To: Robert S. Galen

A gift from the AACC History Division to an outstanding leader in clinical chemistry.

Samuel Meites

May, 1996

Otto Folin
America's First Clinical Biochemist

Otto Folin

America's First Clinical Biochemist

Samuel Meites

American Association for Clinical Chemistry, Inc.

Library of Congress Cataloging-in-Publication Data

Meites, Samuel
    Otto Folin : America's first clinical biochemist / Samuel Meites.
    p.   cm.
    Bibliography: p.
    ISBN 0-915274-48-5
    1. Folin, Otto, 1867–1934. 2. Clinical chemists—United States—
  Biography. I. Title.
    [DNLM: 1. Folin, Otto, 1867–1934. 2. Chemistry, Clinical—
  biography. WZ 100 F6605M]
  RB40.M32 1988
  616.07'56'0924—dc19
  [B]
  DNLM/DLC
  for Library of Congress   88-7418
                                                                      CIP

Copyright © 1989 The American Association
for Clinical Chemistry, Inc. All rights reserved.
No part of this publication may be reproduced,
stored in a retrieval system or transmitted, in
any form or by any means, electronic, mechanical, photocopying, recording or otherwise
without permission in writing from the publishers.

10 9 8 7 6 5 4 3 2 1

0-915274-48-5

Printed in the United States of America

This book is dedicated to the memory of Teresa Folin Rhoads (1908–87), who did so much to bring her father's biography to fruition. There is a measure of consolation in the fact that the book was nearly completed at the time of her death, and that she had seen all of its contents.

Contents

		Part I, His Life	Part II, Methods
Foreword		xi	
Notes from the Author		xiii	
Acknowledgments		xix	
1.	Early Days in Minnesota (1882–88)	3	361
2.	University of Minnesota—Sanford's Delight (1888–92)	10	
3.	Chicago University—Organic Chemistry, With Love (1892–96)	31	
4.	Uppsala and Åseda—A Biochemist's Birth (1896–97)	55	
5.	Berlin—Clinical Orientation (1897)	72	367
6.	Marburg—Quest for Independence (1897–98)	80	
7.	Chicago Revisited—Survival (1898–99)	96	
8.	The Columbus Laboratory (1899)	111	
9.	Marriage and Morgantown (1899–1900)	142	
10.	McLean Hospital—Tooling Up (1900–3)	153	370
11.	A Classical Period (1903–7)	169	380
12.	Harvard—Teaching and Profession (1907–12)	190	386
13.	Willey Denis—The Prodigious Year (1912)	213	389
14.	Developing Clinical Biochemistry (1913–18)	225	402
15.	Hsien Wu—A Major Leap Forward (1919–21)	267	410
16.	Metabolic Studies: MLIC (1922–26)	282	418
17.	Modifications, Micromethods, Unlaked Blood (1927–31)	313	
18.	The Final Years (1932–34)	341	
Afterword		421	
Chronological Highlights of Otto Folin's Career		423	
General References		427	

Foreword

Many of us, when first introduced to clinical chemistry, quickly became familiar with the preparation of the Folin-Wu protein-free filtrate, a technique for removing proteins from whole blood or plasma that resulted in water-clear solutions suitable for the determination of glucose, creatinine, uric acid, nonprotein nitrogen, and chloride. The major active ingredient used in the precipitation of protein was sodium tungstate prepared "according to Folin." Folin-Wu sugar tubes were used for the determination of glucose. From these and subsequent encounters, we learned that Folin was a pioneer in methods for the chemical analysis of blood.

My first experience with the determination of uric acid in serum was the Benedict method in which protein-free filtrate was mixed with solutions of sodium cyanide and arsenophosphotungstic acid and then heated in a boiling water bath to develop a blue color. Even though the heating step was carried out under a hood, there was always a slight odor of cyanide in the air as the final color was measured in a photometer. A thorough review of the literature revealed that Folin and Denis had published, in 1912, a method for uric acid in which they used sodium carbonate, rather than sodium cyanide, as the alkalinizing reagent. Further study and modification of this reaction led to a "carbonate" method, which largely superseded the "cyanide" method in many clinical laboratories.

This review of the literature marked the beginning of my acquaintance with the personalities of Folin and Benedict and with their discussions (sometimes intense) of the relative merits of methods for glucose and uric acid, but even more significant for me, it marked the beginning of an enduring interest in the history of clinical chemistry.

Unfortunately, most textbooks tell us little about the historical evolution of clinical chemistry and the individuals who contributed so much to its development. It is therefore most appropriate to recognize that this biography of Otto Folin fills an important gap which, it is hoped, will set an example that future historians may follow.

This book is a tribute to the memory of Otto Folin and to the diligence and commitment of Samuel Meites. The author has spent much time and effort in collecting background information, correspondence, photographs, and conducting personal interviews to provide a basis for the preparation of a warm, personal, authoritative, and in-depth view of the life of one of our earliest and best-known clinical biochemists.

Wendell T. Caraway
Flint, Michigan

Notes from the Author

How I Came to Write This Biography

Having been appointed chairman of the Committee on History (later renamed the Subcommittee on Archives) of the American Association for Clinical Chemistry late in 1980, I set about the task of fulfilling my mission, fully aware that my historical knowledge was derived largely from my longevity and from little else except a few articles written or inspired by my predecessor, Wendell T. Caraway.

Clinical chemistry has been traced back through the ages, particularly to the ancient practices of urinalysis, but modern clinical chemistry began with the application of 20th-century quantitative analysis and instrumentation to measuring constituents of blood and urine, and relating the values obtained to human health and disease.

In the United States, the first impetus propelling this new area of biochemistry was provided by the 1912 papers of Otto Folin. The only precedent for these stimulating findings was his own earlier and certainly classic papers on the quantitative composition of urine, the laws governing its composition, and studies on the catabolic end products of protein, which led to his ingenious concept of endogenous and exogenous metabolism.

In 1912, Folin began using analytical methods on blood that he had previously introduced for urine, though he had already determined blood ammonia in 1902. This work preceded the entry of Stanley Benedict and Donald Van Slyke into biochemistry. Once all three of them were active contributors, the future of modern clinical biochemistry was ensured. Soon a relatively large group of the early biochemists began to develop the clinical field, primarily because the jobs were available in medical schools eager or constrained to adopt the new science in their curriculum. Those who would consult the early volumes of the *Journal of Biological Chemistry* will discover the direction that the work of Otto Folin gave to biochemistry. This modest, unobtrusive man of Harvard was a powerful stimulus and inspiration to others.

Because clinical biochemistry is a relatively new science, its history should be relatively easy to discover and relate. This science, among other things, includes people, methods, instruments, and the professional aspects relating to publications, organizations, and meetings, as well as teaching and education. Among these topics, frankly, the easiest to write about, and the one most interesting to all of us, is people. The early efforts of the Committee on History were consequently aimed at writing sketches about pioneers, living and dead. Several were published in the journal *Clinical Chemistry* and more are on the way. Other aspects of history are now receiving attention as well.

I chose to write about Otto Folin because he antedated Benedict and Van Slyke. My intention was to write a biographical sketch limited to about four pages in *Clinical Chemistry*. A search of the literature, however, showed that there were already many excellent sketches of him, and that one, a biographical memoir written by Philip Shaffer for the National Academy of Sciences, was definitive. What could I, a latecomer, add to these? Most of the writers had known Folin personally, and at that, for a long, long period of time. Frankly, this was discouraging. Therefore I thought of preparing a limited portrait from the standpoint of one who had entered clinical chemistry during World War II and had extensively used Folin's methods in the laboratories of three military hospitals.

Several people changed my mind. First and foremost, Mrs. Teresa Rhoads, daughter of Otto Folin, not only granted me a seven-hour taped interview on Nov. 7, 1981, but entrusted me with more than 400 letters written by her parents to each other, mostly from 1894 to 1900 before they were married. Not even Philip Shaffer had read them! On top of that she provided some invaluable pictures, four bound volumes of Folin's publications, and other memorabilia. At the time, I was nonetheless doubtful that I could glean more than a biographical sketch from the material that Mrs. Rhoads had placed in my hands, even though that biographical sketch would be replete with highly original, even fascinating information.

Mr. Joseph Benotti, the retired clinical chemist and former co-owner of the Boston Medical Laboratory, made the first contact with Mrs. Rhoads in my behalf. And Mr. Leonard Sideman, clinical chemist with the Pennsylvania Department of Health, made the arrangements at a motel in Philadelphia and helped with the interview.

Within a week of my meeting with Mrs. Rhoads I went to St. Petersburg, Fla., to interview Otto Folin's son Grant in his home. At the outset of this interview, Mr. Folin said to me about his father, "Don't put him on a pedestal. He was a simple, hard-working man, but not a saint." And I replied, "I will try to report his life as I find it. He will have to put himself on a pedestal." Despite my growing feeling about my subject, I was still uncertain that a full-scale biography was possible, or even warranted. How can one write of a man's early life in Sweden without knowledge of the Swedish people, the history of life in Sweden during Otto Folin's boyhood, the causes of his emigration, the education and customs there? How could I learn about his young manhood in Minnesota and his education there? Chicago? West Virginia? McLean Hospital? Harvard? I decided to gather all of the material I could before making a final decision.

In mid-December 1981, I went to Boston. There I accomplished several objectives. With Mr. Benotti's help, I held a four-hour interview with Olive

Watkins Smith, who had obtained her doctorate degree under Otto Folin's guidance in 1928, and who had been a friend of the Folin family. She provided much useful information, mixed with unabashed hero worship, from the viewpoint of one of Otto's former students. Then Joe Benotti and I visited the McLean Hospital where Otto Folin had worked for seven years. Finally, with material obtained through the solid support of Mr. Richard J. Wolfe, Curator of Rare Books and Manuscripts at the Countway Library of Medicine, I was convinced that a comprehensive biography was possible. What I lacked in details about the man could be redeemed by what then seemed most important: reviewing his scientific output in terms of its impact on modern clinical chemistry.

I followed up this thinking with a trip to the University of Chicago, where I examined records at the Regenstein Library, in the hope of finding tidbits about Otto Folin's days as a graduate student there and about the department of chemistry as he knew it. This trip was fruitful.

In October 1981, I received a mailed advertisement from the Immuno Nuclear Corp. of Stillwater, Minn., Folin's home town as a youth. Immediately I wrote a reply stating that while there was no need for their products in my laboratory, could they help me look into the background of one of Stillwater's distinguished sons, as well as send me some of Stillwater's history? Very benevolently, a secretary phoned and said that while they could not further my quest, they could put me in touch with someone who would, namely Mrs. Sue Collins of the staff of the local public library and an active member of the Washington County Historical Society. She not only began to supply me with appropriate historical material, but introduced me to the keen, gracious historian of the nearby Swedish community of Scandia, Mrs. Anna R. Engquist, and through Mrs. Engquist I entered the portions of Otto Folin's life about which so much was vague, his early life in Sweden and in Minnesota. As a result of the help of these women, I was now able to piece together the childhood of Otto Folin. Soon I was in contact with historians in all of the sites where Otto lived in Minnesota, and had contacted several people in Sweden, most notably, Carl-Werner Pettersson, who lived in Folin's birthplace, Åseda. What a find! I began a widespread correspondence then, with people anywhere that a clue on Folin's past occurred—more than 300 letters—and bit by bit, Otto's life began to come to me in the mail.

After a trip to "Swedish" Illinois, Wisconsin, and Minnesota with a group of Swedes from the old country under Pettersson's guidance (he had come to the U.S. to sightsee and to visit relatives), another trip to the Minnesota Historical Society and to the University of Minnesota, and a talk with Otto Folin's niece Mrs. Eva L. Pederson and her nephew John Folin, I was ready to assemble the mosaic of Otto Folin's life. By October 1982, I began writing this biography.

The Problems of Uncovering Otto Folin, the Man and the Scientist

Those of you who are GADDERS, or believers in the Grand American Dream, will appreciate Otto Folin. His life is an old tale touching the roots of most families in our nation. The penniless, half-grown immigrant youth arrives in an alien land whose language and customs he must assimilate. He must change the ingrained pattern of old country ways, absorb the new culture, while he earns his daily bread as an unskilled laborer.

He would not undergo the proverbial evolution from "rags to riches," but metamorphose from the artless odd-jobber just getting by, into a dedicated, far-seeing scientist, laboring in his laboratory of clinical biochemistry to leave a legacy that stands firm 50 years after his death. His dream was the grandest of all: to benefit mankind.

The reader must allow my apparent reading between lines on occasion. The totality of a man cannot be grasped because what we can learn of another is riddled with gaps. The biographer learns that the child perceives his parents only as adults, and vaguely at that. The parents sometimes can perceive their offspring to a certain age short of adulthood, but rarely beyond. The teachers see a growing child among many, and often recognize aptitude and potential and help develop it, but the records of this are sparse. Colleagues see him from the plane of their own understanding and judgment of his work, from the limitation of their own background and interests. However flawed, these multi-faceted views from children, colleagues, friends, and teachers can be brought together as resources to help the biographer approach the problem of picturing the complete man.

A second problem I faced in writing Folin's biography was the lack of information about certain parts of his life; an even flow, a continuum of events or detail was thus not always possible. But even in those portions of his life that were sketchy, Otto Folin's character and mind were revealed.

Fortunately for me as his biographer, Otto had one admirer who not only cherished his memory, but who collected a great deal of his correspondence, published work, pictures, and put into writing all that she could relate of him that seemed appropriate—his wife, Laura. But Laura could not, understandably, grasp her husband's significance as a scientist, the quality of his work, and its influence. This aspect fell to two of Otto's friends: Harry Trimble and Philip Shaffer.

Trimble, one of Otto's closest friends at Harvard, made an effort to catalogue all of Otto's scientific work. (It was Trimble who made the funeral arrangements after Otto's death based on a "pact" that he and Otto had made during Otto's hospitalization in September 1934.) To Philip Shaffer, Otto's student and friend from the University of West Virginia as well as his first assistant at the McLean Hospital, fell the honored task of writing the early biographical sketch for the *Journal of Biological Chemistry* and along with Henry Christian, for *Science*. But it was highly significant that though Philip was asked in 1935 to prepare the comprehensive sketch for the *Biographical Memoirs* of the National Academy of Sciences, he was unable to do so for another 15 years. Then it was made possible because Laura Folin supplied him with the basic facts in her own typewritten version of Otto's life, and Trimble provided a complete list of Otto's publications. But the passage of time had also had a beneficial effect. It allowed Philip's hindsight to expand and a more complete picture to emerge despite some loss of detail.

My opportunity as a historian was unique: to reveal not only Folin's life's work, but to assess its pioneering impact on clinical biochemistry.

Quantitatively, in the years of his scientific productivity, 1897–1934, Otto Folin published 151 (±1) journal articles including a chapter in Abderhalden's handbook and one in Hammarsten's *Festschrift*, but excluding his doctoral dissertation, his published abstracts, and several articles in the proceedings of the Association of Life Insurance Medical Directors of America. He also wrote one monograph on food preservatives and produced five editions of his laboratory manual. He published four articles while studying in Europe (1896–98), 28 while at the McLean Hospital (1900–7), and 119 at Harvard (1908–34). In his banner year of 1912 he published 20 papers. His peak period from 1912–15 included 51 papers, the monograph, and most of the work on the first edition of his laboratory manual.

The quality of Otto Folin's life's work relates to its impact on biochemistry, particularly on clinical biochemistry, and ultimately to its benefit for mankind. From the vantage point of the 1980s an overview may be gained, and a broad assessment proposed for his work before the specific highlights are offered.

At the outset, Otto's two brilliant collaborators, Willey Denis and Hsien Wu, must be acknowledged. Without Denis, Otto could not have achieved so rapidly the introduction and popularization of modern blood analysis in the U.S. And one cannot objectively weigh Folin's or Wu's individual scientific input to the development of the tungstic acid filtrate and the blood sugar method. It would be pointless to conjecture how far Otto could have progressed without this pair.

Otto Folin's work provided the basis of the modern approach to the quantitative analysis of blood and urine through the introduction of new or improved chemical methods that greatly reduced the body fluid volume required for analysis. He also applied these methods to metabolic studies on tissues as well as body fluids. Because Folin's interests lay in protein metabolism, his major contributions were directed toward measuring nitrogenous waste or end products. His work on blood and urine meant that systematic chemical studies of people in health and disease could be made and the quantitative tests used for diagnosis, monitoring of treatment, and therapy. This was most dramatically illustrated by Folin's studies of blood nitrogen retention in nephritis and gout.

Folin introduced colorimetry, turbidimetry, and the use of color filters into quantitative clinical biochemistry. He initiated and applied ingeniously conceived reagents and chemical reactions that paved the way for a host of studies by his contemporaries in biochemistry. He introduced phosphomolybdate for detecting phenolic compounds, and phosphotungstate for uric acid. These, in turn, led to the quantitation of epinephrine and of tyrosine, tryptophane, and cystine in protein. The molybdate suggested to Fiske and SubbaRow the determination of phosphorus as phosphomolybdate, and the tungstate led to the use of tungstic acid as a protein precipitant. Phosphomolybdate became the key reagent in the blood sugar method. Folin resurrected the abandoned Jaffé reaction and established creatine and creatinine analysis. He also laid the groundwork for the discovery of creatine phosphate. Clinical chemistry owes to him the introduction of Nessler's reagent, permutit, Lloyd's reagent, gum ghatti, and preservatives for standards, such as benzoic acid and formaldehyde.

In an era when pure chemicals were notoriously unavailable, he provided procedures for purifying, among others, creatinine, creatine, uric acid, picric acid, Nessler's reagent, gum ghatti, and more than any other, sodium tungstate. Folin's work on protein-precipitating agents led to the creation and use of tungstic acid, as mentioned above, and the introduction of sulfosalicylic acid, metaphosphoric acid, and picric acid.

His studies on urine were classic: the chemical composition of urine, the quantitative relationship between the total nitrogen and its principal nitrogenous constituents, and finally the concept of endogenous and exogenous protein metabolism that remained the accepted explanation until the advent of isotopic labeling and analysis modified it primarily in terms of the dynamic metabolic pool.

Folin pioneered the use of clinical chemistry in hospitals, clinics, and life insurance laboratories. He issued the first call for U.S. hospitals to hire clinical chemists. By his personal example he helped establish the modern role of the biochemist as a teacher of basic science in the medical school. He set standards for the curriculum and the teaching of biochemistry to medical students, including in their training a manual of laboratory methods containing quantitative clinical chemistry.

Finally, Otto Folin created a center for the graduate training of clinically oriented biochemists, and among the most distinguished were Bloor, Doisy, Fiske, Shaffer, SubbaRow, Sumner, and Wu. These young investigators were drawn to biochemistry by Folin and inspired by his personality, teaching, and laboratory skill.

Always unbroken throughout his life, however, were his research themes that built the new discipline of clinical biochemistry, step by step, over and over, the incessant innovations and modifications. First he was alone, then joined by the few, followed by a host, but his remained an unbroken effort exemplified by the recurring uric acid problem: analysis and metabolism.

Sketches on Folin's Life

Following his death, tribute was paid to Otto Folin by his friends and admirers in three (arbitrarily divided) periods of time. First came the obituaries when he was still clearly remembered in close detail by his colleagues (1934–35). After a longer lapse of time (1936–51) and the details of his life were receding, there appeared a few sketches of him. Finally, Shaffer's definitive study in 1952 came when only general impressions remained.

At the end of the Afterword are listed the obituaries and sketches of Otto Folin in the three arbitrarily selected periods. They are incomplete because they do not include newspaper reports and all of the obituaries that unquestionably appeared in foreign journals. Nor are all the possible dictionaries and directories of science and scientists covered.

About the Organization of This Book

Part I of this book chronicles Otto's life and the significant contributions he made to science. Part II presents synopses of many of his published scientific papers for those readers who may want more details of Otto's highly original methods.

Samuel Meites

Acknowledgments

Thanks must first of all be extended to the American Association for Clinical Chemistry, which provided me with the opportunity to conceive and execute this project, and to the Children's Hospital of Columbus, Ohio, where the author has spent three decades as a clinical chemist. In both instances the gratitude is to those making supporting decisions and to those offering encouragement and help. In this, the people of my own section in clinical chemistry, headed by Mrs. Karen Saniel-Banrey, are most thankworthy.

The people and institutional-related sources of my information are many. Some provided absolutely essential documentation. They are presented, however, without any implication as to their relative importance, and in no definite order.

Mrs. Jane S. Price, daughter of Prof. Philip A. Shaffer, and Dr. Paul G. Anderson, Archivist of the Washington University School of Medicine, St. Louis, Mo.; Mr. Harold Forbes, Associate Curator of the University Library, West Virginia University, Morgantown, W.Va.; Ms. Maxine Clapp and her assistants at the University Archives, University of Minnesota, Minneapolis, Minn.; Mr. Alexander Armour and Mr. Terry Alan Bragg, Archivists at the McLean Hospital, Belmont, Mass.; Mrs. Eleanor Sikes Peters, daughter of Madeleine and George Sikes, of Peoria, Ill.; Mrs. Maxine H. Sullivan, University Registrar, and the Archives of the Joseph Regenstein Library, University of Chicago, Chicago, Ill.; Ms. Brigid Shields, Bonnie Wilson, and colleagues, the Minnesota Historical Society, St. Paul, Minn.; Mrs. Cynthia Goldstein, Rudolph Matas Medical Library, Tulane University, New Orleans, La.; Ms. Bonnie Chernin, Archives Librarian, Metropolitan Life Insurance Co., New York, N.Y.; Mr. Clark A. Elliott, Associate Curator, Archives, Harvard University Library, Cambridge, Mass.; Mr. Robert B. Porter, Center City, Minn.; Mrs. Hsien Wu, New York, N.Y.; Dr. Irvin S. Danielson, Pearl River, N.Y.; Dr. Emeroy Johnson, St. Peter, Minn.; Ms. Frances Goudy, Special Collection Librarian, and Ms. Lisa Browar, Curator of Rare Books and Manuscripts, Vassar College, Poughkeepsie, N.Y.; Mr. Theodore A. Norelius, Lindstrom, Minn.; Smith College Archives, Northhampton, Mass.; Carnegie Institution of Washington, Washington, D.C.; Superintendents of Schools, Stillwater, St.

Paul, and Minneapolis, Minn. Special thanks to the Swedish contributors: Dr. Anders Kallner and Dr. Fredrik Berglund, Stockholm; Mrs. Elisabeth Thorsell, Linköping; and Mr. Gustav Samuelsson, Åseda. Others unlisted here are cited in the text and in the Notes from the Author.

I owe much for the skillful services provided by Ms. Mary Pat Wilhem, Librarian, and her assistant, Mrs. Sherry Hay, Children's Hospital Branch, Ohio State University Libraries, Columbus, Ohio.

My wife, Mrs. Lois P. Meites, has served as the typist of this manuscript, and has patiently endured the agony and only partly subdued outbursts of ill humor (exceedingly rare) of its author. I am grateful for her essential, unruffled support.

Samuel Meites

Part I: His Life

Chapter 1. Early Days in Minnesota (1882–88)

On an early August afternoon in 1882, Axel Folin waited on the wooden platform of the Stillwater Station for the train that was to bring the brother he had not seen in 11 years. Axel's memories of Otto were vague. After all the child had been just four when Axel had left Sweden for the United States. Now Otto was nearly an adult, eager to emulate his older brother's success in America.

Otto was strong and sharp-witted, their mother had written Axel. He had been the top student in his class in school and had been one of the few boys selected to receive extra education from the parish curate. Yet the opportunities in Sweden were few then, even for a bright, industrious young man such as Otto.

Thus it was that their mother Eva had asked Axel to provide passage money for Otto to the United States and her sister-in-law Ingrid and her husband Nels Peter Johnson to provide a place for Otto to start his new life, as they had many years earlier when Axel had come to the United States. The Johnsons had a farm in Minnesota that always needed the help of a strong young man.

Axel too could have provided Otto with a place to stay and even found him full-time work. After all, Axel had been successful in his adopted country. As head sawyer at the East Side Lumber Mill, just across the St. Croix River from Stillwater, he was appreciated for his competence and intelligence. On the sawyer's skill hinged much of the financial success of the mill, and Axel was shrewd at deciding rapidly how logs approaching the saws should be positioned to yield the maximum amount and quality of cut lumber.

Axel, however, had discovered that learning English could be the key to an even more successful life. His own English, learned mostly from English-speaking coworkers and self-study, was enough for his job, but beyond that was limited.

That first day when Otto arrived, Axel pointed out a particularly successful Swedish acquaintance. The man had amounted to little in Sweden, but, Axel emphasized, had learned English soon after he arrived. Now just a few years later he was not only a salaried officer in the fire company, but a pillar of the community.

Working full time at the lumber mill with Axel would leave Otto no time for school. So the decision was made. Otto would go to the farm with Aunt Ingrid and Uncle Nels where he could earn his keep and still get an education.

Meanwhile Axel welcomed his brother to the thriving town of Stillwater, Minn. Stillwater with its population of 15,000 was a major hub of the great lumber industry of the Northwest, which was providing immigrants from all over Europe with a chance for a new and better life. Young men were in demand for skilled and unskilled jobs—as lumberjacks, laborers on the log boom, mill workers, road builders. The immigrant could buy land cheaply, farm during the summer, and after harvest, fell trees for the lumber industry. Or he could work at the boom, the state prison, or in one of the emerging industries such as wood products, grain mills, and farm machinery.

Opportunities abounded and the area was flooded with immigrants from Scandinavia, Germany, Ireland, and other European countries. North of Stillwater, in Washington and other counties, ethnic groups settled almost exclusively. But Stillwater was too small and its turnover of population too rapid for significant segregation into ethnic-based neighborhoods.

While Stillwater was home now, Axel and his Norwegian-born wife Anna Marie dreamed of the day when they could settle on a farm further north in Chisago County near Axel's relatives in the heart of the Swedish-settled area. Although Axel spoke enough English to get by, Anna Marie was isolated by her limited knowledge of the language, and her childlessness deprived her of the chance to learn English through her children as many immigrant women did. Dreams aside, Axel had a well-paid position—almost four dollars per day and board, twice what the average mill worker earned—and their home would remain in Stillwater for some time.

For Otto, however, the Swedish area in Chisago County would be home for a good part of the next two years. Ingrid and Nels Peter Johnson, who had settled in the area in 1867, were poor, but among their countrymen they were comfortable and by the standards of their former home in Småland, Sweden, they were solid landowners. Their 80-acre farm with its few beef cattle, a dairy cow, and a horse made them self-sufficient, and during the winter months, Nels worked as a carpenter, making and selling furniture and building and improving houses. Along with his cousins, Hannah (Johanna), 17, and John (Johan), 21, Otto would help Nels work the farm.

The Chellberg school at which Otto would begin his education in his new land was three miles from the Johnsons. Typical of that period, it housed eight grades in one room. Instruction was entirely in English and stressed the proverbial three Rs. Once out of the schoolhouse, however, the children lapsed into Swedish, the language their parents spoke at home.

In Otto's initial struggles with the English language, he undoubtedly had help from John and Hannah. His cousins, who had probably attended the Chellberg school for several years after their arrival in Minnesota, could both read and write English. John, in fact, was a bookworm once he had mastered the language and he must have been able to help Otto with grammar and vocabulary.

Otto's intention to master the new language was clear. He spent what little money he had for an etymological dictionary that traced the meaning of each English word through several languages. Otto soon realized that his general knowledge obtained from his education in Sweden was already beyond that offered in this grade school, particularly in arithmetic and science. However, his grasp of the new language came slowly, and thus the one-room schoolhouse was to his definite advantage, because he could work at the level of his capability.

Records show that the Chellberg school was opened in the 1882–83 school year for four months, two in the "summer" (beginning April 15) and two in the winter (beginning Nov. 1). After the harvest, Otto probably attended the school until heavy snows and biting winds made the three-mile walk impossible.

In the spring of 1883, Otto returned to Stillwater to live and work with Axel, because in addition to an education Otto wanted money to pay his debts to Axel.

The job Axel obtained for Otto at the East Side Lumber Mill only earned Otto a youth's wages, 50 cents a day for light work, plus board and bunk. But Axel did not want his brother to perform the heavy, somewhat dangerous labor of men, even though he would have earned at least triple the wages.

Life in the lumber mill consisted of workdays from sunup to sundown, with shorter hours on Saturday, and Sundays off. The food was hearty, but the bunk room was crowded both with people and bedbugs and as Otto recollected, " I had a rather nice man for a bed partner; but occasionally on a Monday I would find him wrathy because I used to soak the bed with kerosene before going home to my brother on Saturday evenings."

In those days the sawmills were opened from the time the rivers thawed in the spring to the late autumn when the waterways began to freeze. In the autumn of 1883, the weather turned very cold. "No amount of clothes could keep me from freezing and that late fall and winter I suffered a good deal from rheumatism in my feet."

From late November when the mill closed until after Christmas, Otto lived in Axel's house. He celebrated the holidays with his sister-in-law Anna (Axel was working up north in the woods) and then returned to the Johnsons so that he could attend school. Unfortunately, the rheumatism that he had experienced in the fall left him stiff and sore and made the walk to school difficult. In all, he did not attend more than four or five weeks and he benefitted little.

A real step forward for Otto in learning the new language was brought about by events in the spring.

> In the spring I returned to Stillwater, but work was getting scarce and my brother I learned later did not really want me to get work in the mills so I got none. I scratched up a bit of work on the river myself putting ties together into crates as they came down the river to the "Boom." This lasted only ten days, however, but paid me a dollar a day and my board. So I did not get any more work until after the fourth of July when I made my second trip in the country to seek work among the tillers of the soil. This second I had luck and obtained a

position on the evening of the first day with a young Irish farmer who asked me to ride with him and had me hired within a few minutes. Fifteen dollars a month for four months.

At the end of that period I remained to do the chores for my board while I attended the district school—four months. This Section, Erin Prairie (township), St. Croix Co., Wis., is as distinctly Irish as Chisago Co. is Swedish; but still it was more advantageous because we all spoke English. For eight months I did not meet a Swede and when the spelling down came on the last day of school I came within one of winning the book which was given as a prize to the best speller—an art which I have since developed very little, I fear.

The Irish family was kind and generous, and even if the wheat farmer's erratic finances did not always produce Otto's wages, living with an English-speaking family (albeit with a brogue) diluted the language barrier and allowed Otto to make up ground rapidly.

Aside from a brief visit to Axel in the spring, Otto continued to work with the farmer until the end of the summer of 1885. However, a growing determination to prepare for high school took him back to Stillwater in the fall. In the three years Otto had been in the U.S., he had attended only two four-month sessions of school, one of which had been incomplete. Now he enrolled at the Lincoln School, a nine-month grammar school.

Probably through Axel's efforts, Otto obtained a job as a porter in the Daniel Elliott Boarding House, which like other boarding houses of the era provided economical housing for travelers and the semi-transient single men who provided labor for the area's industries. By doing chores around the boarding house, Otto earned his keep and a little extra, and he had learned to make a little extra go a long way. From his various jobs, he had already saved enough to repay Axel all but eight dollars of his debts, including his ocean passage money.

But most important about his new job was that he would now have the chance to go to school for the full nine-month session and to become better integrated into American society. He had grown tired of being considered a "foreigner."

> I was a half grown boy when I first came to this country. I had been brought up in a rather large village and had always imbibed the idea that I was just as good as anyone else in that vicinity and you may perhaps know that the idea of caste pervades all society there—very strongly—from the lowest to the highest so that idea meant a great deal more than the idea of "equality" does here. Then when I came over and began to school I began to feel the stigma which the less intelligent Americans and especially the American boys attach to a "foreigner." I could see the thought reflected in every eye that met mine and whenever I heard a subject discussed it always seemed to be related to the "dangers of our foreign population." One day I heard the most prominent guest at the hotel where I earned my board express this sentiment. . . . "I don't like a Swede and I don't like anybody that does like one"—he knew I heard it—I was only nineteen—he the most stylish, pretentious man at the hotel. I was very sensitive naturally I think and there was no one who seemed to dream even that such might be the case and I hid it always. But it has made me abnormally

sensitive and has always made me shun any society where I was not like the rest.

The year Otto spent at the Lincoln School (in the Third Grammar School) was equivalent to the modern eighth grade. He could supplement the English, history, and science of this curriculum with the mathematics and German he had learned in Sweden. Although no records exist of how Otto did in his classes that year, he must have passed the compulsory state board examinations given to seventh graders in geography and spelling and to eighth graders in history, arithmetic, grammar, civics, and physiology, because in the fall of 1886 Otto was enrolled in the high school.

In those days few could afford to go on to high school and those youngsters who did needed to be highly motivated and strongly disciplined if they were to succeed. So difficult was the Stillwater High School curriculum that nearly two thirds of the students dropped out by their senior year. The school offered only three courses of study in preparation for university admission or teaching school: Latin, German, and English. Much work outside the classroom was expected, from the preparation of two daily recitations to reading of supplementary materials. Each course annually required the student to read and give written reports on three works by different authors.

Otto chose the English course of study, which was difficult enough for those to whom English was the native tongue. But the somewhat startling fact is that Otto received permission to attempt the four-year curriculum in two years. That he achieved this feat was a credit not only to his intelligence, but to the work and study habits he had developed. (His brother Axel once wrote that Otto was a constant reader who would prop up a book to read even when he was chopping logs for kindling.) To make his accomplishment all the more amazing, "inflammatory rheumatism" caused him to miss the entire first month of his second year in high school.

How much influence each of his teachers had on Otto's development is uncertain. Each of Stillwater's five high school teachers were university graduates (in contrast to the grammar school teachers of the day who could qualify to teach with a high school degree), who taught their courses with the intent of preparing students for admission to a university. With only 20 students in the high school, each teacher had time to give students individual attention and to perceive and help develop their abilities.

School records do not distinguish the courses for which Otto attended class from those that he passed simply by taking the examination, although his accelerated curriculum must have required him to do much work on his own. However he completed the coursework, his grades for his junior and senior years showed improvement over the first two years, probably as a result of his growing fluency in English. Five of his 15 grades in the upper level courses were in the 80s and three—for two chemistry courses and solid geometry— were in the 90s.

On the Minnesota State High School Board examinations, where a minimum score of 60 was required to obtain a certificate, Otto received 10 certificates, which would have qualified him to enter the teaching profession. Although his scores were not remarkable, he did receive a 93 in chemistry (*Stillwater Daily Gazette*, Friday, Sept. 7, 1888).

Graduation from high school was a major achievement, and the ceremony was a community event. So large was the expected crowd that the ceremony was held in the Stillwater Grand Opera House. Among the audience must have been Axel and Anna Marie and Otto's older brother Alfred, who had immigrated to Stillwater in the summer of 1887.

The commencement program was elaborate. According to school board regulations,

> Pupils shall be required, upon graduation, to write an original production upon such themes as may be selected by the principal, and to perform such part in the commencement exercises as may be assigned by the Superintendent.

Otto delivered an oration on capital punishment and his classmates delivered addresses on topics ranging from Egypt to the mechanical properties of air.

An article in the *Saint Paul and Minneapolis Pioneer Press* on Saturday, June 2, 1888, which described the commencement, made special note of Otto's remarkable achievement:

> The high school commencement exercises at the Grand Opera house last evening proved a matter of as much interest to Stillwater people as usual, and the house was filled until there was scarcely standing room. It was the only graduating class for two years. . . . The class of last evening . . . contained one member whose career is the most remarkable in the history of the Stillwater Schools, and probably has not a parallel in the state. . . . Last evening this courageous young man graduated with high honors—did in less than two years what many American boys with wealthy parents have found it impossible to do in four years.

Four years later in the *Ariel*, a publication produced by the students at the University of Minnesota, Otto recalled his high school commencement address:

> The borrowed thoughts and stereotyped expressions and gestures which go to make up the ordinary Commencement oration were perhaps good enough when we took leave of the high school. We were then still young and enthusiastic, still susceptible to every new impression, foolish enough to believe that the world need but hear a truth to accept it and to act upon it, conceited enough to believe that we could tell that truth as it had never been told before to that particular audience, eager to lay our newly acquired ideas before them, yet afraid and nervous perhaps, while waiting our turn to come forward and fire. Having successfully given the enemy our broadside, however, we sat down amid cheers and a profusion of flowers—as proud as kings. And the audience good-naturedly applauded the young Demosthenes. They understood the situation, but were pleased and satisfied, because their boys and girls had done all that could reasonably be expected of them, and deserved a little encouragement. (*Ariel*, Feb. 20, 1892, p. 229)

Although the high school degree qualified Otto to teach grammar school, his ambition was to receive a university degree. So in the fall of 1888, Otto matriculated at the University of Minnesota. How he acquired sufficient funds for this step in his education is not known. However, had he worked in the mills during the summer of 1888 at a job that netted the full pay of $1.50 per day, he would have earned enough money to start at the university in the fall.

An article in the *Stillwater Daily Gazette* (Monday, Sept. 10, 1888, p. 3) noted his entrance into the university in glowing terms.

> Otto Folin, the young Swede who graduated from the high school with the class of 1888, and made such an exceptionally good showing in his studies, by mastering a four year's course in two years, entered the state university yesterday, from which he will graduate in due time. He proposes to work his way through this school as he did the high school here. It is of such stuff as he that the sinews of great nations are made. His motto seems to be "Never Know Defeat."
>
> He has been steadily at work during the summer and has earned a considerable sum of money which will help him along. Those who know him best, and are acquainted with his ambitions can but at least wish the worthy young fellow success.

Otto Folin had worked very hard to reach the university level, but he also had been lucky. The high school he had attended had not only performed its educational duties conscientiously, but had challenged his intellect with extracurricular reading and writing assignments, particularly in chemistry. The very small size of the school had also allowed him to receive personal time and attention from his teachers.

Chapter 2. University of Minnesota—Sanford's Delight (1888–92)

The University of Minnesota at which Otto arrived in September 1888 was a young, struggling institution. Although founded in 1851, it had not managed to graduate any students until 22 years later and then only three. Financial disaster had been staved off only by an infusion of private funds from its devoted benefactor and regent J. S. Pillsbury, who was a Minnesota entrepreneur and public servant. Huge tracts of land provided by the Morrill Act of 1862 in return for the university's obligation to teach agriculture and mechanic arts were also a boon to creating financial security.

The university's struggles to establish itself were not just financial, but educational. Instrumental in propelling the university toward its proper role in higher education and research was William Watts Folwell, a scholar who had both the vision and worldly wisdom to see the institution as a state resource.

> He had the satisfaction of knowing in the end that his stubborn determination to create a carefully articulated system of statewide education gave Minnesota the distinction of being the first state to provide free secondary instruction to public schools for all qualified pupils within its borders. (2-1)

Folwell, who became the university's first president in 1869, put his stamp on the new school, but not without opposition from the "old guard." After squabbles with the dissenting old guard faculty on matters relating to his ideas about the university's mission, Folwell was able to replace six of them in 1879 with the "first" faculty of professional teachers, young men and women educated in famous universities.

However, the university's existence was still shaky as enrollment dropped from 386 in 1880 to 253 in 1882 and to 223 in 1883. Folwell stepped down in 1884 and was succeeded by Yale professor Cyrus Northrop. Northrop, through both a series of measures designed to gain public and student support and the charm of his personality, reversed the downward trend in enrollment and finances. He continued to recruit and promote outstanding teachers, some of whom would ultimately contribute in a major way to Otto's education.

In 1888 when Otto arrived, the fledgling university was housed in just a handful of buildings located on 45 acres high on the bluff of the eastern side of a north-south bend in the Mississippi. The original buildings—"Old Main," mechanics arts, agriculture, law, and military—were supplemented in 1889 with the Pillsbury Science Hall and in 1891 with the chemistry and physics building, erected on the site of the agriculture building, which burned down in 1889.

Dormitory space on campus was available only for students in the School of Agriculture. Room and board, however, was available from private families for $4 a week and up, the 1888–89 university catalog notes. Students could also board in clubs at a cost of $2.50 to $3 per week.

In addition to room and board, expenses for students included an annual fee to the university of $5 for incidental expenses, the cost of their books, and laboratory charges, which depended on the amount of material used.

For a young immigrant whose ambition it was to further his education, the University of Minnesota was ideal. The tuition was free and the fees were low, and evidently opportunities were good for young men like Otto who wished to earn their living while they attended school. The catalog stated,

> The University cannot promise employment to those desiring to earn their own living. The public bounty stops at furnishing free instruction. Many of the students support themselves in college, and a young man who really wants work, and will look for it, can generally find it.

(No mention, of course, of a young woman who wanted or needed to work.)

The University of Minnesota provided Otto with even more "breaks" than the high school. He earned his board and room during his first two years tending horses and performing odd jobs for a Minneapolis civil engineer, and he worked as a night clerk in the St. Charles Hotel and at the Hotel Windom during the other two years. He probably spent his Christmas holidays with Axel and Anna Marie, and returned to the lumber industry at Stillwater to generate a little reserve fund during the summers. One can picture this new freshman on campus, good looking, tanned, manly, tall, preoccupied, with the faint aroma of the barn in his clothes, and perhaps a bit of manure adorning his lone pair of shoes.

What was Otto's freshman year like? His transcript reveals two complete sets of grades, one for freshman classes and one for subfreshman classes. (The subfreshman classes were for students not considered adequately prepared for freshman coursework. Otto's scores on the state examinations had been mediocre for the most part.) Why he was involved in both programs at the same time is not clear. However, the scores in his freshman courses for the three terms pointed toward a promising future. Only one grade was below an 80 (one term in rhetoricals). Remarkably his English grades were in the 90s, and his science and mathematics grades were comfortably in the 80s.

That Otto's grades were considerably higher than those he had received in high school shows that he was flourishing under the new academic challenge. But more than academics, the university offered Otto a milieu where learning was the raison d'etre. He could cultivate friendships both with fellow students who were his intellectual peers and with teachers on occasion. He could fill his spare time in clubs and organizations in which he could make both social and intellectual contacts. Best of all to Otto was that the university represented democracy in action. Otto's penury, the continual need to support himself, was shared by many others. In this environment, Otto's humble, laconic, and good-humored outlook on life could be appreciated and admired by his classmates.

In his four years at the University of Minnesota, Otto had highly qualified, dedicated instructors for most of his courses, but none so influential as Maria L. Sanford and James Albert Dodge, two of Folwell's appointees. Sanford, professor of rhetoric and elocution, worked her magic on Otto, especially during his first two years, and Dodge, professor of chemistry, initiated Otto into the advanced aspects of chemistry and the possibilities of further education and research, particularly during the last two years. (A third instructor, Norwegian-American Professor Olauf J. Breda may also have had an impact on Otto's education. Possibly inspired by his friendship with his Norwegian sister-in-law or his desire to know more about his Scandinavian heritage, Otto studied Norwegian and masterpieces of Scandinavian literature with Breda for five terms.)

No question exists about the importance of Sanford's influence on Otto. A teacher gifted far beyond her formal training, Sanford had developed oratorical powers that could "bring down the house." Her greater gift, however, was a teaching prowess that could turn a bright introverted student into one who could speak extemporaneously, debate effectively, or participate in oratorical contests. While she was producing this transformation, she promoted the writing of good essays and appreciation of the arts, culture, and the good life.

An 1855 graduate of the Connecticut State Normal School at the age of 19, Sanford taught school for more than 20 years before coming to the University of Minnesota. In that time she had served in a combined role as a superintendent of schools and principal of a high school and as professor of history, elocution, and rhetoric at Swarthmore College.

Rhetoricals, which consisted primarily of elocution and some composition, were required of students in all four university years. Hour credits toward the degree, however, were given only for one course during the sophomore year. Of 12 possible terms, Otto had rhetoricals in 11 of them. For each term, he had to present one oration or write two essays.

A gifted teacher like Sanford is capable of extracting an extra load of work from a student and making him enjoy it in the process. Otto must have been the raw material that Sanford delighted in molding—a student much in need of help, but willing and gifted. Undoubtedly, he still needed help in pronunciation, for his English must have been strongly accented. But she did more for him than reduce his Swedish accent to a minimum. She taught him to overcome his stage fright and speak calmly and comfortably in front of a group. She helped him spice his speech with humor and reduce his dependence on written material when he spoke. The art of exposition, the writing of essays, the logical presentation of a thesis with abstract and summary, the use of proper grammar—all these she helped him study. But beyond study Otto and her other students had need for practice and more practice, and she devised ways to help there also.

To make elocution more interesting, Sanford helped establish debates between various student groups. Debates in small groups of two or three allowed the instructor to work critically with small groups, and debating allowed the students to practice, work with one another, add and discard ideas—essentially to think out loud.

Sanford encouraged students to join the various clubs, societies, and associations on campus that held or participated in debating contests. As an added incentive for students to develop their oratory skills, her rhetorical department offered three prizes annually for the best public oration by a student. In addition to the prize money of $30, $25, and $20, which was donated by the Honorary J. S. Pillsbury, the winners also received the attendant publicity, another incentive not to be dismissed.

Although the societies stressed intellectual pursuits for the most part, they also served as the principal extracurricular and social activity for most of the 700-odd students in the university. Recreation could also be found in the Student Christian Association or class events, but not in organized athletics. In fact, intra- and intercollegiate sporting events, hampered by poor facilities and lack of funds for equipment, barely existed during Otto's stay at the university.

Otto joined the Hermean, one of the two literary societies on campus, which was formed to promote "culture and literary activity, especially oratory and debate and the cultivation of a spirit of friendship among its members." To fulfill its purpose, the society staged debates within the group and with other societies, and once a year it debated the other literary society, Delta Sigma.

In early 1891, Otto helped organize another society, the Investigator, to study Christianity from a scientific standpoint. With a membership of nearly 50 students, the group met rather infrequently, but apparently mostly on Saturday afternoons. The *Ariel* of March 31, 1891, makes mention of their first meeting at which each participant presented an essay and a short talk on one of the religions of the world. Otto chose the religions of China.

During his sophomore year, Otto followed the prescribed scientific course, choosing chemistry over physics. The first term he took an extension of general chemistry from the freshman year and a brief survey of organic chemistry and the next two terms he took qualitative chemical analysis. The rest of his curriculum included botany, a laboratory course like chemistry; English; rhetoric; history; and rhetoricals.

With the novelty of collegiate life wearing off and his academic efforts showing continued success, Otto began to develop friendships, some of which would be of lifelong duration—John and Anthony Zeleny, George Sikes, Madeleine Wallin.

The Zeleny brothers were both Hermeans and in the Investigator with Otto. In fact, Otto and Anthony teamed up for debates or prepared talks for the same program on almost every occasion.

Like Otto, the Zeleny brothers bore witness to the immigrant's need to succeed. Their Bohemian-born father, a tailor in Hutchinson where the boys grew up, was so intent on his sons' having a university education that he moved his trade and residence to a spot near the university campus, and remained in business there until all four of his sons obtained degrees. The father's devotion to educating his children was richly rewarded. Both Anthony and John were outstanding students. John, bent on physics from the beginning, was a Phi Beta Kappa. But Anthony started as a pre-law major and switched to physics as an upperclassman. Anthony and John both became full professors of physics at the University of Minnesota, with John eventually moving to Yale University. A third brother, Charles, later became head of the department of zoology at the University of Illinois, and the fourth became a practicing engineer.

One of Otto's closest friends in his four years at Minnesota was Madeleine Wallin, who transferred from Smith College in her junior year. Madeleine was brilliant (in fact, had she not been a transfer student, she would have been the valedictorian of the class of '92), vocal, outgoing, sharp, and sophisticated in spite of having been raised in the wilds of Minnesota and North Dakota.

Notwithstanding the differences in their personalities, Otto and Madeleine were to form a friendship out of their mutual interests in the Hermean and later the *Ariel*. In fact, though their friendship remained forever platonic, it became deep enough that Madeleine, noticing Otto's scepticism toward religion and preachers, set out to make him a true believer. Otto went to chapel with her—he appreciated a good sermon as he would appreciate any good piece of oratory—but he held his own counsel when it came to his doubts about God and religion.

Whether Otto had friends within the Scandinavian community (on or off campus) is uncertain. We know he did not belong to the Scandinavian Student Society, although that was probably because he lacked the time, but his admiration of Professor Breda and the growing Swedish population of Minneapolis makes it likely that he acquired friends among them. The number of Scandinavians on campus, however, was small. In fact, an editorial in the *Ariel* (Jan. 15, 1890, p. 52) lamented the fact that there were fewer than 50 Scandinavians on campus, whereas in proportion to the state population there should have been 300 enrolled.

Edwin and Frank Batchelder, two of his university friends who had been Otto's classmates at Stillwater High School, served as his character witnesses as he was sworn in as a United States citizen in November 1890. Otto had met the five-year U.S. and one-year Minnesota residency requirements long before he applied for citizenship. What finally spurred Otto into becoming a citizen may well have been an essay awarded second place in a preliminary contest for the Pillsbury prizes and published in the *Ariel* (Jan. 15, 1890, p. 53). Entitled "Our Greatest National Need," it stated, in part,

> The average immigrant is unskilled, unlettered, unable to earn a competent livelihood, unfit to become an American citizen. He knows not why he comes. He has heard that here all men are equal; mercenary steamship agents have deluded him into believing that he need only come to America to gain wealth and power. He has no money, no friends, no trade. He is fortunate if he can find employment of any kind. He is doomed to a life of poverty and servitude; his fanciful dreams of this marvelous land are never realized.

In the fall of 1890 as Otto entered his junior year, his life both outside and inside the university blended well. He was working in the local hotel as a night clerk for which he received his room and excellent board. The job also provided him sufficient time to do his homework, read, and think. His days and evenings were free, particularly on weekends, so he could adjust his hours of sleep to his activities.

The junior year marked a major dividing point in Otto's college education. More emphasis was placed on advanced courses related to his major course of interest and he was given more latitude in his choice of subject matter. On reaching this point in his education, Otto certainly decided to become a professional chemist, prompted not just by his own maturing interest in the field but by the guidance and teaching of James A. Dodge. Otto enrolled in all the advanced chemistry courses offered by Dodge to the exclusion of courses in the other sciences beyond the introductory level. (Of course, he may have steered clear of physics and zoology because these departments were then in a state of transition and development.)

Dodge, the second major influence in Otto Folin's college career, had been brought by Folwell to the University of Minnesota in 1880 from Baldwin College (Berea, Ohio) where he had been a professor of natural science for two years. Dodge's academic credits were excellent. After having received a bachelor's degree (1869) and master's degree (1872) from Harvard, he went abroad in 1873 to study chemistry for six months with A. W. Hofmann at the University of Berlin. Following his stay in Berlin, he moved on to the University of Heidelberg to study with Bunsen, Kirchoff, and Kopp for six months, and then on to Owens College (Victoria University) in England to study with Professor Roscoe for eight months.

Dodge returned to the U.S. for one year in 1875 and taught in a high school in Omaha, Neb. In 1876, however, he journeyed to the University of Leipzig to work with Professor Kolbe in the chemical laboratory. In the spring of 1877 he went back to Heidelberg and Bunsen to complete his doctorate thesis, which he obtained with honors in 1878 (2-2).

Along with his continued success in academic subjects, including a 93 in analytical chemistry and a 98 in organic chemistry, his junior year brought Otto the crowning achievement of his college career, election as managing editor of the *Ariel* and with it the presidency of the sponsoring organization, the Ariel Association.

The *Ariel* had been started in late 1877 to serve the students in several ways.

> [The students] need a paper which shall be an exponent of their interests, which will keep them informed upon passing events in other colleges and upon matters of scientific and literary interest. They also desire some means, other than writing, of informing their parents and friends at home of the incidents of their college life, and we doubt not that this desire is reciprocated by their parents and friends at home. In response to this want the Senior and Junior classes have elected, from their number, six editors to conduct a paper devoted to the interests of the Minnesota University. ("The First Ariel," *Ariel*, April 28, 1900, p. 361)

The first issue of the journal appeared on Dec. 1, 1877, and it was published monthly thereafter for the 10 months of the academic year September through

June. Because it was only published monthly, it could not be particularly timely, but no other source of campus news existed.

The *Ariel* was purely a student venture, created and operated by students with the blessing, but not the control, of President Northup and the faculty. Other than providing the physical plant, the administration took no part in the publication. Faculty was, of course, a major source for material, both from articles written by their students and their own contributions. Maria Sanford and other faculty members may have served unofficially as advisers and consultants to the board of editors.

In 1878 the Ariel Association was incorporated to act as publisher. The association was at first made up by any junior, senior, or second-term sophomore who signed the articles of incorporation, although membership was later confined to subscribers in those classes. The board of editors consisted of six officers, four from the junior class and two from the sophomore class, elected at a spring meeting of the membership (*Ariel*, April 3, 1878, p. 65).

As the only student publication on campus, the *Ariel* had a wide and vital audience—students, regents, administration and faculty, and growing number of alumni—and its importance was enormous. A position on the board of editors became one of prestige, and no other leadership position approached the managing editorship as the prize plum on campus. Most officers of societies were known only by their limited memberships, class officers were recognized only by their classmates, if then, and athletes had not yet begun to receive the acclamation they would in future years.

Thus for Otto to have obtained a position on the board of editors was a remarkable achievement. After all, membership on the board was by election, and election presupposes that a candidate be well known and popular. By this time being well known on campus was not an easy task because the student body had grown to nearly a thousand. Otto's contacts with the older seniors and large number of sophomores was limited. Nor did Otto belong to one of the fraternities (15 fraternities and sororities existed by then), which might have provided backing. Fraternities and sororities represented the more affluent students who could readily subscribe to the *Ariel* and become members in the association. (In fact, Otto was strongly opposed to these "secret" societies.)

Nevertheless when the election was held on March 21, 1891, the 200 students attending did elect Otto to the board of editors.

> Ten candidates were nominated by acclamation, six from the Juniors and four from the Sophomores. Of these eight got majorities on the first ballot, and the six highest were declared elected. Those successful were Messrs. Folin, Kirk, Sikes, and Miss Madeline Wallin from the Junior class and Messrs. Spear and Washburn from the Sophomore. (*Ariel*, March 31, 1891, p. 106)

How had Otto managed to achieve this triumph? In spite of the fact that Otto's busy life had allowed little time for politicking, chitchat, and socializing—during his first two years he worked at the home of the civil engineer with whom he boarded, and during his junior year at the hotel—he had made a favorable, lasting impression on his fellow students. They had seen him excel at his studies and participate frugally in extracurricular activities while he continued to support himself. He had capably managed the affairs of his life, yet worked well with his associates and groups. Above all he had an attractive personality—modest, thoughtful, unassuming—to which his fellow students could relate. He had the winning combination of dignity spiced with dry humor. In addition, he must have had forceful advocates among his friends and the faculty, particularly Sanford.

The final victory came when the new board of editors met on March 26 to elect officers and assign places. Otto was elected managing editor (editor-in-chief), editorials were given to Sikes, literary and personals to Wallin, notebook to Kirk, "home hits" to Spear, and exchanges to Washburn. The board dubbed Otto the "Chief," a title he held the rest of his life.

Otto's election as managing editor from among the four eligible juniors—Folin, Sikes, Wallin, and Kirk—may have been boosted by Sikes and Wallin. Sikes, who worked as a compositor for the *Minneapolis Tribune*, had too many irons in the fire to be managing editor himself, with part-time work and extracurricular activities that included athletics, the campus YMCA, and a new fraternity on campus. Although Madeleine was also an excellent writer, it was she probably more than anyone who perceived Otto's potential as the "chief" and wanted him to take the job.

The first issue of the *Ariel* under Otto's editorship led off with an interesting and sound statement of what its hopes and purposes for the year would be.

SALUTAMUS—It is with pleasure and pride, commingled with fear of irksome though profitable toil, that the new board takes charge of what it considers one of the neatest and best college papers published. In light of this opinion it is with some misgivings that any set of editors would take upon itself the maintenance of a paper of the standard of the ARIEL. Yet we can but do our best. In this attempt we shall take it for granted that the world in general and the University in particular is in pretty fair condition, after all. We shall be content to let them go in their steady and conservative course. We shall not attempt to reform the journalism of today; nor to oppose the faculty on general principles; nor to bruise our editorial head by using it too vigorously as a battering ram to storm the ramparts of the marking system and examinations: neither are we so sure about the feasibility and advisability of entirely altering the present plan of conducting the commencement exercises of the University—in short, we shall not expect the world to make too big a lunge, just because we have hold of the lever.

However, we shall raise our voice and humbly but persistently urge the football team to maintain its position as champion of the Northwest; we shall find fault with the baseball spirit if the University does not do better in that line than it has in the past; we shall advocate a gymnasium, or at least some gymnasium facilities; we shall call the attention of the regents to the fact that sandburrs on the campus are a thorn in the flesh and not conducive to interest in athletics; we shall advocate a good field-day; decent behavior and less whispering in chapel; proper use of the privileges of the library; more interest in oratory and the Oratorical association; good will and kindly relations among the students; good feeling among the departments of the University; department representation on the ARIEL. We shall ask for the abolition of pie at ARIEL elections and, in fact, of treating at all elections; we shall favor giving the Freshmen a vote in the ARIEL association hereafter; we shall remind the professors not to delay the preparation of the list of books they are to use the coming term so long that the students shall have to wait for them a week or two; we favor a biennial visit from the legislature, and also hope these same legislators will cease threatening to disturb our institution by removing the Agricultural school; we shall suggest to the students who are forever carrying on a conversation in class that they are an annoyance to others and a pest to the professor who is trying to conduct a recitation; we may inadvertently make bold to remark that there is a petty evil under the sun, viz. that staid old professors are occasionally inclined to discriminate in favor of winsome students of the fair sex; we shall advocate the hearty support of all worthy enterprises and organizations; better intercollegiate relations with Western colleges. In short, there is need for more unification of student interest in University affairs. And lastly, we advocate making the ARIEL a weekly. The present board is fortunate in being brought into existence at a time ripe for improvement. A monthly, however good, is no longer adequate to the needs of an institution like our own. We must have a weekly, and a first-class one too.

What is the purpose of the ideal college journal, anyway? We believe it is not primarily to set forth literary matter. It is, first, to give a picture of university life—to reflect the life of the school as it is. Thus news—a comprehensive record of what is transpiring—and comments upon the same should be the principal material for publication. Secondly, the college paper is, or should be, the mouthpiece of the students and should represent them.

The afore mentioned are the two principal aims of college journalism. In addition to these the college paper represents the school abroad, and should give a good impression of the institution from which it is issued. It should help to keep the alumni in touch with their alma mater. A good live paper serves to increase the college spirit and to unite the students. (*Ariel*, April 29, 1891)

The conversion of the *Ariel* to a weekly took place soon after Otto and the rest of the editorial board assumed their positions.

Mr. Folin remains managing editor. The Literary and Note Book departments have been abolished. Hereafter the managing editor will attend to the securing of literary matter and will expect contributions for that purpose from the student body. Mr. Kirk, who was the Note Book editor, will edit department news and Messrs. Spear and Manuel will be the local writers, or "Home Hitters." Mr. Washburn will edit the Exchange Department as formerly. Mr. Brabee will represent the Medical department. When the departments of Law and Agriculture select their representatives the staff will be complete. It is hoped that these departments may thus be induced to take a keener interest in the paper and in the general University affairs. Miss Wallin, who had charge of the Literary department under the old system, will be the associate editorial writer. (*Ariel*, Sept. 12, 1891, pp. 1–2)

Meeting the financial requirements of the new weekly paper was a major task. The appointment of a business manager had been a step in that direction. Although not a board position, the business manager handled the "nuts and bolts" of the publishing process: advertising, contacts with the printers, circulation, materials, purchasing arrangements for meetings, billing and collecting, and record keeping. With student enrollment topping 1200, the *Ariel* board hoped to obtain 800 subscriptions to the new weekly paper, the annual subscription rate of which had been raised to $1.50. The addition of associate editors from the law, medical, and agricultural departments and a freshman as local editor also made the board more representative of the entire student body, an important point in selling subscriptions.

Nevertheless, more had to be done to make the *Ariel* an effective paying weekly that would stir up student interest and involvement. The paper of which Otto and his board assumed leadership had been mousy, small townish, apolitical, churchy, often groveling. It had slumbered along without even a marginal grasp of things that truly mattered to the students: the faculty, the city, the nation, economics, employment, the future, science, people, industry, even current literature. It had been the epitome of dilettantism, although it must be said that in an era when few colleges taught the elements of journalism, it was on the par with most college student newspapers.

In the spring when the new board took the reins, the transition issues seemed to continue in the *Ariel*'s previous vein. However, during the summer vacation of 1891, Otto must have realized some of the paper's shortcomings. The paper needed something more than prize essays from the sophomore rhetoric class. Otto was no iconoclast, but he had some long-standing resentments and a few pet peeves about the university that could serve to stimulate discussion among the students.

One of the major broadsides hurled at college life took the form of an attack on the fraternity system. However, long before the *Ariel* began its campaign, Otto was debating the merits of fraternities. In the Oct. 31 issue of the *Ariel*, a debate on the issue was reported:

> "Resolved that the existence of fraternities among students is unfavorable to the best interests of the college." Messrs. Folin and Gjerset for the affirmative held that fraternities corrupt college politics, that socially, fraternity men set an unfair disadvantage; also that since fraternities select their members on the basis of some supposed merit, the men who have had the best advantages, natural, social and educational, are taken up into some society and receive the benefits of close companionship with its members while the students who have been less favored before coming to college are left to shift for themselves when here. Messrs. Head and Philips on the negative claimed that fraternities were necessary to the life of college students and dwelt at length on the fact that the students who join them are benefitted thereby—the judges and a popular vote gave the decision for the affirmative.

Nothing bone-jarring appeared in the *Ariel* until after the Christmas holidays. However, a letter published in the Nov. 7 *Ariel* gave Otto a royal ribbing.

CITI, NOV 2 - '92

MESTER EDDITUR

    Medical Collesch
    DIR SIRR!

    I tink I is not won ov dem fellows wat do not take the "Airiel", for I read itt and hav pade won daler and a halv to Mester Edditur, so I dount hav to borrow anniboddies others papir.

Dis was not wat I wudd spik about this time how evver.—
I notised in the last ischu of the "Airiel" dat the highli learned and edducated yung men acrosst the river kannt spel to Sweden, dis is the way dey do itt.— Sweeden.
Now, will Mester edditur plise inform dem over there dat wen I wants to reed a Junivursity-paper, I wisch to do so for mine mental kullture and eddificaschun and dount kare to feind anni rong spellings in itt.

                                          Racepectfulli
                                          [MED. STUD]

The first broadside at the college fraternities came after the holidays. Up until this point Otto had been painstakingly stockpiling his ammunition and had begun to gather signatures on a resolution to be discussed among the non-fraternity students, the independents, or as termed by the Greek letter societies, "the barbarians" (barbs).

Under the pseudonym of Erasmus whom Otto had studied in one of his sophomore courses in history, he wrote a series of four articles entitled "College Fraternities." The first appeared in the Jan. 9, 1892, issue of the *Ariel*. In this essay he introduced in general terms the topic of secret societies, their merits and demerits in history. Most secret societies, he wrote, even when they were begun with good motives, eventually were corrupted so that they provoked hostilities that ruined them. They became combatants against democracy and Christianity. Fraternities (Greek letter societies) were secret societies and may have begun as a response to the hated and feared enemy, the faculty, and to fill a need for symbolism, mystery, and ostentation, which promoted bonds of union and friendship, but their role on campus had deteriorated.

> A university or college is an organization in itself. A college fraternity is an exclusive organization in the very midst of that organization—a class of students having an existence apart from their fellow students, not identified with them, not incorporated with the general society of the college; and consequently tempted on all sides to conspire against it, to prey upon it, and to keep it in disorder. (*Ariel*, Jan. 9, 1892, pp. 159–60)

In his second essay, Otto got down to the first of his specifics, perhaps the weakest of all of his arguments, namely that fraternities destroyed interest in the two literary societies at the University of Minnesota. He recalled that as a sophomore he had in one evening visited both societies. One society had only seven members present.

> I went over to the other building and counted six persons, one of whom was a lady. Now allowing that one gentleman went to the society simply to escort the lady it leaves only five who (in the language of a noted comedian) went to the society with honorable intentions. (*Ariel*, Jan. 30, 1892, pp. 195–96)

Otto maintained that only members of second-rate fraternities joined with the barbs in the literary work of the societies. "The members of the leading fraternities here consider it below them to take part in any literary work." While he admitted that poor literary societies joining with poor fraternities was not the fault of the fraternity, he nevertheless maintained that the frats substituted pleasure for the opportunity to become eloquent debaters.

> It is not only the fraternity man who neglects literary work, but I claim that it is due to the influence of fraternities that so little interest is taken on debates. A Thanksgiving reception where the ladies may exhibit their arms and the gentlemen may vie with one another in displaying social polish attracts a crowd, but a joint debate between Delta Sigma and Hermean literary societies has not in the last three years attracted enough people to fill the pit of the chapel. (*Ariel*, Jan. 30, 1892, pp. 195–96)

In his third article, Otto considered the effect of fraternities on its members and on non-members.

> We have read about conceited and unbearable freshmen being reduced to very modest and unassuming people. Men aware of their true ability and proper station. But it is safe to say that this never happens where fraternity influence predominates. Such a system is more prolific in snobs and miniature "tin gods" than in men who have gained a reputation through ability and industry. In fact a man of strong individuality and character is not an ideal fraternity man, and if such a man perchance gets into a fraternity, it needs all the smoothing power of a machine to keep that individuality from asserting itself. (*Ariel*, Feb. 13, 1892, pp. 218–20)

The qualities needed to enter the leading fraternities were wealth, social standing, and personal appearance, he continued. Good students were sometimes taken into the second rate fraternities to "bolster them up." Look at the frats, Otto said, they ignore students, orators, and athletes, and grab men who are thought to have social standing.

> To say that a man is a good student because he is a fraternity man is nonsense. Fraternities impose friendships on their members, and in advance for their high school pledges. In fact, to finish college, some members must stay for a year or two beyond their expected graduation time.

Non-members were socially affected by the fraternities, because the latter took those with social polish. "The men who come here from the farming prairies of Minnesota and Dakota whose chances for social development have been slight, are not only not helped socially, but they are actually ostracized from society." The non-members were practically exiled, socially quarantined. "It is an utter and unequivocal opposition to every peculiarity of free institutions. It not merely lays half our students under the social ban, but it deinstitutionalizes (I may use the expression) our university."

His concluding essay on college fraternities added a few points and recapitulated. Men joined fraternities because they were considered a promotion based on their exclusivism, out of a desire for popularity, to be "in" with the girls. It was the fraternity system that seemed to justify the assertion that the university was an institution for the rich man's son. The frats also interfered in college politics. Their votes were ever ready to be bought, sold, or traded. While the barbs nominated a man on his merit, a frat man was run because he had the 15 to 20 frat votes needed to get the nomination, without regard for his abilities. Be a frat man and get ahead! Exclusion? Even a freshman fraternity man would not sit with a freshman barb in the chapel. The fraternities caused dissension among students and lifelong bad feeling. The frat men were more loyal to their fraternity than to the university. The barb became bitter about his Alma Mater. "It appears to me that there ought to be no place in American Colleges for organizations so undemocratic." Democracy and Christianity were at war with exclusivism. (*Ariel*, Feb. 27, 1892, pp. 243–45)

Two rejoinders in the March 12 issue and three in the March 19 issue indicated that Otto had succeeded in stirring up the campus somewhat. The first of the pro-fraternity responses came from M. Bryo. Entitled ERASMUS ERRATUS, it ridiculed Otto by writing that though Erasmus died in the 16th century, the Great Creator returned him in the latter part of the 19th century to a world that had deteriorated.

> He gathered his bones together, placed a pen in his hand and the Book of Truth before him and told him first to write a letter of warning, forgetting in His haste that the eyes were gone and the skull was empty. Even mute matter must obey His commands and the letter was written. Do you wonder at its content?

He defended exclusivism on the basis that the Christian church was at one time the greatest secret society in existence, a fact which did not prevent its members from obeying the Ten Commandments or living as nobly as nonmembers. Christ chose from among his followers a few with whom he associated closely. Democracy too allows choices among those with whom we associate. Fraternities, he asserted, were one form "of the great plan of the survival of the fittest, and who would not complain if he did not survive?"

Bryo defended the literary societies, pointing out that attendance was generally between 20 and 90 and fell only before examinations. "In an institution where non-fraternity men outnumber fraternity men, the wail that the latter are responsible for poor literary societies comes with poor grace from the lips of the former."

The second pro-frat article, ONE OR TWO SUGGESTIONS, argued that the University of Minnesota was democratic, making possible things that were unavailable in private schools. Students were not excluded on the basis of race, religion, politics, fraternal inclinations, dress, tobacco, or "beer." Everyone looked to his personal affairs. No fraternity was directly immoral. They were off campus and should not be subject to more discussion than the churches and political parties. Social circles were natural. Barbarians had the same rights as the Greeks, but were slower to avail themselves of them. Attaining stability of organization was the reason for limiting membership and unanimous election.

Like the second reply, the third pro-frat response was unsigned. It presented little real argument, but ended with an ironic note. "But bitter experience has taught the world that all men were not born free and equal, and that as long as nature shall continue to make men after different molds, they will walk in different paths."

Support for Erasmus's viewpoint came from an independent, signed Non-Survivor, who attacked Bryo's ridicule of Erasmus and his misrepresentation of Erasmus' views.

> Many of us have believed that the fraternities were thoughtless rather than intentionally unkind. We are now forced to believe that their attitude towards the barbs in general is that of intentional slight and contempt... unprejudiced men whose opinions are worthy of respect, and who have had wise opportunities of observing, consider fraternities detrimental to literary societies.

Attributing part of the decline in strength of the Student Christian Association to the neglect of the fraternities, the writer concluded that if there were more of the Christian spirit in the secret societies there would be less lonely, friendless people on the campus.

> We believe that in some colleges where there are suitable restrictions, fraternities may have their proper place. But as they exist here—totally unrestricted, growing more and more exclusive and arrogant, those of doubtful position scrambling for surer social footing—we are unable to see that they carry out the teachings of the scripture as our worthy Greek friend would have us believe.

In the March 19 issue, George Sikes attempted to discuss objectively all the arguments provoked by the fraternity question. The articles had attracted much attention, he noted, and had probably caused some strained relations among the students, which needed to be avoided. To this end, he believed that passion and prejudice should be eliminated in favor of reason and sober judgment.

Sike's analysis of the articles by Erasmus credits them with the most substantive arguments in the debate, albeit containing several untruths and omitting facts that did not support the arguments. In addition, Erasmus "exhibits so much of bitter prejudice at times and even of misstatement that one seeking the truth would be unwise to allow himself to be much influenced." To say that the best frat men were those with more conditions than credits, he continued, was untrue and ridiculous because a student who cannot pass his courses is a disadvantage to any society.

Sikes found little that was worthwhile in Bryo's letter, lashing out at him for being a poor spokesman for the fraternities. His facetious tirade against Erasmus comprised too much of his argument, and his assertion that fraternities were a form of survival of the fittest could not be condoned.

Otto's independent supporter did not miss Sikes' attention. Although some parts of the argument were good, Sikes contended, portions were "more objectionable than any that had previously appeared," because they misrepresented the religious life at the University of Minnesota. The fact was that religious life was gaining strength.

Perhaps the most objective and useful response to Otto's articles came from Grace M. Rhoades in *Ariel*'s LITERARY column (April 30, 1892, pp. 326–28). She congratulated Eramus for courageously inaugurating the discussion. His treatment of the subject was ably presented, deserving of respect and courtesy, she asserted, and his arguments should be met with arguments, not with ridicule and insult.

Our years in college should be happy, she declared, and for that we need friends. This need for friendship underlies the formation of the secret society. In the process, however, the barb is cut off from that society and deprived of his rights. Why doesn't the barb find friendship among the barbs? Sometimes he does, she continued, but because the secret societies select out people, those left might not suit each other. The barb ranks are constantly changing by those leaving the university, or those going into the secret societies.

The reason for the barbs' bitterness Rhoades found easy to understand.

> It is hard for people to be charitable and well disposed when they are lonesome and dissatisfied. . . . It should be everybody's business to see that all new students are successfully launched into University life, regardless of fraternity distinctions. There is much need of more *general sociability* and *general college society*, to promote good feeling and University spirit. (*Ariel*, April 30, 1892, pp. 326–28)

Otto's last issue as managing editor of the *Ariel* contained his outstanding journalistic triumph, an article under his own name, entitled "Our Secret Fraternities." By this time, Otto had transformed his anger and despair into an undeniably logical and clear argument that the independents were not getting a fair shake socially. The fraternities had defended themselves poorly, choosing the "contempt of silence." No evidence had been presented to show that the evils complained of by Erasmus and the Non-Survivor were imaginary or not caused by the existence of fraternities.

The pièce de résistance of Otto's argument came in the form of a resolution that he had circulated among the independents. For the first time Otto asked the university administration to confront the problem created by fraternities. The resolution read as follows:

> We, the undersigned students of the University of Minnesota, hereby declare that we consider the Greek Letter Fraternity System as represented at our University a detriment to our best interests and to the best interests of this institution. We believe that such secret, exclusive, permanent, intercollegiate organizations infringe upon the rights of those students who are not members.
> That they create artificial social barriers and thereby produce a series of distinct social classes in which series non-fraternity students, as a class, are considered outsiders and in social standing below the lowest fraternity;
> That the principles upon which social organizations are founded are wrong and contrary to the principles of PURE DEMOCRACY and FREE COMPETITION and that for these reasons they have no legitimate place in a state institution.

For obvious reasons the signatures to the above resolution were not printed. It has been signed as follows:

Postgraduates 4
Total number of barb postgraduates 4
Seniors 24
Total number of barb seniors 27
Juniors 18*
Total number of barb seniors 42
    Total 73 46

*The greater number of barb juniors have not yet been seen. As far as possible their opinion as well as that of the barbs of other classes shall be obtained.

The resolution gave clear backing to Otto's contention that the fraternity system made barbs feel ostracized, and in its design and careful gathering of data, spoke of Otto's solid scientific approach to a problem. The resolution was circulated among junior, senior, and postgraduate barbs, and the number of signers was compared to the total available. (To be fair, Otto had not approached underclassmen, who had not been at the university long enough to understand the situation and to form independent opinions, even though doing so could have easily doubled the number of signatures.) Otto made the list of signatures available for inspection to any with the right to do so.

In subtle irony Otto wrote, "For my own part I am perfectly willing to admit that the fraternity people of our University as a class are better in every respect than the non-fraternity people. Why shouldn't they be when they are *selected*?" Most barbs were so from necessity, but it was an honor to be a fraternity man. Then he took a poke at the faculty. The board of the university's yearbook, the *Gopher*, had added the Greek letters to the names and titles of the professors in the '92 edition "to give full recognition to their attainments!"

Fraternities did provide social polish to their members.

> One of the most potent factors that go to make college-bred men and women superior to their uneducated fellow men is the refinement and culture which they unconsciously acquire at college by associating with their superiors. All are agreed that in this respect fraternity people as a class are superior to the barbs.

The problem was, as Otto saw it, that all students wanted to improve themselves, and the barbs couldn't do so with the fraternities in existence. The State of Minnesota had intended to give all an equal chance for improvement.

> If anything approaching a democratic condition of society can ever be attained it ought to be in a state university where human depravity, ignorance and narrow-mindedness have been reduced to a minimum. And the spirit at our University *is* democratic, but we are drifting into an aristocracy.

Otto spelled out the university's role in fostering the fraternity system. First, the university had failed to prohibit secret societies when it had had the power to do so 17 years before. The administration had assumed that fraternities would disappear on their own. Second, the university catalogue did not mention the fraternities and failed to inform students and parents of the social environment that they should expect for four years if they were or were not members of a fraternity.

> The barbs are certainly not unreasonable in demanding that the parents throughout the state who send their children here to be educated be given some reliable information as to the social environment in which those children are to spend the four most important years of their lives.

This article was probably Otto's finest moment at the University of Minnesota. He had bared a small part of his soul, a resentment against any encroaching aristocracy on his idealistic view of democracy. He had become increasingly aware of his shortcomings in social refinement. He had, however, in a small way, struck a blow at snobbery and pretense in college life, and had thrown the ball in the lap of the administration. Theirs was the power to remedy the situation.

In that year not only did the *Ariel* become a weekly, it became a campus newspaper that covered the entire student body instead of being a literary magazine. In its initial season as a newspaper only Otto's fraternity essays stirred any controversy, but nevertheless the publication had become a bit more peppery and relevant.

Much of Otto's life outside the *Ariel* offices was spent in the recently completed chemistry and physics building, which as mentioned earlier, had been built on the site of the old agricultural building. The two-story rectangular structure of Roman brick with a red sandstone basement provided the chemistry department with 21 rooms on the west side of the facility. In addition to large lecture rooms, offices for the teaching staff, and ample laboratory space for general and organic chemistry, qualitative and quantitative analysis, special rooms were set aside for weighing, microscopy, spectroscopy, water analysis, assaying, and furnace work. Provisions had been made for 80 student work tables although in the class of 1892 there were probably few chemistry majors besides Otto to make use of them. Oddly no library had been specified, indicating the heavy emphasis on teaching undergraduates. (For the historically curious, equipment available in the new building is listed in Note 2-3.)

Otto took 24 credits of chemistry in his senior year, comprised of two four-hour courses in each of the three terms. The first term consisted of organic and theoretical chemistry, but no records remain to indicate what subject matter was covered in the last two terms. Because he had exhausted all of the courses formally offered, in all probability he studied special topics in chemistry under Dodge's guidance either in reading the literature or in experiments in the laboratory. (See Note 2-4 for more information on Otto's grades.)

In this period when he spent so much of his time with Dodge, the professor must surely have stimulated Otto to consider postgraduate training and research. The 16 credits of German he took in the first two terms would indicate that Dodge had him considering that direction, for German was the language of the most preeminent chemists and journals in the world.

Of considerable importance to Otto's future was to be another professor, Harry Pratt Judson, professor of history and lecturer on pedagogy. Otto had taken Judson's course on American history in his junior year and had both enjoyed it and scored well.

Like Sanford and Dodge, Judson had a long teaching experience in secondary schools, but not at the college level. He had obtained both the A.B. (1870) and A.M. (1883) from Williams College, and joined the Minnesota faculty in 1885.

One of the most popular teachers on campus, Judson was praised in the *Ariel* for his teaching prowess.

> Professor Judson is said to have a remarkable faculty of making students do a great amount of work without their realizing it. Whatever be the secret of his power, he is certainly succeeding in inspiring a deep interest in the study of history among the students here. All enjoy the opportunities for research and original investigation which are offered. Well selected volumes of reports and documents as well as reference works are provided, from which the student prepares his work. Thus he is not bound to the ideas of other compilers, but is enabled to form opinions for himself from the fountain heads of history. That these privileges are prized will be seen from the fact that a considerable proportion of the senior class is electing United States history with the juniors. (*Ariel*, Jan. 10, 1891, p. 45)

During the summer of 1891, Judson was recruited by the newly appointed president of the planned, but unbuilt, University of Chicago, William Rainey Harper, and a tentative agreement had been reached for Judson to serve as Harper's chief of staff and surrogate in the overwhelming task of organizing the new university. In January 1892, he was appointed professor of history and head dean of the colleges at the then remarkable salary of $7000 a year. About this appointment Judson wrote:

> In the winter (1891–92) it became time to settle definitely whether I should or should not come to Chicago, and some matters had taken such shape that I sent word to Dr. Harper that, on the whole, I didn't think it advisable for me to leave Minneapolis. However, he had anticipated me by having the appointment made by the Board of Trustees, and an announcement made in the press before my letter reached him. Of course that committed me so that I could hardly do otherwise than accept the original proposition.
>
> I must confess that the new scheme, while very attractive to me in many ways, seemed in many other ways quite visionary. There was much in the air, but not much in the ground. When I came down that winter of 1891–92 to look over the plant I found a wilderness adjoining the projected site of part of the World's Fair. Still the possibilities were made to appear much like probabilities by Dr. Harper's enthusiasm. The matter being settled I sent in my resignation to the Board of Regents of the University of Minnesota. They thought that the Chicago plan was even more visionary than it had, at any time, appeared to

me. They did not believe that it would ever materialize in such a way as to make it a permanent institution. Therefore they voted me a year's leave of absence, without pay, to give me a chance to try the thing out. They were quite confident that at the end of that year I should return to Minneapolis.

In the final organization Judson was named acting head and then head of the department of political science instead of professor of history.

When word of Judson's new job reached the Minnesota campus, the following article appeared in the *Ariel*.

> The Chicago University, which is to open its doors next fall, has honored the University of Minnesota by selecting Prof. H. P. Judson for a high position on its faculty, and at the same time given us a severe blow in taking from us one of our best and most respected professors. . . . It has been known for some time that this offer was open to Prof. Judson, but only lately was it definitely known that he would accept. The Chicago University will be organized on a unique basis—different from that of the ordinary American college or university and modeled somewhat after the German. The work of the college and of the university are to be differentiated. The first two years of the course will correspond to the work of the freshman and sophomore classes generally. This will be known as the collegiate or undergraduate work. The years corresponding to the junior and senior years will be given to University work proper and the student will be allowed great freedom here. Those who are familiar with the early history of our own University will recognize in this a scheme very similar to one which was outlined to be adopted here as soon as practicable, but which has evidently been given up. There is to be a dean of each of these departments and under-deans for each college. Prof. Judson is to be dean of the under-graduate department, equal in rank to that of the dean of the graduate department and is a position second only to that of President Harper himself. (*Ariel*, Feb. 6, 1892, p. 206)

At least seven members of the class of '92 followed Judson to the University of Chicago that autumn, probably due in large part to his practical advice and encouragement.

One of those who enrolled in Chicago's graduate program was Madeleine Wallin. Otto and Madeleine, who had shared Judson's American history course, both felt Judson's influence, and his departure must have stimulated long conversations about their academic futures.

Thus though Dodge must have been the one who interested Otto in the possibility of postgraduate training, Madeleine and Judson were probably the major influences in his decision to apply to the University of Chicago. If he applied anywhere else is unknown, but his finances certainly would have prevented him from considering the costly private schools.

The fact is that Otto probably did not consider the possibility of graduate school until his senior year. Otherwise he would have taken more prerequisites in Latin and physics and perhaps more mathematics. His much admired teachers, Sanford, Dodge, and Judson, had acquired advanced degrees by teaching high school and pursuing the degree through summer programs at various universities. Otto may have intended to follow their example, but

Judson had changed his thinking by emphasizing that it was distinctly feasible for Otto to attend Chicago in the fall.

The university that was taking shape in Chicago was based on a plan that was nothing less than that proposed by Folwell for Minnesota, but in this private school, which had the backing of John D. Rockefeller, the possibility of its becoming reality was distinctly better. Great emphasis was to be placed on the university work, the upper two classes, the graduate school, and publications. Judson himself was to head the undergraduate junior colleges, but he was to be a policymaker in all areas of the university administration. While at its outset the university would open under a cloud of disadvantages, these were primarily physical and temporary, because so much construction was occurring and the interim housing of the departments would be poor. To counter this there would be a superb faculty, and a president who was a young genius, a practical visionary who could inspire a great university into existence.

Above all, the real needs of the students were to be given solid consideration. A four-quarter system would keep the doors open to full operation during the summer. Tuition would be $40 per quarter including the library and incidentals fee. The university would provide housing (and control the code of behavior therein) for all students at a minimum cost or approve housing with private families. In addition, an effort would be made through an employment bureau to help needy students find work.

Beyond the students' physical needs, the new university intended to help meet their social and spiritual needs, a distinct difference from the University of Minnesota, which had attended little to the student's life outside of the classroom, social or athletic. To Otto's pleasant surprise, the secret societies that had dominated social life on the Minnesota campus were unreservedly disliked by the administration. The importance of athletics, however, were well recognized by President Harper, and with his passion for excellence, he hired the great Yale athlete, Amos Alonzo Stagg, to become the head of the physical education program.

So the new university was to be both Otto's and Madeleine's destination the following fall. George Sikes too must have been stirred by all the discussion of Chicago, but an appointment as a reporter with the *Minneapolis Tribune* was his chance to launch a career in journalism, and its lure was strong enough that he decided to postpone applying for admission.

Besides all the other events of his senior year, Otto had an event in his family that must have brought him joy. In April of that year, Otto returned to Stillwater to attend the wedding of his brother Alfred George to Mathilde Bengston in the local Lutheran church. Alfred and his wife were to live in Stillwater in the house he had bought from Axel. By then Axel and Anna Marie had realized their dream and moved to a farm near Center City.

Following graduation, Otto may have stayed and worked in the Minneapolis–St. Paul area. His application for a fellowship in chemistry had been turned down, offering no relief to his perpetual need for money (only four fellowships in chemistry were available, and there were 13 applicants). Thus he again had to seek work to support himself during the school year, but the university's employment bureau was there to help him if he had need of it.

References and Notes

2-1. Gray, J. The University of Minnesota, 1885–1951. pp. 1–115. Minneapolis, Minn.: University of Minnesota Press, 1951.

2-2. Carr, J. D., The School of Chemistry. In: *The University of Minnesota and the Men Who Made It*. Chap. XIX. Minneapolis, Minn.: Star Publishing Co., 1907.

2-3. Three analytical balances; three microscopes; a Bunsen's model large spectroscope, a Scheibler's saccharimeter; a photometer; apparatus for gas analysis; a gas furnace for organic analysis; a set of apparatus for projecting spectra and for similar lecture room experiments; a number of technological wall charts; several gas generators; an ice-making machine using ammonia; a number of steam waterbaths; a steam distilling apparatus; three furnaces for crucibles and similar work; and 2400 reagent bottles.

2-4. Otto took eight credits (two four-hour courses) of chemistry in each of his last three terms and received scores of 98, 98 (organic, theoretical); 96, 94; and 80, 80.

Chapter 3. Chicago University—Organic Chemistry, With Love (1892–96)

The train ride from Minnesota to Chicago that fall must have evoked memories for Otto of a similar trip 10 years before, but that trip was from Chicago to Minnesota and his life in his new country had barely begun. He had few memories of Chicago, though. He and a number of his countrymen who had been together on the train from Philadelphia had been gathered in a building after they arrived at the depot. A Swedish-speaking man wearing a cap with a gold band around it—like a Swedish officer's—demanded to be paid in gold for a 35-cent inedible snack. Otto was among those who were daring or hungry enough to venture into another room for the food, but Otto fortunately had already learned to count American coins and paid the man with the correct change. In Otto's case the man had decided not to press the gold issue, but Otto had thought, "Wouldn't it have been fun if one could have been only disguised as an emigrant and after handing him a five-dollar goldpiece demand the change in *gold*?" Now Otto was returning to this huge city of more than a million, an American citizen, a college graduate, a budding chemist, and above all, a man who would now demand his change in gold.

One can imagine the amazement in the eyes of the 742 students arriving on the campus of the University of Chicago in late September 1892. Two vast unrelated areas of construction dominated the scene. To the south and east of the university were the gargantuan earth-moving and building efforts of the Columbian Exposition, the Chicago World's Fair—no less than 200 buildings were under construction, with the completion date for all of them set for the formal opening of the exposition on May 1, 1893—and then there was the university itself rising like a phoenix to the north of the fair's midway.

> Everything was new and everything was incomplete. The site had received much attention from Daniel L. Shorey, one of the Trustees, but in large part was still in its natural state. The western side was flat, but dry and covered with young oaks. The southeast quarter was like it. But these two sides were separated by low ground which was a morass in the spring, being lowest just east of where Haskell later stood, and here there was standing water for much of the year. There were a few board walks, but only a few. There was no gymnasium for Mr. Stagg's athletes, and no building for what was already a

great library. A gymnasium and library building, temporary in construction, was under way and became available at the end of the first quarter. This building, poor and unsightly as it was, was an invaluable addition to the facilities of the institution. Half a dozen other buildings, the Kent Chemical Laboratory, the Walker Museum, Foster, Kelly, Beecher, and Snell dormitories, were being constructed and the campus was covered with piles of earth, and with brick, stone, iron, lumber, every kind of building material, and swarming with workmen as well with young men and women going to and from their recitations. The professors made their way about as well as they could, dodging teams, avoiding derricks, but rejoicing in the promise of increased facilities. They needed these badly. The scientific departments had none whatever on the campus. A four-story brick building on the southwest corner of Fifty-fifth Street and University Avenue, divided into store rooms below and apartments for flat-dwellers above, had been rented for them and into these narrow quarters the biological departments and Physics, Chemistry, and Geology were crowded, and here they tried to do their work through the whole of the first year. As one of the professors said at the laying of the cornerstones of the four biological laboratories: "Our earlier days in the University were spent in the garrets and kitchens of a tenement house." But somehow the departments were housed, and the great enterprise was got under way. (3-1)

A student's view of the opening of the university is provided by Madeleine Wallin in an article she wrote for the *Ariel* in April 1893.

FIRST IMPRESSIONS OF THE UNIVERSITY OF CHICAGO

"The way to resume is to resume," said Horace Greeley, a good many years ago, in regard to an important step in the national finances. Dr. Harper, of the new Chicago University, has acted upon the same principle in opening the great institution under his charge, evidently believing that the way to begin is to begin. Nothing could have been more informal than the opening exercises of the University, which took place on October first; and yet there was about the whole institution, even in its first days, an air of quiet strength, a consciousness of power, that was marvelous. The vastness of material resources is apparent, not only in the extent of the grounds and the number of buildings already up or in process of erection, but in the smaller details of administration, and in the general atmosphere of the place. The bigness of the plan makes itself felt everywhere. Nothing is done on a small scale. Great sums of money are behind every enterprise that is undertaken, and the stability of this financial foundation goes far to insure success in all departments of endeavor.

The intellectual equipment is no less rich and strong than the material. As everyone knows, the faculties of the finest universities of Europe and America have been searched to make up the teaching force of the University of Chicago, and a hundred and twenty men and women are registered on its faculty roll. Next to Dr. Harper, probably the most influential man on the faculty is Professor H. P. Judson, head dean of the colleges, and professor of political science and constitutional history. His winning personality has already made him a favorite among the students, as his fine abilities have given him his important status in the faculty. Without doubt he is the most popular instructor in the university; and we who admired and loved him in Minnesota, are proud of the ready recognition he has won in his surroundings.

A. A. Stagg, the noted Yale athlete, is the director of the department of physical culture. He made his first appearance before the students at a meeting called on Saturday afternoon, October first, to choose a "college yell." A goodly company of students, including a number of "coeds," were present, and Mr. Stagg, dressed in a football suit and sweater, and looking decidedly tough, led the "trial yells," starting the boys off each time with a wonderfully infectious vim and "go" all his own. The next morning, Sunday, at nine o'clock, behold Mr. Stagg, in black coat and irreproachable tie, leading the first student prayer meeting and starting the gospel songs with the same spirit that had sent off the college yells the day before—for Mr. Stagg does what has been irreverently called the "heavy pious," combining athletics and religion in a masterly manner, with equal success in each. The football team organized by him this fall, played very well for a young team; but it will probably be some time before they will be able to compete successfully with the "Minnesota Hustlers" who did such terrible execution in the field last season. . . .

In many ways the University of Chicago is wholly unique. It is at once an experiment and an assured success. The members of the faculty have come here with entire faith in the future of the University, because of their faith in Dr. Harper, and in the creation which his mind and Mr. Rockefeller's money are to evolve upon Chicago soil. They all seem to have given themselves with complete enthusiasm and without reserve to the work before them; but despite that fundamental feeling, there is an upper current of surprise and some little amusement at finding themselves here at all. Most of the eastern professors (and the tone of the faculty is decidedly eastern) have come here against the vigorous remonstrances of their friends in the older communities, who cannot conceive of anything so rash as to leave the established intellectual centers of the east for the raw atmosphere of pork packing Chicago. The memory of these protests, and perhaps of their own up-rooted traditions and beliefs, hovered like a mist about the real enthusiasm of every member of the faculty, and gave at first an air of individual uncertainty and tentativeness to the institution, while collectively there was no such feeling at all. Moreover, when it is remembered that the faculty are entirely unaccustomed to working together, and have not had the time to become acquainted with each other's methods and purposes, it must be conceded that the progress of the university, even in these short months, has been little short of marvelous.

The University of Chicago is unique also in the proportion of its graduate students. Out of the five hundred and twenty-six students who had matriculated up to Saturday night, October first, one hundred and twenty-six were graduates of other colleges. The fact only emphasizes Dr. Harper's thought in an informal address to the graduate students on the day before the university opened:

"It would have been comparatively easy," said he, "to gather together a faculty capable of instructing undergraduates. The work of organization would have been definitely more simple had that been our aim. But we are determined that our instructors shall be capable of directing the studies of advanced students and that the emphasis of this university, both in faculty and equipment, shall be placed upon that grade of work. At Yale and at Harvard, undergraduate work gives color to the whole university, owing to the history and traditions of those Institutions. It shall not do so here. Our record shall be from the top down, and the frame of this university shall rest primarily upon the quality of its graduate work. We intend, therefore, that the spirit that characterizes advanced intellectual work shall prevail throughout the whole university, from the highest student to the lowest. If I were asked to give the key note, the prevailing character of the work to be done in this institution, I should say that it was the individual element. We are not here primarily to teach you what other men have declared to be true; we are here to help you to find and declare truth for yourselves. In every class room of the university, individual and original effort will be encouraged in every possible way. If there is no provision for an especial department of study in which any student feels himself qualified to excel, let him make that lack known, and there will be every effort to provide the opportunity he seeks. The cultivation of peculiar and original gifts—the growth of the individual along the lines marked out for him by natural endowment is the particular province and aim of this university."

Nothing about the institution, perhaps, is more striking than the disparity in age among the graduate students. In nearly every class there is a sprinkling of gray-haired men and women, some of whom have been teachers and professors for years, and others to whom the advantages of special training have been denied, and who eagerly embrace this opportunity to broaden their knowledge.

A great deal might be said about the social and religious features of the university. Mrs. Alice Freeman Palmer, head dean of women, is to spend twelve weeks each year in residence at the university. Hers is a charming personality, and all things gracious and womanly seem to bloom naturally in her presence. She is ably assisted by Miss Talbot, her second in command at the "Beatrice" as the Women's Dormitory is called. Afternoon tea, served every Wednesday by Miss Talbot, assisted by a number of the university girls, is one of the attractions afforded by the "Beatrice." To these gatherings the professors are invited; and not all of them are beyond the age where young ladies might be supposed to feel an interest in them—in fact, there are several of a strictly eligible age, and some of them are quite socially inclined. . . .

Another development of the university life, which may be classed as both social and intellectual, is the formation of clubs in almost every department. Every week or fortnight, groups of investigators in the classics, the sciences, political economy, political science and history, the romance and the semitic languages, meet to listen to the reading of papers or to the discussion of disputed points by the members themselves, or by outsiders engaged in the same lines of work. These clubs not only serve to carry the interest of particular lines of study beyond the limits of the class room, thus making them a part of the independent thinking of the students, but they draw together in close intellectual sympathy teachers and learners; for in these informal meetings the teachers are often learners, and the learners, teachers—both meeting on the common ground of scientific investigation. In order that these clubs, while each pursuing its work along individual lines, shall still feel the guiding hand of university direction and allegiance, the whole number are joined together in the university union, or federation of all the clubs. The council of the university union is composed of one representative from each of the clubs, and this council forms a sort of executive committee, whose duty it is to attend to any business that may affect the common interest of the clubs.

The religious life of the place is organized on the same scale as all other undertakings. There is a marked similarity between the religious and intellectual organizations of the university, except that in the former case, the central body was formed first, and smaller organizations emanate from it. The central body is known as the Christian Union of the University of Chicago, and under its broad principles of liberality it is hoped that every shade of religious aspiration may find shelter and support. It has no creed, and no condition for membership except the signing of a constitution which outlines in the simplest manner the purpose of the union—the promotion of religious feeling without the sacrifice of individual religious conceptions. The plan seems to afford a solution for the difficult problem of religious organization. . . .

When Otto arrived on campus to register and pay his $5 matriculation fee, the apartment building that was to house the science departments was not yet available and would not be for another two weeks. The chemistry department was to have laboratories, lecture rooms, and storerooms in three large rooms on the first floor where it would remain until New Year's 1894.

Selected to be chairman of the chemistry department was John Ulric Nef, who came to Chicago from Clark University. The Swiss-born chemist had been an outstanding student at Harvard and was awarded the Kirkland Traveling Fellowship that allowed him to spend three years in Europe where he received his doctorate in 1886 in Munich under the great organic chemist Adolph von Baeyer. Baeyer considered Nef the most brilliant of all the students he had had in Munich and they maintained a friendly correspondence throughout the rest of Baeyer's life. After receiving his doctorate, Nef spent another year with Baeyer. Upon his return to the U.S. in 1887, Nef taught at Purdue University for a while before moving to Clark.

Nef wanted to model his department on his own solid background as a graduate and postdoctoral student in Germany. As he had written President Harper in May 1892,

> I am really convinced that it will be possible to develop a school of Chemistry at Chicago comparable with the best in Germany. That is my great ambition and to that aim and object I am willing to devote all the strength and enthusiasm that I possess. There is not the slightest reason why this country should not develop men who love chemistry for its own sake and who will be willing to sacrifice much for it. . . . (3-2)

As his statement indicates, he believed a scientist should be married to his science (although he later departed from his own dictum by marrying one of his students).

To achieve his goals for the department, Nef became a stern taskmaster, driving himself with as much intensity as he did his graduate students. He chose outstanding faculty, who in anticipation of better things to come, came to the university at pitifully low pay and rank.

Nef insisted that his graduate students be trained in all aspects of chemistry.

> Special stress will be placed on thorough preparation and symmetrical development of the student's knowledge. The object of the courses will be not so much to train specialists as to prepare the student to undertake intelligibly any and every kind of work of a chemical nature.

To be accepted as a candidate for the Doctor of Philosophy degree, a student had "to satisfy the instructors in the different chemical branches that their previous training has been a broad one, and no degree of proficiency in a special branch will be accepted as the equivalent of this." (3-3)

For Otto this stipulation meant that, before he could even begin his research, he had to complete two years of course work, the first year of which included courses in general inorganic chemistry, organic preparations, and organic chemistry. (How this variety of courses, and all the others, were offered in three rooms on the first floor of an apartment house remains a mystery since there are no pictures extant.) In addition, he may have been required to attend the weekly or biweekly chemical seminars (journal meetings) in which instructors and advanced students participated.

The nature of Chicago's chemistry department differed radically from the one Otto had experienced in Minnesota. Minnesota had only three faculty members and only one of those owned a doctorate. In contrast, Nef had brought in six young doctors—Henry Newlin Stokes and Edward A. Schneider as professors; Massuo Ikuta as an assistant; and James A. Lyman, Julius Stieglitz, and Felix Lengfeld as docents. Five of the seven faculty members had received doctorate or postdoctorate training in Germany and two had received training in other European countries as well. That Nef chose European-polished faculty exclusively must have been due in part to his personal experience with the high standards of chemical education there, but also to the dearth of American schools that offered advanced degrees in chemistry.

Otto's life beyond his studies and whatever job he had must have been limited. While postgraduate training in chemistry is not unique among the physical sciences, the student who aspired to emulate his professors did in effect become the "property" of the department. The faculty set standards for his specific performance, within the general requirements of the university. It was not by a quirk of character that Nef regarded chemistry as his bride. This was the universal attitude toward the field by those who were making extraordinary research contributions, particularly in Germany. Only through long hours of devoted study and laboratory trial could important research occur. The preparation for the graduate student had to be arduous.

One occasional activity for Otto outside the rigors of his chemistry classes was meeting with the group from the University of Minnesota. Madeleine's article for the *Ariel* tells of their meeting.

> The "Minnesota Delegation," as those of us who came from Minnesota are known at the University of Chicago, numbers nine at present, counting Professor Judson, who is, of course, our head and chief. We expect soon to form a Minnesota club, where we may meet more regularly than we have done, to congratulate ourselves upon having graduated or studied at the University of Minnesota, and to express in words that sentiment of affection and loyalty towards President Northrop, the faculty, and the whole institution, which never ceases to glow in the hearts of the Minnesota Delegation. No university life will ever seem so peculiarly our own as that of the University of Minnesota; no teachers will ever be more faithful and earnest; no president more respected—for though we have migrated to a new country and cast our lot with new people, we are Minnesota's loyal children still. (*Ariel*, April 29, 1893, p. 425)

In fact, it was at a meeting of the Minnesota group that he first met Laura Churchill Grant. Madeleine had invited Laura, who, although not from the

University of Minnesota, was a St. Paul native. Laura, who was a graduate of Vassar, was like Madeleine a student in political economy and political science.

Otto's and Laura's first meeting was not the material on which storybook romances thrive. In a letter written to Laura some five years later, Otto recalled their first meeting.

> I remember so well the first time that I saw you. That was in the Fall of '92 at the Beatrice. Prof. Judson had the U. of M. students for a reunion and you were brought in extra. I remember what I thought about you before I heard you speak. A delicate, nice but not particularly pretty young person; then you talked considerable in the course of the evening and you laughed in such a free and beautiful manner that you began to look quite pretty; only your eyes I think looked very weak then. Before I went home I thought that you must be a very nice girl and as we went home I heard Soares say that you seemed to be a very bright girl and that rather confirmed me in my own judgment.

At this first casual meeting, Otto may have been mildly interested, but he certainly was not overwhelmed. Laura was perhaps even less impressed with Otto. She recalled with disdain his eight-dollar suit, his generally untidy appearance, and his lack of gentlemanly polish.

The time was not yet ripe for a relationship to develop between Otto and Laura. For one thing, Otto's feelings and loyalty were with Madeleine and would be for some time to come. A letter he wrote Madeleine in December 1892 makes clear the depth of his feelings.

> I have been thinking of you a great deal during the holidays and the more I think of you the more convinced do I become that you are the noblest, the best, the truest, woman that I will ever know and the more glad am I to believe that you consider me a friend. I have never had a friend whose influence I have felt so strongly as I do yours; none to whom I will be so much indebted for whatever of success I may attain; none whose image would haunt me more if I should deservedly fall by the roadside, and none whose success or misfortune will affect me so keenly as will yours.

While Otto did not openly state his love for Madeleine, he was "ripe for the plucking" had Madeleine so wished. However, even had she thought of Otto in romantic terms, religious beliefs, or rather Otto's lack of them, were a major stumbling block. In the months to follow she made every effort to help Otto "see the light."

With the arrival of the spring of 1893 came the opening of the Columbian Exposition in all its pomp and glitter. Aside from the overwhelming distraction that the fair offered to the students grinding their way through their studies, it also provided welcome jobs to many of the students struggling to survive financially. Otto was offered a job in the Minnesota building—an exhibition extolling the leading roll Minnesota played in agriculture—and since there was no summer quarter at the university, the opportunity allowed him to remain the summer and garner some desperately needed funds. In fact the incomplete credit he took for three courses in the spring quarter indicates that he may have started working when the fair opened in May.

Otto's personal relationship with Madeleine may have continued by letter over the summer, but that is uncertain. However, when she returned in the fall, she resumed her attempts to fire his religious ardor and even briefly succeeded. But once fired up, Otto would lapse into an argument that would drive her to despair. If people could always be good and love their neighbors, he contended, then they would not rest contented while there was misery and want of every description. They would have so much to think of and do, so much pleasure and happiness of the enobling kind, that they would not need the vague, indefinite hope of Christianity held out to the forsaken and unhappy people. The "cause of our social ills is selfishness (in the ordinary sense) in part of the people who are comfortably good and respectable, the ruling 'classes,'" he asserted. The better classes are the cause of social ills, but nevertheless, improvement *must* come from them, because if the evolutionary process is believed, leadership for the good must come from the most developed group, the better class.

The development of his argument from this point showed little faith in humanity. Facts show, he said, that the better class is *hypocritical*. What its members need is the true spirit of religion that goes beyond just subsidizing missionaries to the poorer classes. Conversion should thus begin with the better classes, not the worst ones as Madeleine wanted. Although most good people are only good until challenged, nevertheless reform must begin with the better classes, because it is much harder for one "deeply sunk to become good than for one who is already a good deal higher and who *has* good surroundings."

Unfortunately for their relationship, Madeleine and Otto were both too busy to find a common ground in the matter of religion and social justice. She gave up a second time at trying to change Otto's thinking and with it any romantic notions of him, if she had any.

Although she continued to see Otto (he even taught her to ride a bicycle that spring), Madeleine indeed was a young woman with very little unplanned time. She had already completed an advanced degree (she had been among the first five to do so)—a Master of Philosophy (Ph.M.) in Political Science in June 1893—and in the 1893–94 year she had returned to Chicago for further graduate study. She was involved in the Club of History and Political Science, and she was a "prime organizer" of the University Settlement League of Women that provided an outlet for students to participate in the sociological work of the nearby settlement house. Funds were raised to establish a University Settlement in the stockyards district, several miles from the university, and this Settlement House was an imposing outside interest of the Christian Union. The Hull House of Jane Addams was a model for its activities. The League of Women was particularly active in raising operating revenue for Hull House, and would remain so for many years to come. In 1894, Madeleine left the university to teach history at Smith College.

In all probability, Otto chose to keep his job until the fair closed in October, because records indicate that he took only one course in the autumn quarter. Had Nef appointed Otto a fellow at any time during his four years as a graduate student, it would have lightened his economic load and provided a much needed boost to his morale, but sadly that appointment was never made.

Numerous jobs were available in the chemistry department from preparing reagents to working in the storeroom, and Otto may have earned some expense money that way. But the move from the apartment building into the newly finished quarters of the Kent Chemical Laboratory in January 1894 must have created temporary work aplenty.

The history of the new building deserves special mention because it was finished not only in time for Otto to take his more advanced graduate courses but to work on his research thesis in comfortable surroundings.

We cite first from T. W. Goodspeed's article on Sidney Albert Kent (*3-4*).

It is said of Mr. Kent that he had determined, early in his career, to become rich. It was, perhaps, this purpose that led him into those great ventures that made him known, not only as an ordinary business man, but as an extraordinary speculator. The interesting and rather remarkable fact is that he was equally successful in both these lines of activity.

Mr. Kent's purpose to accumulate large wealth, as wealth was reckoned before our day of enormous fortunes, did not prevent him from being a man of unusual liberality. It has been said of him: "The list of Mr. Kent's public benefactions would be too long to recount. There was hardly a charity in Chicago to which he did not subscribe and no one can ever know the approximate of what he modestly gave to relieve private want." He was particularly interested in the needs of his native town. To its Literary Institution, now known as Suffield School, he made contributions, as did Mrs. Kent after his death. His great contribution to Suffield, however, was the Kent Memorial Library. For the erection of the building, the purchase of books, and the endowment of the library he provided nearly or quite $100,000.

But the greatest of his contributions was to the University of Chicago. The University was being founded while Mr. Kent was preparing to retire from active business and make Suffield his place of residence. This makes it the more surprising that he should have conceived so liberal an interest in this new Chicago enterprise. . . .

More than six months before the new institution opened its doors to students Mr. Kent informed the trustees that he had "decided to erect and furnish a building to be located on the University grounds and to be known as the Kent Chemical Hall." He wished to give the University not a sum of money, but a building. His purpose was to build a laboratory and present it completed, fully furnished, and perfectly equipped. This he did. The plans were laid before him for approval. The details connected with the work of construction were submitted to him. He paid the bills as they came in authorizing and approving all expenditures. The laboratory was dedicated in connection with the fifth Convocation, January 1, 1894, the service being held in the Kent Theater, the auditorium of the building. A letter was read from Mr. Kent in which he said: "I hereby give this building, fully furnished and completely equipped, to the University of Chicago as a chemical laboratory, for the use of this and succeeding generations.". . .

[At the convocation when President Harper spoke of the University's indebtedness to Mr. Kent, Mr. Kent sent the following note to be read to the audience.]

"If in any small measure the work of my life can contribute to the advancement of knowledge and the greater happiness of men; if this can be done in the city where my busy days have been spent and where my heart is; and if, as I believe, we, who have aided in the work of erecting this great University, have helped to lay the foundations of what can never be destroyed, I feel in this work a pride and happiness that have never been equalled in my life."...

On the wall of the entrance to the Kent Chemical Laboratory is a bronze tablet in the center of which is a bust of Mr. Kent in bas-relief with the following inscription below:

> THIS BUILDING IS DEDICATED TO A FUNDAMENTAL SCIENCE, IN THE HOPE THAT IT WILL BE A FOUNDATION STONE LAID BROAD AND DEEP FOR THE TEMPLE OF KNOWLEDGE IN WHICH AS WE LIVE WE HAVE LIFE.
>
> SIDNEY A. KENT

The Laboratory is a three-story and basement building about one hundred and eighty feet in length, with an addition in the rear, known as the Kent Theatre. It fronts south on the central quadrangle of the original University group. It is commodious and attractive structure of blue Bedford stone, like the other buildings of the University of English Gothic architecture, and built to endure for centuries....

Kent and Ryerson were the first of the University's laboratories and they set a standard which could not be lowered.

It must be remembered that this was at the very beginning of things. The only buildings under way were the divinity dormitories and the classroom building which came to be called Cobb Lecture Hall. There was no money for any others. The University was absolutely dependent for the character of its future buildings, whether, indeed, it was to have other buildings of any sort, on the generosity of donors.

We must also cite the report written by Julius Stieglitz in behalf of Professor Nef (*3-5*).

THE KENT CHEMICAL LABORATORY

To the President of the University:
Sir:
I submit herewith a sketch of the history of the Department of Chemistry from 1892 to June 30, 1902.

THE LABORATORY

[The first part of Stieglitz's report summarizes the department's move to the Kent Chemical Laboratory and the dedication ceremonies.]

In the course of nine years the work of the Department has developed until now all the laboratory space has been occupied. The following is, in the main, the present distribution of laboratory space: two large laboratories for General Chemistry, for 190 students; one large laboratory for Qualitative Analysis and General Organic Chemistry, for 110 students; one large laboratory for Qualitative Analysis, for 30 to 40 students; one large laboratory for Research Work, for 11 to 22 students; seven smaller laboratories for Preparation Work, Physical Chemistry, Furnace Work, Gas Analysis, Spectrum Analysis, Combustions, etc.; five private laboratories for the staff; two balance rooms, two store-rooms, and a number of storage rooms; a large library room.

Otto's one course in the autumn quarter of 1893 was organic chemistry, evidently taught by Professor Nef. During the winter quarter, with his department now ensconced in the Kent Laboratory, Otto continued with organic chemistry (Nef), special chapters of inorganic chemistry (Schneider), and theoretical chemistry (Lengfeld). In the spring and summer quarters, Otto completed required lecture and laboratory courses in general physics, along with advanced inorganic chemistry (Schneider). In the spring, Otto finished his third straight course in organic chemistry, again taught by Nef.

Although a reading knowledge of both French and German was required for some of the advanced courses that Otto took, there is no record of Otto's receiving non-credited training in French to meet this requirement. The likelihood is that he did not in view of his later attempt to learn the language. No indication exists on his transcript of credits to confirm that he met the language requirements.

Otto performed well, judging from the grading system used. For some chemistry courses he received As and Bs, whereas an equal number were recorded as Ps (passing). The latter marking was solely present in his final two years, hence was probably restricted to graduate-level courses, not available to undergraduate students.

During that autumn, Otto accepted another job besides his fair job, one that would ultimately affect much more than his finances. Laura Grant asked Otto to tutor her in physics and he accepted. In all probability, Laura intended a "strictly business" relationship. She had a genuine need to learn physics, prompted by the fact that she was soon to join the faculty of Ascham Hall, an elite Chicago school for girls, where she would teach mathematics in the collegiate department.

However, strictly business though the lessons may have been, by that winter Otto's interests had been peaked by their contact, and his calls on Laura were no longer just to give lessons. Because Laura like Madeleine was interested in the activities of the Christian Union, Otto asked to join her in attending services and lectures. "I am very indolent and very wicked—seldom go without some extra inducement," he wrote her on one occasion when he had missed her at her residence. "If I had seen you this evening I should have asked you if you wouldn't do a little missionary work in the way of offering to go with me some Sunday."

Trying to court Laura must have involved more than a little frustration for Otto. Laura was busy with the programs and projects of the Christian Union and absorbed in the work of the League of Women, and she must have had a schedule as full as Madeleine's with her classes and outside activities. And perhaps at this point in her life her interest in Otto did not match his interest in her. Whatever the reason, the many notes that Otto left at Foster Hall, her residence, showed that finding Laura at home was no easy task. If he couldn't find her at home in the winter, read one message of frustration, he must assume that he would never find her in when the weather was warmer. "You will be out bumming and I don't suppose that you will let me go bumming with you."

The summer of 1894 brought a temporary halt to Otto's pursuit of Laura. Although he saw her briefly in St. Paul, probably on his way back to Chicago from seeing his brothers, their paths then parted. Laura went off to the University of Missouri to tutor in mathematics and Otto remained in Chicago for the summer session.

Several events occurred in the autumn that were of major importance to Otto's future. He obtained a job as an instructor of chemistry at the McKillip Veterinary College; he began research on his dissertation for a doctoral degree; and Laura Churchill Grant began to consider him seriously as a beau.

The McKillip Veterinary College, which was opened to receive students in October 1894, offered a course of three collegiate years of six months each. Otto's task was to teach a course in elementary chemistry, probably to first-year students. The class not only provided Otto with a means of livelihood at this critical period of his career, but with an initiation into teaching.

Otto's thesis adviser was Julius Stieglitz, one of the brilliant young scholars Nef had lured to Chicago. Born in the United States but educated from the age of 14 in Germany, Stieglitz had obtained a doctorate in organic chemistry at the University of Berlin by the time he was 22. After several months with Victor Meyer at Gottingen, Stieglitz left Germany to work with Nef at Clark University. Nef and Stieglitz had much in common—broad training in Germany in chemistry and organic chemical research, a special interest in mechanisms of organic reactions such as tautomerism and the Beckmann rearrangement, and a love of music. Thus it was not unexpected that when Nef moved to the University of Chicago that he would choose Stieglitz for his department.

From his incoming rank of docent—a position borrowed from the European university system and distinguished from other ranks mostly by its low pay—Stieglitz had by this time risen to instructor. With Felix Lengfeld, another faculty member in the chemistry department, he had published a series of papers on derivatives of nitrogen halogen compounds, imidoethers of carbonic acid, and the action of phosphorus pentachloride on urethanes. Of particular interest to Stieglitz was the cause of the Beckmann rearrangement.

As Stieglitz's first doctoral student, Otto not only had the benefit of Stieglitz's guidance and leadership, but of Lengfeld's as well. Lengfeld's academic background—a doctorate from Johns Hopkins at age 25 followed by postdoctoral research at the Zurich Polytechnium, the University of Liege, and the University of Munich—fit the pattern of the other young faculty who had been attracted to the university's chemistry department.

In addition to beginning his research in the fall of 1894, Otto took the following course work for the 1894–95 school year: advanced inorganic chemistry, two courses (Lengfeld); the carbohydrates and complex hydrocarbons (Stieglitz); organic nitrogen derivatives (Stieglitz); and gas analysis (Stieglitz). The third major requirement of Otto's academic year was attending and participating in department seminars that reviewed current literature and other topics of interest and heard papers on current research in the department.

While Otto was committed to obtaining a degree in organic chemistry, his true interest was growing in an area just beginning to be appreciated in the United States, physiological chemistry. This field as a separate, distinct entity did not exist at that time except in the Sheffield Scientific School (New Haven, Conn.) under R. H. Chittenden. In Europe, the organic chemists and the physiologists, who were primarily physicians, joined their interests in chemistry so that chairs in physiological chemistry in universities were practically non-existent. The development of departments of physiological chemistry in the United States occurred within medical schools, and gained their greatest impetus after the appearance of the Flexner Report of 1910, when the quality of students and their training in the medical schools improved.

Piqued by his interest and realizing his need for a stronger background in physiology, Otto turned his attention to that department and took his first course in the spring quarter. His previous background in botany and biology proved useful, but his weakness in zoology, particularly in anatomy, was apparent. Unfortunately he was in no position at that time to make up the deficit.

One added bonus came out of his foray into the physiology department and that was the beginning of a life-long friendship with Jacques Loeb, chairman of the department and Otto's teacher.

His other major academic commitment was, of course, the departmental seminars. In addition to reviewing the current literature, hearing papers on current departmental research, and discussing other topics of interest to chemists, Otto had to present a topic of his own, at least once during the three quarters.

While Otto's life in the academic world was reasonably ordered and predictable, his personal life was not. In mid-September 1894 shortly before the autumn term started, Otto wrote to Madeleine that he was going downtown on Sunday to meet the St. Paul–Minneapolis train in the hope that Laura Grant would be aboard; she was to begin teaching at Ascham Hall on Sept. 17. She had advised him against meeting the train, because she was not sure she would be on it, but Otto, not to be put off, was prepared to spend the day in the library or art institute (he wrote Madeleine) if Laura was not on the train.

That letter to Madeleine indirectly revealed much more about his feelings for Laura.

> I shouldn't like to see her as a physicist [he had mentioned to Madeleine that he would continue tutoring Laura in physics] because she was evidently not intended for one but then I shouldn't particularly care to see her become a specialist of any kind—would you? Don't you feel like saying of her what she said of you that she ought to marry—only she ought to have a truly nice, cultured, well to do man that could always both satisfy her and appreciate her and would always be able to keep her free from the common, mean worries of how to get along with less than she is accustomed to.

Otto no doubt would have liked to become that man, but he was in no financial condition to take himself seriously enough to approach her.

In the absence of rivals, however, Otto continued his vigorous pursuit. His presence at Foster Hall had become so frequent that the women were beginning to talk, Laura informed him in a note. Hurt, Otto replied in a huff. He addressed her not as Laura but as Miss Grant. He would not go to the theatre with her next Friday because her friends would be there "looking askance at you and disapprove of me." He would continue helping her with her physics on Saturday afternoons, but if she preferred, he could find her another tutor. He still wanted to be her friend.

Otto must have quickly reassessed his response to her note, because he sent Laura an apology the next day. His letter implies that her note had mentioned several nasty remarks that her friends had made about him. Had she really known him, he tried to explain, she would have understood that he was highly sensitive to being deprecated as a foreigner. He described his early confrontation with bigotry. He had seen the disapproving gaze of Myra Reynolds when he had asked Laura to the theater, and he admitted that he should have left Foster Hall then, rather than wait for an answer. But he concluded on a more positive note. She should forget the whole matter and he would go on teaching her physics as before and not let "any tragic ghosts in the closet" disturb them. (A long encouraging letter from Madeleine that he had just received may also have made him feel a little less bitter.)

In addition to the physics lessons, Otto helped Laura set up demonstrations for her physics class at Ascham Hall. One of his notes written in November indicates that he stopped by on his way to the veterinary school to help Laura with a demonstration of the pressure-volume relationship of gases. Their social relationship, however, had evidently not resumed. In the same note, Otto complained that they had not seen each other socially since the university year began and offered to take her to the theater, as well as escort her some Sunday to hear a sermon by Rabbi Hirsh, so that he could have a long walk home with her.

In that other major part of his life, completion of his graduate work, Otto launched his research thesis. Entitled "On Urethanes," the work was to consist of three parts: the preparation of urethanes from acid bromamides, the action of phosphorus pentachloride on urethanes, and the action of phosgene on urethane. Details of his preparatory work and his thesis are found in Part II, Methods, p. 361.

During the spring, Laura continued with some reluctance to teach physics at Ascham Hall, and she not only relied heavily on Otto's help in setting up classroom demonstrations, but she began to accompany him regularly to various campus affairs, and finally to bicycle with him. In short, they became moderately steady companions.

Otto apparently sent few written messages to Laura during this period. He did, however, write to Madeleine about the progress of his relationship with Laura. "Miss Grant has agreed to go bicycling with me this spring but just how much that will mean I don't know."

To Madeleine's query on how often Otto saw Laura, he replied that he would see her two to three weeks in succession and then not at all for two or three weeks. "I suppose that I should consider it about right if I saw her three times every day on week days and four times on Sunday," he continued.

Of the letters Otto did write to Laura, one was addressed to her at Ascham Hall probably to avoid the prying eyes of the women at Foster Hall, especially those of Myra Reynolds who acted as a surrogate parent for the young women away from home. Myra apparently frowned on "dating" and short-lived romances, and she was also Laura's personal friend. The fact that Otto was a mature man of 28 and Laura was 24 did not seem to enter into it.

What Otto did during the summer of 1895 is unknown, though he must have found some kind of work in Chicago for at least part of the summer, possibly assisting in a session offered to undergraduates during the first six weeks. Madeleine invited Otto to come to her parents' summer home in Minnesota (Wallinwood), but he was not sure he could. (Her motive for the invitation is pure speculation, although it is possible that she had not entirely dismissed Otto as a matrimonial candidate and wanted to see him in the comfortable surroundings of her home, with her parents about.) He did ask if he would see her during her trip home when school was out.

In July, Otto visited Axel in Almelund. From there he wrote to Laura at her home in St. Paul, and asked if he could visit her the following week since he was stopping at a friend's in Minneapolis.

By Aug. 25, Otto was writing to Laura from Chicago. He had seen Madeleine, he told her, and waxed "a little confidential" with her about his plans after the next academic year.

He may have discussed his long-range plans with Madeleine, but he made it clear in the same letter that he wanted his short-range plans to include Laura. Already he was asking to meet her at the train station when she came to Chicago and expressing his fear that they would see little of each other in the fall. He asked her to bring her bicycle so they could at least ride back to campus together from Ascham Hall where Otto also was now involved. One cryptic remark in his letter—"I ought to be there when she [Miss Martin] is going to have the basement fixed up"—could mean that a science laboratory was being prepared at Ascham, and it is not improbable that Otto taught physics there in 1895 and 1896, as well as chemistry at the veterinary college.

His final letter of the summer (Sept. 8) covered a spectrum of topics. He told Laura that, if she wished (because Foster Hall was closed in the summer), she could stay at his place of residence, the Delta House, which was used by graduate students and the more penurious faculty. Madeleine, he noted, had stayed there during a summer term while she served as a reader in history and he had once talked to her until 11 p.m. in her room. Madeleine was also the subject of the letter's main news item. The previous week Madeleine and George Sikes had gone to Milwaukee to get married, but, he wrote without further elaboration, they got into a quarrel on the boat and didn't go through with it. Madeleine was now gone and very unhappy. Finally, Otto, the apathetic churchgoer, mentioned that "the Rabbi preached his first sermon this morning and I don't believe that I ever in my life heard a better sermon. I shall send you the newspaper report of it if there be any."

For both Laura and Otto, the 1895–96 school year would be their final one at the University of Chicago. Laura would receive her master's degree in the spring and Otto was far enough into his research to see its conclusion by spring or at least by the end of the summer.

For his final year as a graduate student, Otto's schedule included his research; courses in elementary spectrum analysis, carbohydrates (Stieglitz), organic nitrogen derivatives (Stieglitz), and physiology laboratory (Loeb); and the departmental seminar.

He had completed a good part of his research by the autumn quarter of 1895. In a letter written from his laboratory to Madeleine, who was back at Smith College, he stated that he was earning all he needed and "I ought to center all my energy this year on my research work. So far I have added four new compounds to the list of known ones and I shall get a great many more (perhaps five or six) unless I am much mistaken." He found the work interesting and he hoped that he would always have time to do a little work on his "own hook."

Otto experienced the organic chemist's joy of synthesizing and characterizing new compounds, but although he knew he could contribute original work in the field and appreciated the enthusiasm of his professors, he nevertheless knew that something was missing. Pure research without palpable, practical consequences was not what he wanted, even if it led to a major finding. Stieglitz, Lengfeld, Nef exulted in widening the field of organic chemistry, but the course in elementary physiology and the work of Loeb had stirred Otto's interests.

Loeb believed strongly in the mechanistic basis of all living things—that is, that life could be explained by its chemistry and physics. The physiologists, trained medically for the most part, were urgently in need of collaborating with chemists and of learning all they could of chemistry. Otto even toyed with the idea of studying medicine, but he did not pursue it seriously, for he had neither the time nor the money.

With their stays at the University of Chicago coming to an end, Otto and Laura had to consider where their relationship was going. Both young people had a major dilemma. If Otto wanted something to offer Laura, he needed to obtain his degree and look for a job so that he could consider marriage. However, he was torn by the growing desire to train in physiological chemistry, available only in Europe. The call of Europe was made still stronger by the fact that he had not seen his parents and relatives in Sweden for 13 years and he wanted very much to visit them. But if he opted for further study in Europe, would he have any right to ask Laura to wait for him?

Laura understood Otto's needs and ambitions, and in light of them, she felt that she should remain uncommitted. All was simple and all was complex; fresh love and frustrating patience have always been difficult to grind together in the same mortar.

Early in December, after a long, chilling discussion with Otto, Laura asked him to consider what he expected from their relationship. We present his reply because it illustrates poignantly Otto's predicament, his strength of character, his manliness, and his humor—the very reasons why Laura could not dismiss him from her mind.

> Dear Laura,
> I did not sleep much Friday night and I was blue yesterday but I have now thought the matter over cooly and seriously and I want to write and tell you of it. Of course I want very much to marry you. It seems to me as though it would be a perfectly ideal existence to me—to live in a busy and somewhat responsible outside position and then have you as the center of my non-professional

thoughts and speculations. If you then only liked me and were content I don't think that I should fear much for the reverses that will be sure to come to almost everyone. I am sure that my life would in some respects be quite difficult with you than it will be without you because I shall probably otherwise marry sometime under very different conditions and probably to some girl of a very different type—and of course I would far, far rather marry you and would much prefer to wait a very long time for you so long as I thought that I had a reasonably good prospect of finally marrying you at all. But on the other hand it does not pay to be subject to such days as I had yesterday and it does not pay to look upon the prospect of marrying you in a different manner from that with which I would consider any other good prospect—as only a possibility that is too risky to stake much upon; and neither do I think that there is any reason why you should see to it that I go home unhappy and miserable whenever I say anything about that which I want more than I want anything else. I will under all circumstances have a great deal to do and to think about for a long time to come and I ought to be able to get along for some time yet as I have in the past without trying to see the future in a too definite form.

I have just now started to make a little preparation intended to make life rather more tolerable for myself this morning. In a few minutes I shall have some strictly first class extract of coffee prepared according to true chemical ideas of purity and cleanliness. What is the use of marrying a girl who doesn't even *drink* coffee! I should then always have to make it myself, anyway, or do you think that you would and could occasionally make a little coffee as indicated above, if I taught you how? You could make chocolate by the same method, you know.

After all dear Laura it is rather unnecessary for you to see the case as much from my side as you often seem to do when we talk about it. Marriage is of more importance to you than it is to a man and you better remember that; she is more dependent upon the kind of marriage she makes than he.

It seems to me as though no other woman could ever mean as much to me as you would if you were to marry me and yet I am sure that I should hesitate if I should be obliged to choose between a life with you without anything to do and the life that I expect to live without you. And you are such a sensible and philosophical little girl and ought to be able to live a married life without getting down to the plane of the average married woman; but for all that marriage is a far more momentous affair with you than it is with a man who has a profession and, dear Laura, I don't want you to do one decisive thing one way or another any faster than you feel certain that it is the best thing for you to do under the circumstances.

And now I am coming to what I really want to say. There is no reason why you should think so awfully seriously about your relation with me when I am with you and between times and then make me suffer for it just as soon as I say the least thing that would tend to mix you up with any plans of mine. You needn't think that I am so stupid that I do not appreciate the uncertainty and the difficulties of my case.

I think that I do and I don't suppose that I should feel any too sure of you even if you promised to marry me, until you had actually done so and there is no reason why you should make yourself restless and me miserable about it until you have to. I was better off and didn't worry half as much before we talked about the subject as I have since—and yet I have said comparatively very little about it. It is especially unnecessary for you to think as you do about it so long as my future is a matter of the greatest uncertainty. It may be years and years before I could press my claim on you very much and you needn't think that I wouldn't have sense enough to know it; and in the meantime it seems to me that it would be much better to just make the most of the present as I did on my own hook last year.

I like very much to see you and to be with you but have always been ready to see the first sign that I could understand of your being bored—yet Friday evening and yesterday I was practically determined not to go to the banquet or to see you any more this year—and you suggested that it would be better if I didn't see you any more til next July and all to gain—nothing.

Dear, dear Laura, I think that I am sensible just now yet I am sure that I should much prefer that we keep on indefinitely just as we have done until either you begin to like some one else or find out for certain that you never would marry, or until I should get tired of it or fall in with someone that I would and could marry. There is nothing unfair in that, either to you or to me. You may think that the latter possibility is none at all but I think that in that you are mistaken. It isn't this year and it won't be while my prospects with you are at all reasonable, but I am fully aware of many of the difficulties and obstacles that are in my way with you; and while I might let considerable time and perhaps some opportunities go by unheeded so long as there seemed to be some reason for hope, I am sure that I would shift for myself sooner or later according to circumstances as I am accustomed to do, and just as soon as I shall have become convinced that it would be useless to think any more of you.

And that is all of my little speculative sermon this time. I hope that it will be all that will be required but I will be sure to give you more unless you be nice to me after this.

I had two, three different propositions to make to you Friday evening for this week but I lost all heart for anything of the kind when you began unnecessarily to be "more disagreeable" to me than you ever were to anyone else. First I thought that we ought perhaps to go and hear Jos. Jefferson in his two plays of this week, and then I thought that we might either go skating at Tattersalls or go out sleighriding one evening this week—i.e., only one thing of these three—you to decide which it should be.

I have one of my new photographs in my table drawer where I keep yours but Mr. B⎯⎯⎯ was a little too ambitious about it and put it on such a large mat that I wouldn't send it to you. I shall very probably soon get the rest of them.

And now, dearest, dearest Laura, please like me as well as you can. I know I have many, many shortcomings but I love you so well and you would at least always have the very best that is in me. Will you write me something in answer to this? I never get any but the stingiest little notes from you—yes, I have received two.

"Knute"

In arguing over and contemplating their relationship, Otto and Laura defined their goals. Otto's grandest ambition was to become a physiological chemist, a teacher in a university with an opportunity to carry out research projects. His least expectation was to teach sciences in a small college, do at least a bit of research, and hope to improve his lot.

He saw his professional future mirrored in the lives of his teachers. All of them had been thoroughly grounded in the European tradition of research, and to a man they had not compromised their training or ambitions, even when forced to make their own way financially. Dodge had been a high school teacher, and then followed his doctorate by teaching in a small college. Stieglitz had given up a solid position as a commercial analyst to accept the docent's rank, the rock bottom of exploitation by the university. Lengfeld, Smith, Stokes, even Nef, had laboriously carved careers in chemistry, refusing to be tempted by industry and "administrative" promotion out of the laboratory at the cost of dedication to research.

Whereas Otto could model his professional life after his dedicated teachers, Laura had no such models that she was comfortable imitating. She appreciated the causes of leaders like Jane Addams, but by nature she was not an ambitious trail blazer. She was content to observe rather than contribute. With her future options limited both by her temperament and the times in which she lived, Laura was ready for marriage. She was already well beyond the age at which young ladies of her middle-class background married.

Laura's comfortable circumstances were a result of the American dream fulfilled. Her grandparents had emigrated from England and Scotland to Nova Scotia where they had been farmers. George Grant, Laura's father, had begun his career as a carpenter in Truro, N.S., where he had married Laura's mother. Then he worked as a ship builder, in housing construction, in railroad building, and finally as a road builder in Winnipeg, Man. When the work in Winnipeg proved unprofitable, Grant moved his family in 1880 to St. Paul where he became a very successful industrial building contractor. By this time, Laura was nine.

Thanks to Grant's success—with an income sometimes reaching $40,000 a year—Laura lived with her parents and her five siblings in near luxury in an elegant home, a household that included a cook, at least two maids, and a coachman.

Mrs. Grant raised her children in aristocratic style, and George indulged her. She did not permit Laura and her sisters to do the menial chores of cooking, sewing, laundering, cleaning, and the like, but, on the other hand, she made sure they were well educated. After high school in St. Paul, Laura and her older sister Kate were admitted to Vassar in 1888. Although Kate never completed Vassar because she returned home to manage the household while her mother was ill, Laura majored in mathematics and graduated in 1892. A top student, she was made a member of Phi Beta Kappa retroactively when a chapter was instituted at Vassar in 1899.

Laura's interest in the University of Chicago was kindled in much the same way as Otto's—through a popular instructor who moved to the new university. In Laura's case the instructor was Myra Reynolds, who was not only Laura's teacher, but her friend.

Laura's time at the university had been educationally rewarding, but had not improved her future prospects. Still, though marriage remained the most viable option for the future, she was patient enough—fortunately for Otto—to avoid chasing marriage for marriage's sake. Once Otto and she had clarified their relationship, they became openly and unabashedly steady companions.

In the spring of 1896, Otto began writing his dissertation under the watchful eye of Professor Stieglitz, and with the end in sight, he began to make written inquiries about jobs. However, Loeb with whom he was finishing the course in physiology encouraged Otto to consider continuing his education in physiological chemistry. As a field, physiological chemistry still did not exist in the United States, but it was now well established in Europe where organic chemists were elucidating the structure of natural products and publishing their results in *Hoppe-Seyler's Zeitschrift für Physiologische Chemie*, which already was achieving a respected place among chemical journals. Loeb was convinced that he would have to get a physiological chemist in his department.

As no job materialized out of Otto's inquiries, the idea of going abroad to study physiological chemistry grew. Loeb held out a solid hope of an academic appointment in the physiology department if Otto had the training. In addition, Otto had a strong desire to visit his parents and his relatives. Thus Otto wrote to Professor Olof Hammarsten of Uppsala, Sweden, asking for the opportunity to work and study with him for six months. Hammarsten had published *A Text-Book of Physiological Chemistry* in German in 1883, a book which impressed Otto very much, and Hammarsten was also an editor of *Hoppe-Seyler's Zeitschrift für Physiologische Chemie*.

By the end of May Otto knew that he could not complete his thesis until sometime in the summer quarter. For the past two years, he had taken only laboratory courses in chemistry, but no "book work," so he had to begin reviewing those areas in which he had forgotten much, particularly in inorganic and analytical chemistry. He also had to be prepared in physics, in which he had only taken two courses, because his examiners would include a physicist.

By July, Laura had completed her master's thesis, "The Theory of Values," and received her degree. The completion of her degree, however, left her without a goal. If she were going to teach, particularly in high school, she could do so in her own home town of St. Paul. And indeed she was under no financial or social pressure to get out on her own. Her decision was therefore to return home, at least for the summer.

Otto's letter of Aug. 23 to Laura was full of anxiety as well as some good news. He was now clear of all requirements for the doctorate except "that dreadful test, the examination . . the uncertainty of it puts me in a most disagreeable position. I want the degree now and I feel as though I must get it, but I am at the same time afraid to try for it."

The tension produced by his upcoming Ph.D. examinations was alleviated somewhat by the reply that Otto had received from Hammarsten.

> My letter to Professor Hammarsten brought a beautiful answer. He wants me very much and will give me exactly the attention that I want. The University year there begins the first of September, but he tells me to come, by all means, even as late as the first of December, because my work will all be under his special personal direction. . . . I need pay no tuition, nor any other fees while I am there so that my expenses would be practically only the cost of living.

Following receipt of Hammarsten's letter, Otto had a long talk with Professor Loeb that he related to Laura.

> Dr. Loeb wants me, I am sure of that and I am rather determined to carry out my end of the arrangement. We have already been talking research work together . . . the idea at present is that I go over to Upsala and remain there till the end of the year (first of June) and get at the methods of the routine work, first of all, so as to know how to handle a class on the subject. Then I shall go to Germany for the summer semester and my chief point there is to learn a certain kind of physiological gas analysis which seems to have been developed to a high degree of perfection in one of the German Universities. . . . He wants me and he wants me at the end of a year and he told me that he did not expect that he would meet any opposition and I have left the question at that point.

As soon as Otto's thesis had been accepted with some modification, he wrote Professor Hammarsten to expect him by the second or third week of November. To finance the trip, Otto had again turned to Axel for a loan and by this time he had probably been assured of Axel's help.

Laura's letters of this period indicate that she was torn about being left behind while Otto went off to Europe to study. She even went so far as to relate to Otto a fantasy that she, her younger sister Jo, and Madeleine had created during one of Madeleine's visits to the Grant home. In this scenario, Laura would be married to Otto in the Grant home

> without the knowledge of any of the neighbors, would immediately leave for New York, and sail as soon as possible—would learn Swedish on the boat, and teach English on the other side to help pay for my board and lodging, would ride my wheel around foreign countries, and master the German language incidentally. I would also cook soup in flasks and improve you. To carry on this foreign expedition you would borrow money—as much as necessary and pay it back at your leisure. I would contribute the enormous sum of three hundred dollars and if possible, get Papa to give me what would naturally be the price of a wedding on the grounds that that function would be omitted.

While the proposals were seemingly made in jest, Laura gave serious consideration to the thought of their hasty marriage and then a year abroad with Otto. She weighed the pros and cons in her letter to Otto. Otto could not afford the added debt. Her papa would oppose it. Marriage would not further Otto's interest. She did not like Otto *that* much.

> Now you will be sensible enough to attach no importance to my foolish remarks, which, in the first place, should never have been made, and which certainly should never have reached you. You will work away and take your examination, and go abroad, and be thankful that you have no one on your hands to amuse and provide for.

In their letters, the possibility of their marrying and going abroad together was turned over and over again, sometimes seriously, sometimes lightly. Otto's primary objections to the plan were financial. He could barely borrow enough for his own needs, how could he borrow for two? Her chances of supplementing their income by tutoring in Europe were rather remote because of the scholars already there. And Otto did not want to be in a position to need her father's financial help because "I know well enough that Mr. Grant would be very seriously opposed to my marrying you even under the best of circumstances."

In his talks with Dr. Loeb about his future prospects, Otto did not discuss potential salary but as he wrote to Laura, the best he could expect was an annual income of $800 the first year. He could not pay his debts and be married on that income, although the income would grow. "I believe in my own future. Once in the University it is only a question of a reasonable time—the moment the medical department becomes a reality I am sure." They must, therefore, wait although Otto did not like the idea that Laura might not necessarily remain favorable to him.

Laura was still not ready to give up the dream. She could raise enough money for a year abroad, she wrote. Then upon their return they could settle somewhere south of the university and live on the $800 income, although she admitted that this prospect was unreasonable for them. "It is good that Dr. Loeb likes you. I think that eight hundred dollars is very good for you to start on, if you start in such a promising place as Chicago."

Otto wrote back that he couldn't sleep because of the irresistible idea of marriage. He had written three letters to see if he could raise more money. "I cried just a little as I read your letter, and I don't mind telling you, though I know you consider it very unbecoming to a man. However, I am happy." In a tender note near the end of his letter, Otto gently tucked their dream away.

> It is quietly raining tonight and I do not feel like working. I am not blue, only a little wistful. It was such a wonderfully beautiful prospect. I felt that it was so much more than I had earned and that it did not belong to me; but it was impossible not to contemplate it, or to vaguely hope that it would be all right if I did my very best to earn it hereafter. Even as it is I have very, very much to be thankful for.

In Otto's next letter to Laura he remarked that he would be taking his written examination for his thesis on Saturday morning, Oct. 31, and the oral on the following Monday afternoon. Otto had, in fact, plotted to make his orals fall on a Monday when Professor Albert A. Michelson, head of the physics department, had his day off and would not be present. (Otto's fears about being examined on physics probably had foundation, because he was not trained beyond the fundamentals.) "I wish it were over," the nervous Otto confided. "However, I am going to try to remain cool and deliberate till the end though I am afraid that I shall lose my head when the committee gets at me."

On Oct. 26 with examination nearly upon him, Otto wrote to Laura again.

> I am still perfectly master of myself and I shall probably get through—but I won't do very well and Dr. Nef, who is no friend of mine, is going to cover all of organic chemistry with me himself so that I am going to have a pretty severe siege, I guess. . . . I received a fine letter from my father Friday. He wonders very much whether I am never to get through with studying and he wishes very much that I could come home once while both he and mother are alive, because he says "I have long ago given up the hope of ever seeing any of the

others anymore." I shall answer that letter from Upsala, I think, and tell him that if he will prepare a fine Christmas tree I might come home to see it. That will be fine, but very, very queer.

The rest of his letter was filled with details of his coming departure for Sweden on the steamship *Campania* on Nov. 7.

Laura's return letter assured Otto that his second-class accommodations on the *Campania* would be comfortable, if cold. But the letter included more than assurance.

> To keep you from freezing to death, to ward off seasickness, to insure a safe passage for the "Campania," to make you have a good time in Upsala and Germany, to bring prosperity to yourself and all your friends, to secure for you a good position in the fall when you return, and above all to give you good luck next Monday afternoon, I am going to send you an amulet. . . . It has my name on the back, but nobody need see that, and only a very few persons would recognize it as a Vassar pin. Anyway, you must have it, as you are now so very much in need of good luck. . . . You must return this to me next Fall.

She elaborated on this subtle love token. All medieval scholars believed in amulets. Losing it was a bad omen. It should never be lent to anyone and it should be worn inside the coat like a fraternity pin.

His written examination, Otto wrote Laura on Nov. 1, had lasted from 8:45 a.m. to 2:20 p.m. He had spent nearly three hours writing on Nef's questions in organic chemistry and "did fairly well." But he had spent his best mental energy after he had briefly "but correctly" answered Lengfeld's question on theoretical chemistry. He did not do well, he thought, on the questions in general inorganic and analytical chemistry. He had not been confused, but tired. "If I only could have taken one hour off for lunch I should have done better."

On Nov. 2, the following telegram was sent via the Western Union Telegraph Company:

> Hyde Park, Ills. Nov. 2nd, 1896
> Miss Laura C. Grant
> 462 Holly Ave.
> St. Paul, Minn.
> Yours Truly Doctor Otto Folin. . . .
> No Sig.
> 6:44 PM

Laura immediately replied:
Just a few words of congratulations to the new Doctor before the lights go out for the night. I am very, very glad for you that the strain is over. You have done an enormous amount of work since you came to Chicago, and it must be a real satisfaction to have the degree to show for it. You are a success, and success brings success ad infinitum. Now you must try to make up for your hard pull by having just as good a time as it is possible over the ocean. You must enjoy life for the year, and not be bothered by any unnecessary worrying. Don't regard yourself as poor, but remember that you will be reasonably well off

financially in a very few years and bank on the future enough to make the present as happy as it ought to be.

How does it feel to be a Ph.D.? I wonder whether I can muster the energy necessary to become one of the noble rank. . . . Why don't you call on Miss Reynolds and tell her you have the degree and are going abroad. I know she would like it.

When Otto boarded the train for New York, he was not officially a Doctor of Philosophy, because he could not wait for the convocation ceremony to occur at the end of the autumn quarter. During the ride to New York, he wrote of the hectic events of his final days in Chicago.

I must tell you that the telegram I sent you was almost a fraud. I was not very happy when I sent it but I am pretty well over it now. The fact is, you know, that I made a complete failure at that final test. Dr. Nef took me up to begin with on a small point in my written examination which was not quite right and I lost my head and never recovered except to some extent in the physics. As a consequence I no more than got through. . . . Dr. Stieglitz told me that had my oral examination been as good as the written one they would probably have given me magna c.l., but it cannot be helped now. Both Dr. Stieglitz and Dr. Loeb told me that the grade in a doctor's degree makes very little difference.

Stieglitz's view of Otto's oral examination comes indirectly through Prof. A. Baird Hastings, Otto's successor at Harvard (3-6):

Shortly before I left the University of Chicago for Harvard in 1935, I received a telephone call from Professor Stiegliz asking me to come to his office. Since I barely knew Professor Stiegliz, I wondered what I had done. When I got there, he said: "Sit down, Mr. Hastings. I hear you are going to Harvard. I want to tell you a story. A year or so after the University of Chicago opened in 189(1), I was assigned a young graduate student. In time he came before the chemistry faculty for examination for his Ph.D. degree. The examination of my student was in the afternoon, following the examination of one of Professor Ne(f)f's students in the morning. Professor Neff's student had done brilliantly, answering all the questions put to him; but my student did miserably, answering hardly any of the questions. However, since I was young and this was my first graduate student, the faculty voted to pass him. Professor Neff's student was never heard from again, but my student was named Otto Folin. Good day, Professor Hastings.

References

3-1. Goodspeed, T.W. *A History of the University of Chicago—The First Quarter-Century.* pp. 248–49. Chicago: University of Chicago Press, 1916.

3-2. John Ulric Nef File. Archives. The Joseph Regenstein Library, University of Chicago, Chicago, Ill.

3-3. Department of Chemistry. Annual Register, July 1895–July 1896. XX. Chicago: University of Chicago Press, 1896.

3-4. Goodspeed, T.W. In: *The University of Chicago Biographical Sketches.* Vol. 1. pp. 94–96. Chicago: University of Chicago Press, 1922.

3-5. Stieglitz, Julius. In: *President's Report, University of Chicago, From 1892–1902.* p. 453. Chicago: University of Chicago Press.

3-6. Hastings, A. Baird. *Ann. Clin. Lab. Sci.* 4, 217–18 (1974).

Chapter 4. Uppsala and Åseda—A Biochemist's Birth (1896–97)

On Nov. 6, 1896, Otto boarded the *R.M.S. Campania*, which was scheduled to sail at 6:30 a.m. A bon voyage came in the form of a letter from Madeleine. "Dear Old Chief," it began.

> I feel as if we hadn't made half fuss enough about your degree, dear chief. It isn't a thing to be passed over lightly. . . . You are the pluckiest, and the grittiest, and the truest-hearted, and most unselfish old philosopher who ever said he didn't believe in altruism, and then acted against his own theory every day of his life. . . . I believe you are destined for more than ordinary success in the work you have chosen. . . . Good luck, dear chief, both in what lies before you on your journey, and in what will await you on your return.

The voyage from New York to Liverpool, which took somewhat more than six days, was blessed with ideal weather for the crossing and Otto found the trip enjoyable. Although his two cabinmates were not stimulating company, he had time to stroll, play deck games or chess, or delve further into a textbook by Bunge in which he was now into the chapter on inorganic foodstuffs.

But beyond relaxing and reading, Otto could contemplate his return home. Fourteen years had passed since Otto had left his family and Sweden behind, and he had already decided to surprise his parents by writing them from Uppsala that he would visit them at Christmas.

The visit to his parents in Åseda was not going to be a large family reunion. Although Otto had had eleven brothers and sisters, only the youngest—Gertrud Lydia Amanda—still lived with his parents. Two brothers also lived relatively close to Åseda. Ludde (Ludvig Severius), the brother whom Otto knew the best among his siblings in Sweden, was a barber in Ljungby. Gusten (Gustaf Adolf) lived in nearby Älmhult.

Of the remaining Folin children, five had died, four in infancy and the oldest, Wilhelm, in 1874 of typhoid at the age of 23. Carl Ossian had left Åseda when Otto was nine years old to apprentice as a tanner in Växjö, and Axel and Alfred had immigrated to the United States.

Otto's family history in Sweden is notated mostly in parish records. His grandparents on both sides were land-owning farmers from Linneryd parish.

Nils Magnus Pettersson, Otto's father, was born in 1825, the seventh of eight children (Mrs. E. Thorsell's notes indicate 10 children). The youngest in the family was Ingrid, the aunt with whom Otto, and probably Axel, had stayed when they first came to Minnesota.

Eva Persson, Otto's mother, was also one of eight children. Like the Pettersson's, Otto's maternal grandparents, the Perssons, had a fairly large farm, which remains in the family to this date.

Those used to the American system of surnames may find it strange that Otto's paternal grandparents had a different last name from Otto's. However, in that era, farmers tended to use their father's first name for identification, and were thus conveniently recorded in the parish records. Craftsmen and town dwellers resorted to the family names, hence the name Folin (actually Fohlin).

Eva and Nils Magnus were fortunate to grow up in Linneryd parish, which had a progressive pastor and a gifted teacher. Because the primary purpose of education in Sweden in the early 1800s was to ensure that young people had sufficient religious instruction in the Lutheran catechism to be confirmed and to receive Holy Communion, education was the responsibility of the church. The Linneryd church council made school attendance compulsory as early as 1825 whereas in Sweden as a whole it did not become compulsory until 1842. "... all children in the parish both rich and poor ... after one term shall be able to read Swedish in a book." The notes of Mr. Carl-Werner Pettersson indicate that by the time Otto's parents attended school, the term had been extended to two months and could be prolonged if needed.

In 1841, Nils Magnus was apprenticed as a tanner, possibly with his mother's relatives in Ålmeboda. Eventually, he came as a journeyman tanner to Bro Hakånsgård, a village near Linneryd where handicrafts were taught. By 1850 when he married Eva, he owned his own tannery.

During the eight years in which the couple lived in Linneryd, Eva grew interested in midwifery. In late 1856, the Stockholm Palace issued detailed instructions for midwives, including educational requirements, examinations to be passed, duties, and how the midwife was to be paid, and in the same year schools for midwifery were established, first in Stockholm and then in Gothenburg. The schools, however, only accepted midwives with special capabilities.

Eva's chance to attend a school on midwifery did not come until four years later. In the meantime, possibly because his business in Linneryd had failed or he had experienced a major fire at his tannery, Nils Magnus shifted his tanning business and his growing family (he now had four sons) to Langhult Mellangård in Ryssby, about 75 kilometers away.

Thus it was to the Ryssby church council that the provincial doctor wrote the following letter.

> The wife of tanner Folin in Långhult, Ryssby Parish, 30 years old, has declared to me her wish to train in the Maternity Hospital and Educational Institute in Stockholm, to pass the examination as a midwife. As the aforesaid wife, of sound intellect and intelligence, can read and write clearly, and has small hands, I consider her fit and proper for the desired appointment; and should she now be accepted by the Ryssby parish at wages determined for a period of at least 5 to 10 years, I will try to obtain free education for her.

In June 1860, the Ryssby council established a contract with Eva to be the midwife. The congregation granted Nils Magnus 50 Riksdaler toward Eva's midwifery education, even though he was evidently so well off financially that he offered to pay her expenses during her education, as well as the cost of travel and obstetrical instruments. Her midwifery fees under the contract were to be mostly in grain plus a special fee for each delivery of 1 to 3 Riksdaler depending on the financial condition of the family. Travel and housing at each delivery were to be provided.

Because there was no vacancy in Stockholm, Eva began her training the following October in the Midwife Training Institute in Gothenburg. The nine-month course consisted of three months of lecture material, eight weeks of practical training, eight weeks of assisting at deliveries, and lectures to complete the course. To obtain instrumental training required an additional three months.

Eva's training, which she probably undertook out of desire to help people—certainly the family seemed to be sound financially in this period—had two major obstacles, a fifth pregancy and a prolonged separation from her family in Ryssby, 180 kilometers away. In fact, just one month before she completed her final examination, she bore her fifth son, Oscar George, whom she had to care for during those critical days in her training. After she took her midwife's oath on June 28, she did not take the additional three months of training for instruments, probably because of the birth.

While Eva was away at school, life could also not have been easy for the family left behind. Even with household help provided, the period must have been a strain on Nils Magnus, although the strain must have been more in his personal than in his business life. Records do not indicate that he suffered any financial hardships in his business this time.

A second fire in Nils Magnus' business or possibly Eva's impending certification as a midwife caused the family to move in late 1861 to the tannery of Tuna Bengtsgård, which was also in the parish of Ryssby. There the Folin's sixth son Alfred George was born.

In September 1863 Eva again went to Gothenburg, this time to receive three months training on the use of instruments, following which she passed her second examination on Dec. 5, 1863. The training allowed midwives to use instruments under specified conditions in the presence of witnesses; afterward, the midwife had to file a detailed written report.

With income coming from the tannery and Eva's midwifery, the Folins should have lived comfortably after 1861. However, Nils Magnus "was erratic, dissipated, generous to a fault, and assumed little responsibility in the rearing and support of his rapidly growing family. He was brilliant, but lacked stability. . . . It fell upon the young wife to support the family," according to Otto's niece Hildur.

In this period, Eva's midwifery must have become more important to the family income than the tannery business to have justified their next move to Åseda. In the early 1860s the Council of the Åseda parish, about 97 kilometers northwest of Ryssby, discussed ways to get a new midwife, their present one being either too expensive or too old. By November 1863, the need for a new midwife became acute and Eva was appointed their midwife at the end of February 1864.

The Folins had difficulty finding adequate housing. Initially Nils Magnus stayed in Linneryd, while Eva, who was again pregnant, moved with her five children to Åseda. When Ludvig Severius was born early in December, the problem still had not been resolved and would not be for some time after that. In October 1865, a farmer in a nearby village offered a house to rent. While renting, Nils Magnus evidently was building his own house and tannery north of the center of Åseda, which opened in early 1866. On Feb. 4, 1867, two months before Otto was born, however, the tannery was destroyed by fire.

Nils Magnus sold the house and rebuilt tannery in July 1872 and the family, which had been increased by the birth of Otto on April 4, 1867, and Gertrud Lydia in 1871 (triplets born in 1869 did not survive), became renters again. From 1872–75 the Folins suffered hard times brought on partly by much illness in the family. Nils Magnus appealed directly to the parish council for financial help, and Eva asked for an increase in her salary because their rent was high. However, by 1876 the family must have resolved their problems, because Eva applied for a three-month leave of absence to take another course of study at Gothenburg.

After 1872, no information exists as to whether Nils Magnus made further efforts to operate a tannery, but records do show that by 1875 he had made a commitment not to build a tannery in Åseda, probably because there was already one in operation. The mention of him as a baker in the records may indicate that he developed another occupation, but even so it seems fairly certain that in this period, Eva did, in fact, become the family breadwinner.

In 1879, the Folins—Eva, Nils Magnus, Ludde, Otto, and Gertrud—moved into the parish house in Åseda. Otto had happy memories of his childhood in Åseda, particularly of his mother. In spite of several children under foot, almost perpetual pregnancy, cooking, washing, cleaning, midwifing, and the burden of an improvident, often helpless spouse, Eva remained a lively, fun-loving woman, thanks in part to a robust constitution and the ability to handle the problems of a large household.

Although Otto's respect and love for his mother was deep, his esteem for his hapless father was low. Nils Magnus and Eva must have had happy years early in their marriage, but by the time Otto was born his father could no longer cope with his misfortunes and was on a downhill slide.

Otto began school in Åseda at age seven in a room in the parish house near the church, but when he was about nine, a new schoolhouse was built to relieve overcrowding. The clerk and organist for the congregation, Anders Gustaf Bergqvist, was a licensed teacher and served as the schoolmaster as well.

The parish curate from 1872–87, Lars Johan Lange came to Åseda after serving as a missionary in East Africa. He was well educated and fond of books, so that he exerted a profound influence on the quality of education in general as well as on the religious training of his congregation.

As early as 1853 the school term in Åseda was 10 weeks per year, but probably by Otto's time in the 1870s, the term had lengthened, at least within the village itself where farm duties did not intervene. Evidently, the number of years a child attended school was not set by law, so that the parents' attitude and the child's aptitude were deciding factors. For many parents, the objective of school was to teach the children enough for confirmation, just as it had been for them.

A school day lasted six to seven hours. The day began with the singing of a hymn and the reading of a prayer by one of the children. Christianity was the first subject studied and lasted for two hours. Lessons in the Lutheran catechism were read and then the older pupils read the Bible aloud. All students were taught reading, beginning with the ABC primers and tables of letters on the walls and continuing with the New Testament and the hymnal. Those who learned to read (and spell) were then taught writing on paper with a quill pen and later with a pencil. Those who could read were also taught arithmetic, as well as instructed on weights and measures. The students' progress was monitored at the end of the term by Reverend Lange and the teacher, who questioned them on catechism and the Bible. Later the examination broadened to include arithmetic when the school became obligated to teach reading, spelling, writing, and arithmetic, in addition to religion.

Otto enjoyed school, and when he learned to read and write, he took pride in being the best student in the class. Though the school term was short he developed an early love for reading. Much later he would write,

> It is curious, but a strong, exciting story can hold me as perfectly now as it would and often did almost twenty years ago. Mother laughed very much this summer as she told me several times how I used to be on my knees on a chair and hang over a book on the table for hours and hours and she would call on me for something, perhaps to do something, and I would never hear until she would finally lose her temper and make me do so.

In the autumn of 1880, Otto, who was by then living at the parish house, was given the opportunity to enter private school that Reverend Lange had created in the parish house for teaching the older, brighter students some advanced subjects that could help prepare them for professional work. This extra schooling, which lasted for about two years, included mathematics, German, and probably some natural history and was crucial to Otto's successful education in America. Otto owed an eternal debt to his Swedish teachers, Bergqvist and Lange, in spite of the fact that Otto disliked the vicar for ridiculing or shaming him in front of the class on more than one occasion.

The demarcation line between childhood and adulthood for the Swedish child in Otto's time was confirmation in the Lutheran church at about age 14 or 15. Not only was it the culmination of a year of special religious instruction, it signaled, for the boys at least, entry into the work force and departure from home. Following their confirmation, all of the Folin boys were sent off to other communities to learn a vocation, with Wilhelm being somewhat exceptional because he went to Växjö to study.

By the time Otto was confirmed, the stagnant economy of rural Småland, and of Sweden in general, was at so low an ebb that a vast emigration had begun. "Vaccine, peace and potatoes," had caused acute overpopulation, particularly among the primarily agrarian populace of southern Sweden.

Although Otto was the top student in the pastor's select group, the family had no money for his further schooling in Sweden. As Otto reached confirmation age, the talk was about America. Ingrid and Nels Petter, already in Minnesota, owned more land than the wealthiest farmers in the parish! Axel, at such a young age, owned his own house and was expecting to buy a large farm without debt.

Few families in the parish had not been touched by emigration. Still it could not have been easy for Eva or any loving mother to send her son to a land thousands of miles away, knowing that she might never see him again. Eva's gift of love to her son Otto was the opportunity to develop his sharp mind as she arranged with his Aunt Ingrid and his brother Axel for his journey to the United States.

When the passage ticket (steerage class) to America arrived from Axel, Otto remembered his euphoria. Suddenly everything for a week or two became a fantasy, and he moved about trancelike in Åseda, unable to grasp the reality of events. It may have been the most exciting moment of his life. He would leave at the end of July, too wound up and too young to reflect upon the dear ones he might never see again, and the sacrifices of the mother he left behind.

* * *

After landing in Liverpool, Otto took an evening train to Hull, where on the following day he boarded the small vessel, *Romeo*, for a three-day trip across the rough and cold North Sea to Gothenburg. Otto spent the night in Gothenburg and the next day took an all-day train ride northeast across Sweden to Stockholm where he found overnight lodging in a cheap room at the Continental.

> The following day I went to my cousin's barbershop [Johan Folin, probably the barber from whom Ludvig had learned his trade] and he shaved me. Then he turned me over to one of his assistants (a girl) to cut my hair and during this operation I began to talk and finally told him that I thought that we were cousins. Well, it was fine. My cousin was a sailor in his younger days (he is now 56) and he is an unusually intelligent man. I really enjoyed his company.

That afternoon Otto took the train for the two-hour ride north to Uppsala. The scenery on the way reminded "[me of] a trip I once took from Duluth to Tower up in the pine district of Minnesota." On the next day in Uppsala, Otto called on Professor Hammarsten, who welcomed him graciously, and wasted no time in getting down to work.

> A few words about Dr. Hammarsten. He is a Jew, a Swedish Lutheran Jew. He is, however, a fine man and I like him. He made me feel at home at once and he will give me exactly what I want. He is a very important man here, the fifth in rank, and he has made me acquainted with Rector Fries who is fourth in rank and who comes from my province and who also made me feel good. Hammarsten is quite proud to have me, I think, and he wants me to do some research work with him and I probably shall. . . . My Swedish is rather broken I am told. It is not a dialect but it is the Swedish of an Englishman and I am therefore a foreigner everywhere.

Professor Olof Hammarsten (1841–1932), who welcomed Otto so graciously to Uppsala, was one of the leading pioneers of physiological chemistry. The son of a shipyard owner, he was entirely a product of Sweden. Following graduation from the lyceum in Norrköping, he virtually rooted himself in Uppsala for the rest of his life where he became not only a major scientific figure, but a capable administrator, both in the university and town affairs. He had obtained a bachelor's degree in medicine in 1866 and a doctorate in medicine after presentation of a dissertation in 1869. He immediately joined the faculty as a lecturer in physiology. At the time Otto came to Uppsala, Hammarsten was vice president (1893–1901) of Uppsala University as well as professor of medical and physiological chemistry.

Later on in his distinguished career, Hammarsten became chancellor of the university (1901–05). In 1905, he joined the Nobel Prize committee and served as its president from 1910 to 1926 (4-1).

The earliest physiological chemists, particularly in Europe, were establishing this new discipline as a branch of physiology within the medical schools, and Hammarsten was no exception. Simultaneously they were laying a solid foundation for the purely clinically oriented field of clinical chemistry. Hammarsten's laboratory would not only provide Otto with the proper direction—indeed, inspiration—for his career, but the next year would attract another young founder of modern clinical chemistry, the incomparable Norwegian Ivar Christian Bang (1869–1918) (4-2)

Hammarsten had shown himself a scientist of insight as well as a gifted teacher. When Svante Arrhenius' 1887 theory of electrical conductivity—ionization as a cause of chemical reactions—had fallen on deaf ears and he was unable to find work in any other department at Uppsala, Hammarsten offered him a position.

Hammarsten was a master of all the existing physiological chemistry of his day, as evidenced by his comprehensive, popular textbook that Otto had studied so diligently. The book, published first in 1883 in German, would undergo 11 editions, including many in English.

Hammarsten's pioneering research occurred primarily, though not exclusively, in protein chemistry. He was the first to demonstrate the proteolytic activity of rennin on casein to produce paracasein, which forms an insoluble curd (coagulate) in the presence of lime salts. Rennin, found in the abomasum (fourth stomach) of ruminant animals, was not found in other mammals or in birds or fish, although rennet-like proteases were present. Hammarsten demonstrated convincingly that pepsin and rennin were distinctly different enzymes (though it is now known that they have similar structures and origin) and that the action of rennin produced not only paracasein but a second proteinaceous substance, a whey protein that he characterized as an ill-defined proteose. This work in turn paved the way for the discovery by others of additional milk proteins such as lactalbumin and lactoglobulin.

Hammarsten also made fundamental discoveries in blood coagulation showing that fibrinogen, which he isolated in pure form, is the sole source of fibrin and pointing out that both thrombin and fibrinogen are essential to blood clotting. He also noted the important role of calcium in blood coagulation.

In addition to these areas of research, Hammarsten isolated, purified, and determined the elemental composition of mucin substances that he later classified as glycoproteins and made extensive studies on bile formation and bile salts and pigments.

All in all, it was Otto Folin's grand luck that he should bridge the gap between organic and physiological chemistry under Hammarsten's aegis. With this master, Otto found not only the wellsprings, but the forefront of the knowledge he needed to enter the field of his ambition.

Life in Uppsala in which Otto would spend the next several months of his life was centered in the ancient university—created by papal sanction in 1477 and endowed in the 1600s by Gustavus Adolphus IV—a university which had produced great scientific thinkers like Carl von Linne (Linnaeus, 1707–78), nature's systemizer, and Torbern Olof Bergman (1735–84), considered a founder of analytical chemistry (4-3).

And outside the university walls in Uppsala great scientific learning had also taken place. There the impoverished apothecary Carl Wilhelm Scheele (1742–86) had discovered fluorine, chlorine, and, with Priestley, oxygen; helped identify tungsten, barium, molybdenum, and manganese; determined the composition of inorganic substances such as cyanic acid, hydrogen sulfide, and compounds of arsenic; and revealed many new organic compounds, particularly the organic acids. (Many years later in 1930 Otto would be honored by the Stockholm Chemical Society with the Scheele Medal named for the man who had brought such great distinction to the soil of Uppsala with his prodigious, unheralded research.)

Laura and Madeleine were soon receiving letters describing Otto's impressions of and experiences in his native land. The university had about 2000 students, mostly male, with an "undesirable reputation," he observed to Laura. Their moral tone was below that of students at the University of Chicago and they were coarsely familiar with the dining room girls, he continued. "The girls are queer things; very strong, very pretty, very meek and splendid cooks!"

To Madeleine he wrote,

> I am distinctly a foreigner. The cut of my clothes is very different from that of the short sack coats used by everyone here. Three kinds of headgear are used by the University people; a semi-soft hat, a low crown stiff one and the "stove pipe." Mine is a Dunlop and would be correct in Chicago, but it looks very different from the stiff Swedish hat. . . . Then nearly all wear boots or shoes that are very heavy and they are very broad in the toes. In the way of neckties, 99% wear the small black "bow" that I used to wear while I was in high school, and the remaining 1% use the long tie (black). I sport rather gayly colored bows; one blue and black, and another red and blue. Nearly all here also have their hair simply standing on end (as I used to have it) while mine is parted very near to the middle and combed sideways. Consequently I am an altogether different looking creature and the difference is of course very much emphasized when I speak with anyone. I enjoy it, however.

Uppsala was a distinctly university town in which everything depended on the university people, he told her, and everything was old as if in the Middle Ages. He described the "Chemikum."

Uppsala and Åseda 63

> Imagine a very narrow and very dark and winding stairway with your hat scrubbing against the ceiling and the walls of damp, clammy stones several feet thick. It is almost as different from Kent laboratory as it well can be. There everything is distinctly modern; here everything is distinctly ancient, except the professor. This is not altogether desirable. I miss a good many convenient devices although I do not think that the lack of them will materially interfere with my work.

Otto had called on Birger Lange, the son of the former Åseda preacher, who, with one of Hammarsten's associates, Holmgren, tried unsuccessfully to find Otto a private room so he could move out of the private hotel that was costing him 1 kr. a day ($1 then equaled 3.67 kronor). Shortly before Christmas, however, he moved to a temporary room in a private house with a permanent one available to him after Christmas for the five-month term.

Otto had written his parents after he arrived in Uppsala that he was coming to visit them on Dec. 22 and stay until he had to return for the next term, which began Jan. 15. Once he received a reply from his parents, he began to make plans for his journey home. He related plans for the family reunion to Laura.

> I will have a three or four hours sleigh ride before getting to Traheryd. I have one married brother, a barber (Ludvig) in Ljungby. My father is also to meet another married brother of mine there that morning, a coppersmith (Gustaf) who is coming home over Christmas and who has not been informed of my being in the country. My father was of course very astonished when he received my letter and I can see that he will be very glad to have me come home.

On Dec. 21 accompanied by an assistant to Hammersten, he journeyed to Stockholm and then the next day to Ljungby, the closest train station to Träryd (Traheryd) where Otto's father and Ludvig met him. Otto recognized them at once, but they did not recognize him.

> Yes, Papa would scarcely believe that I was the one he was looking for. He has been observing me very much ever since and he says I am so tall and so quiet and thoughtful that he can still scarcely understand how I can be "Otto." And Mama says the same, and so do my brothers. They don't seem to be able to understand how these last fifteen years could bring about such a difference. My father is extremely proud of me, however, and I think that he is quite satisfied. He usually sits in a corner and says nothing while the rest of us are talking, but he takes it all in.
> I have had quite a happy Christmas. My parents have not changed very much and it seems very nice to find mother in such good condition. She is said to have been very good looking when she was a girl and she is a very fine looking old lady now, especially when she laughs. It does one good to see her laugh and she does laugh almost as soon as I look at her. We have had a beautiful time recalling old incidents.

Otto had some more pleasant surprises waiting for him. He had last seen Gertrud when she was eleven and had been worried that he would not like her.

> Nearly all of us inherited father's bad temper and I had thought that it, together with lack of training, would render Gertrud disagreeable; but I have found her very nice and it seems quite fine to have a sister. Her personality was originally very much like my own, except for my taste for study and her exceedingly quick temper, and now I find it very interesting to be with her because I understand her so well. She is really the head of the family and she seems to have developed into a splendid business woman and I am not afraid about her future any more. She is twenty five and is fairly good looking, only her mouth is in a perpetual pout and she is too fat and she does not know how to make up her hair.

A recurrent and perhaps troubled theme of his letters from this period to Laura was alcohol.

> The more I see of the Swedes in different stations and conditions of life the more I see what a horrible national curse the drinking habit is. Traheryd is only a village of a dozen houses; but a railroad is being built and scarcely a day passes that I do not see one or more drunken, crazy fools on the road. It is a very sorry spectacle.
> Drink ruined our family in the early times; it has already ruined one brother; and I am afraid that it, together with a rather unhappy marriage, is going to be the ruin of another. . . . Father is about what he always was in such matters, only he is now harmless, because mother and sister run everything.

While Otto had multivariate experiences to relate in his letters to Laura, her life was routine. Laura eagerly accepted Madeleine's invitation to be bridesmaid at her upcoming marriage to George Sikes in February. Laura's letter to Madeleine accepting the invitation indicated that all was not well in St. Paul.

> [Father has] lost heavily for a man of moderate means and has a very expensive family for his present financial circumstances. The ranch in California and the home expense make it necessary that he should earn a sum each year for at least three years which it is not likely that he can, although he is unusually good at making money in his own business. There is an oil mine on the market which Pa has shares in, and if it should sell, all debts could be cleared up and things would be straight. But that can't be counted on just now. Times have been dreadful since '93 and it is greatly to his credit that he had kept up so well under such frightful pressure.

In a letter to Otto written just after New Year's 1897, Laura also described the family affairs for Otto, including her father's financial difficulties and the severe rheumatism that had confined her mother to bed. She told him she was considering teaching and had tried unsuccessfully to a get a job at a school for girls in the area of St. Paul.

Laura's letters expressed curiosity about every detail of his life, but he, too, wanted to know everything about her activities and her family. "Give a reasonably full account of yourself from your waking to your sleeping hour for a week, won't you? It would interest me." He was aware that the Christmas season was perhaps the busiest time of the year for the Grant family, as all of the family would be home. Laura would be busy with that and with plans for attending

Madeleine's wedding. Otto commiserated with Mrs. Grant, because he knew the anguish of rheumatism.

> I have had it in my system for a dozen years; but it does not seem to affect me except upon exposure. It is much worse in cases like your mother's where it is not a matter of cold or fatigue.... Some rich Nobel ought to give several million dollars to promote the investigation of rheumatism. It ought to make a most interesting chapter in the development of the science of medicine; only, it must be studied on a large scale, with a hospital full of rheumatics and with a large staff of scientists to collect data and make experiments.

Otto returned to Uppsala in a happy frame of mind. He had enjoyed his family, in a sense becoming a part of them again. He had discussed plans for his future with them; and Gertrud, surprisingly, had offered to help him financially. None of them, naturally, understood what it meant to become a physiological chemist, but they did grasp that he wanted to be a professor at a university and to find new things that would benefit the practice of medicine.

In late January 1897, thanks to Professor Hammarsten, Otto was given the use of a private laboratory, for which he was grateful. It was adequate but not up to the quality of the Kent Laboratory. Otto thought that his proposed research problem was good, but difficult; that he would be able to clear up some "preliminary questions"; and the work was worthy of publication in the *Zeitschrift für Physiologische Chemie*.

> This will be sufficient to secure the field and I believe, anyhow, that only a few physiological chemists are capable of working out the last part of the problem because it is a question of pure chemistry and involves the identification of a complex and probably hitherto unknown compound.
>
> The problem in brief is as follows: Mucin is a slime of quite wide distribution in the animal. It is best obtained by extracting the large salivary glands of oxen with water. It has long been known that mucin can by proper treatment give a substance which shows some of the characteristic properties of sugars; but all reasonably exact work hitherto on the subject has simply been to observe the absence of characteristic reactions of sugars after first obtaining pure mucin and also to find an intermediate substance the composition and properties of which is a matter of dispute. My work it is hoped shall clear up the question of this intermediate product first and then I shall attempt to obtain pure and if possible identify the sugar like compound which is supposed to be there. Next Friday, I hope to be ready to make a preliminary analysis of the intermediate product.

For about two weeks, Otto wrote in February, he had been working like a clock: up at 8 a.m.; in the lab by 9 a.m.; work to 3 p.m. Then two hours off. Work from 5 to 7 p.m. Then after supper at 9 p.m., read to midnight.

The social life of the Uppsala University student was not particularly to his liking.

> I am already tired of the famous Upsala student life. It is not to be compared with life at our own universities, I think. It is much more monotonous and I miss my occasional evenings with you and my chats with Miss H⎯⎯⎯ and

the others. . . . We have only about thirty lady students at the University and the best family society seems to be absolutely shut to students. The general effect of this is not good. . . . Prof. H is alone, a widower, and besides no one seems to think of ladies' society. There must be some fine women here but I don't know.

Otto's progress on his project was slow, compounded by the temporary absence of Professor Hammarsten whose help he needed on the use of an unfamiliar method. Holmgren was also working on mucin and wanted to collaborate with Otto, but Otto was too independent for it.

By the end of February Otto had made plans for his immediate future. He decided that if he could not get a job lined up in the U.S. before his year in Europe had ended, then he would stay an extra year to study in Germany. Otto wrote to President Northrop at the University of Minnesota to let him know what he was doing, and that he would be looking for work upon his return. And he kept in touch with Professor Loeb.

Meanwhile, he informed Laura that if two years were to pass before he returned and even longer to get a job, it would be unfair to her. He felt that she should act as if he did not exist and meet more men. But he pledged, "I want to marry you and I hope that I may do so someday; but the above is my sober thought and you can count upon it, even if I should occasionally write sentences which may indicate various things."

Laura agreed with Otto that if he failed to get a suitable job in the U.S., he should remain another year. The University of Chicago would grow. A great medical school would become possible in a few years. Perhaps he would be able to find something in the fall of 1898, even if it were not what he most wanted.

> You mustn't take me into consideration when you plan. Plan for yourself. That is the only way to do. And if I ever should marry you, what is best for you will in the end be best for me. And if I never marry you, I certainly should not be a factor in your decisions.

One letter from Otto to Laura written about this time reflects his surprise that she was attempting to get a teaching job, because he knew she did not like to teach. Was she not on good terms with the superintendent in St. Paul—Dr. Curtis? Was he not the individual who was once superintendent of schools in Stillwater and who went to Massachusetts? If so, Miss Minor [Otto's math teacher in high school] should know him well.

A later letter from Laura indicated that Miss Minor had indeed introduced her to Mr. Curtis, but she had still been unable to land a teaching job. However, she knew of an impending resignation that would leave an opening that she would try to fill. At the end of March, Laura did some substitute teaching of algebra in the high school, and she enjoyed it. She definitely wanted the position the next autumn. Perhaps Laura was now genuinely interested in not only relieving some of the financial stress her father was facing, but had taken the only course open to a woman of her education.

As spring approached, Otto was working intensively in the laboratory and having a "sharp succession" of failures and successes. "My work is taking all my thoughts at present and I really think now that I shall have a little something to publish before I get through."

Professor Hammarsten gave Otto a research problem about which there was at the time a conflict of findings: the composition of so-called animal gums. Several years previously Hammarsten himself had worked on an "animal gum" obtained by alkaline cleavage of a mucin extracted from the shell of the snail, *Helix pomatia*. He thought the gum was free of nitrogen, but he had not confirmed this. On the other hand, the gum he obtained from the mucin of bovine (ox) submaxillary gland contained nitrogen. But this work and that of several others conflicted with the generally accepted work of Landwehr, who had prepared and characterized the "animal gum" from the submaxillary gland and other sources and not only declared it to be nitrogen-free, but assigned it the formula $C_{12}H_{20}O_{10}$. Landwehr's concepts and analytical methods were generally favored but knowledge of what later was termed glycoproteins (now proteoglycans) was scanty. There was a profound need for research, to put it mildly, particularly by skilled organic chemists. From Hammarsten's viewpoint, Otto must have been a plum! Here was the gift of a Swedish-speaking organic chemist, trained in the chemistry of carbohydrates and organic nitrogen compounds, versed in analytical chemistry, and interested in learning physiological chemistry. He could join Holmgren on this fascinating problem and aptly use his scientific background.

Otto had learned well with Stieglitz, Lengfeld, and Nef how to characterize organic compounds. The problem that Hammarsten had, after all, was one primarily of characterizing the mysterious substance, an animal gum, that had to be isolated in pure form like any other organic substance. Was it a carbohydrate? If so, how was it possible that Schmiedberg had obtained from the mucin of cartilage (chondrin) a gum-like substance that contained both nitrogen and sulfur?

Otto's first task was to prepare the animal gum from bovine submaxillary gland by the use of Landwehr's method. Much to his chagrin, he found that the extraction method described by Landwehr in the literature was vague and flawed. He was forced to make little side experiments that took up too much of the little time he had, even with his working a seven-day week. For example, the procedure called for extracting the finely ground glands in water by heating in an autoclave, but failed to give the temperature and the length of time. Then it called for neutralization of the liquid with acetic acid, but the liquid was already acid. In the final stages of Landwehr's procedure, the animal gum was precipitated with iron chloride and after the excess iron was removed with carbonate, the purified gum was extracted and dried. It contained about 2% ash as iron oxide and about 10% nitrogen on an ash-free basis.

The gum gave a protein reaction (biuret) that Landwehr had not reported and upon boiling with Föhling's solution gave a qualitative test for reducing sugars. What was the source of the nitrogen? Landwehr had missed it probably by use of an inadequate method for its analysis, and he had not performed an ash analysis. Otto made several modifications of Landwehr's method but in no case could he produce a nitrogen-free substance. On the other hand, Otto could explain the high nitrogen content of the product on the basis that the original extract contained "mucin-albumose" and that the albumose was carried along in the steps of the Landwehr procedure. Knowing this, Otto believed that his task was now to separate the albumose from the "gum." He had definite ideas about how to do this, but unfortunately his time was running out, and Hammarsten wanted him to report specifically on the fallacies of the Landwehr method.

Otto's letters to Laura in April dealt a great deal with his research. Hammarsten had wanted Otto to write his findings in English to keep his thoughts clear and publish in an English journal of physiology. Otto, however, preferred to publish in German, because in modern parlance, "that is where the action is." In addition, Otto was preparing to read a "short extract" to the Upsala Lökore Forming (Uppsala Doctor's Society), but although Hammarsten wanted the presentation made in English, Otto preferred Swedish. Otto had read about 2000 words of his paper to Hammarsten, who was quite satisfied, but Otto himself was in one respect not happy: He would have to attack Landwehr's work unmercifully, an injudicious approach for a neophyte. Hammarsten, however, insisted that Otto include points that Otto preferred not to mention.

> Yesterday afternoon I came upon a new idea of how to obtain a certain substance, but my time is so short now that I dare not follow it up. I also met with an unexpected obstacle in the work that I was really trying to do. A certain phenomenon which I have before often observed under almost the same conditions as yesterday now failed to appear and as it was one of the essential facts in that particular experiment I left the things going overnight in the hope that I shall today find what I am looking for. If I cannot obtain it I must return to exactly the older conditions next Wednesday when I get another batch of glands. Time, time, time that is what I need and when I get it and feel reasonably settled and have a snug little corner where I like to be between times then I am going to do some work.

Despite Otto's high esteem for Professor Hammarsten, he preferred to work alone on his project. He felt that, in spite of his failures, he would get at the facts.

> I was quite a little dissatisfied some time ago because I was prevented by his line of thought from following up a discovery which I had made without his knowledge; but I finally let him see my dissatisfaction and since then he has been very fine. He surrendered completely. I did, however, not want that and the result is a compromise. I shall prove the preliminary proposition which is all he wanted me to think of but in addition to that I shall have enough on the main problem to show it is really *it* that I am after. [My results, however, show that the extra work] is not going to be good enough to serve the double purpose for which I wanted it, but it will serve one. The whole thing is a substance which the German never saw and for which he is going to get a drubbing. The substance is probably a nukleoproteid (you understand) and I want on the one hand to get it out from the other substances and on the other hand I want to investigate it a little.

But Otto's work had too many ramifications; and Hammarsten had sent him on several "wild goose chases." He wrote that his work was so intensive and so demanding that he could do nothing else, not even write to Laura.

> One reason for this has been that I have not had so much confidence in the *planning* of my professor as I had in the plans of Dr. Stieglitz and I was afraid that I should have nothing to show for my winter's work. It made me work and think very hard. The fault was not that he does not appreciate the whole question and he knows how it should be attacked, but he is altogether too easy going and I have wasted a great deal of time and energy on excursions that were interesting enough, but they were not to the point.

Never again would Otto work on what was basically another man's project. By now he had gained confidence in his critical faculties and creativeness in research. He had become the feisty young scientist in an era when opportunities were scarce, and careers were often delayed or failed to materialize at all.

In late March, Otto wrote Professor Ernst Salkowski of the Pathological Institute at the University of Berlin that he would like to spend some time working in his laboratory. Uppsala University closed on June 1, whereas the German university closed on August 10. This would give him about two months there. No doubt Otto had discussed this with Hammarsten, or perhaps Loeb had recommended Salkowski. At any rate, he was now planning the next step of his postdoctoral study.

Otto had now formulated his plans for going to Berlin. Otto wrote Laura that Salkowski had now replied that he would reserve a place for Otto. Otto thus planned to leave Uppsala on May 20, spend a few days in Stockholm either alone or with his sister, and then go home for a few days before going to Berlin on June 1. He planned to spend two months in Berlin, return home during the vacation period, and on Nov. 1, go back to Berlin for the winter quarter or go to some other man, perhaps Voit in Munich.

About Laura's elaborate suggestion that he write an autobiography detailing his experiences abroad, he expressed astonishment. He assured her he liked the idea, but that it would not pay for him to scatter his energies, especially when his time in Europe was so short. Perhaps he could write a little thing now and then, particularly during summer vacation home in Åseda.

Otto spent the weekend of the Easter holiday in Stockholm, and he described in detail the royal castle he visited and the beauty of the city for Laura. Being alone took some of the joy out of his sightseeing.

When Otto returned from Stockholm, his work necessarily had to wind down because he was now writing his manuscript, and time to follow up his own ideas was not sufficient. He busied himself and perhaps earned some money by testing a patent medicine for Dr. Mörner, one of the faculty members in physiology.

At Hammarsten's insistence, Otto spent his last few days in the laboratory at Uppsala doing C, H, N analyses on the substances he had obtained from mucin, but he considered the attempt a waste of time except for one product that he purified. Elucidating the structure of the compounds would have taken

Otto a lifetime of work had he decided to pursue the project, but the problem was not to his liking and he never returned to it.

Otto's article was to be published in German in *Hoppe-Seyler's Zeitschrift für Physiologische Chemie* and he was thankful that Hammarsten himself wrote it in the language. Otto's paper—received on May 20, 1897, and published on Aug. 28, 1897—may have been useful in debunking Landwehr's approach at characterizing proteoglycans, but offered no new insight, although this lack reflected on Hammarsten's skill as a chemist, not Otto's. Otto meanwhile had learned a great deal of physiological chemistry, thanks to Olof Hammarsten.

With the arrival of spring and his work drawing to a close, Otto took to his bicycle on the weekends, exploring the area. Otto found the road to "Old Upsala," about three miles from Uppsala, and he described for Laura the sights on the way, a castle, and nearby, the Upsala Cathedral. His explorations took him to the burial mounds of Old-Upsala for the kings of the "heathen" Yngling dynasty (about 500 B.C.).

May Day in Stockholm was an experience that Otto described to Laura in great detail. "Tomorrow is the first of May the greatest day in all the year here in Upsala," he wrote. On May Day, the students took over the town. Spring clothes emerged, especially the beautiful caps, which were worn by all the students and even the professors. At 8:30 p.m., all the students (1500) gathered in the central marketplace and led by the singers and flagbearers

> marched up the hill to the castle. The whole of Upsala was one waving, white sea for two blocks, and on the sidewalks were the common people. It was a magnificent display.

By Tuesday, May 11, Otto wrote, "My work is done or will be after about one hour more of writing and two more hours of H^{ms} lectures." He was eager to attend one last, major event taking place that Friday, the annual spring concert. Tickets were in great demand, because the concert was to be attended by royalty, namely Crown Prince Gustaf, a Swedish princess or two, and also one or two Danish princesses. Fortunately, Otto got a ticket through a friend because, although tickets did not go on sale until Wednesday, the line of students and hired ticket buyers had formed early.

After Hammarsten's final lecture, Otto bade him goodbye and left him with the research paper to be sent to Berlin.

> Then I went home to put on a black pair of pantaloons and a white bow which was a necessary outfit to make it proper to be among those students who (in a body) paid their respect to Royalty as represented, not by our Crown prince who was unable to come; but by his princess and by prince Carl and by the crownprince and crownprincess of Denmark together with two quite young and very kind-looking princesses from Denmark.

That evening after he enjoyed the spring farewell festival, he had a last supper with Holmgren and a few others. In the morning he left at 8:30 a.m. on a boat for Stockholm. The trip down the Fryris River and through Mölaren, the most beautiful lake Otto had ever seen, brought Otto to Stockholm by 2 p.m.

Uppsala and Åseda

On his bicycle that he had brought with him, Otto chose to explore Stockholm rather than visit the exposition that had just opened. Stockholm was not so large that he could not travel all over it. On Sunday morning he biked into the countryside and then returned to the steamer to begin his journey inland to Jönköping that evening. The steamer *Primus* traveled by inland waterway giving Otto a leisurely day-and-a-half trip through magnificent countryside. "Sweden is called the land of a thousand lakes and the lakes are connected by a famous system of canals and locks which make it possible to travel in this wonderful fashion."

From Jönköping, Otto journeyed by train to Ålmhult where he went to see his brother Gustav.

Gustav is the happiest man in the family. He learned the trade of coppersmithing and for many years worked, worked, worked, trying to get a living at it; but fate was against him. Copper utensils have rapidly been driven out of use by glazed ironware and the coppersmiths as such are now practically exterminated. The very man from whom he learned the trade and who for almost a generation has been established in a little city (Wexiö) [Växjö] now works all alone in his shop (he used to have several men to help him) mending old pots and kettles of the town, but never making any new ones. Gustav has managed to learn the tricks of another trade, that of putting on iron plate roofs and other similar work, and this has helped him considerably. He has also secured the position of Chimney Sweeper for the village for which he gets a salary of 100 kr. a year and 25 ore for every chimney that he cleans. He is without debt by the end of every summer; but as regularly, gets a little behind every winter. He is extremely quick-tempered; but his wife has learned to oversee that and they are living very happily. I knew very quickly last Christmas that Gusten is a fine man and later experience and investigation has only strengthened that opinion although my recollection of him since fifteen years ago was not at all to his advantage. He was not sober then, he swore awfully, he did not write his own love letters (I was his scribe) and he didn't pay me as he agreed for work that I occasionally did for him in the shop. . . . My father came after me the following day and by half past seven I was once more at home. I never expected to feel "at home" but I really do. It is by no means an ideal one; but the certainty that they really want me to stay and that they are very much interested in everything that concerns me and also want to do everything that they can for me produces essentially a "home" impression.

Thursday evening my father must take me back to Elmhult. Friday morning at 5:34 I expect to take a train from there to Trelleborg the southernmost sea town in Sweden, and by 8:45 in the evening I shall then hope to be in Berlin. I have two acquaintances there, one a Minnesota classmate [John Zeleny] to whom I have written and asked to meet me that first evening. . . . It is going to be queer to get into a really foreign land once again.

References

4-1. Blix, G., Olof Hammarsten (1841–1932). *Svenkst Biografiskt Lexicon* **18**, 207–9 (1969–71).
4-2. Van Slyke, D.D., Ivar Christian Bang. *Scand. J. Clin. Lab. Invest.* **10** (Suppl. 31), 18–26 (1957).
4-3. Moore, F.J. *A History of Chemistry*. 3rd ed. pp. 63–73. New York: McGraw-Hill, 1939.

Chapter 5. Berlin—Clinical Orientation (1897)

The trip to Berlin went as planned. The boat for the ride from Trelleborg to Sassnitz, Germany, across the Baltic Sea was large, and though the sea was not very rough, all the passengers were sick. At Sassnitz, Otto boarded a train to Berlin where he was met by John Zeleny.

> Zeleny, a classmate of mine was waiting for me at the Station in Berlin, and that night I slept in his room. He studies physics here; his roommate, a Chicago man, studies brewing chemistry. The latter, Mannhardt, walked me around Berlin yesterday forenoon until eleven o'clock when I was to meet Prof. S. Berlin is a magnificent city. By eleven I went to the Pathological Institute and met the professor, a small intellectual looking man with keen, blue eyes which seemed rather friendly. My German was awfully poor just then and I found it very difficult to converse with him. Prof. Hammarsten seemed to have written that it might be a good idea for me to continue the investigation of animal gum here, but I am not inclined to do that because I am perfectly capable of doing that alone and if I am to stay at this costly place I want to get something out of the professor.

The two months that Otto intended to be in Berlin was a very short time in which to do a research project. If no project could be formulated, Otto preferred to work through Salkowski's course in physiological chemistry. Meanwhile, with Zeleny's help, he found a room very near Salkowski's laboratory, which was housed in the Charity Hospital.

The next two weeks Otto spent working on milk and then on lean meat, evidently carrying out some established procedures with which he wished to become familiar. His laboratory was smaller than the one in Uppsala. The spirit here was also much different. Seven or eight men were doing original work under Salkowski, and Otto obviously would not get as much attention from Salkowski as he had from Hammarsten. Otto noted that Prof. Salkowski's specialty was urine.

> ... next week I shall begin with it [urine] and shall probably work with it as long as I am here. The professor has given me a special question to investigate in connection with it. To determine the exact quantity of each constituent of urine both in connection with diseases and in connection with scientific experi-

ments is very important and the professor would now have me test a new method which has been introduced in rivalry to his for determining one constituent [uric acid]. My taste does not at all lie in that kind of work. How much I shall do with this is therefore uncertain. I think that I shall rather aim to get a satisfactory experience in the whole subject and simply accept the methods which we consider to be the best. I am rather anxious to get to Voit in München. I think that he is my man and if I like him I want to stay by him perhaps the whole of next year and work upon *living* organs. That is the chemistry I want.

What Otto wanted was to get into the "meat" of physiological chemistry. He did not wish to be stuck with a purely analytical problem related to a "waste" product in urine. The core of chemistry was in the living cell, and *there* was the proper place to apply one's skill. There, for that matter, was where disease was often manifested, the chemical changes that could be revealed by careful research. But Otto grossly underestimated the importance of Salkowski's petite project on his future life's work.

Ernst Leopold Salkowski (1844–1923), an outstanding physiological chemist of his generation, was a founder of this new discipline in Germany, along with Hoppe-Seyler, Baumann, Hofmeister, and Huppert. He studied medicine at the University of Königsberg, obtaining his doctorate in 1867. He then spent two years of further study in Vienna and with Hoppe-Seyler in Tübingen before returning to his hometown to work as an assistant under Ernst von Leyden in a medical clinic until 1872. After six months with Kühne in the Institute of Physiology in Heidelberg, he accepted an appointment to the staff of the Institute of Pathology in Berlin. In two years, he was promoted to professor (extraordinary), and in 1880 he was named head of chemistry. His research prior to Otto's arrival was particularly strong in the development of new analytical methods suitable to clinical practice—in short, clinical chemistry.

Contrary to Otto's thinking, urine was a readily accessible body fluid that had been used since ancient times to provide a great deal of information for diagnosis of disease as well as monitoring health. Salkowski's interests in chemistry, however, were wide. He contributed much through his studies of uric acid and urea formation; he developed tests for the presence of sugars and peptones, the antiseptic action of chloroform-water, the excretion of phenol, autolysis, the effect of heating on protein, and pentosuria. At the time Otto was in the Berlin laboratory, some of the work going on was the followup of Salkowski's discovery of pentosuria, the effects of antiseptics on toxins, and alloxan bases in urine. Salkowski had published many articles in the German journals and two useful textbooks, *The Study of Urine* (with W. O. Leube, 1882) and *The Practice of Physiological and Pathological Chemistry* (1893).

Salkowski's laboratory served as a training center for students from many countries. He not only taught them the basics of chemistry but also introduced them to the use of this specialty in clinical diagnosis and treatment. He was described as rather unassuming, polite and correct, deeply concerned, and kind. He enjoyed most of all working with his students (*5-1*).

As he had in Uppsala, Otto soon established a routine for living and working in Berlin. The room he had found was fairly comfortable, if not as nice as the one in Uppsala. Things were more expensive in Berlin, so the only meal he ate out was lunch. He rose at 7 a.m. and made breakfast: milk, coffee, bread and butter. His evening meal was oatmeal and eggs, together with bread and butter. At lunchtime, he sometimes joined his Minnesota colleague, C. P. Lommen, '91, who had been one of the editors of the *Ariel* and was now professor of biology at the University of South Dakota. "He has his family with him and tomorrow afternoon I am going out to visit him at his home in the suburb Adlershof. He is an old fraternity friend of Mr. Sikes and is a fine solid man. He has been here a year and returns home next August."

From 9 a.m. to 1 p.m. and from 2:30 to 5 p.m., Otto worked in the laboratory. He then returned to his room, rested a bit, read the paper, and fixed his supper. At 7:30 p.m., he went for a walk, but felt limited in his sightseeing without a bicycle.

A letter from Laura had come and made Otto so happy that

> I had to go home where I could be alone and then I read the letter again and thought about the writer. I felt as though you were, after all, really falling in love a little yourself. Do you think you are? . . . Goodnight my dear little wife-to-be. I love you very much. I hope that I shall not have to give you up in the end.

Like most of his letters, Otto's letter to Laura contained not only a reaffirmation of his love, but many details about his life and work. He grumbled that he was not getting along very fast with his work (June 12), which was supposed to be an evaluation of the relative merit of two methods (for determining uric acid). When he had decided which was preferable, he wanted to end it there. "Prof. Salkowski is a fine physiological chemist; but he is so nervous and irritable that one cannot go near him half the time." Otto had learned that the best results were obtained without hurry, by careful deliberate work. Prof. Hammarsten thought that the Germans hurried to publish so that their discoveries were seldom reliable and had to be revised. Otto had just heard from Hammarsten about their little "arbeit." They had to withdraw a point to avoid a contradiction from another investigator's work. Otto promised to send Laura a reprint of the paper when it appeared in the *Zeitschrift*.

Berlin in Otto's view was the foremost city of the world in civilization and progress, and it made him feel very small. "Science, science, everywhere; the wonderful university, the bookstore, the science stores, the newspapers, the men, all seems a part of a crusade more wonderful and perhaps as fanatical as the crusades of old." This was, however, a lopsided world. "I should say that science is producing a most wonderful external civilization but individuals are all going to the devil."

Otto worked in the Pathological Institute, a part of the Charity Hospital. The whole occupied a square in the center of the city as large as that occupied by the University of Chicago. It had five to six times as many buildings as Chicago, yet it was only one of several sections that made up the University of Berlin. Otto had not yet seen much of Berlin or the departments of the university.

Otto's next letter on June 17 indicated that he was working hard on testing the two methods, but was faced with a problem. He had found the "Hopkinsche" method for determining uric acid as accurate as Salkowski's method, but shorter and simpler to execute. He had told this to Salkowski whose method had become the standard one. It would hardly do for a paper to leave the professor's laboratory acknowledging the superiority of another method. What a dilemma! Salkowski did not yet believe in the accuracy of the Hopkins method. Otto would have to proceed slowly. "Salkowski is unusually touchy and extremely jealous of his reputation."

Life in Berlin had its drawbacks. Otto complained that the street below his room was noisy but the climate made him sleep as a counterbalance. There were ringing street cars, steam-puffing trains, the clatter of hoofs on the asphalt pavement—all intolerable until he got used to it. It grew quiet between 2 and 5 a.m. About 6 a.m. the racket began with a brass band! Berlin gave the impression of having as many soldiers in training as existed in "our whole republic." One of the Kaiser's military companies marched by every morning, either at 6 a.m. or earlier (75 soldiers led by six musicians). Climbing four stories to reach his room deterred Otto from coming to his room during the day for coffee and rest.

It was almost astounding how rapidly Otto warmed to the simple research task that Professor Salkowski had given him. In view of Otto's short tenure in Berlin, Salkowski's choice was astute, amazingly germane to Otto's needs. Two features of seminal importance to Otto's future materialized. In the atmosphere of experimental pathology of the hospital and the medical school, he now sensed the immediate significance to medical practice of using chemical tests. For physiological chemistry, readily available urine was a resource to study metabolism in the human body as well as to diagnose disease. Then too, his simple research problem provided an adequate challenge for an organic chemist with training. In fact, it gave him a unique opportunity to release his teeming ideas, his creativeness in laboratory experimentation on improved or original analytical methods. Otto was not only discovering his future, but in the spirit of many young insecure researchers, he became frustrated, suspicious, inwardly abusive, jealous, possessive, and defensive. He had a "whopping" inferiority complex that spurred him on to superior, even remarkable achievement—an old phenomenon for scientists and other creative people.

Otto informed Laura that he was tremendously absorbed in his work now. He wanted a little rest, and a different kind of work for next year, but his present work was suggestive and he wished he had more time to pursue his ideas. "It is analytical; but correct and speedy analysis of the waste products of life is absolutely essential to progress toward the chemistry of the same." He was about to test a scheme of his own, and if it worked, it would be a fine thing. He had prepared three substances any one of which he hoped might react as he wished. The first one tested, however, failed. (More than likely, Otto was trying out new substances that could react or precipitate with uric acid quantitatively.)

As his work ran into difficulties, Otto's feelings about Salkowski grew more negative. "The professor is a narrow, selfish, unsympathetic stick. He is of absolutely no use to me and simply does not want to get results because they would then lessen the value of his." Otto hoped that he would never be like Salkowski towards enthusiastic young men who worked with him. Salkowski had only opened his eyes wide when he heard Otto's plan, instead of offering suggestions as to how Otto could obtain quickest the substances he needed—but not one suggestion. Salkowski had pocketed Otto's registration fee instead of letting it go to the university, "and not being registered I cannot search the literature because I must get the books from him and I get only what I insist on having."

Otto's analytical schemes failed. All three substances tested were ineffective. But the work was not lost.

> It was a double-acting-Stieglitz-sort of a scheme, and by refusing to act both ways it will act with double force on one side and right on my main point, too . . . the point is, you see, to get a crucial, quantitative distinction between uric acid and these other things. Salkowski's method, though acknowledged as the standard, is laborious and unsatisfactory and I am certain that a proper touch of ingenuity and luck should produce a very much better process. There is not a touch of genius or originality about the old Salkowski method. It is simply the work of a mind insisting on making an idea go by means of a system of corrections for its deficiencies.

Otto's attempt in mid-July to improve the method failed. His results seemed low, but he still had to compare them with those from the older process that he applied as a criterion. He hoped to know by week's end if he had "a really good thing." If not, he had one more idea. If it too failed, then he would get more data and report that the Englishman's (Hopkins) method was reliable and should be so recognized, and he could meanwhile add his own data, and correct a point or two in the method. There was not time to work out a good method. "Yesterday, I did the most satisfactory one day's work that I have yet done in Germany. It was on my last idea. The other one may yet work." Otto was now working with some success on his own modification without Salkowski's knowledge. Otto's new Italian acquaintance in Salkowski's laboratory agreed that the professor's method was inferior.

During Otto's last week in Berlin he wrote that his "schemes have worked to perfection on normal urines," but he had found disagreeable difficulties with pathological urines. He would, however, have a "very neat little paper, I think" provided he did not antagonize the professor, who did not yet know the chief point of merit in Otto's work, "nor of my attitude toward his method."

By July 30, Otto was still trying to overcome a very disagreeable difficulty with pathological urines (probably end-point detection in icteric urine). Although the method already had some merit, overcoming the problem would have resulted in a good method. The professor was rather stunned by Otto's results and praised his work. Salkowski, Otto believed, had changed his attitude toward him and recommended that Otto publish his findings. Otto wanted to publish in the *Zeitschrift für Physiologische Chemie*, but Salkowski had fought with its editors three years previously and had not published there since. However, Otto wanted it published there because it "practically knocks his standard 'silver' method on the head." To Salkowski's credit, he agreed. Otto would write the article in English and then it would be translated. Salkowski preferred it that way as had Hammarsten.

As it turned out, Otto was in Berlin until the evening of Aug. 17, at least a week to 10 days longer than he had expected. On Aug. 5, he was putting in "the last finishing touches on a really pretty piece of work. Yes, I have had a series of little inspirations and I believe that the Folinische Methode will prove to be a beautiful thing. Within five years I shall expect to see it described in every reputable textbook on the subject." Although Salkowski might try to disprove it, the method would be sure to have its defenders. If he could write the method up so that its main points were brought out in proper perspective and find a translator to do it justice, then he could go home in peace to Sweden for two months.

By Aug. 8 he was writing the paper, which was to be about the same length as the one from Uppsala, but would have 40 times the audience. He had worked and worried through his laboratory difficulties, but, in time, he had succeeded despite the pressures. He also had another idea that he wished to follow up, and to avoid someone else's grabbing it, he intended to mention it in the current paper. He did not think that this was a work of genius or unusual. Any Ph.D. from Kent Laboratory would have settled the question as he had, though at a slower pace. "It is straight scientific chemistry only it is not commonly applied in this science and that is why the subject was left for me. I have done good work, but nothing more."

Finally on Aug. 13, he finished the paper and put it into the translator's hands. He was satisfied with his work. Indeed it was the only publication among the eight people in Salkowski's laboratory that semester. "Piece of good luck, you see. . . . I don't like Salkowski." If Salkowski cut things from the paper, as Otto felt he might, Otto intended to publish without Salkowski's sanction.

The translation did not go entirely smoothly. The translator, Dr. Kruger, arrived too late in the morning. He smoked three cigars, drank four glasses of Otto's horribly strong liquor and also some coffee, within a space of the two hours that it required to translate just six of the 31 pages. In addition, he left out the smoothness and fine points of the work. "Now I sit right by and explain the meaning to him." Yet Dr. Kruger was a good man, a well-trained physiological chemist.

By Aug. 15 the translation was nearly finished, but Dr. Kruger's failure to show up on Aug. 16 stranded Otto in Berlin one more day, and he could not return to Sweden. Otto would have left without his manuscript, but he did not trust the Berlin Germans. They showed dishonesty, he said, and stupidity when it concerned the affairs of others. By the next day, however, the translation was finished just in time for Otto to catch the 7:10 train to Stettin-Sossnitz for the trip home.

Otto's next destination to continue his professional education was undetermined as he left Berlin. He had written to Voit in Munich, but because of expenses had decided against making a personal visit. A brewery chemist whom Otto had met earlier was in Munich and wrote Otto that Voit was a very nice man. Dr. Wakeman, an American acquaintance that Otto had made in Berlin, had offered to check for Otto on conditions in Munich. (Wakeman was a physical chemist from Yale who was studying physiological chemistry to run a private research laboratory in New York.) Otto had written Voit to determine what subjects he could work on, and if these were not satisfactory, he would not consider going until he had written Kossel in Marburg. "I want to get out a good piece of work next year and if I do get another Salkowski I want to have something to say in regard to the subject I am to investigate." He felt "lucky" that he had a fruitful subject in Berlin, and he now believed that he could get results from most subjects in physiological chemistry.

"Physiological chemistry is simply the 'wild and wooly west' of science and there are 'vacant claims' everywhere. I am very much satisfied with it." Dr. Wakeman was a queer example of the men who worked in the field, he commented to Laura. Two years before, Dr. Herter, a physiologist in New York, asked Prof. Chittenden at Yale (by far the most prominent American in this subject) to recommend someone to be his assistant. Chittenden could not recommend anyone, but suggested that the assistant instructor in physical chemistry might do. Dr. Wakeman knew almost nothing of organic chemistry and nothing of physiology, because his field was closest to physics. After two years with Herter, he still knew nothing, and he ran around a good deal looking for a tutor. Otto predicted (wrongly, as it turned out) that Wakeman would never accomplish much in physiological chemistry.

Otto received a friendly letter from Voit, but Wakeman advised Otto to write Kossel with whom he had dined. Voit had not mentioned any subjects for research. Kossel was only 45 and the editor of the *Zeitschrift* that Otto "patronized." Otto decided to write Kossel and ask him if he could continue his work on uric acid.

Otto's second paper resulting from his European work appeared in the December 1897 issue of *Hoppe-Seyler's Zeitschrift für Physiologische Chemie* (**24**, 224–45) and was entitled "Eine Vereinfachung der Hopkinsche Methode zur Bestimmung der Harnsaure im Harn" ("A simplification of the Hopkins method for determining uric acid in urine").

In the future, Otto would produce research of greater significance to biochemistry, but it is difficult to comprehend how in any future research, he could have put in more intensive laboratory effort than he did in the two months in Berlin. He probably worked 12 to 14 hours a day, seven days a week. Unlike his Uppsala experience, he had little recreational time, saw little of the beauty and culture of the city, and except for the Italian Biffe knew his coworkers only slightly.

Uric acid analysis was his maiden voyage into clinical chemistry, and he attacked the problem with all the intellect and skill that he could muster and mixed in the vigor, thoroughness, and tenacity that would characterize his approach to all of his investigations.

The work itself is not a classic, but would bear studying by any aspiring clinical chemist who may be seeking a modest entry to research in a clinical laboratory, to harness his skills into something productive. The project that Salkowski had proposed—comparing his own method for uric acid determination with the Hopkins method—was an easily fulfilled, low-budget project that certainly was not intended to last more than two months or to result in sufficient data for publication. Within the two-month period Otto had to set up and master the details of two analytical methods, perform a large number of analyses on pure solutions of uric acid and on urines from both normal and pathological sources, and assess the significance of the differences in values obtained as well as the complexities and pitfalls of the two techniques—tasks to accomplish that made the odds very long against his writing a paper. Combining these tasks with the need to get any paper written in German before the term ended made publication seem out of the question.

However, not even Otto anticipated how all-absorbing the problem would become. In setting up the two analytical methods for uric acid he decided almost at once that Salkowski's silver method was too cumbersome, already too much modified, and thereby less challenging as a project. On the other hand, the much simpler Hopkins method had more appeal to his chemical sense, and left open several doors for improvement. Because this paper was profoundly important to Otto's scientific future and illustrates his approach to clinical chemistry, we present in Part II (p. 367) some of its features in greater detail than would otherwise be merited.

Laura's letters throughout the summer indicated that her life in St. Paul had changed little. Her family's prosperity had continued to wane to the point where the Grant household was reduced to one servant, and Laura was now helping with meals. Her brother Hedley in California had grown increasingly ill and was not expected to recover. In September, she intended to work as a volunteer cataloguing books, papers, and reports for the Associated Charities of St. Paul. However, her letter at the end of August indicated that she still had a chance to teach in the fall.

Reference

5-1. Neuberg, C, Ernst Salkowski. *Biochim. Zeit.* **138** (1923); Ernst Salkowski on his 70th birthday. *Deutsch. med. Wochenschr.* **40**, 1870–71 (1914).

Chapter 6. Marburg—Quest for Independence (1897–98)

Otto undoubtedly arrived in Träryd exhausted and probably somewhat undernourished. A letter to Madeleine at the end of August indicated that he was doing no real work, just reading a little. His sister was trying to fatten him up, he told her.

Within the letter to Madeleine was an explanation of his financial plans. She had agreed to loan Otto $100, and Gertrud had offered a much larger sum for his continued stay in Europe.

> [I want to get] the whole hundred from you before the fifteenth of October. My fare to Germany and the fee for the University [Marburg] will take quite a sum at the very start of the year, but with a hundred dollars I think that I would be safe until Christmas. I can get a little more money from my sister; but I would not like to take any from her at the beginning of the year because I reach her by mail in two days and it is an advantage to know when I am over here that I can get money quick and sure from her in case I get into any kind of a pinch.

Otto's brother Axel had also offered more help. He served as a permanent safety valve, but farmers never had much money and Otto wanted to spare him. Otto had one other unnamed, promised source of help in the U.S., he told her, but he had not used it. However, he would write everyone to determine how much he could count on. "It was not half as nice to live on borrowed as on earned money."

On Sept. 3, Otto began a long letter to Laura just before setting out on a bicycle trip with Gertrud. He had received a satisfactory reply from Professor Kossel in Marburg and had decided to go there for the October opening.

> I wrote him that I am a chemist and preferred to get a subject that is not purely chemical and I indicated further that I would like to have something involving investigation of some chemical work done by the animal system. He wrote in answer, "I will vigorously strive to be helpful to you in your work, and will gladly suggest a project from the field of metabolic science."
>
> Prof. Loeb is a good friend of Kossel's I believe, because he offered to write me a letter to him. It is probably not going to hurt my interests to go to Kossel if I do reasonably good work.

Otto intended to write to Loeb to ask him to write Kossel, actually to remind Loeb of Otto's existence. He hoped that Salkowski would not cause him any difficulties.

Otto had not yet taken action on future jobs, though earlier he had corresponded with President Northrop at the University of Minnesota. Northrop had nothing to offer, but thought highly of Otto. Otto was also going to ask Loeb for suggestions on future jobs. One job possibility was a job with Dr. Herter in New York through Wakeman.

Otto received a copy of his Uppsala paper and an "Honorar" of 5.50 kr.—his first bit of money as an author! He had no expectations of getting rich on his scientific writings, at least not directly. Otto averred to Laura that he did not like the article much. The best point was left out to please Hammarsten. The point—he had cleaved mucin with acid and, following alkaline hydrolysis, had obtained a crystalline osazone-like compound with phenylhydrazine—had been in the last two or three lines of the article (excluding the paragraph devoted to the professor). These lines, which represented two weeks of work plus four more weeks to show the presence of a hitherto unreported component (the aminoglucose) of the glycosoaminoglycans, would have given the paper an altogether different character. Who knows what he might have accomplished?

Otto was thinking about his future in even more concrete terms. He asked Laura if she would get the high school teaching job. If so, would she salt away some money? After discussing saving money for the future, he figured that to live in New York, if he got work there, would take $2000 a year and a constant increase in salary. It would be like getting a mortgage on happiness if he could find a place with Dr. Herter. Herter had been very generous to Wakeman.

Laura wrote that she was intoxicated by the thought of living in New York, whether or not he had a job with Herter. It was a thousand times better there than Chicago. A European city in the U.S. Hospitals; big city life; cheap living; pleasanter than Chicago; old friends there; Barnard College, where she could study.

Indeed that fall Laura may have needed to fantasize about escape to New York. Her father's finances had hit a 17-year low (a capitol building contract had gone to another firm, and there was no work); her brother Hedley's condition deteriorated and finally he died; and lastly she did not get the teaching job, although she was to do some substitute teaching of history in the manual training high school.

Otto left Träryd for Marburg in mid-October.

> I arrived and called at once on Prof. Kossel. He received me in a most cordial manner and I was quite satisfied with him and will be with my subject if I can find a way to attack it.

His first few days in Marburg, Otto settled in quickly. He found a pleasant room overlooking the botanical garden of the university, only a two-minute walk from the physiological institute. Thursday morning he spent in the library and then talked more with Professor Kossel. On Friday, Otto spent a little time in the laboratory where he met Kossel's friendly, obliging first assistant, Kutscher, who promised to be a treasure to Otto. Kutscher proposed a walk with Otto after dinner when the laboratory was closed. Otto expected to be led to a drinking establishment, but instead they took a stroll through a beautiful promenade.

Marburg was an old town built at the sides and foot of a very high hill on top of which lay an old castle. As they walked down to the foot of the hill through a curious, ancient-looking village, the sights were picturesque. Kutscher, whom Otto found uniformly interesting, took him to a tavern for coffee and a cigar. While Kutscher and Otto were enjoying their coffee, in came Professor Kossel with his wife and young son—going out with the whole family to enjoy a couple of hours was typically German. Otto and Kutscher joined them at their table for more coffee, cigars, and conversation.

He spent his second Saturday in Marburg with his newfound friend Dr. Kutscher, and again they walked over the hills and back into the valley, finishing with coffee, cigar, and good talk. He shared a Saturday supper in Kutscher's room.

Otto was as eager to understand the student life in Marburg, as he had been in Sweden. The fraternity men preserved the dueling system in Germany, and Kutscher promised to let Otto witness such an affair.

Otto did not formally register in the university to save the 40 marks required, but he attended Kossel's lectures free and paid for his laboratory privileges at 40 marks per semester. He was working on a trial experiment for a month as a possible basis for a project. "My work will probably be rather costly anyhow. I am already the owner of nineteen white rats and one starving dog." He had already fed one dog six pounds of good meat and then sacrificed him for an experiment. However, in spite of these expenses, he was financially secure until mid-January, thanks to Madeleine, who had sent him $50, and to his brother and sister.

Encouraged by Laura's letters to be more confident of their future relationship, Otto began to consider his future in terms of teaching and research in physiological chemistry. A letter to Laura at the end of October spelled out some of the possibilities. He calculated that it would take 10 to 15 years to master his subject. Beyond that, it was a matter of keeping up with the literature. Marriage could cause a falling off of scientific output, he speculated, but he was uncertain of that and he wanted a home life and felt that it would promote his best work. What jobs were available? If Loeb could get him a job it would not pay well the first year. But if that were all that he could get, Otto would take it. After he had heard from Loeb and Herter in New York, Otto intended to communicate with Dr. Abel at Johns Hopkins and with Professor Chittenden at Yale. He did not, however, expect much from either of them. There was an American from Johns Hopkins already in Germany, working as an assistant to Professor Schmiedberg in Strasbourg, so he would get any position that became available at Johns Hopkins. Otto also would write to other places such as Ann Arbor, the medical schools. Otto dismissed the possibility of applying to the University of Minnesota, because he felt that Northrop did not like men from the University of Chicago since Harper had taken Judsen and others from him.

Otto's laboratory work had scarcely begun. His first experiments were failures. He did not want Kossel to think him incapable of doing difficult experiments, for Kossel might write this to Loeb. Otto had expected to isolate 10 grams of a substance (protamine), but got only half a gram from 20 herrings and about 2 gallons of solution. Yet Kossel's procedure should have given at least 10 grams. He had lost a week's work. He had 20 more herrings and would try again. But maybe the experiment would not work on "salt" herrings as the procedure had called for fresh ones.

His first six weeks in Marburg, he reported on Dec. 1, 1987, had considerably reduced his purse, and his trial experiments ended in a fizzle. He was not downhearted and would start a "new point of attack."

> Kossel is in almost all respects the opposite from Hammarsten and especially in that he will suggest operations and experiments that have not as much of chance in their favor as against them. He is full of ideas, but some of them are indifferent and some are too visionary to be realized. He let me work with salted herrings without having made more than one—and insufficient—test whether they could be used as a source of protamine and now I am at last to begin with another fish from which protamine can be obtained. I don't complain. I think it is right enough that a mature student should be set at problematical questions and I very much prefer a man who gives lots of ideas to the man who sets you on one little point that he knows will come out right.
>
> Kossel gave me tonight an outline of the work he wants me to pursue and it is going to be very valuable work to *me* because it is going to give me experience with many different substances of importance in physiological chemistry. I am strongly inclined to believe that one half, and the more important half at that, are not going to be found. This makes, however, not so much difference because I shall get valuable experience and I shall get enough results to be able to publish a paper toward the end of the year.

Otto's judgment of his three eminent professors in Europe was based on his youthful feistiness and impatience, his narrow field of vision, and limited experience as a graduate student at the University of Chicago. However, he was hardly in a position to assess the impact on physiological chemistry of the three men or to appreciate the significance of their work. Hammarsten and Salkowski were outstanding protagonists for this new discipline born of organic chemistry, physiology, biology, and medicine, but Kossel was brilliant, a "great one," who would become a Nobel Laureate in 1910.

Albrecht Kossel (1853–1927) was born in Rostock and attended the gymnasium there where he was the top student of his class. He displayed an early interest in science, particularly biological, and was so well versed in plant taxonomy that during a meeting of the German scientists and physicians he was entrusted with leading them on botanical excursions. Though he wanted to pursue botany as a career, his father objected for economic reasons, and Albrecht, instead, studied medicine in the Imperial University in Strasbourg in 1872 and finished up in Rostock. While Kossel was in Strasbourg, Felix Hoppe-Seyler (1825–95) stirred his interest in physiological chemistry, and he became Hoppe-Seyler's assistant from 1877 to 1881, then a privatdozent in physiological chemistry and hygiene until 1883. Then Emil DuBois-Reymond appointed Kossel the director of the Physiological Institute and assistant professor of physiology. He remained there until 1895 when he was appointed to the directorship of the Physiological Institute and a professor of physiology at the University of Marburg, where he was when Otto came for his postdoctorate experience. At Marburg, collaborators came from around the world to work with Kossel whose pioneer studies on nucleoprotein had received widespread attention.

After Hoppe-Seyler's death in 1895, Kossel became editor of the *Zeitschrift* in which his own first work on the dissociation of salts in aqueous solution by means of diffusion appeared in 1878–79. In 1879, while in Strasbourg, he first reported on what became the major area of his life's work, nucleoprotein (yeast). He examined both the protein and nucleic acid portions and had reported the occurrence of hypoxanthine and xanthine. He discovered adenine in pancreatic nucleic acid and incidentally in tea leaves, where he also found theophylline. With Neumann, he found in the nucleic acid of the thymus gland, thymine and cytosine. One of his collaborators, P. A. Levene, would pursue the carbohydrate (ribosyl) fraction to its complete identification. Later with other coworkers, Kossel found uracil. He also worked on the protein portion and reported several of its amino acids. This led him to studies on histones and protamines, the action of nucleic acids on bacteria, the formation of urea, tryptic digestion, and to ideas on the remodeling of proteins via histones and protamines and on fundamental protein structure. Kossel left Marburg in 1901 to take a similar chair in Heidelberg, where he completed his distinguished career (*6-1*).

An advanced announcement of Otto's article that was to appear in the next issue of *Zeitschrift* provoked Otto's irritation. Salkowski had changed the title from "Simple Method. . ." to "Simplification of Hopkins' Method. . . ." The change was unfair, Otto vowed, and made without his permission. Otto wrote the publisher to get hold of the page proof and see if other changes had been made. (Author's Note: Otto's problem, by the way, remains to this day in clinical chemistry. When is the modification so important that the original method need not be titled? A fair compromise for Otto's paper would perhaps have been to leave the name Hopkins out of the title entirely in favor of "A Simplified Method for Determining Uric Acid in Urine," so that modification is implicit. Those who used the method would then have referred to it as the Hopkins-Folin method or vice versa.)

Otto's progress on his current work was slow, but he was getting valuable experience and facts so that he felt comfortable about physiological chemistry. He had read an article by an American on the work that he had done in Salkowski's lab. He considered the article poor and immature. He theorized that although Germany was the place to come for physiological chemistry and other sciences, the U.S. could do as well if the universities paid professors more and let them have time for research.

An American woman was also learning physiological chemistry in Kossel's laboratory. Otto's discussion of her presence clearly shows his male chauvinism.

Miss Langenbeck works pretty hard and is undoubtedly learning a great deal. . . . Women will never do much in science, I believe, because it requires both head and a good physique and that we find in only exceptional cases like that of Miss Felton, perhaps. Of course I have not sufficient knowledge of women to be a competent judge. By head I don't mean that they may not have intellect enough but it is generally not of the kind that naturally deals with science. Miss Langenbeck is a very fluent talker, however, and that is a wonderful help when one wants to get along; we shall probably hear of her as holding some pretty good position in America by and by probably a professorship in one of the women's colleges. [Author's Note: She returned to Cincinnati College and eventually got a Ph.D. at Bryn Mawr.]

By his Dec. 17 letter to Laura, Otto was somewhat disenamored of Kossel. He thought the work was progressing rather queerly, and Kossel, who came into the lab four or five times during the week between 6 and 7 p.m., didn't seem to give a thought to his students between visits. He had no direct responsibility in the student's work and no planning for the year's work. Otto thought that he had showed Kossel that night his lack of enthusiasm for the problem with the result that the outline of the work was made more satisfactory. However, Otto had already wasted a long time

> in getting a substance out of salted herrings that could not be so obtained, then he allowed the season to pass by when I could have obtained fresh fish of another kind that would give the substance, and now it appears that he can let me have about ten grams (1/50 lb.), and then we can probably obtain none until March.

Otto expected to talk to Kossel the following morning to let him know plainly that he was dissatisfied with such a project. He wanted work that could be completed in a continuous, systematic manner, and he did not want to be brought to a standstill in the best season of the year for want of material.

But Otto confided that he was really not in bad humor. He believed he had made "another beautiful little find," good for another "effective bit of a paper." About half a dozen "strong" men, including three German professors and Chittenden of Yale had for 12 years been studying the properties and methods of preparing a certain class of substances, but the methods were crude and unsatisfactory. Otto needed one of these substances for his work. The substance could not be prepared, however, by the direct way. Kutscher had also found difficulty that way. Now Otto had successfully tried a little variation, so that he now obtained

> a most beautiful substance, three times as large a quantity and made in a day, while the old method takes a week. My substance is either not identical with this corresponding product or it probably is much purer and in either case I have scored a point and secured a field of work I hope.

Kutscher approved of the new method, but Kossel did not know yet. Otto would tell him, but meanwhile work his on the side as his own. It would make Professor Chittenden "aware of my existence," and show in others a lack of the caliber of training in chemistry such as he had received at the University of Chicago.

Otto found that Salkowski had not altered the uric acid paper, so with a few minor changes, Otto completed it.

> On the whole I am tired of professors. I am perfectly capable of going on alone now and I do hope that I shall find a position in America where I am scientifically free. I feel quite confident that I shall leave a good plain toe mark behind me if I get a chance to follow the path that I seek.

Jobs in Europe were not to be had. There were already too many men for the positions. The man who translated Otto's uric acid paper had been well known for 10 years, but could not make a living by chemistry. Also one had to be an M.D. to get a chair or a class in physiological chemistry. That was why pure chemical skill had not been used enough. There was a fight to establish physiological chemistry as an independent department. When that happened the work would proceed at a tremendous pace. Even in Germany the physiological chemists grew into their positions as M.D.s! "I have a fine advantage if I get the right kind of position!"

Considering that the year was 1897, Otto was very perceptive. He would be one of the foremost to establish an independent footing for biological chemistry in the U.S., with the principal sponsorship in the medical schools (6-2).

Kossel presented Otto with some substance for his research, possibly protamine, on Dec. 20, with the promise of getting him more if he needed it. Otto described his method for preparing protalbumose, which others had found difficult to do. Kossel was very much interested and gave him suggestions for extending the application of the method. Otto thought that Kossel was slyly mixing himself in the procedure so that Otto would not be able to publish it without giving Kossel more or less credit. But Otto did not care. He felt it showed his worth.

> . . . every time a man finds something like that, down goes his name into the great book, the literature of science, and the name will live and the work will live, and the work will be remembered as long as humanity has any use for science. Such are the joys of scientists.

Professor Kossel, Otto assured Laura, was a very friendly man, easy to get along with, and he did not mean to indicate otherwise.

Otto's second Christmas in Europe had arrived. He described his Christmas Eve to Laura.

> The janitor scrubbed the floor and the desks all around me this noon and it was high time to leave; but I held out until my experiment was finished and the substance secured which proves beyond a doubt that my method for preparing albumoses is simply a wonderful improvement upon the old cumbersome method of Kühne and Chittenden. "Solch ein Menge Deutero-albumose habe ich nie gesehen" said Dr. Kutscher as he left the laboratory about half an hour before me, and he has worked with these things for half a year. The result was my Christmas present and just now the bells are ringing in Christmas, a

monotonous dolorous ring that somehow suggests "sacred" history and is not in harmony with my present frame of mind.

Otto's thoughts ran to Laura and what Christmas must be like in her house. How would they spend the holiday together in the future? He continued,

> I had for years a notion that then nearly everybody was unusually lightminded and happy and consequently I was more or less vaguely depressed in spirit on those Christmas Eves when I seemed to have nothing in particular to remind me of Christmas time.

On New Year's Day 1898, Otto wrote,

> I thought that I should be able to do some work in the laboratory, but it was altogether too cold and disagreeable there and besides I am aware of the fact that Kossel for one reason or another rather discourages work during such vacation times.

Otto reported that late New Year's evening he was trying to write his little article in German. ". . . of course I cannot write a sentence correctly because I don't know the grammar." However, he mentioned, Kutscher would edit it. This was a beautiful paper, as good as the uric acid one. Otto was justifiably proud of having two papers in such a short time. The new paper would contradict and criticize one that had just appeared.

Kossel was not interested in his paper, Otto felt. He was too absorbed in his own ideas to see anything else, and wanted his own ideas worked on. Both Otto and Kutscher felt that Kossel's ideas were not as "important for the science" as "our own," and he stuck to his ideas altogether too much. Otto thought that he would get little to publish from working on Kossel's ideas, so he would leave Germany early, especially if he could find a job in the autumn.

> I have now myself an idea that I had already last summer and that it will take me perhaps years of work to exhaust. This idea I shall outline as a conclusion to my paper because it fits in there and then I shall go to work on it just as soon as I get the necessary opportunity.

Otto had received from E. Merck, Darmstadt (probably the largest dealer in chemicals in the world), some deutero-albumose to test its purity. Otto found it horribly impure, yet it cost $45 a pound. Otto wrote Merck that the product was a swindle, and that he could prepare it at about 1/15 the current price. Though he and Kutscher could use some monetary recognition of their discoveries, nothing much was to be expected because of the limited market for the product.

Otto had received no word about a job. He had written again to Loeb and would soon write to Chittenden at Yale. "My uric acid paper is out and I shall send copies of it to the physiological chemists of America as I already have sent copies to a few acquaintances and friends. Twenty two marks as honorar!"

By late January Otto had begun a new line of research and had told Kossel that he could not do any more work on Kossel's subjects for some time because Otto wanted to dig a little deeper into his own problem.

It turns out not only that I have found a fine method, but that before me people have never had pure products and I shall therefore practically have to revise this whole chapter in physiological chemistry.

Kossel, of course, did not like this, but Otto stood firm. He had wasted too much time on "visionary schemes."

Otto's interest was related to what he felt was false or worthless information that Kühne, Chittenden, and Neumeister had been spreading for 15 years. Otto had now written Chittenden about his own new method and mentioned his job hunt. The question in Otto's mind was whether Chittenden would fear Otto's results or be willing to help him. After all, Otto's pursuit of purified albumoses was in the area that Chittenden had made his reputation. Otto believed that Chittenden's work was all nonsense and that his own was the new beacon light. "[Chittenden] seems to be a man totally innocent of all exact scientific ideas, and what is more, does not even seem to be aware of his own limitations."

Otto was obviously and rightfully getting worried about finding a position in physiological chemistry. His insecurity about Chittenden's reaction to his work was understandable; indeed, he added Chittenden to the growing list of phantom incompetents who stood in his way. But he did not want to leave this newly emerging field to teach or work in "pure" chemistry. He felt that he could contribute a great deal to physiological chemistry, as much as anyone living. It was exactly what he wanted and he craved a break.

Otto had come up with a new idea, he wrote Laura, that was an inspiration, almost a revelation. If he suggested it to Professor Kossel his interest would be aroused "to the highest pitch." However, Otto planned to take the idea back to America and let it incubate for awhile. He did not even intend to tell Kutscher.

> Kossel is interested in a certain substance found in the male reproductive cells of fishes and a paper turned out of here last year by another American boldly asserts that this substance though relatively simple for an animal product is *the* substance which is vital to reproduction and to transmission of inherited qualities. He attempts to prove it by an argument of exclusion.
> There is however a large amount of another substance in the same cells—which Kossel and his scholars as well as all previous investigators have considered without any other than nutritive importance. Now I have almost proven today the idea which came to me last night that the two products so obtained from the same cells by the rendering of their solution alkaline with a little ammoniak at once combine—giving of course a rather complex protein product. This reaction then appears to me as the first step, the very beginning of the process by which a new individual is produced!! The contents of these male-cells have an acid reaction while the yolk of eggs have an alkaline reaction and therefore easily could bring about the above wonderful combination.

He realized that the idea was "far-reaching but probably has only a grain of truth in it," but he couldn't test it until March when he could get fresh ripe herrings. In the meantime he returned to the other work "where I am on solid ground."

Otto read in the *Zeitschrift* that G. Ritter had announced the publishing of an "erwiderung" (reply) in the next issue, probably a reaction to his own paper on uric acid. Otto intended to "stand by his guns," but he wished that he had been more diplomatic in his choice of language. There was no need to antagonize people. "I must try to adopt a peaceful style rather than the pugilistic one that I am apt to get into when I detect too many flaws in the work of others." The next paper he published would come in conflict with several, active, "dangerous" individuals. He had to be fair!

Otto was again explaining his attitude towards women. In the lab, Frau Langenbeck came up against male chauvinism—though not from Kossel or him. She came to Otto for scientific suggestions. In his letter of Feb. 11 to Laura (which he addressed *My dear lecturer* in honor of the lecture series she was doing at her church), he encouraged her to become a philosopher, a career woman. Marriage need not stop her, although he advocated that she should not plan for a Ph.D. Too much drudgery and supervision.

He noted that his work was going slowly, but "I think a lot and the ideas do come and the paper will undoubtedly be the best that I have yet contributed to knowledge."

Otto had had a friendly letter from Professor Loeb. Rush Medical College had affiliated with the University of Chicago, but had not yet opened a department of physiological chemistry. Loeb would not know Harper's intention until Harper returned April 1, but Loeb had little power at the university, and Harper would not pay an adequate salary. Loeb himself wasn't being promoted though none did more scientific work. If a department of physiological chemistry were created they would probably bring in a man from Johns Hopkins.

As February 1898 closed, Otto felt that his next paper on protein digestive products would be his best. Although he would treat Chittenden "mildly" (a concession), he intended to "shatter" Chittenden's ideas of what took place when meat and kindred foodstuffs were digested.

Otto was now actively searching for a job in the U.S., and the time was passing too rapidly. The man he had just ridiculed for his scientific work, Russell H. Chittenden, wrote Otto a friendly letter, and had sent one or two letters of inquiry on Otto's behalf. Otto felt that it was essential that he continue his work, because it would make him an authority in one of the most important subjects in physiological chemistry, protein metabolism. Otto described his work to Loeb, just to maintain his contact with him, and wrote to Professor Howell at Johns Hopkins. He planned to contact the Armour Packing Company in Chicago about the commercial possibilities of making "extracts" of beef and such. He had also written to John A. Mandel, the translator of Hammarsten's textbook of physiological chemistry, who had been recently appointed professor of chemistry and physiological chemistry at New York University and Bellevue Hospital Medical College. No word had come from Wakeman or Herter. "I should be quite blue if it were not for the success that I have in my work because therein I think that I have a guarantee for a position a little later." He felt that if he did not have a job, he would lose out for awhile not being able to defend his position once his paper was published.

Laura's letters of late February and mid-March dealt with her new experiences at teaching and lecturing. Her lecture series had been a drain, because she was not a confirmed lecturer and she was relieved to be done with it. She had also begun teaching steadily at Cleveland High School and found that she liked teaching, but this branch school was 70 minutes away. She was teaching plain and solid geometry to two classes of 16 students each and elementary Latin to a third class of 22 students.

By mid-March Otto had handed a first draft of his paper to Kutscher for editing. This time it was written in German rather than English. He had discussed the proper tone of scientific articles with Miss Langenbeck and they had agreed that such articles must not be personal, rude, unduly critical, or pugilistic. Even so, Otto thought his Marburg paper would engender a fight. The leaders in the field of protein chemistry would try to squelch him; so that he would have to get another paper out on the subject or at least work ahead to be ready. But, he must do some of Kossel's work also.

> I hope that there is no fatal error in my work. If there isn't then my paper is one of the very best that have been published in physiological chemistry since I entered the science, I believe. It will undo at a stroke all or nearly all the work that has been done by several men these last fifteen years. And they will be shut out and I shall try to work the subject out on a sound scientific basis.

Otto believed this work would have a strong bearing on his future, that it would lead to something substantial in the way of a position, sooner or later.

Otto's work gave him some delusions of grandeur—as to where that work could lead, not because of its merit. As he had debunked Landwehr's method for preparing animal gums, now he debunked methods for preparing intermediates of protein hydrolysis or digestion. These intermediates were termed by Kühne, Chittenden, Neumeister, Wenz, and others as proto-albumose, hetero-albumose, deutero-albumose, and the final digestion product, peptone. At each "stage" of digestion, they theorized that an anti-substance was formed that resisted further digestion—for example, for proto-albumose there was anti-albuminate; for deutero-albumose, antideutero-albumose; for peptone, anti-peptone (6-3). Through a series of chemical manipulations that demonstrated that Otto had learned his organic and analytical techniques well, Otto succeeded in obtaining what he thought were superior products of both proto-albumose and deutero-albumose, but he found that they differed in some respects from the expected physical properties; and he was absolutely convinced from its sulfur content that the deutero-albumose was impure and contaminated with proto-albumose. In short, the intermediates were to be considered only theoretical and not factual.

To find some of the real answers, as with his work on mucin, would have required many years of work. Otto would have been better off working on Kossel's problems, for much had to be discovered and invented before the answers were obtained.

Ritter's expected "erwiderung" appeared in the journal in defense of his work on uric acid. The last sentence aroused Otto's bile, "Die Art Folins, Kritik zu üben, ist wohl der beste Maß stabe für die Beurtheilung dieser selbst."

My answer is going to be short, decisive and sharp. Had Ritter not used the above gruff sentence I should probably not have answered because he had in a way a right to complain of the fact that I had said that he had not made his experiments in a certain way which it now appears that he has. The fact that he has does not change the result one bit however, because his result is false and if he through a faulty manipulation of a right procedure obtained a false result it is no better than the same result. It is rather good that I get a chance to do a bit of fighting just at this time because it will show the other gentleman before hand that I know how, and they will be apt to be a little careful and peaceful.

Otto's short and effective reply appeared in the *Zeitschrift für Physiologische Chemie* [**25**, 65 (1898)] with the title, "Die Hopkins'sche Harnsäure-Bestimmung." Ritter had made three errors that could not be denied by using the wrong conversion factor for calculating uric acid content from permanganate consumed, a wrong correction factor for the solubility of free uric acid, and he had completely overlooked the solubility of ammonium urate.

Otto did not feel particularly combative. He had spring fever, "I suppose; but it, while it lasts, is as serious as typhoid." He was losing his interest in doing more work in Marburg. He would not work on the Kossel's subject, nor would he register for the next semester. Instead, he would leave in a few weeks for Sweden, to rest and then return to Chicago for the July convocation. The uncertainty of his future was intolerable. But even if no job was promised, Otto had had enough of his postdoctoral work.

Dr. Kutscher constructively criticized Otto's new paper as being too long-winded, and Otto respected his opinion. Kutscher had also helped Otto in making his reply to Ritter. Otto's latest article, entitled "Ueber die Spaltungsprodukte der Eiweisskörper" (On Protein Cleavage Products) and further labeled "Erste Mittheilung, Ueber enige Bestandtheile von Witte's Pepton" (First Communication, On several constituents of Witte's peptone) was to be published in the *Zeitschrift für Physiologische Chemie* [**25**, 152–63 (1898)].

"I was quite pleased that it passed muster as well as it did." Kossel had added no criticism. However, Otto bemoaned his lack of ability at German grammar, though he had a good command of vocabulary and idioms. But Otto had not at that time finished writing the article and he wondered what Kutscher would say when he summoned one big man after another into court and then let them out "on parole."

Otto advised Laura not to come to the July convocation. He would be unhappy without a job, but hoped his next paper would bear fruit. Stieglitz had written a "nice" letter assuring Otto that the department would help if any opening occurred somewhere.

On March 26, Otto wrote that he had handed his paper to his editor (Kossel) who would look it over and not change anything but language. Kutsher was excited about obtaining a few crystals of something that "supplements mine most beautifully," so they would celebrate the event. Otto thought that Kutscher was as careful a worker as himself. He would like to work with him, but it was out of the question. "I must keep my subject for myself as much as I can." Otto thought that he was more original and had a better grasp of methods than Kutscher, and added to this that he was less conservative and would probably turn out more important work as well as make more mistakes. As to Kossel,

> [he] . . . is brilliant enough, too brilliant in fact and hereafter I shall ever take his publications only at considerable discount. He is too uncritical and he would get into heaps of trouble if it were not for the fact that his specialty is an out of the way subject which does not engage the attention of other men.

Mandel wrote from New York that there were no local jobs, but suggested Cornell and the University of Pennsylvania. The tone of his letter was not very friendly. Otto had forgotten to mail the letter to Armour's and decided not to send it. He was certain that he wanted a job that gave him opportunity to carry out his own research. He was waiting to hear from Loeb and from Johns Hopkins.

On Saturday, April 2, Otto wrote his last letter from Marburg. He would leave for Sweden on Tuesday, April 5. His leaving Marburg was a little precipitate. But the laboratory would be shut for 10 days due to the Easter holidays, and there was insufficient time left to bring Kossel's project to a definite stopping point. Furthermore, Gertrud had sent a "remonstrating" letter for him to return for the holiday. Kossel was already gone, so Otto would write him and Kutscher would explain his situation. Otto felt that he had gained two good friends in Marburg, Dr. Kutscher and Miss Langenbeck. He would write her from Sweden, and hoped that he would meet her again some day.

He had the last of many walks with Kutscher. Prospects for Kutscher were good. He would become a docent at the end of the semester and was satisfied with his work. Otto felt that his own work covered a larger territory and could not be settled faster, so "the brunt of the coming fight will fall on me." Otto's and Kutscher articles, if they came out soon, could hurt chances for a professorship for the Leipzig man—unless he could prove Otto's work was wrong as to fact or interpretation. "It is queer; a ten page bit of work may down one man over here and may possibly make another man [himself] on the other side of the water if it comes out in time."

Otto enclosed a satisfying letter from Dr. Loeb:

Pacific Grove, Cal. March 14, 1898.

My dear Dr. Folin,
On account of an inflamed eye I am obliged to dictate this letter to you. A few days after the arrival of your letter I heard from Prof. Welch of Baltimore that they were looking for a physiological chemist and that you were one of the candidates. I recommended you very strongly and I should not be surprised if you would get the position. It is very fortunate that they appreciate research work. Prof. Welch is a perfect gentleman and you can trust him implicitly. I should advise you to send reprints of your physiological papers and if possible of your thesis to the following members of the medical faculty there at once: Prof. Wm. H. Welch, John Abel, William Howell, F. P. Mall, William Osler, Simon Flexner; Johns Hopkins Hospital, Baltimore, Md.

I am very much interested in the contents of your last publication and I still hope . . . you . . . back in Chicago one day. The conditions are decidedly better than you imagine, only I have to move slowly.

I am yours sincerely,
Jacques Loeb

Please give my kindest regards to Prof. Kossel.

In Otto's April 13 letter from Träryd, he indicated to Laura that he wanted to stay home for about two months. While this was prolonged, ". . . it may be the last time I am with my parents so I don't mind." Otto extolled his mother who had come into his room for a nap while he was writing. "She was away yesterday and we all felt it at once. Nothing seems right when she isn't at home although the machinery was just about the same. Now she is already asleep." (6-4)

Now that his work in Europe was over, Otto's thinking about Laura and their future began to be practical. He wondered how he would get along with her family. He expected they would be friends. "It is very good that everything has been done so slowly that your father has had lots of time to think about it. Now he will probably not have so much to say if we are satisfied." But what would their own reaction be when they met? Otto confessed that he now gave the bodies of the young women the eye—much more than he once did, and that he was curious to see her. He added that he was certain that she would look as nice as ever. If they were both satisfied, then they would be married as soon as finances allowed it.

Although Otto still had no promises of a job in the U.S., he felt that his trip had been a "fine thing" for him. "In pure chemistry I could never have dreamt of getting in at any Eastern University at least for many years, but now I think that I can at least apply for any position that appears."

Laura wrote back, "I would far rather have you come to St. Paul than go myself to Chicago. It won't be half as awkward." In Chicago, they would have to account for their behavior to Madeleine and George, and everyone would know that she had come just to see him, but it would be less pleasant for him in St. Paul. Come to St. Paul, she teased, and they could get acquainted again. Though Otto should not stay at her house, he could eat all of his meals there, and be there nearly all the time. They could take their bicycles and ride to Minneapolis, White Bear, Fort Snelling, and other places.

Perhaps, she mused, it was too much to expect an appointment at Johns Hopkins, but she was proud that Dr. Loeb had recommended him. It showed what he thought of Otto as a scientist. Otto would have a good place in a few years. "I have always believed in your success—now more than formerly. But it may be that you will get such work as you want sooner than I had thought."

The Spanish-American war had come, but Otto was unable to follow events because local papers were inexact. The Swedish press was not friendly toward the U.S. in this cause. Nor was the German press. Otto thought the war unwise and that the government seemed hasty and immature in its actions. What would the effects be at home, he wondered. No doubt money would be diverted from education, and opportunities for finding work would be reduced.

He had no word on jobs. Otto was writing to President Harper to get the exact convocation date. He would stay in Sweden as long as possible. After the convocation, he hoped to go further west to see her, or go first to his brother, then to her.

Otto agreed that he should visit Laura in St. Paul rather than Chicago. He wanted to learn to know her family, but he expected not to be there more than a couple of days at a time. "I shall be a little in Stillwater and most of the time in Center City either of which places it does not take long to go into St. Paul." Otto was a bit afraid of being with Laura more than about a day at a time. He did not want to wear out his welcome, unless he had a job.

Otto had received reprints of his Marburg paper and mailed about 15 copies to friends and colleges.

Laura had found out that Otto's convocation would take place on Friday, July 1, probably in the afternoon. She pleaded with him to come to St. Paul before going to Center City. He must stay at least two days so that they could talk, preferably a week. Certainly not a short stay! It would hardly look right for him to rush off. They must make plans.

Madeleine had written to Otto, he related in his next letter to Laura, and asked him to stay at her house when he came to Chicago, and sent him $50. Otto lamented his large debt. He owed his brother $800 to $900; $300 to $400 to Madeleine and Sikes; and about the same to Gertrud—in all, a little short of $1600. With his debt and no job, Otto felt that he had to go slow on claiming Laura. His debt was at 0.7% interest and would be hard to overcome, especially when he wanted to get a job that must avoid stress on getting money. But, of course, he would keep his eyes open to earn money. They must have a reasonable living before they married.

Otto was expecting to leave about mid-June, depending on departure dates for the ships. By leaving Gothenburg on June 15, he could sail the *Lucania* from Liverpool on June 18 and reach New York on June 25.

Laura's letter of May 26 to Träryd was readdressed to a Goteborg (Gothenburg) address on June 13, which indicated that Otto was there perhaps a bit earlier than anticipated. Beyond inquiring about his travel arrangements, the letter brimmed with thoughts of their future. Laura was now declaring that they could be married in the fall even if he had a position paying only $600! They could earn extra money tutoring, and in other ways. "I should not be afraid to risk it. The risk on your side would not be so very much more—as I will explain when I see you, if we agree." Board with a nice family in the neighborhood of the University of Chicago would be $10 per week, which she could earn tutoring. They would not need furniture. "If you can find *anything*, it can safely be managed. . . . I should not be afraid to risk it even if you had no position—but naturally you would." She had clothes enough for two years, except for alterations. Otto would not need many clothes. He could get fine business suits ready made in St. Paul and cheap. Even her papa now buys good looking ready-made suits. "This is the last letter I shall write to Sweden for many, many years, in all probability. The end of our letters is near at hand. Are you sorry? I have learned to like your letters."

During his last two weeks in Sweden, Otto shared visits with his brothers Ludde and Gusten and visited more distant relatives with Gertrud. He bought a suit and overcoat for $22 and sold his bicycle for $34. Dr. Wakeman had finally written a friendly letter. Dr. Herter had evidently changed his mind about having another investigator in his laboratory. Otto would try to see both of them when he arrived in New York. He had written again to President Harper concerning his degree and putting his name on the program for the convocation on July 1. He expected to be in New York on June 25 and in Chicago by June 28. He would try to visit his relatives in Minnesota, perhaps on July 4, before coming to see Laura in St. Paul, but he would have to find out first if Axel were home, as he often was called for work during the summer at the sawmill in Afton.

References and Notes

6-1. Felix, K., Albrecht Kossel (1853–1927). In: *Great Chemists.* pp. 1033–37. Eduard Farber, Ed. New York: Interscience Publishers, 1961.

6-2. Kohler, R.E. *From Medical Chemistry to Biochemistry—The Making of a Biomedical Discipline.* New York: Cambridge University Press, 1982.

6-3. (a) Foster, M. *A Textbook of Physiology.* 3rd ed. London: Macmillan, 1879.

(b) Halliburton, W. D. *Chemical Physiology and Pathology.* New York: Green, 1891.

(c) Bunge, G. *Textbook of Physiological and Pathological Chemistry.* Philadelphia: P. Blakiston's Sons, 1902.

(d) Herter, C.A. *Lectures on Chemical Pathology.* Philadelphia: Lea Brothers, 1902.

(e) Hammarsten, O., and Hedin, S.G. *A Textbook of Physiological Chemistry.* 7th ed. Authorized translation by John A. Mandel. New York: Wiley, 1914.

6-4. Strangely Otto never elaborated about Eva's work as a midwife, which she had been performing for over 30 years. She had served in the parish of Träryd for 15 years after having been removed as midwife by the Åseda parish council in 1882. In Åseda, she had been accused of being responsible for many unsuccessful deliveries, and there were written testimonials from 15 husbands in the parish accusing her of having brought on infant mortality by the use of obstetrical instruments to start deliveries several days ahead of the time of natural delivery. The Crown in Council did not reverse the decisions of the Åseda council to remove her, so Eva left Åseda in December 1883 and began her duties anew in Träryd. There was no other hint of a blemish in her record of her 40-year work as a midwife. (Carl-Werner Pettersson, Brinkelid, Åseda, Sweden. Personal communication.)

Chapter 7. Chicago Revisited—Survival (1898–99)

In contrast to the daily account of his shipboard life when he had sailed from the U.S., Otto had little to report during his return. The trip had been smooth. Even the North Sea had been friendly this time.

Otto at this stage must have had mixed emotions about his immediate future. The fact that he had accomplished so much in Europe, primarily in grounding himself in physiological chemistry and in showing that he was adept at research, was counterbalanced by the fact that he had no job offer. What was worse, there were few prospects for people with this new discipline. Important stirrings were taking place in the better medical schools, but this was the era when high school graduates entered medical school, so that their training in chemistry was rudimentary and superficial.

To a jobless young man, a debt of $1600 in 1898 (about $19,300 in 1983 U.S. dollars) must have seemed staggering. He owed money to five people who believed in him and his future and who knew him well enough to trust him unreservedly. But debt and umemployment could only aggravate his despair about his future with Laura. He would now have to be the "sensible" one and caution patience and forbearance. For both of them, this was a bitter pill to swallow.

Laura wrote him on June 25 in care of the Sikes flat in Chicago. She missed writing to her foreign correspondent. She knew that he would feel lonely taking his degree.

> I did when I took my master's. Almost nobody said anything after it was over and I had to go home alone—but you needn't care if you do, for I will sit down in my room and clap at about quarter before five, and then I will meet you at the station the next day.

Laura was anxious that Otto visit her first after the convocation.

Otto wrote soon after his arrival in Chicago that he had received her letters and they were much appreciated.

> I shall leave on the last train Friday probably about eleven oclock. . . . Both my brothers come to Center City on a visit the fourth of July and I shall therefore have to leave St. Paul on Sunday. . . . I shall be very, very glad to see you and

yet I am afraid to. I hope that all may go well. We shall have a great deal to talk about.

But in the early afternoon of July 1, Otto sent Laura a special delivery letter stating he would not come to St. Paul, now! "Mr. Sikes rightly suggested that it would probably be much better for me to remain in Chicago one or two weeks and try to get my bearings especially since Prof. Loeb is at Woods Hole and I have really no hold whatever as yet." Otto wanted to begin working as soon as possible, and it would be impractical to conduct correspondence from Center City. He would stay for a few days with the Sikes and get letters off to Loeb and to Johns Hopkins.

Fortunately, Laura understood Otto's predicament and must have swallowed her own frustration, not to mention a little pride. She could wait 10 days as well as three.

By July 5, Otto telegraphed Laura that he had obtained a job with Stieglitz for six weeks to assist with a class in chemistry for summer students. It paid only $75 for the term, but it was better than nothing.

Oddly Otto, who had so faithfully recorded most events of his recent life, mentioned nothing of the July 1 convocation in which his doctoral degree was formalized. Not one word did he write of the acquaintances he renewed, the students, faculty, speakers, nor of his emotions upon returning to the campus where he had found love while he unlocked the door to his future. Only silence!

On the day previous to sending the telegram, he had written a letter talking about his visit to Stieglitz, who had assured Otto that his prospects for the future were good. Otto had enjoyed a brief visit with Dr. Wakeman and his wife in Connecticut, before he left New York for Chicago. Dr. Herter was out of the city. "I have not yet closed down the Missouri Medical School but it seems to be of no account," which meant that Otto was following up any leads for jobs that were available. He commented sadly that prospects for their marrying in the fall were decreasing. He was optimistic, however, that he could manage to get along for a year and await further opportunities. Madeleine and George seemed wonderfully happy, and it made him eager to wed—but he must not think about it. Sikes earned about $1800 and they spent it all! Laura and Otto could live on less, but not on a third.

On July 7, Otto wrote that he was already working in the laboratory, but that no importance was to be attached to the six-week job. He would live there for the summer session on the second floor next to his friend Dr. Veblen. He hoped that Laura and he would remain "friends" and that they could get together pretty soon.

Laura wrote on July 10 that Madeleine had invited her to come to Chicago, but she wanted Otto to come to St. Paul first—which she thought would happen in five weeks when his job was over.

Otto wrote that it was a trifle dreary to be locked up in one little "furnished room" again.

> My room is on the second floor facing Ellis avenue and the U. of C. grandstand. Dr. Veblen has the large frontroom next to me. Dr. Veblen is the same quietly sociable, interesting sort of a man that we used to know.

Madeleine was now pregnant. Otto had taken a longing look at the $12 flats in the building next door, and saw one he particularly liked. They could not marry this fall, but he thought of it anyhow. Madeleine thought it would be "anything but conservative" to marry this fall even if he were to get a good position and Otto agreed.

Otto hoped that when he went on to Center City after he visited her in St. Paul, he would be able to see her again.

> Quite a number of people spend a part of the summer at Lindstrom or Chisago City. Those places are smaller and less frequented than White Bear; but the lakes and surrounding country are fine. At either of these places I could see you everyday if it seemed desirable.

Otto wanted to get acquainted with Laura's mother. Maybe they could get married earlier than it seemed they could.

Otto now found his work at the lab a blessing. He was readjusting to American life. He had spent a pleasant evening with Dr. Veblen, who was probably working on his first book at this time, *The Theory of the Leisure Class*, which would appear the following year causing a minor sensation.

Otto was busy getting a paper ready to be presented to the Chemical Club, which was the department of chemistry seminar that was open to all the interested campus people and Chicago chemists. He was going to review recent work on fermentation. Dr. Loeb had written that he had asked President Harper to appoint Otto an assistant in physiology, and Otto had gone to see Professor Judson about it. Judson promised to help second the recommendation when Harper consulted him. But it would only be a one-year job paying $300 to $400, enough for a single person, perhaps. Stieglitz, too, could get him that much. Harper was shrewd and knowing Otto's dire need, would not offer more, Otto thought. From Madeleine Otto had learned Laura had a full-time teaching appointment for next year. She had not mentioned it. Otto thought it was wearisome to teach large classes and wondered if it would be too hard on her.

Laura sent Otto a brief letter on July 22. She thought his prospects were improving. She had faith that Harper would come through for him because he recognized manliness and ability. Otto need not think about marriage. That would take care of itself. She had a high school appointment, and it might pay $650. They would talk about it. Mother was opposed to it. She hoped Otto's paper would be a success.

Otto had been worried about giving his paper, so he had arisen at 5 a.m. and gone into the park where he mulled over the subject of fermentation. Then he went to to his room, made a brief outline of the subject, left the manuscript there and tried to forget it. When his turn came at the Chemical Club meeting, he felt that he had given his most satisfactory talk so far.

Otto was concerned about his coming interview with Harper. What approach should he take? Otto's job as a laboratory instructor was getting easier as the men (students) learned now to work from the directions in the textbook.

In early August Otto held his interview with President Harper. Result: "I believe in you Mr. Folin but we have absolutely no money." But Otto was to call on Harper again to see if they could possibly raise $500 to $600. Otto was now writing the director of the American agriculture stations asking about vacancies, but was pessimistic about his chances, and he had written to Professor Abel at Johns Hopkins.

Laura tried to comfort Otto. Don't worry if Harper offers nothing. Get temporary work. Appointments at Chicago U. were made four times a year. Wait until January. Don't get discouraged! She had made up her mind that they could not get married in 1898. When would he come? "I am growing impatient to *see* you."

Dr. Veblen again took a great part of Otto's evening

> in giving me a most interesting discussion of how our very wealthiest railroad men obtained their fortunes. It was very interesting, but he has really taken the time I meant to give in part to you. By the way and before I forget it, you ought to read his article lately published in the Harvard Periodical on "Why Economics is not an Evolutionary Science." I have read it and he has talked to me about it and I suspect that it will create a great deal of stir in the circles concerned. He also took pleasure in telling me (I think) of how Pres. Harper asked him to call today without having anything to say except that he had heard nothing but the highest report of him and that he appreciated it. V$^{\underline{n}}$ made a great deal of fun over it. He has simply no use for H.

Nothing came of Otto's second meeting with Harper, and Judson promised to see Harper about it. Now Otto felt he could not count on going soon to Minnesota. But don't worry; all was not hopeless.

Of course Laura would have none of that! If he did not come by Sept. 1, then she would have precious little time once she began schoolteaching. For that she would leave at 7 a.m. and return at 5 p.m. or later. The days were passing too quickly now.

Madeleine was ill. A doctor and nurse had been with her the whole day. The blessed event expected tonight (Aug. 7). Otto would drop by again tomorrow at noon. He would let Laura know if the baby was a boy or a girl.

Otto was badly discouraged and feeling guilty. He did not dare to come to St. Paul unless he had a job. It was "unreal" with them. They ought to meet and settle the matter. When his six weeks were up, *nothing* was in sight. He would now look among the physicians. There was more encouragement there. Otto wanted to hold out as long as possible to stay in his line of interest. Madeleine's baby was born on Aug. 8—a very large boy, Alfred Sikes. She had had a hard time, but she was all right.

Laura sent her congratulations to Madeleine. Laura wrote that though she could come to Chicago because Madeleine had invited her, she would not do so until he came to St. Paul first. How about Saturday? They should get it over with—but not if it was unwise. Otto could stay at the house. Mother said so, and meant it! She would like, to have him for awhile. He would not be uncomfortable. It was a large house. Bicycles, Saturday—on to White Bear on Sunday a.m., and have dinner with her Aunt at Bald Eagle. Stay to 9 p.m. He should come for even two days. Then she would go to Chicago for two weeks.

What could Otto say to this impassioned plea? "I feel like packing up at once. . . ." As long as there was any immediate and definite reason for remaining in Chicago, he would stay. He had met Dr. Wesener the day before, a professor at the College of Physicians and Surgeons. He was a partner in a large commercial laboratory in the Memorial Building. He would see Wesener again tomorrow. Otto was "reading critically for him a paper which he means to publish." Wesener had made a good personal impression but was a decidedly poor writer. Otto could help him without hurting his ego. Wesener also asked Otto to write a review of a new textbook on physiological chemistry. "The book gives credit to 'Folin's Method' and on that account, if for no other reason, I must be generously fair in my treatment of it." Wesener thought there was a job for Otto in Chicago and that was why he must stay for a few days to see him a little more, and to get some clues on employment. "I shall soon come, however."

At last, the moment came when Otto went to St. Paul and visited Laura and stayed in one of the six Grant bedrooms. He went on to Center City to visit Anna and Axel, and no doubt his Aunt Ingrid and his cousins, John and Hannah; and probably Alfred and his family in Stillwater. How much time Otto spent with Laura is unknown, but he left St. Paul on the morning of Wednesday, Sept. 1, two weeks after he had left Chicago. As we shall note from the letters that follow, they discovered that their bonds of love were genuine, that their growth in maturity since their last meeting in Chicago had brought deep mutual understanding and friendship. They talked out their problems sensibly, and Laura knew that their restraint would be rewarded. By controlling their passions they were building a future together, which was fast approaching. The short period ahead was essentially the last challenge that they would have to meet apart. To them, it would seem a lifetime.

It was clear now that Otto and Laura had reasoned that they must wait another year for marriage. Otto needed time to find appropriate work unhampered by marriage obligations, however desirable they were.

While Otto was trying to get his career launched in Chicago during the months ahead, he found work at two jobs that at least offered subsistence and provided a springboard for his future. These jobs were the result of two people, John Alfonso Wesener (1865–1926) and Fenton Benedict Turck (1857–1932).

Wesener was a pharmaceutical chemist, who had obtained his degree from the University of Michigan in 1888, and his medical degree from the College of Physicians and Surgeons, Chicago, in 1894. He had been a professor of chemistry at the Chicago Post-Graduate Medical School in 1889 and at the College of Physicians and Surgeons in 1891. In 1893 he was elected professor of chemistry at the College of Pharmacy and the American College of Dental Surgery. While he was teaching for his income, he founded and was president of a private commercial laboratory, the Columbus Medical Laboratory. With him in this enterprise were three active associates and a "silent" shareholder. The actively involved included a bacteriologist, a pathologist, and a business secretary, all of whom were physicians. The laboratory, located in the Columbus Memorial Building at 103 State Street, specialized in water analysis, toxicology, food analysis, and analysis of medical products. Wesener was the laboratory's chemist, specializing in dietetics and sanitation, and he served as a consultant in chemistry and pathology in criminal cases. He had been for a number of years the chemist at the Cook County Hospital laboratory. He was said to be thorough and exacting in his work and research, a close student of scientific methods and principles. Wesener published several papers of clinical interest as well as some on food products. During the time Otto knew him he was evidently working on the relationship of indicanuria and oxaluria to gastrointestinal fermentation.

Turck, who was from Milwaukee, was a graduate of the Chicago Medical School in 1891. He had been a professor of internal medicine at the Post-Graduate Medical School in 1893, and a lecturer at the Jefferson Medical College, Philadelphia, in 1896. Although he did most of his scientific work in New York, he had in 1893 devised instruments, including the gyromele, for exploration and scientific studies of the alimentary tract. He worked on gastritis, peptic ulcer, traumatic shock, immunity, cell division, and regeneration, among other things, and published many articles and a book on "Experimental Studies in Biology." He established and directed the Research Laboratory Turck Foundation and was a delegate to the International Medical Congress five times between 1894 and 1913.

On Sept. 6, 1898, Otto wrote that he had talked to Dr. Wesener, who held out certain prospects for his future, no doubt at the Columbus Medical Laboratory. He had also spent some time with Dr. Turck, about whom he noted his superficial impressions.

> He seems to be a brilliant man, very brilliant, but scarcely a man to be trusted . . . full of ideas, schemes, a science enthusiast—but erratic, not well trained scientifically. Others (W., and others) do not speak well of him. Think him a wild, unscrupulous bluffer. Partly true maybe—but it may stem from the fact that Turck mixes science and practice—hence differs from the majority of medical men. . . . I believe that Turck has many brilliant ideas and will do some splendid work between times, but I cannot say that I like his personality.

Otto ate with Dr. Turck and remained four hours in his house. Otto thought that Turck was wealthy, but did not own the house. His laboratory was in the basement and was small, not well arranged.

> His specialty is diseases of the stomach and while I was there I helped him to torture a poor mortal whose appetite or digestion was poor. Dr. T. has invented an instrument by the use of which he can scrub the stomach on the inside and we just scrubbed and rinsed out the above patient in fine fashion—but I had all the time the idea that the doctor did it more for my benefit than for the benefit of the patient.

Before Otto left, Turck offered him $40 a month to work in his laboratory at both assisting Turck in his work and doing his own research. This would certainly help Otto get on his feet. Otto had found that among the medical people in Chicago, "professional Chicago," there were cliques. Once in one of them, he could "make a place" for himself. Dr. Turck was currently a professor at the Chicago Post-Graduate Medical College. Turck offered Otto the choice of working there or in his basement lab, and Otto chose the latter so that he could be near the Newberry Library, only two blocks away. While Otto's assessment of Turck was overly sweeping as it had been of his mentors in Europe, Turck not only hired Otto because of his superior background in science, but also because he and his young wife wanted to befriend him. Unfortunately, Otto apparently misunderstood this desire, probably because Turck was older by 10 years and they could not "hit it off" as friends, and partly because money was involved that Otto needed badly.

Otto hoped that Laura's first day of teaching went well. He had inventoried Turck's laboratory and found it needed about $187 in items and another $50 worth of chemicals. Meanwhile he had found a place to board and room at 348 LaSalle Avenue, North Chicago, two blocks from the Newberry Library, costing $5.50 per week. Otto had agreed to lecture to the Pathological Society on Oct. 10.

Otto, for the moment, felt that Dr. Turck was offering a potentially more lucrative proposal than Wesener. Turck was a much finer man than he was given credit for, Otto had discovered. Meanwhile, he was getting familiar with the Columbus Medical Laboratory. Wesener treated Otto to a fine 65¢ dinner the previous day. Otto had helped him, or rather kept him company for a couple of hours working on a scientific experiment that "he is repeating at my suggestion." Otto would help Wesener the coming weekend. Wesener was quite friendly and was trying to get him a $300 to $400 job lecturing on physiological chemistry at the Harvey Medical School.

Otto's work began favorably in Turck's laboratory. He felt that the outlook was not half bad. He was going "right on with his own investigations just as I would if I were still in Germany." Dr. Turck came down to play with one thing or another, and Otto "plays" with him. Things looked good. The point was that "he is a specialist on the stomach and digestion and my work is just on the pepsin digestion of the stomach so it is of course right in his line and I can see how he probably will make some good advertising for himself out of it which I am very willing that he should." Otto would investigate the products formed in artificial (in vitro) digestion, and then Turck would provide human digestors for duplication. This was interesting, but "showy." Otto could not decide whether Turck was a trifler or not; time would tell. Otto expected to complete a good piece of work by springtime.

Otto was going to help Dr. Wesener the next day. Wesener had proposed that they work together on the pancreas and how to cure diabetes mellitus. No money for that. Dr. Wesener's work was negative, probably from faulty experimentation. Otto remarked that another paper was announced on albumose and peptones in the German *Zeitschrift*, and that it was possible that he would be attacked in it.

The first week of teaching had gone well for Laura. She had picked up a rather interesting side job, that of teaching a neighbor's daughter history, literature, and composition. She would be rich enough to come to Chicago by Christmas, if Miss Reynolds did not come to visit her in St. Paul.

Otto had received a letter from Dr. Wakeman asking for his latest improvements on his method for uric acid analysis. Wakeman was about to start another year's work with Dr. Herter, and was still trying to find something for

Otto. "My work [with Dr. T.] is going awfully slowly and the Doctor is simply a man to be humored not to be seriously taken or consulted about anything."

Miss Myra Reynolds had decided not to visit until spring, so Laura planned to "run down to Chicago" about Dec. 24, though she was not certain of it.

Otto had spent more time with Wesener and was going to his house for a Sunday dinner. They were constantly discussing applied chemistry schemes of one kind or another for the Columbus Laboratory. Wesener had given the Harvey School position to his assistant, but Otto believed that Wesener genuinely wanted to work him into his laboratory. Wesener expected a considerable increase in salary in the next academic year and would need an associate. With more than 400 medical students and the rapid affiliation going on with the University of Illinois, there was nothing improbable in this. He also wanted 10 or 12 students to get fellowships for investigations that could be carried out at the Columbus Lab because their staff included professors of pathology and of bacteriology who were stockholders. A special man (Otto) would be needed to guide this research in chemistry for these students. What a pipe dream! But Wesener already had a married man there who left a $100 job to join him—the one who had the teaching job at the Harvey School. The man was dissatisfied and unsatisfactory for the Columbus Lab.

> I am getting somewhat scared as I read up on all the troubles that the stomach is liable to get into and see the living examples enter my laboratory and anxiously watch me while I take out some of their stomach contents and gravely test it for hydrochloric acid, lactic acid and peptones.

After another week had passed (Oct. 1) Otto wrote that he was becoming unhappy on his income of $40 a month and was doing his best to join the Wesener crowd. He and Dr. Turck had not yet accomplished anything. One of Otto's ideas had proved inadequate. If only T. could think consecutively for 10 minutes on any one topic then they could accomplish something! Dr. Turck was talking (again) of establishing a clinic. He wanted to send out fliers to doctors to teach them "our methods" at a cost of $30 per month. This would increase Otto's income—but he did not like the plan because it would bind him to the place and he did not expect to remain there until spring.

Otto worked again with Wesener on Sunday morning. The project on diabetes they were working on seemed futile.

Laura was hoping that her busy life would get smoother. She had ups and downs in activities. She had given her students a written lesson to do, and it took five hours to correct the papers. She went to the station to pick up Miss Felton, who came for a day, so she took her to school with her. Miss Felton had proposed that Laura join her next year in a course with Sidney and Beatrice Webb at the London School of Economics. Laura said she was not inclined to go unless something happened to Otto. Chicago was good enough for her. But Miss Felton thought that she should go because of marriage so that she could get a start in some line of work in economics. Laura asked Otto to call on Miss Felton at Foster Hall. She hoped to look at flats with Otto at Christmas time.

The following came to Otto like a bolt out of the blue:

Physiological Laboratory
HARVARD MEDICAL SCHOOL
Boylston & Exeter Sts.
Dr. Otto Folin,
   Chicago, Ill.
Dear Sir:—
   Your letter of August 6th reached me while I was in Europe and I referred you to Dr. Pfaff for reply.
   I find that in my department there is a vacancy in the position of Assistant in Physiology for the coming year. The duties of the position are to superintend the practical laboratory work of the first year men, and to engage in original research. The salary is 400 hundred dollars a year, which is rather a nominal sum, the real compensation being found in the opportunity for original work afforded by the Laboratory.
   If you desire to apply for the position please let me know at your earliest convenience.
                                        Yours truly,
                                        H. P. Bowditch

"The enclosed typewritten letter came and disturbed me last evening. I went at once down to the U of C to consult with Professors Loeb and Stieglitz." Loeb was not at home, and S. thought he should take it. "Harvard is Harvard," he said, "The most exclusive and most difficult University to get into." Otto finally saw Loeb, who advised Otto not to take the job, because his opportunity would be better in Chicago. "This relieved me very much, because I rather want to remain in Chicago." Loeb did not place value on the Harvard position. He was trying to get a $500 endowment through Atwater, head of the Agricultural Experimental Station of Connecticut. Perhaps this was a visionary scheme, but Loeb had influence. Otto wrote to Bowditch telling of what he was then doing, and declining the job. Otto told Wesener but fibbed about the amount offered. Wesener also advised him not to accept the offer because there was a good future for him in Chicago.

We must note that Bowditch's offer to Otto was for a job in physiology, and Otto was not a physiologist, though perhaps it might not have mattered. When Bowditch was involved in recruiting Otto to Harvard less than nine years later, the offer would grow by an order of magnitude, plus fringes.

Otto wrote that he would call on Miss Felton when he had her address (although Laura had indicated Foster Hall). "Miss Felton must try to convince me, too, if she wants to get you to follow her to London." Could Laura still be that interested in economics? Perhaps domestic economy!

He had to work on the paper for the Pathological Society the following Tuesday. He had finished 2000 words. Had she seen the review he wrote on the text in physiological chemistry? He had stated that the book reflected a man at home in the subject, but it was labored.

Laura's vote was also a "no." She thought Harvard was great, but not at $400—though it was hard to decide. She hoped Otto would enjoy giving his paper on Tuesday, and wanted to see the review he wrote.

Harvard would be fine, Otto avowed, and Professor Bowditch must have been a very prominent man. He was the only American to receive the honorary doctor's degree at the congress of physiologists in England a few weeks before. But Dr. Loeb said the original assistant's salary had been divided into four parts and would not be increased. Otto would have been looking for a new job within the year, anyhow. Chicago would do for a start.

Otto's paper before the Pathological Society he thought was a fair success. The manuscript hampered him. "My voice is not at all adapted to lecturing and I don't know just what I can or shall do about it." Dr. Turck made some remarks about Otto's paper and what they were going to do, but he was rather curtly treated by the president, Dr. Hektoen.

> Dr. Wesener then got up and also made some remarks and then I added a point which Dr. Wesener somehow forgot to bring out. I stated, however, that I knew from personal discussions which I had with Wesener that he knew perfectly well the point that I brought out, only he forgot it somehow and he nodded approval to the president and so it ended smoothly by my receiving the thanks of the society, through the president, for my paper. Wesener and I left the society and he treated me to a social cigar and a glass of beer!
>
> [I] wrote a short note to Dr. Hektoen today and I almost apologized or explained why I went in with Dr. T. I do not like to be associated with a man who is not particularly upheld by the other men in his profession. This I did not state to Dr. H. but I simply said that I hoped to be able to make some work that should bear repetition even under the present conditions.

Otto had been lately excited by a pair of ideas he had. One was on unmentioned biological processes. He would have needed a month to test it out, but could not do it in Turck's lab primarily because he did not wish to share it. Although publication of ideas alone was not a sound practice, he thought, he was thinking of doing so because he felt that Kossel would not hesitate to publish them, and they would create a stir and, if true, be a feather in his hat. His second idea came while he was smoking a cigar to satisfy a strong urge, and he was reading a long article by Nef that he regarded as "terrific." "I took it up and found one of my most beautiful central points hinted at so plainly that I cannot think of disputing the priority with him although I certainly had worked it out absolutely without knowing about it." Otto wanted to write Nef to see if he should go on with the idea. Six months too late! If Nef did the continued work it would be his best. But the idea referred to was only half of the total idea that Otto had. Otto thought he had better get it into print before he lost it to some other genius. Nef's article [$Ann.$ **298**, 202–374 (1987)] was on bivalent carbon, and the chemistry of methylene. So Otto must have been thinking about something related to bivalent carbon. He may have been conjecturing about free radicals. Nothing on the two ideas was published or recorded.

Loeb asked Otto to present a paper on fermentation to his seminar and Otto accepted. Loeb had not yet written to Atwater about the grant, but would do so. While Otto did not like the grant idea, Nef, Stieglitz, and Lengfeld approved, meaning that Otto had visited the chemistry department and spoken with them.

A few words about Jacques Loeb and his brother Leo would be appropriate here. "Jacques Loeb (1859–1924), brain physiology, tropisms, regeneration, antagonistic salt action, artificial parthenogenesis, duration of life, colloidal behavior." So reads the inscription on his memorial tablet placed in the Marine Biological Laboratory, Woods Hole, Mass. (where this great physiologist was a trustee), as well as in the Rockefeller Institute for Medical Research, New York City (where he headed the experimental biology division from 1910) *(7-1)*.

Three men helped more than others to influence Otto's postgraduate training and career once he had left Minnesota: Stieglitz, Loeb, and Hammarsten. Once Otto had departed from Chicago, of these three he would see only Loeb periodically. He would also see Leo Loeb (1869–1959), a gifted, experimental pathologist and biologist whom Otto had met in Chicago that summer of 1898. Leo was then struggling to establish his remarkable scientific career with his pioneer work on cell cultures, and would later do outstanding studies on cancer; transplantation, hereditary and hormonal factors; and resistance to transplants. Like his brother, Leo worked on a diversity of problems *(7-2)*.

Jacques Loeb never did raise enough money to hire Otto whom he wanted. Under ordinary circumstances Otto could have taken the job that eventually went to A. P. Mathews, but only an unencumbered man could afford to do it.

On Sunday, Oct. 23, Laura wrote that she was thinking of spending next summer with Miss Emily Reynolds at St. Ignace, Mich., where Lakes Michigan, Superior, and Huron come together. Could Otto come for a couple of weeks, too? It could be reached from Chicago inexpensively by boat, in August when he was on his way to Minnesota. What would Otto think of living in town cheaply for 10 months, perhaps near the Hull House, and "then go into the country to some very beautiful (cheap) place in the summer." They could live frugally until they were rich, then get a large "establishment." But their summers would be for the wilds.

Otto liked the idea. Though he expected they would not live there, they could certainly look at flats in the vicinity of Hull House, but being near a library could be an important factor. How did she reconcile the St. Ignace scheme with their marriage? Would they talk things over there, go to St. Paul to be married, then complete the circuit in Chicago in time to get "our place" in order for fall work?

Otto had been to the Weseners and had borrowed a book on physiological chemistry. He was helping Turck write an article as "fast as I can get his ideas sufficiently definite out of him." Otto was giving a dietary preparation to one of Turck's patients who had a poor digestive tract. The patient was doing well, but Otto was saddened by the fact that the patient also had another disease (probably Bright's disease) and would die in two years, according to Turck.

Otto wrote that he and Wesener had gone to the Post-Graduate Hospital to operate on a dog.

> The surgeon of the place was to make one point and Wesener was to make the other but the point actually made went to the servants of the hospital. The dog that was to furnish the points has been kept by the surgeon for over a year and in that time the animal had gotten into the good graces of the servants.

Consequently there was absolutely no dog present when the time for operating had come. The surgeon swore and raged, but the dog evidently was not near enough to hear it and no one knew where he was.

They would try again on Wednesday morning at the Academy of Science Building where no one would probably spirit away the dog either before or after the operation. They had gone to the Columbus Laboratory where Wesener was quite excited because he thought he had found a remarkable cure for consumption. His excitement grew to fever pitch when Otto told him of the wonderful work completed in Berlin with almost the same reagent. It might prove to be a sure cure in the early stages of all infectious diseases.

Otto had received a heartening letter from Kutscher that in part translated as follows:

> On the contrary, right now I believe he (Kossel) would consider you very favorably because he has become rather proud of your work. Initially Hofmeister was the one who wrote to him and praised your work, and lately Kühne also has mentioned your work to him. Kühne specifically recognized the advantages of metal precipitation and praised the clever use of polarization to try to determine whether an alteration in Deutero-albumose is caused by pepsin. Kühne was not too convinced that Deutero-albumose remains totally unchanged and he is going to check out more results of yours with the help of the polarization apparatus.

Otto had stirred the right people with this work, and his opportunity would come; but this was a final period of frustration and delay for him. He was pleased with the scientific judgment and critical power that he had developed in the past two years. He was confident that he would do creditable work if he had the chance.

Otto had been writing for Dr. Turck, doing a little experimental work, and reading a good deal. He noted a paragraph in a journal in which one of the German professors pointed out that there was no conflict between his views and Folin's work.

On Friday, Nov. 6, Otto had dinner with the Turcks. After dinner Otto and Dr. Turck went to a meeting of the Chicago Medical Society, then he went with Wesener and another M.D. to have a pleasant glass of beer at the Schiller. On the following Friday, Dr. Turck took Otto to a meeting of the exclusive Chicago Academy of Medicine, where they had dinner and then heard two papers. There were many fine men in this group, and Otto wanted to join as soon as he had worked up the right kind of paper. Dr. Turck, Otto remarked, was certainly liberal and kind to him in every way, "and I must reciprocate and do the right thing by him." Otto had worked more than a week in writing an article for him, and Turck appreciated it.

They would have a good time at Christmas without quarreling, Otto assured Laura. She was his constant friend now. The College of Physicians and Surgeons was next door to the Cook County Hospital in the direction of Hull House. While Wesener might not be able to get Otto in next year, they could still look at flats in that neighborhood. They could enjoy going to an "auction" store downtown to look at desks. Much to talk over.

Otto was on his way to Turck's to finish his "article," if possible. T. was working on plans to keep Otto because he wanted to write a book with him and he could not do it alone. Wesener also wanted to write a book with Otto. That would be different. They could produce a first class textbook together.

No chance to get more pay. Surely by fall. Turck sent 400 circulars and if they were successful, Otto should make more money. If not he would have to be content with the status quo. Otto had an interesting session at Loeb's seminar. His department needed a chemist. "I like Loeb. He is said to be extremely jealous about his ideas; but is as kind and friendly as a man can be toward his students. I can bank on him to help me when the proper time comes."

Otto sent Madeleine a graduated beaker so she could measure liquids that baby Alfred was to get. She thought Otto should get a room in the neighborhood when Laura visited at Christmas. Thus far, Laura and Otto were scheduled to have breakfast and Christmas dinner with Madeleine.

Laura told Otto to be his own judge about rooming near Madeleine's. Perhaps without his belongings it might not be such a good idea. Laura wanted to "rest up" a little this vacation. She would like to go to one theater production. Chicago was too big to go out in more than once a day.

Although Otto spent a cold Thanksgiving afternoon with Madeleine at the Chicago-Michigan football game it was characteristic of him not to mention anything of the game itself. Though he had known George since his Minnesota days, and they were certainly good friends, George remained "Mr. Sikes" or plain "Sikes" throughout Otto's correspondence, whereas Miss Wallin had become Madeleine, sometime during their University of Chicago days. After the game, Madeleine had Otto sorting cooking recipes from newspaper clippings. Baby Alfred was doing well and looked better. Otto discussed the arrangements for Christmas. He would rent a room near Madeleine's house for a week and would bring reading material for his extra time alone.

Otto's work for Turck at this time was mostly spent at writing. He was preparing a brief resume of Turck's works for a textbook to be issued soon at Johns Hopkins. Then there were 50 to 60 letters to write "of an advertising nature," and Otto disliked doing that. Wesener felt Otto's prospects were good for a job in his laboratory before spring. Meanwhile, his old job teaching chemistry at the veterinary school would be open next year, and he could earn $100 assisting in the summer quarter at the university for the six-week term.

Otto was fortunate in one respect financially. He had long practiced frugal living and spent within his means. He accepted the inconvenience of a cheap boarding house and the company of people he cared little for, so that he could stay within walking distance of Dr. Turck's house and the Newberry Library. His travel by the economical streetcar was to either the University of Chicago or the Columbus Laboratory. For the most part during the period he worked for Turck, Otto kept himself busy by reading scientific literature—a habit that would stick with him for the rest of his life. Of course, he visited Madeleine and George occasionally. The money he earned from Turck was more than adequate to take care of his own needs during this particular period in his career. Otto evidently had abandoned the idea of doing his own research in Turck's laboratory because as he had already well demonstrated, he could not collaborate, particularly with someone he did not respect. If Kossel and Kutscher were inadequate what chance did Turck have? There was no further mention of studies on peptic digestion, and very little about doing gastric analysis on Turck's patients. His schedule at Turck's was reduced to three or four evenings a week for which he was paid regardless of whether or not he had performed any useful function. He was embarrassed by this and the kind hospitality displayed by both Dr. Turck and his wife.

On Thursdays, Otto was a regular visitor to Professor Loeb's seminar. He had heard a poor lecture given by A. P. Mathews, who was to present a course in physiological chemistry, though he was apparently a pharmacologist. (Mathews had also spent time with Kossel and published a paper from Marburg. Kossel had three outstanding American postdoctorate students besides Otto: P. A. Levene, A. P. Mathews, and H. Dakin.) Otto was to present a paper the following Thursday on the recent investigations of Kossel and his students, a topic that Otto felt comfortable with and which was interesting. As it turned out, Otto gave two lectures, the first on Miescher, who, Otto wrote, worked 25 years without publishing anything only to die just as he was about to get his results into shape for publicity. Johann Friedrich Miescher was a pioneer in studies of nucleoprotein and nucleic acids. Two posthumous works were published by his friends, a collection of early papers and lectures and another containing extracts of his private letters; Otto would review these because Miescher's work was confirmed by and antedated Kossel's.

Otto usually went to hear sermons or invited speakers on Sundays after he stopped spending the mornings with Wesener. For exercise, he walked a great deal, and looked at flats. He was getting excited about Laura's impending visit.

Laura wrote that she had visited an Italian couple, and would teach the husband English in one-hour sessions twice a week for about three months. Laura announced that her sister Kate had promised to marry Mr. Duncan, of whom she approved highly. They would be happy, despite one drawback (like Otto's): little money. Laura wanted to take Madeleine with them when she visited to see Mansfield play Cyrano de Bergerac. "Let me treat the round," she wrote. "You take me to something else—not to the theatre." He should get tickets for three! She wanted the two of them to spend Christmas night visiting with Miss Felton after they had been together all day.

Otto was amazed by Laura's newfound vitality. She seemed far more active than she had been. Otto was astounded by the news of Kate's engagement and wanted to know more about it and how it would affect their own marriage.

Laura replied that she would tell him all about Kate when they were together. She did not want a double wedding, so they must wait until fall rather than consider a spring marriage; Kate would marry in June. Laura would be leaving Friday night on the Burlington, but if she could do better on another line she would telegraph him.

For the young couple, despite the bad weather, it must have been an ecstatic week, though the only comment about it after their resumption of letter-writing was from Otto's statement, "You were very satisfactory." Otto had, as planned, moved to a room near Madeleine's and George's flat. And while Laura was introduced to baby Alfred, Otto probably met Madeleine's parents for the first time. Laura returned to St. Paul on New Year's Day, 1899, in what was to be a momentous year in Otto's and her life, the last year of the 19th century.

On the train back to St. Paul, Laura wrote Otto this touching message:

> Dearest,
> It is only half an hour since you hurried off the car, and I am already writing to you! Why not! I have had my supper of baked apples and cream and bread. It doesn't take long to eat when one has such elegant service as the dining car affords. You are probably eating now—and alone. After all, I am writing first. I am sorry for a number of things.
>
> 1st—that you didn't kiss me goodbye, for I know you wanted to.
> 2^d—that I told you to take your hand down when we had only a very few minutes left.
> 3^d—that I did not teach you how to tie your string ties.
> 4th—that I did not sew your glove and the button on your coat.
> 5th—that I talked to Madeleine about you etc.—(all of which you are in honor bound to forget).
>
> I hope you didn't get hurt when you got off the car.
> I think my watch must be slow.
> I hope you will write me tonight.
>
> The End.
>
> The car jolts, and my pen will scarcely write.

Otto's New Year did not get off to an auspicious start. After a week's absence from the boarding house, his landlady had evicted him in favor of a "better" customer. So on the first day of 1899 Otto moved across the street to 357 LaSalle Avenue, North, where he expected to remain for some time. By Jan. 3 he was comfortably settled.

"Your dear little note written on the train did not reach me till this morning" (Jan. 5). He had not expected it, so had not asked for it at the "old" boarding place. The note made lonesome Otto quite happy.

References

7-1. Osterhout, W.J.V., Jacques Loeb. *Biogr. Mem. Natl. Acad. Sci. U.S.A.* **13**, 318–401 (1930).
7-2. Goodpasture, E.W., Leo Loeb (1869–1959). *Biogr. Mem. Natl. Acad. Sci. U.S.A.* **35**, 205–51 (1961).

Chapter 8. The Columbus Laboratory (1899)

Two projects were developing at the Columbus Laboratory that Otto thought could get him on the payroll sooner than he had expected. Wesener was making an arrangement with a local milk dealer, G., to develop a substitute milk for infant feeding. And Otto was talking to a manufacturer of "acid-phosphates," K., who also wanted some developmental work done. Meanwhile, Otto continued to be useful to Dr. Turck. "I felt a little queer Wednesday evening as I heard him deliver the paper of which I had written almost every sentence. He was afterwards told and I was told that it was the best paper he has given here." In this era, ghost writers for scientists and Presidents were unheard of. Otto would help Turck with another paper to be given the following week. Meanwhile, he and Wesener were trying a high caloric, two-meal-a-day, weight-gaining diet that Turck had successfully used on his patients.

Wesener plunked $15 on Otto's desk at the Columbus Laboratory on Saturday afternoon, Jan. 7, and told him that he could expect this as a weekly salary if the G. milk scheme came through—a salary of $900 for the first year. Things looked promising. ". . . we may begin some trial feeding experiments with infants within a week." On Sunday Wesener and Otto would do some scientific experimenting, and after dinner, call on Dr. Christopher, the great baby specialist about their milk plans. Christopher had not encouraged G. (who had called on him yesterday) sufficiently, so they would try to win his support. Christopher, however, was extremely busy and could not talk much. ". . . he is so crowded with patients and duties to patients that he has simply no time that he can call his own." He usually ate dinner between 11 p.m. and midnight! His wife complained about not seeing him enough. Otto felt that being too busy must prevent him from doing more scientific work in his specialty of pediatrics. More men of this caliber were needed, however. Christopher could not talk much about the milk, but at first he did not like the plans. After Otto talked to him, he wanted to know more, so Otto promised to visit him again. Dr. Wesener needed Christopher's endorsement of the product, and Otto felt that Christopher understood the subject of infant nutrition better than anyone in his recollection.

Otto had bet K., the phosphate manufacturer, the price of a new $4 hat that he could readily prepare a "dry" acid phosphate, which he did in two hours, and won the hat. The manufacturer was interested in its commercialization, and Wesener wanted definite financial contracts to be made. K. balked at this. He wanted Otto to work directly on the phosphates at the factory, if they could come to an agreement. K. did not want a royalty agreement with Wesener, but asked Otto if he could spend one day a week at his factory and let no one know about it. Otto did not wish to do that while he was with Wesener. K. made an enticing offer of $8 for one day and $1 per hour overtime.

When Otto saw K.'s factory, he was impressed, although he had the impression that it could be losing money. Otto had meanwhile told Wesener of K.'s offer and Wesener felt than $50 per month was inadequate. He thought that the original offer was better financially because a lot of hours were involved. Any chemist could earn the $50, but Otto's background and the responsibility of the position called for more. On the other hand, if K. would pay more than $50, Otto wanted that to come to him personally. K. did not want any deal involving Wesener, but would hire Otto directly. Otto considered the possibility of leaving Wesener, and just getting income from K. and Dr. Turck. But he backed away from the idea because Wesener's laboratory held more promise for the future. Otto felt that Wesener was treating him fairly and trusted him, and wanted to keep him permanently at his laboratory. This was demonstrated when he asked Otto to prepare a batch of patent medicine made by the Columbus Laboratory for a firm, and for which the Columbus Lab received $12.80 a gallon, though it cost only $3.00 to prepare. The firm would purchase $3000 worth of the medicine the first year. Otto was at the moment temporarily filling the place of the chemist, R., who was now apparently teaching. There was a 50 percent chance that R. would not return in the spring. If he did not return, Wesener would pay Otto $50 per month "with forty percent of what I earn for the laboratory until my income shall amount to $100 a month; but then he does not want to pay me more than $100, for a time at least." If the milk scheme came through, however, Otto could get the $100 almost at once. "We now have the full support of Dr. Christopher about it and G. has seen Christopher about it and is now sending us his milk every second day for some preliminary experiments."

Otto's anticipated income sent him into premature visions of early marriage, payment of debts, and a sizable bank account. If Wesener's plans worked, Otto might not care to take a college position anywhere! If R. returned to the Columbus Lab he would bring with him about $50 per month income (source unstated) so that Wesener felt that the lab could use two chemists. Otto assured Laura that he would not make permanent arrangements until he had first discussed it with Professor Loeb. Otto would not take less money to work at the University of Chicago. In fact he believed that he would require $1500 per year in view of his potential at the Columbus Lab. "It is fully as interesting to work out scientifically how to prepare the best possible substitute for human milk as it is to work out scientific problems without commercial value." For the moment Otto preferred to get money as a recognition of success. Did Laura agree? he asked. Should he seek money first of all? "We want books, you know, and a thirty dollar flat and a hired girl (by and by) and our debts paid and a little bank account."

The Columbus Laboratory 113

Otto discussed in detail with Laura his debts and how to repay them, and his personal needs if marriage were involved. He wanted to pay off entirely his debt to his cousin Miller who had gone to the Klondike gold rush and was now on his way back, evidently none the richer. He was steadily examining flats now. The talk with Loeb revealed that the University of Chicago was out as a job prospect. Otto had mentioned $1500 as a competitive starting salary, and had found that Stieglitz who was the most popular professor in the department of chemistry was just getting that salary now, in his seventh year. Of course, this was Harper's fault.

Madeleine and George had brought Alfred to the Columbus Lab and they had gone to see Dr. Christopher about an eczema that Alfred had developed. The pediatrician changed Alfred's food that could be causing the problem. Temporarily, he would be on condensed milk with water and cod liver oil. As the month of January 1899 drew to a close, Otto was able to put one baby on the new formula, but after a few days on the diet the baby could not stand it, so the doctor and nurse switched him back to a mixture of oatmeal and cow's milk. There was no new word on the phosphate deal. He was dissatisfied with his boarding house and was thinking of leaving it.

Here is an exact synopsis of my indebtedness:

1st	My brother	$867.00	with	7%	from	Sept. 1, '98
2d	" cousin	$100.00	"	8%	"	Nov. 1, '96
3d	" sister	1250 kr [about $340]	"	6%	"	June 12, '98
4th	Madeleine	$50	"	7%	"	Ap. 22, '98
5th	"	$100	"	7%	"	Oct. 1, '97
6th	Sikes	$100	"	7%	"	March 20, '97
7th	"	$100	"	7%	"	Nov. 4, '96

"How in the world can you wish to marry a fellow with that sort of a financial standing to start on!" It kept him from accepting a small salary at a college. It sharpened his desire for money. But he figured that he could pay it off reasonably fast according to his present "indications" in Chicago. Why? Because Wesener must soon guarantee $100 per month. Side earnings would bring in quite a little. He wanted Laura not to consider working when they were married, but to stay bright and happy. Otto liked what he was doing, but not the sessions with Dr. Turck. He suggested Aug. 23 as a wedding date.

Laura responded by comforting Otto about his debts. He could clear them up in a few years. He should not fret constantly about money. They would live cheaply for awhile and be comfortable. During January, she had followed Otto's progress with Dr. Turck and Wesener, but did not know what to advise him about the factory. She advised against taking a share in the business, but a royalty was fine. Holding stock involved a double or triple liability. She discussed paying the debts off.

In early February, Otto thought that his predicament had reached the "pits." The formula failed, there was no word from K., there was no permanent arrangement with the Columbus Lab, and now Dr. Turck had so far forgotten to pay him. What was more, on Monday he had to pay $11.60, his semi-annual life insurance payment, and after receiving a letter of advice from Axel, would pay $55 to $60 to his cousin Miller on the $100 debt. After paying all this plus his board, Otto felt he would be as poor as he had been when he returned to the U.S. in late June. On top of this, he was determined to move and advised Laura to address her next letter to the Columbus Laboratory.

The move was a good idea. On Sunday, Feb. 5, he found a pleasant room with a private family at 462 Oak Street, with room and board at $6 a week. Otto made a deposit of $5. From the new room he would have a view of the lake looking sideways across Lake Shore Drive. One drawback was that the path to the bathroom led through the sitting room, just short of the kitchen. As he used a "cold bath" regime at 7:30 a.m., he would not disturb anyone. But Otto was getting fed up with this type of living and hoped that this would be the last place he would live in as a bachelor.

Laura wrote trying to cheer up the depressed Otto. She wrote about flats and clothes. Laura had her heart set on buying a sealskin coat. She had enough in her savings and was wrestling with the economy of such a move. Otto encouraged her to buy it. There was no hint in his letters that he had the slightest possessive thought toward her savings. Otto and Laura had looked at housing possibilities during their Christmas vacation together, and now there was an increasing tempo of discussion about them. Otto sent Laura clippings from the Sunday *Tribune* so that she could see what was available. They had worked out a system of progressive advance in housing as debts were paid and income grew. Begin with room and board, then on to a flat, from the humbler to the more elaborate ones, then to rental of a small house. Otto would make no arrangement without her judgment. Otto, however, did not want to consider boarding houses because of the likelihood of meeting too many disagreeable people. Another possibility was a mutual arrangement with Miss Felton. Finally, Otto was letting Laura know that his mind sometimes turned toward enjoying the fruits of their future marriage.

Otto stopped using himself as a guinea pig for Dr. Turck's high calorie diet without indicating the final results, but it is safe to assume that were it successful, he would have mentioned it because he was now steeped in reading literature and texts on nutrition. Not only was this related to his projects with the Columbus Laboratory and Dr. Turck, but he was preparing to give a public lecture on food and dietetics.

Otto was now trying the milk formula on three more babies. Then a fourth baby was added, the Turck's daughter, five months old. Several "air castles" floated across his view in February. A potential philanthropist would talk to Evans (pathologist of the Columbus Medical Lab) about founding a medical research laboratory in physiological chemistry and bacteriology. Otto would be in charge of physiological chemistry with a staff of three or four assistants and no students. Wesener told Otto that if H. became editor of a certain national medical journal, Otto would become editor of the physiological chemistry in it, at a salary of $15 a week. Then Otto was to be an expert witness for the U.S. Customs Department. "The question involved is the chemical resemblance between bitter almonds and the seeds of peaches and of apricots. Wesener turned the man over to me." Otto loaded himself with information, but nothing materialized from this or the other two pipe dreams.

Otto wrote from the Sikes' flat. He had been out for an hour's stroll around the park with Alfred. The baby had improved in appearance since he had last seen him. The eczema was almost gone, his skin was getting clear, and his eyes were bright and full of intelligence. But he had not gained any weight in three weeks.

Otto's lecture on nutrition had gone off well enough, though acoustics were rotten for his voice. "I hope that I may have enough sense not to write out any more lectures. It is an awful waste of time and I believe that with proper care I can make what I want to say clear without all that trouble."

Otto had gone with the Turcks last night (Saturday) to the theater, but they left before it was over because it was a bore. They returned to the Turck's home, lit the fire, drank champagne and talked. Dr. Turck was dejected because he had to perform two operations, both of which would end fatally. Mrs. Turck did all she could to cheer him up. Turck surprised Otto by handing him a check for $20, two weeks pay, and Otto commented that he scarcely knew where he stood with him now, as they were not collaborating anymore. "The help I give him is not so much scientific; but I think it is worth to him what he pays for it and he probably wants to be satisfied that he treats me nicely—as he certainly does."

During the last week of February, Otto had a good deal of work to do at the Columbus Medical Laboratory. He analyzed two water samples, one for sanitary and the other for boiling purposes, and also possible poison in a batch of candy that a woman had been mailed anonymously. Otto also prepared formula for six babies. He was concerned with how to prevent curdling or the whole project would fail. Then one of the babies taking the formula died suddenly, but not because of the milk. "Those hospital babies are very bad ones to handle, because they do not get the proper care. They do not use our milk anymore but the doctor in charge of them has consented to give us some of the private patients as soon as she gets an opportunity to do so." Another baby, however, brought heartening results. This baby had received the very best attention for months, but Christopher and the attending physician together had not been able to find a food that the infant could tolerate. At six months the baby weighed only 12 lb. Three days on the formula, however, and the child improved. Though this was too short a period to tell, if the baby were restored then it would be a "great victory" for them and the formula. Turck's baby was thriving on the formula. Otto hoped to become expert in compounding infant mixtures in the next few months.

Otto was not sure he could afford to visit St. Paul in June for Kate's wedding, and he doubted he could earn enough for a new dress suit. He was now paying off interest on his debts. He had paid out $107 since Jan. 1: $60 to his cousin, $35 to the Sikes (between them), and $12 for insurance. He might be able to come in June if they were to delay their marriage from August until October.

Otto sent Laura a clipping from *The Chicago Record* about the public lecture he had given. The lectures were sponsored by Northwestern University, the University of Illinois Medical School, the Parkman Club of Milwaukee, the

Board of Education, and the *Record*, and covered history, art, science, and literature. The lectures were presented in the high schools, and more than 500 people attended and packed the schools to where there was only standing room available. Otto's talk was described as follows:

Dr. Folin Talks About Foods

"Food and Dietetics" formed the theme on which Dr. Otto Folin of the Columbus Food laboratory spoke at the Bismarck school. He said among other things: "We know the elements of which the living body is made up, we also know the exact chemical composition of the various foodstuffs, and since there is a continual breaking down of the tissues of living bodies, the extent of which can be and has been investigated, it is clear that we can know of what kind and how much food must be furnished the body in twenty-four hours in order that it may remain undiminished and in normal condition. To replace such broken-down parts of the body is one function of food. Another equally important is that of furnishing all the energy which the body needs for its activities and in order to preserve the normal body temperature. The disintegration of living tissues does not go on rapidly enough normally to furnish the energy, as some foodstuffs are certainly burned in the system without first becoming an integral part of the living substance."

March began with another air castle for Otto. A professor of physiology at the College of Physicians and Surgeons was fired for being rude and coarse, particularly to the female students. While the dean of the college had made arrangements to fill the rest of the term, he was open to suggestions for next year. Wesener and Evans wanted Otto to consider the position, which involved teaching the first-year course. "I don't feel well enough prepared for it." Though Otto knew much more physiological chemistry than most physiologists, he was relatively weak in those parts involving anatomy.

In a few days, Otto convinced himself that he could take the position if it were offered. After all, fundamental physiology was largely physiological chemistry. Most of the physiological chemists of Europe, Kossel, for example (Otto reasoned), were professors of physiology though their research was in physiological chemistry. But grounding in anatomy was essential as well as in the clinical side and this is what appealed to the would-be physicians. If he were to become a professor of physiology, Otto would try to get an M.D. degree from the college. Wesener insisted that Otto get the degree, and Otto thought it was not a bad idea.

Dr. Dickinson, dean of the Harvey Medical School (for women), asked Otto to teach the freshman class of that school for a six-week course beginning on May 1, "the greatest possible" amount of chemistry, for three hours every evening. Otto had hoped to teach there the previous fall. Dr. Dickinson was a very capable woman. Otto accepted the job, though Wesener advised him not to take it if he got the P & S appointment. If so, then he should also quit Turck, and move to the West Side near the private clinic of Dr. E., one of the best anatomists around, and take a course in anatomy, perhaps two hours each morning, before going to the Columbus Laboratory. He could use all of his "spare" time working up physiology and preparing the course for his students. Dr. E. was a professor of anatomy at the college, so Otto would get all of the anatomy he needed to become an M.D. Otto liked thinking about such prospects and was meanwhile preparing his first lecture on a modern chapter of chemistry to give to the medical students in Wesener's classes in the medical school.

Word came to Otto from a friend at the University of Chicago that Loeb's request for a physiological chemist had been granted and that Otto should go see Loeb. But there was no point in it, Otto felt. He had spoken to Loeb only a few weeks before and only $600 would be forthcoming. The professor also agreed that this amount was not enough in view of Otto's debts and prospects downtown, especially because Harper did not attach much value anymore to high-grade scientific work and would not promote or pay anymore than he was forced to. Otto believed that Loeb would tell him if he had a living wage to offer. "It makes me ugly just to think of the treatment that Dr. Steiglitz is receiving at Harper's hands. Stieglitz published an article a short time ago that is a masterpiece."

Otto languished and pined, feeling a bit sorry for himself. Laura remained 500 miles away for another six months.

> After I get a reasonably good position and income I think that I shall try to settle down to just do my work well and then simply work with the utmost leisure on my science without regard to how to get results quickly or how to attract attention anywhere and then spend considerable time with my family and help my little wife with the worries incident to normal family life.

"The chances are," Laura confessed, "that I shall become domestic, and shall never develop any ambition. You know, I never have shown any so far, and when we are married I shall probably not be so very different from what I am now." She would not be able to do much outside of her home duties after marriage. "I have tried to hold up the other side, but I do it without fully believing my own position tenable. I don't think I shall be unhappy if I should not study at all any more, that is, if I were happy with you."

YALE UNIVERSITY
SHEFFIELD BIOLOGICAL LABORATORY
New Haven, Conn. March 11, 99.

Dr. Otto Folin:—
  My dear Sir:
  I have just heard that there is an opportunity for a physiological chemist— research work—at McLane [sic] Hospital, near Boston. Nervous diseases—It is a magnificent hospital and might prove a good opening for you if you have nothing better on hand.
  I have written concerning you to the Superintendent. Should you care to consider such a place address concerning it:
    Dr. Edward Cowles
    McLane [sic] Hospital
    Waverly [sic], Mass.

Yours truly,
R. H. Chittenden.

"What do you think of it?" Otto was delighted to find that he was on Chittenden's mind for any position that did occur. Otto wrote his thanks to Professor Chittenden and a letter of inquiry to Dr. Cowles. Otto, of course, was full of curiosity.

> I do not see how the facilities can very well be good in the way of a laboratory, apparatus, chemicals and literature or how it comes that they want a research man; but we shall probably find out more about all that before long. At any rate it gives us one more respectable looking piece of iron to attend to.

Otto, with the callousness of the still youthful, struggling, insecure doctor expressed no shame for his recent bitter remarks about Chittenden.

Otto had gained another influential supporter, Dr. B., for his position at the College of Physicians and Surgeons. He "spoke of it openly" at a dinner he gave to about a dozen faculty members.

This was an interesting period in Chicago medicine, Otto observed, because in the next few years one or two medical schools would rise above the others permanently. There would be a struggle for supremacy between the P & S and Rush, and the outcome would be determined by their permanent arrangement with the state university (P & S) and the U of C (Rush). If both arrangements became sufficiently close to the universities, Chicago would have two great schools. All of the medical men talked about it. In 10 years, Chicago would become a great medical center.

Laura was encouraged that Professor Chittenden was thinking of placing Otto, and she thought that this should greatly please him. "Yale has a fine sound." Of course Otto was getting a good start in Chicago and would not be eager to leave it.

R. was *not* returning to the Columbus Lab, so the job was now Otto's. He anxiously awaited word from Dr. Cowles, and reminded Laura that Chittenden had gotten Wakeman his job in New York with Herter.

"I am wasting an awful lot of valuable time with Dr. T.," Otto wrote, "but he has been very generous with me and I cannot desert him yet for awhile."

Although Otto would write of having failed to accomplish anything, it would generally be followed by a description of accomplishments. He was trying to prepare a new meat product that could be dissolved in water to be used by invalids at times when they were unable to digest solid meat. There was one already on the market, of German make, and it sold for $3 to $4 a pound. He had two more of Dr. Christopher's babies to put on the milk formula, and he was preparing his lectures to give to Dr. Wesener's class. He complained steadily now of having to go help Dr. Turck, but he always enjoyed talking to Mrs. Turck.

While Otto had not yet heard from Dr. Cowles, he did not particularly care in view of his brightening prospects in Chicago. He was now evidently drawing regular pay from the Columbus Laboratory and as usual from Turck, and would probably earn $100 from the Harvey School. Otto estimated that he would not earn $1200 in 1899. His reserve was down to $45, and with clothes to buy and marriage ahead, he could not get cash quickly enough, no matter how much his income improved. "I am of a too hopeful disposition and build air castles too rapidly as soon as I have a bit of success."

At the beginning of April Otto's hopes began to rise sharply. The P & S faculty had met and the dean told the faculty that Otto Folin was the man he had in mind for the physiology replacement. Then,

> Wesener has further this week (yesterday) made an arrangement with the American Cereal Company (the manufacturers of Quaker Oats among other things) to do all their chemical work, in fact to be their regular chemists. The agreement begins today and the compensation is to be $2500 for the first year and $3500 for the second. Wesener and I are to be jointly in charge of the work and I shall get an assistant to do the routine analysis for the company. But we, or practically I, shall attend to the experimental work, occasionally visit fairs, agricultural experiment stations, etc., in short be their chief chemist. Monday evening we shall go to the company to discuss the work and to sign the contract. This means that I shall begin to draw a higher salary probably $100 at once!

With this joyful news in mind, Otto went to see Loeb, partly in regard to an article that Wesener wanted to publish in Germany, but also to discuss his own affairs. Loeb did not like them. Otto would be unhappy trying to teach physiology because of the tremendous amount of preparation he would have to make, although it would be so easy for him to teach physiological chemistry alone. As to his other income, Loeb felt that the prospects were extremely good for a physiological chemist at the U of C at $1200 or $1500 next fall. And he would *not* consider anyone but Otto for this position! What should Otto do? He was apprehensive about teaching physiology, but there was more money downtown than they would need to live on. At the University of Chicago there was just enough, but a high grade, pleasant, scientific work and a name as a physiological chemist!

Otto did not have to struggle much to make a choice. "Well, I tell you, we will probably choose pure science provided that we get enough to live on comfortably." A choice, however, could be difficult to make if the salaries were $1200 for pure science vs. $2200 for the Columbus Lab and P & S.

All of these approaching riches might have turned an ordinary mortal's mind, but Otto's temporary euphoria, if any, was tempered by reality. Wesener cautioned Otto about marriage. He would be less productive at this critical state of his career as he took on the laboratory job and the role of a teacher. His need to be with his wife would hamper his creativity. Of course, Otto disagreed. He felt that Wesener was reasoning from his own marriage, which was not conducive to restfulness and peace. Wesener talked of making Otto a member of the firm. For $300 to $400 Otto could buy out one of the present shareholders. Now was the time! But this was another reason not to marry yet! He could draw a reduced salary of $75 per month, and the rest would go to pay for his interest in the business. Otto, however, was not interested in accepting less money now. Meanwhile, Wesener did not mention any raise in salary at *this* time.

Otto's thirty-second birthday was approaching. A real birthday present would be an hour with Laura alone. If he should get $25 a week from now on, then he would come to St. Paul for Kate's wedding and would stay a few days, but only on condition that Laura would not be stingy with her affection.

On his birthday, Otto voted in the city election, then read Laura's letter on the way to the elevated train. He was on his way to give a lecture to Wesener's class. He felt that he was wasting a lot of time trying to answer farmers' letters to the American Cereal Co. Laura could help him out of some of this, as the letters involved the most efficient use of feed mixtures. This could be reduced to a mathematical basis. A certain ratio of fat to protein to carbohydrate must be maintained at 1 to 6.25 to 31.25. Given four feeds with different ratios of the three constituents, a formula had to be created so that the maximum amount of the cheapest feed was used, and the minimum amount of the costliest, the cheapest mixture, in short, using either any three or all four feeds. Otto provided an example, and wrote that he had worked for two hours trying to make the ideal mixture.

Otto sent Laura a letter he had received from Dr. Cowles, along with a report on the McLean Hospital. He thanked her for his birthday present.

"Your letter on the mathematics of feeding reached me yesterday morning but I did not go through the mathematical part of it until after I had gone to bed last night," Laura wrote. It was clear, as she put it. She seemed to know at once what could and could not be done. Otto, however, had found a commonly used, easy though inexact way out of the difficulty, based on the asumption that fat and carbohydrate could be used interchangeably by the animal system. It would be convenient to have a mathematician in the house. Otto wanted to study the subject of animal feeds more.

The previous night, Otto reported, the dean of the College of Physicians and Surgeons and Evans of the Columbus Lab had a "terrible" row at the faculty meeting. Evans represented the progressive, aggressive faction of the faculty and the dean was ultraconservative and dictatorial; the resulting breach could cause one or the other to resign. Bad for the college and its expected affiliation. Maybe bad for Otto's chances as a physiologist, though this did not bother him, as he was less keen about teaching physiology now that he had talked to Loeb. Besides, he should soon be earning $100 a month at the laboratory. On Monday, Otto expected to get a new assistant at $50 per month. Certainly it was time that Otto received more than $60 per month.

Otto had a pipe dream about earnings at the U of C. If they paid him $1200 for teaching three quarters (base pay), then he could earn proportionately $1600 for four quarters; at $1500 base pay he could earn $2000, and still have the vacations available between quarters, amounting to a month off. Harper would not go for this plan, probably.

> In the meantime I shall also write to Dr. Cowles of the McLean Hospital although I scarcely know what to think of it. I do not like that Dr. Cowles did not tell me what salary I could expect although it is plain that he does not himself know. This seems rather strange since the catalogue indicates that there is considerable money to be had for the pathological laboratory. I wonder how the hospital idea will strike you? I would not particularly like it if it were

not so near Boston because I would be apt to be scientifically too much isolated on the hospital grounds; but within six miles of Boston this would not matter I guess. At any rate I shall write Dr. Cowles in a few days. I shall not fix any salary but simply state that the salary I would accept must naturally depend on the situation at the time. Then I shall tell him that my financial prospects in Chicago now indicate $2400 for the coming year (which they really might be, you know, at $1200 each place) that the work (in Chicago) will, however, not be nearly so satisfying as would straight scientific work and that I would therefore accept considerable less in a place where the work would be desirable and other conditions satisfactory, but that I must have a reasonable living salary because I am going to get married in the fall.

Laura had wondered if Otto could combine the U of C work with that of the Columbus Laboratory, but that was impractical. Working four semesters, as well as earning on the side was possible. They could discuss all of this when he came to Kate's wedding. There were a good many dreary days and nights remaining for them until their own planned wedding on Aug. 23.

"My period of trial and suffering is not yet over," Otto complained. Each morning for a week he had awakened with a headache that lasted to 3 p.m., when it stopped or almost did so until the next morning. This periodic pain was severe, and puzzling to the specialist. The doctor found in Otto's right nostril a so-called polypus, an inoffensive tumor, and removed it. Otto hoped that it had been the cause of the headaches and the ringing in his right ear. "I think the doctor did a good job, but it was a very severe bit of experience." For the moment, it did not stop the pain in the front part of his head, but it did stop the buzzing in his ear. Would the headaches stop, too? Perhaps at Laura's instigation, Madeleine called and wanted Otto to visit while he was unable to work, but Otto preferred to nurse his pains alone. George Sikes also dropped in on him (probably to check on his progress). George was now making an enviable salary ($40 a week), and had four to five years start on Otto.

Laura cautioned Otto that he could not teach an "extra" quarter at the U of C. If he taught in the summer quarter, he would *not* teach in one of the other three quarters. Otto agreed, but again mentioned that he really was not hopeful about a position there anyhow.

Otto went through a period of pain and illness for about two weeks after the polypus was removed, not to mention the week before. During this period he evidently ceased going to Dr. Turck's and spent a minimum of time at the Columbus Laboratory. His two lectures at the P & S were given before the "surgery," and he felt that they were above the heads of the students. The second lecture was more satisfactory to Otto because he used no notes or manuscript as he did in the first lecture. He liked the "free-hand" talking method.

Otto's new assistant at the Columbus Laboratory was a hustler, but not an accurate worker. The man had taught at a college and thought he knew a great deal. He spouted some nonsense at Otto, perhaps thinking him an M.D. Otto showed him the inaccuracy of his results, and that the mistake was an elementary one on an essential point. The new man had not been taught to do it right. Although he was a nice fellow and they would get along, he would have to descend from his "considerable experience" position. He needed to be watched for awhile. Wesener would have to keep an eye on him now that his teaching term was over, because Otto would not be able to do it during the first phase of the work for the American Cereal Co. when he expected to be traveling for them.

Otto had received "a good deal more in the way of literature than I sent you" from Dr. Cowles. Otto had not yet written him, but was thinking of asking for a salary of $1800.

As April neared its final days, Otto reverted to his long day's routine. He had talked to Dr. Dickinson, and another air castle almost tumbled down completely. She wanted him to give only eight lectures to her medical students beginning in the middle of June, three hours each, for which she would pay the grand total of $25. He would take it.

The job at P & S, according to Wesener, would probably go to a man from Iowa. How curious the rapid change of things in Otto's expectations.

Laura was troubled by Otto's complaints of headaches and hoped he would get further medical examination. It was good of Madeleine to have asked him to stay with her. He needed quiet and good food. He must take care of himself. In less than 12 weeks they would be riding their wheels out at Como, or Fort Snelling, or White Bear Lake. She began emphasizing Otto's needs in clothes more and more, as she expected him to come to Kate's wedding. He must be stylish.

Tuesday, May 2, Otto prepared a report of the whole month's work—208 analyses—for the American Cereal Co. Laura had agreed with Otto that a university position was best for him, but Otto did not foresee any real opportunity at the U of C. He was not sure he wanted to teach anyhow, but would know better after he taught the medical students at Harvey College. He then launched into a favorite pastime that Laura and he had developed: living within projected costs and income.

Laura had done three days of substitute teaching at Central High School but turned down the opportunity to teach to the end of the term. She discussed their living costs.

Wesener again talked to Otto about his income. He was *not* inclined to pay him $100 a month unless the milk scheme worked out, but agreed that Otto needed more pay. He still wanted Otto to become a shareholder in the firm by buying out the "silent" member who held a fourth of the shares. He paid only $48 originally, wanted $800 to sell out, but would be forced to settle at $300 or $400. Evans would buy it, but Otto could buy it from him at the same price. Otto still would not do it until he was certain that he would stay with the firm.

He and Wesener had a long parley with the two chief men of the American Cereal Co., and everything was favorable. The second year would definitely bring in $3500 instead of $2500. When that point was reached, Otto's salary would rise to $125 a month (April 1900). "I certainly do not expect to be limited to $100 a month for very long." But at the moment, these were all more pipe dreams. He was still drawing $60.

The course that Otto would teach at the Harvey School ended on June 24th. Otto wanted to know when he should come to St. Paul, based on his taking a week's vacation. Could Laura suggest a wedding gift? He wanted to get something useful but unique—not ornamental but lasting. Cost, $10 to $15. He had gone to the Turcks for an hour, not to work, but to see Mrs. Turck, who had been away for a couple of weeks. Otto added that he now felt fine and hoped to regain some lost weight. He had no treatment from the specialist for a long time.

Wesener had talked with Otto a bit. He wanted to know when he would be married, and when Otto said, "this fall," he smiled and said that he would get his "one hundred plunkers" all right; but that he would like to make that G. (milk) deal a go." Otto did not specifically ask when the raise would come, but felt certain that this would happen on June 1, if not before. He wanted it by May 15 because he needed the money now. They would keep the laboratory assistant though the rush was over and Otto could probably do all the work. It would be more profitable in the end if Otto had more time for constructive purposes and let the assistant do the routine work.

> There is no doubt whatever in my mind but that Wesener really means to keep me permanently. In a way he is going to be quite square with me, too, though he will be apt to keep for himself most of the work that increases one's reputation—except in the food line. It will be left for me. We can therefore, I think, consider it settled now that $100 will be the foundation upon which we can build and on the strength of that I move that August 23rd be finally accepted as the day on which you marry me!

With only the Columbus Laboratory as a source of income, they could now consider living in any part of Chicago. Meanwhile, the income was still $60 per month. Otto had written to Dr. Cowles.

Laura wrote that Otto should try to get two to three weeks vacation to get married and settled, but the vacation should come *after* they were married rather than immediately before. They would have to do a lot of searching for a flat, and then prepare it properly. Laura could not tell Otto when he should come for Kate's wedding. "I do not care for a large wedding—especially a few weeks after Kate's." Most of the people Laura wanted to invite lived out of town, in contrast to Kate's friends.

Laura described Miss Felton's plan for mutually sharing housing. She was employed by the Associated Charities and worked near Hull House. She would like to live nearby and eat at the Coffee House. She wanted them to join her in taking a flat with entirely separate establishments, but uniting to reduce expenses. When Otto was away Laura could join Miss Felton in some of her work and gather material for a worthwhile economic study. They knew each other well enough that they could arrange matters so that they would not interfere with the other's private lives and yet be very companionable at times. While the Folins might not wish to live in the slums, Hull House was the cultural center of the city.

Laura did not like the locality. Being near Hull House did not compensate for the unpleasant neighborhood. However, Laura wrote Miss Felton that she would ask Otto about it. Otto had mentioned the possibility of a flat on the fourth floor of the "Irving." She liked the idea, but not the floor.

In mid-May Otto telephoned the Associated Charities and learned that Miss Felton would attend a Hull House conference that evening. Otto found the meeting interesting but when it was over, he almost missed Miss Felton, who had left by a side door. He found her on a street corner and they talked for awhile as they rode on the street car. She asked Otto to call on her the next evening so that she could explain her ideas. What she had in mind was that together they would rent one large or two adjoining flats, or even a house, in the Hull House or in the "Commons" district. They would use a common kitchen, and a competent girl—in short, they would be sociable yet free and private. Otto talked about Miss Felton's social work and his fears that Laura would get too much absorbed in it for her own good. After an hour's discussion they went to look at flats together. Otto liked Miss Felton's suggestion, because it meant living economically without being isolated. There were plenty of nice people in either locality. At any rate, they could try it for a year, and it would be better enjoyed than living on the north side of town, would it not? Otto stated that if Laura agreed, he would be decidedly in favor of the plan. The distances were not inconvenient for him. It would be a streetcar ride of 10 minutes downtown, and the elevated train would take him to the Curan Library. Going or coming in the localities that Miss Felton preferred would not be pleasant, but there were quiet, decent, peaceful side streets there. He went to inspect in detail one house that they liked, a two-story brick one with three rooms on the first floor, four on the second, and a large basement, at $25 per month. Otto thought that, as far as housing sites were concerned, it was a choice of the university, the Newberry Library, or the slums.

As to his vacation, Otto remarked,

> Now I must make a visit with my relatives either this trip or in the fall. I could not go up there twice one summer and not visit them especially as I have not seen my Stillwater brother (Alfred) for four years. If I could go up to Center City on the 4th of July and remain for about two days that might suffice as my Stillwater relatives would probably go up to Center City then. I ought, however, really to give them three or four days, because this is probably the last I can give them much of my time if I take any vacation in that direction. You will probably want to go with me to visit them once; but that can perhaps just as well be postponed to next summer or some other time.

Otto liked the Felton proposition because he wanted to associate with cultured people. He felt that boarding house life had hurt him socially. The discussions he had heard at Hull House were stimulating. Good surroundings had a strong influence.

A classmate and friend of Laura's at Vassar, Cornelia Golay, who had married Francis Gano Benedict, a young Ph.D. and professor of chemistry at Wesleyan College (Middletown, Conn.), had written Laura that an opening was available at the college, but gave no details. She told Otto to write to Dr. Benedict. As to the Felton idea, she did not want "to see the other side" just for the sake of seeing and "I do not possess much if any enthusiasm for the people." She wanted to think it over more. "I hate the thought of the settlement idea degenerating into a fad."

Otto wrote to Benedict and asked Laura to thank Mrs. Benedict for her thoughtfulness. The position might be good from the scientific standpoint, but if it was under the supervision of Professor Atwater, he would not take it, because some of Otto's "friends" did not speak well of him. As to the Felton scheme, he had thought of it in terms of its possibility for Laura's conducting scientific inquiry rather than philanthropic work. Perhaps she would do better on that score, anyhow, with the Salvation Army.

Otto assumed that the Middletown job involved the agricultural experimental station. Such jobs were the best stepping stone into other desirable positions.

> I may be sent to visit the eastern experiment stations this summer or fall by the American Cereal Company. This trip would then include Atwater's station and then I should also, I think, make a call on Dr. and Mrs. Benedict! At present I am figuring a little on getting them (the Am. Cer. Co.) to send me westward the latter part of June, taking in the stations of Wisconsin, Minnesota, North and South Dakota, Nebraska and Iowa. If that could be arranged, you see, I should get my expenses paid, probably very liberally while I am gone.

It was warm at last in St. Paul as May drew to a close. Kate's wedding was now to be either on June 29 or 30.

The job at Middletown was an assistant to Atwater, without faculty rank, and at a salary in the range of $600 to $1200. It would have been joint research between the Storrs Experiment Station and the U.S. Department of Agriculture. Otto replied to Benedict:

> Dear Sir:
> Your prompt reply to my first letter was received a couple of days ago. The information I wanted was not so much in regard to the general character of Prof. Atwater's investigations as I was fairly familiar with his work before; but I thought that possibly the work for which a physiological chemist was required had been more or less definitely outlined for the first year.
> I should certainly not leave my present position to accept another at less than $1200, nor would I accept the last named sum unless I could also see my way clear to do considerable independent research work since my chief reason for desiring such a position is that I would like to follow up some of my own ideas. Prof. Loeb, who is also a friend of mine, was unable to give me any definite information of the kind when I saw him yesterday.
> Thanking you for your letter and for the publications you sent me. I also take the liberty to send you such reprints as I have left of my own investigations.
> Respectfully,
> Otto Folin

Laura wrote she would forget about Middletown. Chicago was more to their taste. She had been busy sewing. Kate and she would go calling five afternoons, and perhaps go to Minneapolis [probably in relation to Kate's impending marriage]. Three weeks from Thursday was certainly not far off. Mr. Duncan had moved into his own home, only two blocks away. His tenant had moved out on June 1.

Otto had received a letter from Miss Felton, but he left the proposition entirely to Laura. No hurry. Otto bewailed his financial status. Not enough stored cash. Then he visited Mr. Sikes' tailor and ordered a suit for $60.

Otto had decided on his schedule. He expected to leave on Saturday evening before the wedding, spend Sunday with Laura, and go to Stillwater or Center City on Sunday or Monday evening. He would be with his relatives until Thursday, then return to St. Paul where he could stay in a hotel nearby until the rush was over. He could arrange for a few extra days if it seemed necessary.

On June 7, Otto wrote that his lectures would begin the next Monday (June 12) at the Harvey School. He was not yet prepared.

Otto had another consultation with the American Cereal Co. It was highly probable that he would visit the Experiment Stations of Wisconsin, Minnesota, Iowa, and Nebraska when he came to see Laura. So he could be disposed of in a St. Paul hotel for two or three nights, if her house was crowded. Perhaps he could come by for meals and see her that way. It would be interesting to see the wedding and he would have to talk over the details with Laura, if they could get time to themselves. Were Kate and Mr. Duncan taking a trip?

Mr. Duncan offered Otto the use of his house, which he would not occupy after the day of the wedding. So Otto could have his own key and eat at Laura's house. Kate and Charles (Duncan) would probably go to Lake Superior, perhaps Madeleine Island, and expected to be gone for a month and return to their house on about Aug. 1. The Grant house would be crowded for a while, particularly when everyone was dressing for the wedding. Madeleine would arrive the morning of the wedding. They would discuss their own marriage and Miss Felton's scheme when Otto was there.

Otto spent most of June 9 taking part in and listening to business talks. One thing was settled:

> Dr. Wesener and I will visit the Mich. Experiment Station (at Lansing) next week, probably Thursday evening. The next trip I shall make alone and will probably include Wis., Minn., Iowa and Neb. In this way I shall at least get my travelling expenses paid and as much for living expenses as seems fair. I shall get an expense book and will charge my expenses up to the Cer. Co. It will save me probably 25 to 30 dollars I should think for which I am duly pleased.

Mr. Sikes wanted Otto to room with him while Madeleine was gone, and Dr. Wesener had suggested that Otto live in his flat when he returned from his first trip as Wesener would be leaving for Europe. Otto was concerned about his cash on hand. They must talk about the possibility of delaying their marriage a bit.

Otto, of course, accepted Mr. Duncan's offer. He would not mind if her brother Wallie also slept in his room at Mr. Duncan's. What time of the day was the wedding? He was not certain just when he would arrive in St. Paul because he had to write letters to the Experiment Stations to find out when anyone was there, and how much time he would spend at each station.

Otto's cash was practically gone. The poor man had lent Mr. Sikes, $20, and his friend Bernhard, $10. Meanwhile, he bought a fine looking pair of tan shoes, and found that he would need another pair to accompany the wedding suit. Should he buy patent leather shoes that he didn't like? Did he need gloves? "I don't know nothing," he double negatived.

It would be pleasant if Otto could come on the Saturday before the wedding. By leaving on Friday night, they would have both Saturday and Sunday together, and then Otto could go to his brothers on Monday. They could perhaps have Sunday at the lake. Laura was against postponing their wedding as long as he was meeting current expenses. They would not need much furniture, and what they needed would not require much outlay. "When will you begin to get the $100 a month? You said once that you thought it would date from June first. . . . I forgot to ask you before to get a copy of Dr. Veblen's book—you know it has been out sometime. I have seen it advertised. 'The Theory of the Leisure Class.' You could bring it with you, could you not?"

Otto gave his first lecture on Wednesday, June 14, to the students at the Harvey Medical School. Two students had told him afterwards that they now had a clearer view of organic chemistry than they had of inorganic chemistry from last year's term. This is understandable in view of the fact that the premedical requirements for medical school then included only a high school education. In eight lectures Otto would evidently prepare the students in both organic and physiological chemistry! It was no wonder that Dr. Wesener, with only a modest background in the actual science, had been a professor of chemistry in a medical school.

The next day, Otto and Dr. Wesener would travel to Lansing, Mich., and return probably late on Saturday. He could not be sure when he would arrive in St. Paul. He was a bit worried that the people he wanted to meet at the experimental stations would be gone to the annual meeting of the Association of American Agricultural Colleges and Experiment Stations taking place in San Francisco, July 5 to 7. The director of the Michigan station, however, was president of the national group, and could advise Otto about whether it would pay him to visit the various stations. Otto would write or telegraph when things were set for his trip.

Otto told Wesener that he expected to get $100 beginning in June. But he was doubtful now that he would get it. A meeting would be held to settle this matter either Saturday or Monday. Otto stated that he would consider it tricky if the Columbus Laboratory did not give him the full amount ($100) for June.

Laura wondered what kind of a school the Harvey Medical School was. Was it a women's school, with a woman dean? Never heard of it, before.

Otto had returned from his trip with Dr. Wesener to the Michigan agricultural college, which he thought had the "finest college grounds I have seen in this country." Although the trip was tiring, he met a few men worth knowing; it was a "valuable experience for me to visit experiment stations and to get acquainted with the directors and chemists of each place . . . the only purpose is to get acquainted. . . . I bought Veblen's book the other day. It is extremely interesting. You will like it."

On Tuesday, June 20, Otto wrote that he would leave for Madison on Friday morning, arrive at 2 p.m. spend the rest of the afternoon there, and take the evening train (about 11 p.m.) to St. Paul, to "reach my little friend by Saturday morning." He had given a two-hour examination the previous night.

Laura replied with a few words on June 21. Glad he was coming. By what road? Would meet him. Just three days!

We assume that Otto was able to stick to his planned schedule for the trip. If so, he left Chicago on Friday, June 23, to visit the experiment station in Madison in the afternoon, then took an overnight train to St. Paul, where he visited until Monday morning, June 26. Then on to Center City to see Axel, Anna, and the Johnsons, possibly his cousin Adolph Miller, lately of the Klondike, perhaps Alfred and his family in Stillwater. Otto returned to St. Paul on the morning of the wedding, Thursday, June 29. He had by now, met the bridegroom, Charles Duncan, and taken his baggage to Charles' home, where he would spend that night and probably the next four, until he left on July 4th.

With so many days free for our couple, we can be assured that they found time to be alone to discuss their own impending marriage. To Otto, the wedding he witnessed must have been a revelation; his interest was sharpened because he had to learn at that opportune moment how to play the role of the groom in the ceremony. He probably last attended a wedding in Stillwater— Alfred's, in 1892—and this took place in a church where Otto felt at ease among his Swedish brethren. The wedding ceremony and preparation for it, in this instance, belonged to the bride and her family. Otto would have little to say about it, and bear none of the immediate expenses.

As he was preparing to leave Mr. Duncan's house early on Tuesday, July 4, 1899, Otto penned the following touching note on stationery of the American Cereal Co.:

> Dear Laura:
> It is about ten minutes after six and I am ready to go. The fire crackers have been exploding for a long time. Perhaps you don't hear them but are still sleeping soundly. I am glad that I did not come over and make you get up so early. I am content to leave now because you have been such a dear little girl to me these last two days. Goodbye.
>
> Otto.

Otto Knut Olof Folin left St. Paul convinced that Laura Churchill Grant would let nothing stand in the way of their marriage, that their puritanical restraint towards each other was ended, and that their love was forever sealed.

It was a tiresome trip to Ames, Iowa, and Otto wrote just before going to bed at 9 p.m. despite the fireworks and other July 4 activities. He had been thinking about Laura and was happy because "at present I feel as though you really think more of me than I otherwise had reason to hope for." Otto had now put Laura on a pedestal.

On July 5, Otto made two trips to the station, where an instructor showed him around. In the evening he was on his way to Omaha, where he would spend the night before going to the station in Lincoln in the morning.

After his visit to the station in Lincoln, Nebr., Otto had skipped going to the Kansas station because of a train delay in Nebraska City. As it was, he arrived in Columbia, Mo., at 4 p.m. after having changed trains in Kansas City. But he was able to have a two-hour visit with Director Waters of the Experiment

Station. This was his final stop on this trip, and he would return the next day (Sunday) to Chicago. "Columbia is a beautiful little place and for a small town could probably be very pleasant to live in. I have just been out for a bicycle ride around the town and wondering at the time where you could have been living while here." On to St. Louis, where Otto got a sleeper for the overnight trip to Chicago. He was practically out of money, but was trying to save enough from the trip to buy Laura a present—something intrinsically useless. He hoped to save $10 to $15 from his trip and $10 from the coming New York trip. "How would Monday 18 of Sept. do for our wedding?" He would work on the previous Saturday, reach St. Paul on Sunday morning, and take that week only for a vacation. They could be settled in Chicago by Sept. 23.

On Tuesday, July 11, Otto was back in the Columbus Laboratory. The trip from Centralia, Mo., to Chicago had been smooth. He had enjoyed the two letters from Laura awaiting him. George Sikes had called and wanted Otto to live in their flat while the Sikes were on vacation, and Otto agreed because he would learn something about flats that way. He asked Laura to tell Mrs. Benedict he would like to call on her because he meant to call on Dr. Atwater in Middletown, during his eastern trip to the experiment stations, but not to mention anything about the American Cereal Co. because Atwater might try to sound him out about certain dealings Atwater was trying to make with the company.

It should be noted that Otto found that he had not been raised in pay to the $100 per month that he expected. He would now be on the lookout to find a job offer somewhere, more to use it as a lever for a salary increase than anything else. At this moment, he was not enthusiastic about the idea of leaving Chicago, despite the reneging by the Columbus Laboratory. After all, Laura and he had just put in so much detailed discussion about their future there!

With her parents and Kate gone, Laura was busy running the household. Otto's suggested wedding date of Monday, Sept. 18, was all right with her, though it may have been a bit too early to decide. If it were the date, they could be married as late as 6:30 p.m. and still catch the 8:05 p.m. train for Chicago. The ceremony could be over by 6:45 p.m. (15 minutes), leaving 55 minutes to leave the house (half an hour for supper, 25 minutes to dress) and the rest to arrive at the station and not have to wait. Otto could arrive on Sunday morning and there would still be a week in Chicago. Although she did not want a wedding present, if he insisted on getting one, let it be a watch for $20 or $25 (gold). He must again search for housing.

Otto wrote that he was now moving to the Sikes' flat at 215 Jackson Park Terrace, and would look at other housing in that territory for the next three weeks. He would just take his wheel when he wanted to, and go searching, then report to her. If Laura thought it best, she could come to Chicago to look over the suitable housing he found or she could wait until she came for good. The paper had abundant ads for "south side flats." Good ones, $20. Anything between 47th and 65th was suitable if it was reasonably near the Illinois Central. First though, he would look near the university. As to their wedding date it must be suitable to the bride. He was thinking she could use an extra ring beside the wedding ring when he wrote that he wanted her to have a gift that was "intrinsically useless" with the money he saved during his travels. A watch would be fine.

On July 17, Laura asked, "How should you like to live on Cornell Avenue between 51st and 52nd street—about three minutes walk from the Illinois Central Station?" The express could get Otto to the heart of the city from 53rd in 11 minutes—no stops. Laura's friend, Marion Craig, had offered to share a flat with them. The bathroom separated the two areas involved, so that it was the only thing to be shared—hence no cause for friction. The back part of the flat faced the lake. Sunny. Pleasant. Nearby 53rd Street restaurants.

Though unaware of it at the time, Otto received a very momentous message.

> Dr. Stieglitz called here yesterday. There is a vacancy at the U. of West Virginia and Nef has already written down there strongly recommending me for the position. Medical chemistry is wanted. The salary is low and I do not know much about the whole affair as yet. Bernhard wants me to work the opening as much as possible in order to use it as a club with the Columbus Laboratory people. Steiglitz has told him that if I wanted the position I could almost certainly get it. I shall investigate it of course but feel almost certain that I do not want it.

Otto wrote on Aug. 2 that he was not thinking about flats at the moment, but the trip to the East. He would stop at the Hoffman House in New York.

> Tomorrow I am to meet the president of West Virginia University but I know that I cannot consider the place. Yesterday I had quite a friendly letter from Dr. Cowles of Waverley. I have written him a second letter to the effect that I am going east on business and might visit him and the institution.

On the next day, Otto wrote,

> Now we shall just have to let the flat troubles cease for a short time because the conditions here may be a little different on my return from the East. Two interviews with Pres. Raymond of West Virginia University today resulted in his offering me a position as instructor in Physiological Chemistry at a salary of $1000 which is $200 more than the position was supposed to pay. From what he told me the conditions down at Morgantown are such that I would not consider it half bad as an opening leading to a permanent University career. I shall have to do some assisting but will have independent charge of the medical chemistry courses and have been promised that I shall have time and facilities for doing research work. Of course his offer is not final because it will have to be acted upon by the board on the 14th or 15th of August but there is really little or no doubt as to what the action will be. Well, I shall probably find out while I am in N.Y., about the income and will then write to the Columbus Laboratory of the offer and then they will either have to pay $100 a month or I shall leave them.

Otto then added that if Laura had strong feelings about Chicago, then he could hardly play the position that he proposed, though he really thought they would stay in Chicago. Morgantown was a very pleasant little college town (3500) and they could live there happily for some years, perhaps, until he could get another "call." As he would be within 100 miles of Morgantown during his trip he might go there to look the place over.

Otto sent Laura the following letter that he had received from Medical Superintendent Edward Cowles of the McLean Hospital, Waverley, Mass., dated July 29, 1899:

>Dr. Otto Folin
>426 Oak Street, Chicago, Ill.
>My dear Sir:
>In reply to your letter proposing to visit the Hospital about the middle of August I am glad to say that it will be a convenient time for you to see both Dr. Hoch and myself. The vacation arrangements are now made so that he will return from his absence a little before the middle of the month; and my vacation has been postponed to some time after that.
>It will give us much pleasure to see you and to show you something of our work. We were very much interested in your last communication to me and I have been hoping week by week—indeed it has become month by month—to see some changes in the situation favorable to my purpose. But I am still constrained by the state of Hospital affairs that is not yet quite worked out, and I am not in a position to let any one expect anything of us in regard to the future work I hope to introduce here. Still it is near enough "in sight" to allow us to welcome examination of the matter by anyone who might wish to engage in it. I shall be obliged to you if you will write me a little in advance of your coming and it will give us pleasure to see you.
>
>>Very sincerely yours,
>>Edward Cowles
>
>Med. Supt.

Otto, Aug. 4:

>If I am to carry out my plan of using the W. Virginia offer I will do so from N.Y. as soon as I have received a definite offer. The answer from this laboratory I shall not get till I return about the first of September. Consequently I cannot set a marriage date because I could leave here for good before the 15th of Sept. and in case of necessity I presume I would need to remain the full month. Still in that case we could marry and you could come with me to Chicago and we could board until time to leave for W.Va.

This was all hypothetical and highly improbable, he thought, because his desire was to remain in Chicago, and the Chicago lab would come to terms. If so, he could take only a week's vacation due to his long absence. They could be married on either Sept. 11 or 18th, the difference being that on the 18th he would have $25 more in his pay than on the 11th.

Laura's next letter was sent to Otto's traveling address, c/o the Hoffman House, New York City, N.Y. It was a masterpiece of logic, mathematical precision, and subtle irony.

>In considering the Virginia plan I think that we should regard little else than your probable future career. Whatever tends toward your advancement is without doubt best for me also—that is, if conditions for living are fairly

comfortable. And in a small University town I should suppose that the life would be very pleasant and healthful. The climate would certainly be better than that in Chicago, and the country in many parts of West Virginia is very attractive. I have little doubt that we could live happily in such surroundings. I know I had a pleasant three months in Columbia [Missouri], and I am inclined to think that I should be easily content under similar circumstances again. But what you need to think about is whether it is going to be to your advantage professionally to accept a position in a Southern institution. You know almost all of them are much below the standard of even our Western State Universities. And men from the best Southern families, I think, are educated in the East. If it were a subordinate position in Johns Hopkins, or Harvard, or the College of Physicians and Surgeons in New York, even, that was under consideration, I should look upon it as the probable beginning of a University career, and although the salary would be small I should consider the opening a good one.

But in West Virginia—while I do not know the University—I should fear that there would be little or no stimulus to do research work beyond that which your own ambition would furnish. I should think the chances about what they would be if you were teaching in a preparatory school, and had an equal amount of leisure. Then as to the leisure you would have, it will be difficult to know in advance. The grade of students in the University of Columbia, Missouri [University of Missouri] is decidedly below that of the undergraduates in Chicago, and I presume that in the medical school of W.Va. you find students who had come up fresh from the country schools of the State. Of course I do not know. I should think, too, that it would not be easy to get a position from the University of West Virginia—I mean a position in one of the leading universities—unless you were to wait until you had published considerable, and had made a reputation in spite of adverse circumstances.

One thousand dollars in West Virginia for the first year would scarcely be as good as $900 in Chicago, on account of the additional travelling expenses. And the second year in Chicago you would almost without doubt make more than in West Virginia. Then, too, you have just got a start in Chicago. You have worked very hard to get a footing, and you have got it. If you are satisfied to work for the Am. Cereal Co., you will soon have become very useful to that company and can probably take a responsible and well paid position. If you want to do purely scientific work you may have an opportunity through your hold with the men of the departments of chemistry and physiology in the University of Chicago to get a really desirable opening in some first class institution. Then it is just possible that Dr. Cowles may have such a position as you desire after another year or so. I should not like to see you leave Chicago for anything less than a good position. By that I mean either a large salary or a chance for making a reputation in science.

Of course there may be advantages in West Virginia that I have not thought of, but I should fear that the work in undergraduate science would scarcely come up to the standard of first class high school work. Of course your own courses might be excellent but they would of necessity be elementary. The only advantage I can see to you is the promise of leisure and facilities for work. About the leisure I should want to be pretty certain before I would consider the situation. There may be a new clean laboratory and apparatus, but would such a University probably have the journals? And if it is so poor as to make the difference between $800 and $1000 a point, would the board be likely to supply money for journals and other things you might want, even if the president were broad minded and generous? Of course I do not know that the institution is poor. It may not be, but so many second grade State institutions are in a struggling position.

Now I do not want to influence you unduly. You do not need to do anything just to please me. I should hate the thought of that. I shall be willing to go where your judgment decides it is best to go.

I can not possibly know just how you feel about the Columbus Laboratory. It may be that for some reason you would feel relieved to get away from it. Of course I understand your desire to have your salary increased, and your feeling that when Dr. Wesener promised the hundred you should have got it at once. But I would not let impatience for money stand in my way. You have been patient so many years that you can afford to wait even longer. Unless you *want* to leave the laboratory, I should think that to leave it for the position in West Virginia was foolish. Of course I have not the remotest idea how safely you can play this new offer if you do not desire to accept it. The men in the laboratory may be willing to advance your salary, but it is only a very few weeks since it was advanced. Perhaps they would gladly pay the $100 at once if they understood that you would leave otherwise. But if you do not acquaint them specifically with your offer they may take it as a bluff, and may think that you will probably take the same means to increase your pay again soon, and *that* they may not care to have you do. And if you do tell them that you are offered only $1000, they will see that you are practically as well off or better financially with them considering the opportunities you will have to make money in Chicago. Some men you can force to do what you want them to, and others cannot be worked that way.

Now please try not to let me influence you too much in this matter. I have only tried to say what I think (after I have expressed my opinion, but not in accordance with my opinion because I hold it).

I have written all this rather hurriedly, but I have had the matter in mind over Sunday.

Even though Laura's reasoning was solid, it was nevertheless missing two ingredients. She had not talked to President Raymond and she could not gauge, as she mentioned, Otto's feelings about the Columbus Medical Laboratory and his professional future. The latter was understandable because Otto himself was not sure of them. His trip to the East crystallized his thinking and settled all doubts.

A preliminary word about the president of West Virginia University is appropriate here. Jerome Hall Raymond was only 28 years old when he assumed his new office in 1897. He had been a professor of sociology and secretary of the extension department of the University of Wisconsin. Raymond held two degrees from Northwestern University. He was only two years out of the University of Chicago where he had obtained his doctorate in sociology. He was highly recommended for his new post by none other than William Rainey Harper, who came to his inauguration along with two other prominent educators, E. B. Andrews of Brown University and Chancellor W. J. Holland of the Western University of Pennsylvania. Theodore Roosevelt, assistant secretary of the Navy, sent regrets that he could not be present.

Raymond brought with him youthful vigor, ideas, ambition, the "know how–can do" spirit that for awhile far outweighed his lack of experience, impracticality, and capacity for long-range planning. With Raymond at the helm,

> West Virginia University successfully moved into the 20th century. . . . During his presidency WVU became one of the first universities to develop the elective system permitting students to select their own courses. He abolished compulsory chapel attendance, strengthened the social sciences, established summer school, developed a modern library, and provided graduate fellowships. Dr. Raymond was responsible for great increases in state appropriations and student enrollment. His four-year leadership from 1897 to 1901 brought about the University's greatest and most progressive changes and prepared it for the years to follow. (8-1)

Obviously, Raymond had chosen the University of Chicago as a model and brashly pushed reforms that led to conflicts with faculty and regents. Patience and taking wise counsel with his peers were evidently not among his virtues.

Raymond recruited new faculty successfully and personally. In Otto he had found not only a potentially productive scientist and teacher, but a man with whom he shared much in common, the new University of Chicago and what it represented, as well as their personal experiences, goals, and youthfulness. In short, they spoke a common language that could easily lead to a permanent friendship. With a man of this caliber heading the university it was no wonder that Stieglitz and Nef had recommended that Otto join his faculty. And Raymond went after Otto knowing full well that he might not stay long, that one of his potential jobs was at the McLean Hospital. No doubt he thought that Otto would find a very promising future there in Morgantown that would bind him to West Virginia.

Having arrived in New York, Otto spent three to four hours on Monday, Aug. 7, with representatives of the American Cereal Co. The rest of the time he spent getting acquainted with the geography of "this remarkable city." At night he would be on his way to Ithaca (Cornell) for the experiment station there, and would return to the Hoffman House on Thursday. He had decided despite all of his effort, not to pick a flat without Laura's presence. After all, she would live in it more than he; and she was certainly more capable of selecting one.

From Ithaca on Aug. 8, Otto wrote that he had spent the whole day at the university and accomplished much for the American Cereal Co. In the evening he would be on his way to Geneva, then on the same day after his visit, he would return to New York City, reaching the Hoffman House about 11 p.m. Wednesday. After a second visit with the company people he expected to go to New Jersey. He enjoyed it all, and felt he was personally developing strong ties with the American Cereal Co.

Otto preferred marrying on Sept. 11. He had worked on the ceremony a couple of times. The words somehow meant more to him now than when he read them in St. Paul.

Laura's brother Percy wanted to go to Andover Preparatory School and join three of his friends, so he would now be tutored. Andover, founded in 1778, had a fine reputation. It was about 30 miles from Boston. Laura was glad that Otto was seeing New York. She recommended that he visit the art gallery in Central Park. Mrs. Benedict could be at her old home in Maine now, but Otto should call anyhow in Middletown. Laura had not seen her since Vassar seven years ago.

Otto wrote on Aug. 14, and reviewed the activities of the past several days. He had not had good luck since leaving Ithaca. In Geneva he had missed seeing the director of the station, but talked profitably with the chemist. The past Thursday he was back in New York City and went to Middletown. No luck there, nor the next day in New Haven. All were gone, including Professor Chittenden. On Friday, he saw Dr. Wakeman (from Otto's days in Berlin) at Green Farms. He stayed overnight, and both of them set out for New York City on Saturday by way of a day steamer up the Hudson to Albany. It was a fine trip. Laura, of course, knew the locale, the scenery. From the boat, Otto did not see much as they passed by Poughkeepsie. He assumed that the red brick buildings north of the railroad bridge were Vassar College. At Albany he and Wakeman went to a poor vaudeville show, and the following morning (Sunday) Dr. Wakeman returned, while Otto continued his journey northward to Port Henry and Lake Champlain. He ate his Sunday dinner at Port Henry.

> I had figured on getting the Ticonderoga-Burlington boat at 2:25 and so to reach Essex, NY at four o'clock. Essex is just opposite Burlington where the Vermont Experiment Station is located. True, the Washington director of all the experiment stations spends the summer at Essex and Atwater was to be there visiting him—up to Aug. 14th. By the time I got to Essex the bird was gone. There were no boats on Sunday, but I took a livery rig at Port Henry driving 22 miles in the afternoon along Lake Champlain. It was a fine drive. The Lake and the Green Mountains of Vermont to the east, and in the west the Adirondacks.

That morning he had met the Washington director and had a short talk with him before the director left. The professor was gone. Otto hoped to catch up with Prof. Atwater by taking a stage the next day going west into the mountains, 27 miles to Lake Placid. Now, Otto had come to Elizabethtown, N.Y., and wrote this letter from there.

> I am sitting on the highest peak of one of the Adirondack Mountains. The horizon on all sides is bounded by an irregular chain of other mountain peaks. Elizabethtown is in the center of the nearest valley below, seemingly in the juncture of six different valleys along which I can see roads (with here and there a house) and in one (directly below the mountain) a pretty little river—all winding themselves irregularly in more or less definite directions—something like the process of evolution, I fancy, because the roads are quite as irregular as the river.

Then Otto wrote a significant passage, indicating his current concept of his position with the Columbus Medical Laboratory:

> I am glad you wrote as you did in regard to Morgantown. I think that I could have forced the Columbus people to give me the $100 all right but the occasion

is scarcely mature enough. Dr. Evans would be pretty sure to take advantage of any later opportunity that might arise for getting even with me and it is better not to get into such a relation with him till I can do so at a smaller risk than is just yet possible. I have therefore written to Morgantown that I cannot accept the position.

They would have to get along on Otto's present income, and try to supplement it with small jobs and wait for a good opening. They would choose their housing accordingly.

Percy's catalog from Andover had arrived. He would have his entrance examination on Wednesday, Sept. 20, so he must leave no later than Sept. 17. Laura wanted Percy at their wedding, so they should be married on the 11th, or any day of that week. They must decide soon, so that the announcements could be sent without hurry and trouble. They must be engraved, which took a few days.

In mid-August, Otto was at Lake Placid. "Caught up with Prof. Atwater on the stage coach today and had quite a chat with him. . . . He is a suspicious, wary individual and did not like to talk very much about his relation to the Am. Cer. Co. because I did not have authority." Atwater thought Otto wanted a job, but Otto showed no curiosity. "I gradually got him to come to terms a little more in regard to the A.C.C. business." Otto would try to see him again the next day, though he might not wish to see Otto. Atwater had gone to a private club so Otto would have to cross the lake to see him again. "I have had a fine coach ride through the Adirondacks and I am now staying at perhaps the finest of all." Otto remarked that he was spending a lot of money. But while it was a fine thing for him, he was trying to benefit the A.C.C.

Laura was sorry to read that Otto had such a chase after Atwater. "I am glad that you have decided against West Virginia. We can live this year without difficulty on your present income, and I am inclined to think *that* will naturally increase this coming twelve months."

Otto had written a letter on Aug. 16, but had not mailed it. He had, after all, had a second and more satisfying meeting with Prof. Atwater for about one and a half hours, then remained at Lake Placid until evening before going on to Plattsburgh and Burlington, Vt. He had written up his interview with Atwater for the American Cereal Co.

Otto's next letter on Aug. 17 came from Burlington, Vt. He had slept on the Lake Champlain steamer at Plattsburgh, then ridden on the steamer that morning to Burlington. He had a very pleasant time with the director and the chemist at the Burlington station. "I shall have a pretty rough time of it tonight trying to get to the N.H. station at Durham. The train connections are poor." Otto regretted the duration of his trip. It could seem like a vacation to the Columbus Lab, and they might begrudge him the week he would need in September. He wished Laura could have accompanied him through the Adirondacks, but she would not have enjoyed the traveling. It was very expensive. He would not return to New York for almost a week so he would write the Hoffman House and ask them to get his mail forwarded to Boston.

On Aug. 19, Otto was in Bangor, Me. "This work is doing me a great deal of good and after the salesmen have made their next round and visited the stations I hope that they will get such a report from the station men as will materially strengthen my position in Chicago." Although Dr. Benedict was within three miles of Bangor, Otto had no time to pay him a visit. His wife was not there. Otto had a letter from Waverley. Both Drs. Cowles and Hoch would be home the following Monday.

On Sunday, Aug. 20, Otto wrote, "I have loafed about Boston all day." He planned to see representatives of the A.C.C. early in the morning, so that he would waste no time in going to Waverley for a visit before setting out for Amherst either at 1:35 p.m. or at 4 p.m. He had written an unidentified long and quite important business letter, and had reread a lecture given by Dr. Cowles last year. He liked it. "There seems to be a true scientific spirit running through it and I am rather curious to meet both him and Dr. Hoch." Otto was now tiring of his rat race. He was thinking of Laura and how little she was going to get in their marriage. Later that night he wrote he had been using the prayer book for the wedding service. He asked that a phrase be left out or changed: "In the name of the Father, and of the Son, and of the Holy Ghost. Amen."

Suddenly, Otto received a letter and a telegram from President Raymond, which he enclosed and which probably prompted the important letter mentioned above. The telegram stated:

> Elected twelve hundred Morgantown soon can arrange matters satisfactorily when will you come? Wire me Morgantown, Jerome H. Raymond.

The letter:

August 16, 1899

My dear Dr. Folin:

I returned this morning from the meeting of our Board of Regents at Parkersburg, and find awaiting me your letter of the 11th which reached Morgantown yesterday, the 15th. As I wired you from Parkersburg last night, you were elected Assistant Professor of Physiological Chemistry, in accordance with the agreement between us in Chicago. I found that I could do a little better for you than I agreed, and so your salary was placed at $1,200 [about $14,500 in 1983 dollars] instead of $1,000, as we agreed.

I am somewhat disturbed at reading your letter. I had understood the whole matter was settled in our interview. After talking with you, I made no further effort to look up a good man in this direction. As for the matters to which you refer, I am confident that they can be arranged. I think you under-rate our University. It is true we have not done any work in Physiological Chemistry here heretofore, but if we embark in this work, we shall expect to have it done properly and to supply necessary tools. It is late now for me to look up another man. Moreover, our Board of Regents will not have another meeting this fall, and I have no authority to employ anyone without the action of the Board. If you withdraw now, therefore, it would mean delay in the inauguration of the work.

I hope you will decide to come to Morgantown soon, so that we may make arrangements. Before the beginning of the fall quarter, we must issue a circular in regard to the Premedical course. In order to do this, it is necessary for us to be advised fully in regard to the Physiological Chemistry.

Hoping to see you here soon, I am,

Yours truly,
Jerome H. Raymond.

"As the situation now stands," Otto wrote, "I think that I must go unless the Chicago people come to terms. The president means to make the University a success and I will not hesitate to go."

Otto found Dr. Hoch, but not Cowles at the McLean Hospital. He liked Dr. Hoch and the appearances of the place very much. It was only a 12-minute ride from Harvard. Otto would go back for a visit upon his return from Amherst.

Laura received Otto's letter from Burlington on Aug. 21. They would get married even if the lab missed him, would they not? Otto could leave on Saturday night and be back in Chicago, Tuesday morning, and if necessary, go to the lab on Wednesday. It would not be so nice for Otto that way. Laura could see to getting the flat in order.

"Madeleine was very urgent in her invitation for me to stay with them while we were getting our flat settled—and of course it will save a great deal of money for us—still I should scarcely like to do it just to save money." It was certainly kind of the Sikes to invite them. She wanted Otto to answer frankly. She also asked Otto not to buy her an extra ring, reemphasizing that she cared little for jewelry.

From Amherst, Otto noted on Aug. 22: "In an hour I return to Boston. I shall get there about six and after supper am going out to the McLean Hospital to stay with Drs. Hoch and Cowles (pronounced Coles) overnight." If he could get 12 or 13 hundred, he would be willing to work there. However, even more salary was possible. Dr. Hoch had impressed Otto very favorably and advised him on Dr. Cowles so that he could more readily understand him.

"Last night I wrote to Dr. Evans of the Columbus Laboratory. I also sent the following telegram: Have been appointed physiological chemist University of West Virginia. Shall accept unless Columbus Laboratory can pay me 100 a month." In the accompanying letter, Otto demanded "100 a month from the first of Sept. to April, 1900 and $125 after that, and in addition 20% of the revenue obtained through any new combination or commercial products that I am chiefly instrumental in bringing out." Otto had a copy of this letter in case the Columbus Lab had not notified him by next Friday that they agreed to his proposition, then he would telegraph his acceptance to Morgantown, and "I shall run down there when I visit Pa. . . ." Because Raymond's telegram had reached Otto in the office of the American Cereal Co. in Boston, he had explained to the head representative its importance, and his present relationship to the Columbus Laboratory. "If my connections with the laboratory should be severed I shall send the company (A.C.C.) a copy of the letter I wrote this morning so that they cannot blame me." Otto still expected the Columbus Lab to accept his terms. If so, he would try hard to gain "good understanding" with Professor Evans as rapidly as possible. Otto had also asked for a vacation during the week of Sept. 10–17, and in case of separation he set his final date at Sept. 9. "In either case we can marry Sept. 11th and then either go to West Va or to Chicago."

We must note two developments in the undertow of the last few letters. Raymond's offer of $1200 must have swayed Otto's thinking toward West Virginia. This was the first concrete offer for him to begin functioning as a physiological chemist, and Otto had made a serious pledge to come. To continue with the Columbus Laboratory required that they show better faith than they had in the past; so he raised the ante. Unfortunately for this negotiation, but not for Otto, the main power in the ownership, Dr. Wesener, was evidently still out of the country.

On Aug. 23, Otto noted, "I am sitting at the station waiting for my train to Kingston, R.I." He had returned the past evening to Boston, but did not go to Waverley because of the rain. This morning he had gone to the McLean Hospital and stayed for dinner (lunch). He had a pleasant time with Dr. Hoch but spoke very little with Dr. Cowles. Otto felt that his two visits had converted Dr. Hoch from a lukewarm attitude toward him to one strongly in favor, and he had said as much to Dr. Cowles within Otto's hearing. Dr. Hoch told him that though he had no influence on whether or when a physiological chemist can be procured, he had much to say about who would be hired for the place. Otto could feel certain of getting the place if he wanted it at such a time.

"All this makes it decidedly possible that you and I will be in New England within a year from now—and if I do not go to W.Va. it may happen quite soon." This would have some bearing on their flat arrangements in Chicago after they had learned whether they could be in Chicago or West Virginia. Otto would write Dr. Hoch and "simply state the case and ask whether he can offer any opinion in regard to when an arrangement can be made or whether it would be advisable for us to consider this in our flat arrangements." In short, Otto needed Dr. Hoch's view on *when* they expected to hire him, as this could influence whether he would stay in Chicago for the time being.

Laura wrote to Otto at the Hoffman House. She was no doubt amazed by the turn of events, and particularly by Otto's notes on Morgantown.

> I shall be content either there or in Chicago. If we do go to Chicago it will mean simply that we shall be on a much better financial footing than if this offer had not come up. I hope that whatever turns up you will be well satisfied. The University of West Virginia may be much more promising than I supposed, and if it is not a very good institution you need not stay more than two years. By that time you can go to Waverly [sic] perhaps, or to the University of Chicago.

When he wrote from State College, Pa., on Aug. 26, Otto was beginning to have second thoughts about his recent actions. "Am rather sorry about it, but if the Columbus Laboratory or rather Dr. Evans cannot be fair there is no point in remaining there. I am not sure but that a year in a small place might be good for me."

On the next day, Otto enclosed President Raymond's letter (cited above). Otto belatedly (he thought) remembered that Laura's birthday was nearing, and he avowed that he had been so preoccupied that he forgot their impending marriage. However, both the birthday and marriage could coincide on Sept. 11.

> I am not entirely happy over the outlook. I had to act too rapidly and may not have taken the most judicious course. If I could have waited until after I had seen Dr. Cowles I should no doubt not have written to Dr. Evans as I did but rather waited for a better opportunity.
> [I] resent his $900 decision and the $1200 offer together with the urgent letter from Pres. Raymond and after discussing the situation with the American Cereal representative in Boston made me decide to force the issue at once. Evans made a curt answer as I should have known that a man of his pugilistic qualities would, especially as my letter was not very politic, but on the contrary rather challenging. When I received Evans' letter I showed it and a copy of mine to the Am. Cereal representative in New York. He has written to Chicago and asked me to delay acceptance of the Morgantown position till Monday evening. The matter will therefore be taken up Monday in Chicago. The right head men are, however, unfortunately not home from Europe yet and it scarcely looks as though the Am. Cer. Co. could demand of the Columbus Laboratory that I be retained.

This letter from Otto revealed that just before he left on this trip, the meeting of the owners of the Columbus Laboratory had decided to raise Otto's salary from $720 to $900, rather than $1200 per year, as Otto had been led to believe by Wesener. This was a raise that Otto had expected in mid-May, and definitely by the first of June. We must bear in mind that the Columbus Laboratory was owned and operated by physicians who, with the possible exception of Wesener, were contracted to do analytical chemistry and research work that was apparently outside the area of their competence. Their decisions were purely monetary, and income of this source had nothing to do with their medical background.

Otto's regret was that he might as well have stayed in Chicago because the Waverley job would be available within a year. On Aug. 27, he wrote from State College that he had been working on a report of his visit to the last three stations. He expected the Chicago matter to be settled by tomorrow night. Then he would visit Morgantown before going to Washington, D.C., or to New York. If Morgantown were as pleasant as State College, then he was inclined to be philosophical and think that a year there would be good for him, though "I feel a little guilty on your account." Laura would be happier in Chicago. He asked her not to worry. Regardless of what happened, they would be married on Sept. 11.

Otto expected to part company with the Columbus Laboratory on Sept. 9. He hoped they would pay him the salary due for the past two months. They owed him $83; and Sikes owed him $30. His cash on hand was $30, so "I shall be exceedingly poor and can buy nothing." That was why he had pushed for $1200. He was frustrated by seeing men of lesser training who earned two to three times his pay. They could, however, live economically in Morgantown.

Back in New York on Aug. 29, Otto wrote:

> I have accepted the Morgantown position because no word came from Chicago and I could not hold off any longer—and I have thought it over, and while I sometimes regret it some I am on the whole quite satisfied. I shall really be more in touch with scientific men from there than as a commercial man and I am inclined to think that we shall find it a pleasant little town. I am almost sure that a year at such a place will be good for me.

After mailing his letter to Evans, Otto received one from Wesener telling him that he was wanted by the Harvey Medical School to teach four evenings a week. He could, therefore, have earned more than $900, possibly more than $1200; but it would be hard work from 7 to 10 in the evenings, and would make it impossible for him to keep up with the literature.

References

8-1.(a) Dawson, J. *WVU—An Early Portrait*. Limited First Edition. Morgantown, W.Va.: West Virginia University Archives, 1971.
 (b) Ambler, C.H. *A History of Education in West Virginia*. Morgantown, W.Va.: West Virginia University, Archives and Manuscript Division, West Virginia Collection, West Virginia Library, 1951.
 (c) Doherty, W.T., Jr., and Summers, F.P. *West Virginia University—Symbol of Unity in a Sectionalized State*. Chap. 4, pp. 59–78. Morgantown, W.Va.: West Virginia University Press, 1982.
 (d) Catalog, 1899–1900. Morgantown, W.Va.: West Virginia University.

Chapter 9. Marriage and Morgantown (1899–1900)

On the last of August, Otto was traveling.

> I am now on my way to Chicago via Morgantown. A telegram from Pres. Raymond asks me to. It will cost me eight or nine dollars extra to do it but then it will be very satisfactory to have seen the place and the men with whom I shall work. I shall also not forget to inquire what arrangements can be made for living comfortably and economically. A N.Y. acquaintance told me it is a very pretty but sleepy place. Goodbye.

As Otto was moving out of Pittsburgh on the westbound train to Chicago on Sept. 2, he wrote about his visit to Morgantown.

> The town is small (3500) nor does it look attractive from the train. Once there, however, it is not half bad. It is useless to go into details on a shaking train but it is not bad. The university is also relatively small, but growing rapidly under the present administration. The buildings are also new, but the chemistry department has exceedingly good quarters and moreover is astonishingly well equipped. I am decidedly pleased with it. One reason for this condition is I presume the fact that the chemist, Prof. Whitehill, is also treasurer of the University. While in Morgantown we outlined the chemical courses to be given next year and I am very well satisfied with the arrangement. I shall have full charge of qualitative analysis and of the medical chemistry. I shall have medical chemistry 2 hours a day, 5 days in the forenoon and the analytical classes 2 hours each afternoon, four days a week. I shall therefore by no means have a very hard year though some or all of the work will require considerable preparation the first year—as any course really does the first time you give it. Prof. Whitehill estimated that I shall have not over 15 in each of my analytical divisions and probably only 7 or 8 in my medical courses.
>
> I shall get practically all the literature that I need and shall have about $350 at once to spend for such chemicals and apparatus as I shall need for my courses. Really for me personally everything looks very much better than I expected. The literature I ordered will probably cost at least $250 possibly $350. Prof. W. means to be exceedingly congenial and is I believe a very kind man.

Marriage and Morgantown 143

But Otto was dejected when he thought of Laura's interests. What program of activity could she make for herself? Keeping house was out of the question. A building boom was on. Too many people; too few houses; no hired help available. A large number of small families boarded, particularly university people. The board was fine. The best hotel in town gave table board at $4.00 per week. Rooms were costly and difficult to find at any price, especially rooms with modern conveniences. But Otto had one "strictly fine" place in view, only he had to decide at once whether to take it because of other parties' wanting it. Professor Whitehill had found the place and had accompanied Otto to see it.

The place that Otto would choose had five furnished rooms on the second floor all of which would be rented out. Otto could rent one *very* large and very pleasant room and a smaller one across the hall, the latter a very sunny room, for $20 a month. On the same floor was a perfectly fine bathroom (to be shared). The rooms were all heated by open grate fires fed by means of natural gas. Otto was quite pleased. The bedroom or rather the smaller room had two windows on the south, the larger room one window north, the other east. The furniture of the whole house was fine, but the beds were heavy, clumsy, European looking things.

The place was expensive. Two of the other rooms would be taken by another young instructor and his wife (from Ann Arbor) although they were going to try hard to find a house. Otto was referring to their future friends, Dr. John Black Johnston and his wife Juliet. John had obtained his doctorate in zoology just that year, after six years as instructor at the University of Michigan, and Juliet had gotten her M.S. in the same field, the previous year. They were newlyweds, as Otto and Laura would soon be, and about the same in age.

If Otto were to rent the two rooms, and they boarded at the hotel, they were really getting about the best that money could buy in town. It would cost them a total of about $55 a month, somewhat high but there was little else around to spend money on, so they could save almost all of the rest! Otto could have rented the larger room alone at $12 or $13, but he did not like to start out in one room. It would be good to arrange things so that they were not compelled to be in one all the time. Did she agree?

There would now be no need for Otto to return to Morgantown before Oct. 1. His letter to Dr. Evans had placed his *latest* resignation as Sept. 9. He did not know yet what the reaction of the Columbus Laboratory would be. They could demand that he leave at once, if they felt ugly about it. This did not matter so much, unless she wished to change the date of their marriage, provided the announcements were not already made. If not, Otto could come on Sunday, or perhaps visit his brother in Center City, and they could marry at their leisure later in the week, spend the rest of the week in Chicago, and on the following Monday, proceed to Morgantown. Otherwise, Sept. 11 was fine and they would plan how to spend the following two weeks.

Otto wanted Laura to write him at once whether or not he should rent the rooms in Morgantown. By the time he received her letter in Chicago, he would also know when his last day of work would be, and he could telegraph her, if necessary, of any point she might wish to know.

Once back in Chicago, Otto had a long talk with Wesener, who had returned from Europe. Wesener regarded Otto's proposition as a bluff and still

> ... expects and wants me to come back. He insists that my demands were put in such form that no firm would have agreed to them. He now wants to give me $75 and a commission on what I do outside the American Cereal and wants me to see the dean of the evening school the first thing tomorrow morning, i.e. before seeing Evans. Well, I will do that to hear what she can offer me for fun but I can anyhow scarcely change my mind now.... Wesener thinks I will virtually get $100 but the appearance must be that Evans won.

Otto did not want to change his mind about West Virginia but it could be done, if warranted, because President Raymond had agreed that he would assist Otto should he get an atttractive offer elsewhere. He would not blame Otto, in other words, for looking elsewhere, but would recommend him. This did not really mean that the president would release him from his present obligation to take the position in Morgantown. But Otto, of course, had signed no contract.

The next day, Sept. 4, Otto stated,

> Evans, I and Wesener talked a short time ago. Evans said that my insinuations of bad faith on his part was a lie. I said there was bad faith somewhere and afterwards told Wesener that it lay with him which he practically admitted. This closes the affair as far as Chicago is concerned. I think and I hope that your announcements are already printed.

On Sunday, Sept. 3, Laura wrote that she had delayed as long as possible but finally ordered 425 wedding announcements. A neighbor who had been in Morgantown to sell law books at the university told her that it was a very pretty place.

> Mother likes the idea of our going there very much. The thought of hills and a river with a steamboat pleases her. ... As for myself I cannot but confess a lingering fondness for Chicago, but I doubt not that rural life will be much better for me physically, and I shall like a University town well enough, I am sure.

They would be married on Monday the 11th so Otto *must* turn up on Sunday the 10th—just a week from today. "I want to stop over in Chicago a few days if it can be managed as well as not—or rather if it will not be very inconvenient or expensive."

> We should stay with Madeleine while in Chicago. I should certainly think so. ... When does the University of West Virginia open? Probably you will tell me that before you answer this. Of course we could not start earlier than the eleventh. But if it opens so soon or not to allow a day or two or three—it might take three—in Chicago, please let me know *at once*, for in that case I should get another pair of glasses here. I can not get along without them at all. And it always takes a couple of days to get them made—or nearly always. [Laura had broken her glasses and had expected to replace them in Chicago].

Laura was packing several boxes and trunks for shipment by freight. She urged Otto not to worry about money. Rather than borrow, it would be much better to use hers until he got his first pay. She gave Otto detailed instructions on what to write the minister.

Otto was staying with the Sikes following his return from the eastern trip. He had written Morgantown and asked for the two rooms he had reserved. "I shall now sever my connections here as rapidly as possible but was told at the Am. Cer. office that of course the managers want to talk personally with me in addition to reading my written report." He was now presumably writing the report that had been practically finished before his return.

Otto's last premarital written communication to Laura was on Sept. 7, 1899. "Unless you get further notice to the contrary I shall reach St. Paul Saturday morning." The announcement of Otto's and Laura's marriage appeared on page 4 of the St. Paul *Pioneer Press*, Tuesday, Sept. 12, 1899.

Folin-Grant

Otto Folin and Miss Laura Churchill Grant, daughter of Mr. and Mrs. George J. Grant, were married at 6 o'clock last evening at the home of the bride's parents on Holly avenue. Rev. Maurice D. Edwards, of the Dayton Avenue Presbyterian Church performed the ceremony. Only relatives were present. . . . Mr. and Mrs. Folin have gone to Germantown [sic], W.Va., where Mr. Folin is on the faculty of the university.

The newlyweds took the evening train overnight to Chicago on their wedding date of Sept. 11 where they went directly to Madeleine, who expected them. Madeleine had not come to the wedding, but more importantly, provided them with economical hospitality in a setting they both knew, the Sikes flat. More than likely, George was again in Springfield covering news of the state legislature. They had 10 days before they arrived in Morgantown on Friday, Sept. 22, eight of which were spent in Chicago, a city of fond memories to them for the rest of their lives.

Fifteen days after her marriage on Tuesday, Sept. 26, 1899, Laura Folin, lately of St. Paul, Minn, wrote her friend, Madeleine Sikes, the following detailed letter. We cite it all because there would be no other from her for the next year, and moreover because it gives a precise picture of the first housing the Folins would share, and a glimpse of the life that they would lead (*9-1*).

Dear Madeleine:
Otto and I are sitting at a big table in a large corner room with a gas fire burning in the grate (for it is a rainy day). We are both writing letters. Yesterday I wrote six letters or notes, some reasonably long, and this morning I have written one. After I write you I am going to copy three which Otto and I "composed" Sunday evening. He helped on all the difficult ones. He still continues to be *very* useful, and I believe that he is fast becoming ornamental as well. Yesterday he made a towel rack of this beautiful description only it stands upright which the picture does not. [Laura provided a diagram]. We had a hard time arranging for places enough to put our clothing as we had only two exceedingly shallow closets—not room enough in both to make one as large as that in the room I had at your house [flat]. There were some hooks on the back

of the enormous bed which I occupy, and we pulled the bed nearly four feet from the wall and Otto put in a strip of wood to which were attached a dozen hooks. My trunk stands in the extreme end of the improvised closet and most of our every day clothes hang there, also laundry bags. The towel rack already referred to also stands just within the closet behind a large wash stand which, with its high glass, screens in the closet up to the headboard. This arrangement is so good that I am very proud of it. Here is a picture of the front view.

Our sitting room we cannot fix up until the end of the week. Our landlady provides nothing in the line of book cases, so we have ordered one wooden—$2.25 it will cost without paint or staining. It will doubtless come to $3.00. This I think a good bargain because the bookcase is to be 6 ft. 9 in. long and is to have about 27 ft. of shelving, and is nearly, as we could measure our books which are piled on the floor, about 30 ft., but there is a small arrangement of shelves to be set on a table, so we shall get along all right. I do not like the paper in our sitting room so we are going to put burlap on the wall above the book shelves. That arranges for one wall, after considering that there is a door in it. (The picture tells all we have decided on about the sitting room so far).

We did not reach Morgantown until Friday morning. Our train was too late Thursday to allow us to catch the 11:15 train from Benwood Junction where according to our plans we should have changed cars. So we had to go on to Wheeling, where we spent five hours. Wheeling is an exceedingly picturesque place. We walked an hour and a half across one of its hills. The view was wonderful. After three hours more on the train we reached Fairmont where we spent the night. The next morning we proceeded to Morgantown where we arrived at 8:30 after an hour's ride. The country through which we passed during the last few hours of our journey is remarkably beautiful—hills on every side. And Morgantown itself is very attractive. It is built on hills. We have walked about quite a little already and like the place. The board at the hotel is strictly first class.

So far you see we have lived a peaceful contented life. May it continue!

It is just twelve o'clock and as I must do a little more writing before lunch, I shall say goodbye for a short time to you.

I hope you know how very good you were to us while we were in Chicago. We both thank you very truly for all you have done for us—individually and collectively.

Write soon, if you have time. With best regards to Mr. Sikes.

As ever,
LCGF

The *Record* we read religiously.

"The Opportunity of the Democratic Party" might also have served as an occasion for an expression of approval addressed to Mr. Dennis. (This last sentence is Otto's. He liked the article [probably George's] very much). With much love. Laura

On Oct. 23, 1899, Otto wrote the following letter to Professor Loeb (from the Manuscript Division, Library of Congress, Jacques Loeb):

Dear Prof. Loeb:

I had no opportunity to see you before leaving Chicago or before accepting the position offered me in the University of West Virginia. The offer from this place came rather suddenly and I did not at first mean to accept, but intended, in fact, to use the offer simply as a means of forcing the Columbus Laboratory people to live up to their early agreement of paying me $100 a month.

As it turned out, however, the laboratory did not want to pay me $100 but expected me to earn the extra money by teaching evenings. President Raymond of this University on the other hand proved to be a very satisfactory man and made it possible for me to give most of my time to physiological chemistry by making it distinctly my department. I have already ordered some of the more essential literature, such as Maly's Jahrsbericht der Thierchemie and sets of Hoppe Seyler's Zeitschrift and of Schmiedberg's Archiv f. Pathologie u. Pharmakologie. And as I shall have access to the "Berichte" and several other chemical journals at the experiment station, also located here, I am not badly off as to literature.

I have further a well arranged and quite well equipped laboratory, so, considering everything, I am not sorry to have left the commercial laboratory in Chicago.

I shall give two courses, both extending through the entire year; one in analytical chemistry calling for two hours work during each of four afternoons a week and the other in physiological chemistry occupying eight fore-noon hours per week.

The latter course is intended for such premedical students as may want to prepare themselves here in certain laboratory studies, so as to get credit for these in the leading medical schools of the country.

I have ten students taking my physiological chemistry. In addition I shall probably later have a certain number of "domestic science" students but their work will be made to fit in with the other and will simply mean that I shall go rather extensively into the chemistry of foods, digestion and nutrition.

On the whole I am fairly satisfied. I may not be able to do much research work the first year but I shall hope to keep somewhat in touch with considerable of the current literature and to work up two sets of lectures—one on physiological chemistry in general and another one on the special chapters of foods and their relation to animal metabolism.

My purpose in writing you this letter aside from letting you know how I am situated is to ask whether you would consider it feasible or advisable to try to get an opportunity to give either or both of the above lecture courses at the University of Chicago during the next summer quarter? The reason why I ask you so soon in regard to this is that it would naturally make my lecture work here all the more interesting if I thought that I might give the same courses to another set of students in Chicago.

I have now a living salary and am reasonably pleasantly situated so I shall not be in any great hurry to leave, but of course I shall still be looking for better things.

Very sincerely
Otto Folin

Laura's apprehension about the probable lack of modern facilities in Morgantown was ill-founded, fortunately for her. She was hardly the backwoods type. What she found was doubtlessly a pleasant surprise. Not only were there paved streets, sewerage, gas, electric lights, telephones, and a copious supply of filtered water, but there were commendable hotels, homes, churches, and several public institutions. A greater surprise came to her when President Raymond asked her to teach mathematics to two "overflow" classes of freshmen.

Because of the rapid burgeoning of faculty, students, employees, construction workers, and new buildings at the university during the Raymond regime, there was consequently a severe housing shortage as well as a building boom in town. But an additional factor significantly stimulated the growth of the community. Morgantown was (and remains) the county seat of Monongalia County, a region holding some of the best oil fields and coal veins in the country. Strengthened by the presence of great fields of cheap natural gas as fuel, and nearby sand and limestone deposits, it was becoming a manufacturing center of bricks, woolen textiles, and most particularly of glass products. The population in 1899 was about 3500, but within three years it had doubled (9-2).

The university was beautifully situated on the bluffs above the Monongahela River. The campus consisted of about 25 acres, and the university also owned an experimental farm of about 100 acres.

Six brick buildings comprised the teaching facilities in the "square" area known as Woodburn Circle, and included the Agricultural Experiment Station.

Otto worked in the department of chemistry, on the second floor of the Science Hall, built in 1893. The four-story building also housed the departments of physics, geology, civil engineering, and drawing and painting, as well as the president's office. Under construction on the campus were the Armory, the Library, Mechanical Hall, an addition to Woodburn (University) Hall, and the beginnings of an astronomical observatory.

The chemistry facilities consisted of two large laboratories and a lecture room, with an associated weighing room, preparation room, and a darkroom. The laboratories provided for 50 students. Otto mentioned his delight at finding how well equipped the laboratories were with modern apparatus.

The faculty of the university had grown from 30 in 1896 to about 69 in 1900, and the student enrollment from 465 to 885 (9-3).

Professor Alexander Reid Whitehill (1850–1921) had joined the university faculty in 1885, in the chair of combined physics and chemistry. The following year the two sciences were separated and he became the head of chemistry. He had received excellent training in the physical sciences. He received his A.B. from Princeton College (later, University). After graduation, he won an award, the Experimental Science Fellowship valued at $600, based on an examination in chemistry, physics, and geology, and delivered the commencement oration on geology. Whitehill then worked for his Master of Arts degree which he also obtained at Princeton, on an "honorary" basis, in 1877. During a good part of this time (1874–76) he was abroad for travel and educational experience. He was a graduate student in Germany at Leipzig University and the Freiburg School of Mines. Despite his fine academic background, he did not obtain a doctorate degree but was awarded an "honorary" doctor of philosophy in 1887 by Washington and Jefferson College.

Whitehill assumed his position at the University of West Virginia in 1886. Although not research-minded, he was fond of writing on science, and he kept up with progress in the chemical field. He had assisted in organizing the West Virginia Agricultural Experimental Station and authored its second bulletin in 1887. He had written a history of education in West Virginia in 1889. For 15 years he served as treasurer of both the university and the experimental station (9-4).

As Otto noted, the chemistry laboratories were surprisingly well equipped, and there is every reason to believe that he was too busy at this time to tackle a research problem. President Raymond had already taken strong action to encourage graduate work. He had appointed nine fellows for the 1899–1900 academic year and each fellow was awarded $300. For the following academic year there would be a fellow in chemistry, and Otto, had he stayed at West Virginia, would have had his first graduate student.

For a mere two-man chemistry department, one must be astounded by the variety of subjects offered, or promised at least, in the 1899–1900 catalogue. Not only were the usual elementary courses listed in inorganic and organic chemistry, qualitative and quantitative analysis, but specialized courses were also available. Professor Whitehill taught courses in analysis of coal, coke and water, iron, ores, pig iron, and steel, and probably the chemistry of everyday life. Otto offered qualitative and quantitative analysis for students making chemistry a specialty; toxicology especially designed for medical students; water and food analysis, analytical chemistry applied to domestic science, an extension of toxicology, but including analysis of water and foods; medical organic chemistry, special subjects of organic chemistry such as carbohydrates, fats, proteins, and foods in general, together with certain organic amido and other nitrogenous compounds; physiological chemistry, a continuation of the above, with lectures and laboratory work on ferments, digestion, respiration, nutrition, and general animal metabolism; medical analysis, a study and practice of all the essential analytical methods as used in the modern practice of medicine; organic chemistry, open to students of pharmacy only, and consisting largely of laboratory courses in organic preparations, and including writing of abstracts and discussion of recent chemical literature. Finally research work was offered for the various chemical specialties, with credit to be given in proportion to the amount of work accomplished.

Otto must have been hard pressed not only to prepare his lectures at once for medical chemistry, but also to put the students' laboratory exercises into immediate use. Fortunately, the texts by Bunge, Hammarsten, and others gave him a comfortable, familiar reference base for didactic material. His course in quantitative analysis should have been somewhat easier to give because of the well-equipped state of the laboratory. With so much need for preparation of his lecture and laboratory outlines, it is doubtful that Otto had any thoughts of launching a research project during his first year. Besides he undoubtedly wanted to come home to his wife, who was now pregnant.

Otto's classes were small by today's standards in state universities. We can assume that during the fall term his classes in qualitative and quantitative analysis had about 15 students each, as Professor Whitehill had estimated; and Otto had written Professor Loeb that he had 10 students in the medical organic chemistry course. Because there had been little time to order the necessary chemicals and to prepare the reagents for the latter course, Otto was obviously forced to improvise, demonstrate, or merely explain the basis for some of the laboratory exercises. Fortunately, Otto's experience as a teacher in the Chicago veterinary school and at the Ascham School for Girls paid off and the small classes soon familiarized him with each student's capability.

Above the rest of his students stood the youngest and the brightest among them, Philip A. Shaffer (1881–1960). Otto and Philip would become lifelong friends. From publications they shared later it seems obvious that Shaffer was highly adept and painstaking at quantitative analysis; and it was Otto's introductory course in biochemistry that stirred his interest in that new field. More than likely, had it not been for his mentor, Philip would have gone on to medical school and emulated his maternal grandfather, Philip Williams Anderson, for whom he was named.

Philip Shaffer, a native of Martinsburg, W.Va., had been precocious, having begun reading at age three. This led him into later difficulties because of the expected lack of challenge in public school. Fortunately he was taken in hand by a new teacher at the Martinsburg High School, a recent Harvard graduate, who tutored him, and maneuvered him through high school so that he could enter West Virginia University shortly before his sixteenth birthday in 1897 (9-5).

Sometime early in the spring, Otto received the momentous letter from Dr. Edward Cowles, asking whether he would now take the position of research chemist at McLean Hospital. Cowles was now assured that the position would be ready by summer and they had only to determine Otto's salary and when he could start. As it turned out, he was offered a salary equal to that of Dr. Hoch, $2000 (about $24,000 in 1983 dollars) per year. Of course, Otto accepted both the job and the salary. He could not leave Morgantown before the end of summer quarter in fairness to President Raymond. In addition, Otto asked that he be given a research assistant, and if so, he had a bright young student in mind. This was granted.

One can imagine the latent excitement of both Laura Folin and Philip Shaffer at Otto's news. Laura would not only bear her first child, but now Otto's golden moment had come for both research and security in an environment that would assuredly provide both of them with the benefits of urban life in one of the most intensive intellectual centers of the U.S.

But for Philip Shaffer, not yet 19, it must have seemed a miracle. He had talked on frequent occasions with Professor Folin about chemistry, and his vague plans for the future, perhaps as a physician. But the physiological chemistry that Otto was teaching interested him greatly. Phil's academic record at West Virginia University was outstanding by any standards, particularly in the scientific course he was pursuing. During his first year, Phil not only excelled in his class work, but had been a top performer in the military cadets. When the Spanish-American War began, he had hoped to be called to active duty, but his age and his parents fortunately bottled that patriotic fervor. Now he was equally fired up by the vista of research and study that his professor outlined.

Otto was obviously drawn to this talented, outgoing young scholar, whose youth belied his serious desire to achieve his mark in science. Would Philip like to have a position as research assistant to the professor, in Waverley, Mass.? If he liked, he could enter the graduate school at Harvard, only a few miles away.

By taking a pair of courses during the summer quarter at West Virginia University, Shaffer could complete his academic requirements for graduation at the end of the summer term.

Otto advised him to strengthen his language training in German if he wanted to pursue chemistry, and though Shaffer had already taken extra coursework in German previously and also had credits in Latin and French, he concentrated on advanced German during the summer of 1900.

Of his experience in West Virginia with Otto Folin, Philip would one day record the following (9-6):

> In that small institution, perched on a hill overlooking the Monongahela River, Folin gave during one year a course in quantitative analysis and another in elementary physiological chemistry. The writer of this memoir had the good fortune to be a student in both courses. The text used in the second course, the first Dr. Folin had given, was a laboratory manual by F.G. Novy; Hammarsten's book (translated by Mandel) was used "for reference." Many of the exercises in Novy's volume were omitted; the margins of my copy bear notes "no microscope," "apparatus not available"; but the omissions were more than compensated for by the spirit of inquiry aroused in several of the students who voluntarily worked overtime on small tasks they were encouraged to think of as "research problems."

As Laura's confinement drew near, her mother came to Morgantown to help. Having had six children of her own, she knew what was needed to make this moment more bearable.

Joanna was born on Aug. 8, 1900. Whether this occurred in the rooming house where the Folins lived or in some local medical facility is unknown, but the former possibility is more likely. Children were born at home and not in hospitals. As Otto would later describe this period, it was hardly credible—an unreal moment that he could scarcely grasp, despite the fact that his own mother was a venerable midwife back in the old country.

Otto had written the Sikes about his taking the new job at Waverley, and of their immediate plans. Madeleine no doubt promptly answered, asking them to stay with her while they were in Chicago—at least for several days while she became acquainted with the newest Folin, and they all had a solid visit.

References

9-1. Madeleine Wallin File. Archives. Chicago, Ill.: The Joseph Regenstein Library, University of Chicago.

9-2. Grant, H.L. *Greater Morgantown and Its Environments*. Morgantown, WV: West Virginia University, Archives and Manuscript Division (West Virginia Collection, West Virginia University Library), 1902.

9-3. (a) Dawson, J. *WVU—An Early Portrait*. Limited First Edition. Morgantown, W.Va.: West Virginia University Archives, 1971.

    (b) Ambler, C.H. *A History of Education in West Virginia*. Morgantown, W.Va.: West Virginia University, Archives and Manuscripts Division (West Virginia Collection, West Virginia Library), 1951.

    (c) Doherty, W.T., Jr., and Summers, F.P. *West Virginia University—Symbol of Unity in a Sectionalized State*. Chap. 4, pp. 59–78. Morgantown, W.Va.: West Virginia University Press, 1982.

(d) Catalog, 1899–1900. Morgantown, W.Va.: West Virginia University.
9-4. (a) Dr. Alexander Reid Whitehill. *Biographical and Portrait Cyclopedia of Monongalia, Marion and Taylor Counties, West Virginia.* pp. 45–46. Philadelphia: Rush, West and Co. Publishers, 1895.
   (b) Alexander Reid Whitehill. *History of West Virginia Old and New.* Vol. II, Biographical, p. 508. Chicago and New York: The American Historical Society, 1923.
   (c) Whitehill. *Genealogical and Personal History of the Upper Monongahela Valley, West Virginia.* Vol. II, pp. 478–81. New York: Lewis Historical Publishing, 1912.
9-5. Doisy, E.A. Philip Anderson Shaffer. *Biogr. Mem. Natl. Acad. Sci. U.S.A.* **40**, 321–26 (1969).

Chapter 10. McLean Hospital—Tooling Up (1900–1903)

Sometime about the week of Sept. 22, 1900, the Folins bade farewell to their friends in Morgantown. Otto had packed their trunks to be sent on to St. Paul. On the train ride to Chicago, Otto must have marveled at his good fortune since he had left Chicago. A productive year! A family, a developing career, and a living.

After a brief unrecorded visit in Chicago that week, Laura and Joanna left for St. Paul on the Thursday evening train, Sept. 27, 1900. The next day, Otto wrote,

> I do not leave the city until 11:30 tonight and therefore will not reach Boston til 7:30 Sunday. I wish I had known this and kept you until this evening. . . . By this arrangement I save the Saturday night's lodging in Boston and I get the ticket for $15.50 so that I am coming out very well. . . . Mrs. Sikes stopped my paying one cent for my own board but did accept what I had already paid Sikes and two dollars more as payment on the note due in October. I am therefore again richer than we thought. . . . I hope that you arrived safely at 462 Holly this morning. I felt pretty glum last night but was so sleepy that I never woke til a little before six in the morning and today I have been "on the go" all day.

Laura also wrote on Friday, addressing the letter to Otto at McLean Hospital. "Joanna and I are perfectly well, and so the essential thing to tell you is told."

At home, Laura had found that her father had been gravely ill shortly after her mother had reached Morgantown—so ill with "gastric fever" that they thought he might die. He had a trained nurse. Her mother was not notified of this because they feared "it might hurry up the arrival of the baby." Her mother had been ill with dysentery on her return home. At the moment, however, things were going smoothly.

On Otto's arrival in Boston, he was well provided for. He was temporarily given a room in the Service Building of the McLean Hospital with Mr. Shaffer as a roommate. Housing was apparently abundant in the area. The location was beautiful! But it was quiet, probably too quiet for Laura. She would probably prefer to live in Cambridge.

The McLean Hospital where Otto would spend seven of his most productive years was and remains a venerable private institution that from the early part of the nineteenth century had its beginnings conjointly with the Massachusetts General Hospital and is an integral part of the hospital's corporation. The first mental patient in the "asylum" at Somerville was admitted in 1818. Thanks to an adept administration, and in particular to the accomplished, visionary leadership of Superintendent Edward Cowles, the hospital served as a model of enlightened care for the mentally ill. Although patients were selected from the "middle class" for admission and paid for their keep, much was done for charity. From 1883 on, more than 40% of all admissions were voluntary.

The hospital at Waverley was relatively new, having opened in 1895. The grounds then comprised 176 acres, growing within a few years to 317 acres. The patients had been transferred from the 77-year-old Somerville facilities, which could no longer withstand the encroachment of the railroads. At that time (1895) the hospital could house about 180 patients. It bears a strong tradition as a teaching and research institution. In 1900, when Otto arrived, it had a physician staff of seven, including August Hoch, the neuropathologist.

The choice of the hospital site and the design of its buildings were visible testament to the influence, devotion, and skill of Dr. Cowles (1837–1919). When Otto met him in 1899, Cowles was nearing the end of a long, productive career. In hiring Otto he was fulfilling a major phase of his deeply cherished ambition of establishing a research laboratory for pathology, chemistry, and psychology. His farsightedness was matched by his practical sense and experience.

Cowles was a unique blend of practicing physician and administrator. He had been an assistant surgeon on the Union side in Grant's Army of the Potomac. He had run a hospital in Harrisburg, Pa., for six months after the battle of Gettysburg. He remained in the military service until 1872 when he resigned to become superintendent of the Boston City Hospital.

Cowles, a Vermont native, had received a bachelor's (1859) and a master's (1861) degree from Dartmouth College, followed by the M.D. in 1863. Prior to entering the military, he had practiced on his first mental patients for six months at the Retreat for the Insane in Hartford, Conn. His skill as an administrator for seven years at the Boston City Hospital left a lasting impression. He made improvements to the physical plant, established the first hospital-run nursing school, and increased the length of tenure and number of house officers.

Following his appointment in 1879 as superintendent of the McLean Asylum for the Insane in Somerville, he first traveled in England, Scotland, and France to visit mental institutions there. Upon his return he "hospitalized" the asylum in terms of patient care and terminology. "Boarders and attendants" became "patients and nurses." The use of restraints was reduced; screens replaced bars on windows. The patients were encouraged to communicate, to have visitors, and to visit at home. Cowles' second nursing school, first for a mental hospital, was established in 1882, and this helped increase the number of women nurses on the men's wards. The "asylum" officially became a "hospital" in 1892 (*10-1*).

In the early 1880s, Dr. William W. Gannet of Boston had served as a visiting pathologist to the hospital and instructed the medical staff. Following several months of study at Johns Hopkins with the pioneer psychologist G. Stanley Hall, whom he befriended, Cowles returned with a conviction about the need for experimental laboratories. Dr. William Noyes was appointed full-time pathologist in 1888, and he began research work in connection with the patients and was a pioneer in establishing a clinical laboratory. He resigned in 1893 and was replaced by Dr. August Hoch. Of course, Otto's arrival in 1900 to head the laboratory of physiological chemistry achieved the second phase of Dr. Cowles' laboratory plans, establishing what was probably the first hospital "metabolic" laboratory in the U.S. (in today's terminology a clinical study or research center). It was not until after he retired in December 1903 that the third laboratory, psychology, would be opened (*10-1*).

On Tuesday, Oct. 2, Otto wrote,

> My laboratory is only one small room. I could get more if I insisted but I prefer to see what can be done with it as I have full command of good carpenters, plumbers, steamfitters, metalworkers, etc. . . . Dr. Cowles is so far exceedingly accommodating, and will let me spend at least as much money as I had planned to. . . . I am sitting in my library, a very large and commodious room adjoining the laboratory . . . the hospital is having one of their regular home evening dances so Mr. Shaffer is out and having a good time in another part of the building [Entertainment Room, Service Hall]. . . . We have a fine golfground outside the hospital and yesterday afternoon I made one round and a third, playing from four to six. This is a fine place as far as I can make out.

On Oct. 6, Otto wrote that he was extremely busy planning the made over laboratory. Though he was still living at the hospital he would move the next day into new quarters—a very good room in a beautifully situated house in Waverley. The owner was connected with the hospital and was renting the room to Otto as a matter of accommodation.

"I scarcely knew when I proposed to stay away from you and Joanna for several months what it would mean." Fortunately Otto did not have time to remain lonely. He was working at that moment on stocking the library.

Although Waverley was very pretty, Otto was perplexed about where to live. The servant question was very difficult to solve, even more so than in Morgantown. Mrs. Hoch could not keep a girl on $4.00 per week in Waverley. Otto was "seriously thinking at odd moments of writing a letter to Sweden now and inquire so that we could get a girl in two or three weeks notice after you come."

Otto took two walks that day in the Waverley area. The more he saw there, the more he liked it. He wanted to acquaint himself with its housing possibilities before tackling Cambridge. Dr. Hoch lived in a satisfactory house in Waverley, but Otto was surprised by his expenses. Though Hoch's expenses were less than $100 per month, he had no servant. He paid: rent, $25; cleaning woman, $6; laundry, $8; provisions, $30. Otto's interest was heightened because he and Hoch drew the same salary. Hoch was contemplating moving to Boston or Cambridge because his daughter would be entering school. Hoch's house-hunting experience was limited, so Otto thought him "not a practical or resourceful man in matters pertaining to everyday life."

But it did not take Otto, the veteran hunter, very long to see what Waverley had to offer, and on the same day, he spent some time exploring in Cambridge and learning its street system.

Waverley existed because of the McLean Hospital and Otto did not think that Laura would care for it as a source of her social life. He had found that the hospital was not a state institution as he had supposed. It was part of Massachusetts General Hospital, and the latter was simply a very wealthy incorporated institution under control of a board of trustees. "It is as stable as Harvard. . . . The more Mr. Shaffer and I have seen of the institution the more we have wondered over what we have seen."

Otto stated that it would take several weeks before they could work in the laboratory. It would be well arranged and accompanied by a well-equipped library. Otto and Philip had begun by ordering the equipment and chemicals needed for the laboratory and by purchasing the journals and texts for the library. A great deal of this would be shipped from Germany and particularly via Eimer and Amend of New York, the only reliable U.S. source for relatively pure reagents and laboratory ware. Otto had also assigned themselves the task of combing the literature for reports of any chemical tests that purported to distinguish sanity from insanity.

Because of his free time in this period, Otto was able to follow two extra pursuits, house hunting, and for a short while, when the weather permitted, golfing, that he had learned about in Morgantown, to the point of purchasing a set of golf clubs.

> The existence of both you and Joanna and of my last year's experience occasionally seem like a kind of a pleasant dream as I again find myself in a new place and among strangers, socially. I shall be very glad to get my possessions back again.

Otto was glad to learn that Joanna was fine and wanted to know when Joanna's cousin was due; Kate was evidently pregnant. He had written to Percy at Andover to come visit him in two weeks (or whenever he liked). In two weeks Otto could take him to the Harvard–Carlisle Indians football game. Otto would get him a room at the place where he boarded. When did Laura expect to come?

Otto was unquestionably drawn to August Hoch, who was described as charming, open, and sincere, with a lively sense of humor. Hoch (1868–1919) was born in Basle, Switzerland, where he received his primary and secondary education. His father, a clergyman, was superintendent of the city hospital. At the age of 19 (1887) August, oddly, emigrated to the U.S. for his medical education rather than attend one of the great European centers. After two years in medicine at the University of Pennsylvania, he obtained an M.D. degree at the University of Maryland in 1890. He then spent three years at

Johns Hopkins Hospital in the medical and neurological clinics and attended Osler's medical clinic. Having been chosen by Cowles for the position of pathologist at the McLean Hospital, he went abroad for two years of postgraduate study in pathology, psychology, and psychiatry with Professors Recklingshausen in Strassburg, Wundt in Leipzig, and Kraepplin in Heidelberg. In 1895, he returned to become neuropathologist at the McLean Hospital. He displayed extraordinary acumen clinically, and was sharp and clear about his observations and his case histories. From 1902–5 he taught neuropathology at the Tufts Medical School in Boston. He left McLean in 1905 and had an eminent career in psychiatry in New York, where he eventually was director of the Psychiatric Institute of the New York State Hospital and a professor at the Cornell University Medical School. He was very involved with organizations related to mental health, particularly as an editor, and published much in his field. Hoch was extraordinarily helpful to Otto in his clinical orientation to the problems he would shortly tackle.

By Oct. 15, the list of chemicals that Otto needed was complete and it had been sent for bids to four different German firms. It was a sizable order of 5800 marks or a little more than $1400 in duty-free goods.

For recreation, weather permitting, Otto would go golfing in the late afternoon, usually after 4 p.m. Shaffer preferred tennis, so Otto would golf with Dr. Miller. He had lowered his score by nine points, and had made a round in 58 strokes.

Otto had attended a very interesting meeting of the Boston Neurological Society at the University Clubhouse. He had heard a paper about a patient who had at least three distinct, persistent, and different personalities.

On Oct. 19, Otto had gone househunting with Shaffer in Cambridge, and had no luck. Flats were not numerous while double houses were common. Again, on Oct. 22, using his bicycle, Otto spent between 2:30 and 5 p.m. househunting in Cambridge. This time he explored a large district and saw a number of vacant houses. This might be futile for him to do, but he would thereby get acquainted with the town and its rental prices.

What was Cambridge like in those days? A friend of the Folins would one day describe it as follows:

> To a person who has seen the intervening stages of development and decadence, it is difficult to compare the American Cambridge of today with the Cambridge of the beginning of the century. It is only by imperceptible steps that the houses have become grimier, that the traffic has become heavier, that the vacant lots have vanished, and that a community which in 1900 preserved much of the atmosphere of the country town has grown into a great, dirty, commercial city (*10-2*).

Otto was now concerned about getting a servant, so Laura must have encouraged him in this effort. Local girls were difficult to get on account of the factories.

> I wrote to the Swedish family which I had in mind before about a week or ten days ago offering to pay the girl's fare and to pay her $1.50 the first six months

and $2.00 per week the second six. I thought it would be better to put the figure low and then do better, because if she is willing to come at all she would as soon come for $1.50 as for $2.50.

Dr. Hoch joined Otto in the house-hunt, with the possibility of sharing a residence. They found a fine house with 13 rooms and an acre of ground fenced around it, which Otto described as "quite a bit of park." This house rented for $600. With $800 between them, they looked for extra fine homes with separate kitchens and bathrooms. What did Laura think of combining? Dr. Carver had joined them in their search. He had been over the whole territory himself and had obtained a nine-room house in a good locality for $37.50 per month.

Election day had come on Nov. 8 (McKinley-Bryan rematch). It was the chief topic of interest around the hospital, he wrote Laura. Otto was now reading the scientific literature seriously and expected that this would reduce the frequency of his writing. He had ordered $250 worth of literature and still had $250 to spend for current periodicals. The annual allowance for his literature was $200 to $300, a liberal amount.

By Nov. 14, Otto reported that "my laboratory is gradually taking shape. It is going to be a beautiful little shop." Otto's househunting finally bore fruit. On the following Sunday he planned to inspect a small nine-room house in a good district of Cambridge that he would eventually rent. It was the middle house of three and somewhat more distant from the street. He drew a diagram of the layout of the housing. He oddly expressed the thought that it would pay to have a somewhat respectable looking house in case Mrs. Cowles came calling. He had not met Mrs. Cowles, but may have heard about her tastes through the "grapevine" at the hospital, and this concerned him, for some reason. Otto was a master of understatement when he so chose. His "somewhat respectable" actually meant elegant; "small" in terms of a house meant what might now be considered a minimansion; a "nice little piece of work" could border on a masterpiece. On the other hand, his description as a rule was very sharp, e.g., his laboratory was truly a "beautiful little shop."

On Sunday, Nov. 18, Otto had an hour's talk with the owner of the house he would soon rent. He would not be able to see the inside until the following day, but he was quite enthusiastic about it.

> If the place is reasonably satisfactory I shall close the bargain. . . . Mr. Shaffer and I went down and took a photograph of the place about half past three last Friday afternoon. We stood right in front of it on the side of the street and hoped to include both other houses but we caught only the one to the right and in fact one half of it but that you can see just how much of an objection it is. The picture somewhat flatters the place very much and yet I do not see why because it is of course true to nature.

The house had been painted on the outside and would be painted and varnished on the inside and some of the rooms repapered. It had a range in the kitchen, a grate in one of the rooms, and a furnace in the basement. The rent was $30 per month, but if Otto were to rent it from Dec. 1 to April 1, he could get it for $25 per month, with no water tax. On May 1, he could lease the house for one year at $28 per month. Otto was, of course, eager to see the inside of the house. At the price offered, he could think about hiring a good girl without much trouble from a charitable institution, where orphans were trained for housework.

On the laboratory front, Otto had received the bids on the chemicals from Germany and was sending off the first order. He had received $100 worth of books, and the lab was "a little beauty in embryo so that the prospects are good. I am quite satisfied."

In his next letter to Laura, Otto enclosed a detailed floor plan of the house he would rent at 16 Avon St.

> Not very exact but it gives you an idea. . . . I was quite agreeably surprised when I finally saw the inside of the house and yet I have scarcely seen it. I simply walked through, but I saw enough to satisfy me that we could get along very well indeed in it. There are as you see five rooms on the first floor. They are small but taken together certainly make quite as large a downstairs as you and I will need. The hallway in the center is large and the stairway quite broad. There is a good supply of closets on the second floor (which is rather low) and a very convenient closet with a wash bowl in it in the closet on the first floor between the dining room and the back parlor. The dining room is a neat little room with two tiny, pleasing china closets one on each side of the door leading into the kitchen. . . . The paper downstairs is so good that I do not think that we can ask to have them repapered but if we should, what shade would you prefer? I think that a rich red color in the little room next [to the] dining room would be pretty nice if it is to be re-papered.

We have presented the above letter partly to illustrate Otto's faculties of observation and retention following a brief "walk through" the house.

Otto now began to think in terms of how Laura and he could plan their days. He hoped that she could work out her schedule for nursing Joanna so that their daughter slept through the night. Perhaps the last feeding should be at 10 p.m., and then the 6 p.m., night period gradually extended until 7 a.m. "Then I could get up permanently, take my bath, fix the fires and prepare breakfast (if we have no girl) while you nursed Joanna and got yourself dressed." He then discussed a room for Joanna, and the small yard near the kitchen, that could be a potential playground for Joanna. Otto would send Laura a map of Cambridge and a Cambridge newspaper. Otto found that it took 14 minutes for him to walk from the house to Harvard Square.

Otto had enjoyed a play put on by the Radcliffe girls, and the social afterwards. Otto had received his second month's pay, and stated they could afford Laura's fare East from it. Would she come before Christmas? Or did Laura want to see how Kate, who was expecting her child soon would get on, first? He hoped Kate would do as well as Laura had.

> It was all like a dream. I seem to be living as I used to before, and can scarcely realize that you are my wife now and that I have a little daughter. I ought to be very thankful that you were not taken away from me on the eighth of August. Come as you think best.

Having rented the house where they would now live for six years, Otto's thoughts turned toward its furnishings.

> Our house is now empty and clean. I was down there last night. The lady told me that the red paper is now out of style! She said it twice and so I could not well insist. If we get new paper it will be brick red and of the same kind in the dining room and library.

It was now Nov. 29. Otto had finally mailed the order to Germany for chemicals.

Dec. 3 brought the following news from Otto:

> I am afraid that I may very much regret what I have done today. I have spent exactly $90.00! I have bought two beds with springs and hair mattresses to fit. Also the following bedroom pieces all enameled white: one fairly large and good bureau, one small and cheap, one washstand, one small straight chair, and one small rocker and three small tables. . . . All of the downstairs is really yet to be provided for and you will have a hand in that. . . . We will have a good time arranging our little home. I shall have the house warm and the kitchen in good working order when you come. . .

Otto thought that the room for the servant girl "is naturally the most pleasant bedroom in the house and by fixing it up judiciously it ought to prove quite an attraction to the kind of girl that you would like to have."

A receipt from the Association of Collegiate Alumnae was made out to Mrs. Laura Grant Folin on Dec. 13, and mailed from New Haven to Waverley where it arrived on the 14th. Therefore we assume that Laura and the recovered Joanna had arrived at their new home before that date. Certainly the Folin family was ensconced in their Cambridge home well before the Christmas holiday of 1900.

Philip Shaffer described his experience with Otto in those early months at McLean as follows (*10-3*):

> At that time physiological chemistry was a novelty in Boston. Soon after his arrival at McLean in October 1900, Folin went to the libraries in Cambridge and Boston, and to consult a few friends and new acquaintances especially at Harvard Medical School on Boylston Street. Frequently his young assistant accompanied him. One of these early visits was to A.P. Mathews, then assistant to Professor Bowditch. In search of a volume located in the pathology library, Mathews took the visitors to call on Dr. Councilman. When Folin was introduced as the physiological chemist at McLean Hospital, Dr. Councilman asked in a brusque manner, "So, do you know Pfaff?" Folin had to admit that he had not yet made the acquaintance of Dr. Pfaff, a reply that dampened and almost closed the interview. For Pfaff since 1898 had been the instructor in both pharmacology and physiological chemistry at the Harvard Medical School, with a laboratory at the Massachusetts General Hospital. He was perhaps the only avowed representative of either of these subjects around Boston at the time of this incident. . . .
>
> During Folin's first months at McLean while the laboratory tables were being installed and the apparatus secured—most of it imported from Ger-

many—the "chief's" time was devoted to reading about metabolism, the chemical composition of urine, and the claims (mainly in French medical journals) of the presence in urine of insane patients of toxic substances thought to be concerned with their mental states. The first experiment undertaken, as the writer recalls, was an attempt to test the toxicity of normal and other urines, and of their known constituents separately by injection in rabbits. The only marked toxicity observed was that due to potassium and ammonium salts, effects already known. Folin was skeptical of the idea then harbored by some French writers and by Halliburton and Mott in England that "toxins" perhaps related to choline or other nitrogenous bases derived from nervous tissue could be a factor in mental disturbance; yet that idea was probably in Dr. Cowles' mind in creating the chemical laboratory, and the possibility was not wholly rejected by Folin. Finding no evidence for it in the preliminary experiments Folin gave up that approach.

He decided instead to study the protein metabolism of normal and mentally disturbed individuals by measuring as accurately and as completely as possible all of the known nitrogenous and other products excreted in the urine, hoping thereby first to learn the normal range of variation in the partition of the total nitrogen among the known products and residual fraction and then to consider possible abnormal variations. Although the primary purpose was ostensibly a search for abnormal features, they were presumably unrecognizable except by contrast with the normal patterns which were then unknown. To establish norms would alone be an important undertaking and progress could in any case be made toward that.

The first essential would be to devise more and better quantitative methods before any worthwhile surveys could be started. These were the considerations that led to Folin's interest in developing suitable quantitative methods for urine and blood analyses, an interest which held his attention for the rest of his life. The methods he developed enabled Folin, and following him, many others with even better methods to explore normal and abnormal features of metabolism with consequences not then forseen.

By December 1900, though the laboratory and library were not yet fully equipped Otto and Phil Shaffer as his assistant launched their initial projects, only two months after they had opened the laboratory. The most important study, as Dr. Cowles described it in his Annual Report of 1901,

> was a fairly exhaustive investigation of the question whether the toxic or nontoxic character of urine in the insane can serve as an index to the production of toxic substances in certain forms of insanity. This question has been repeatedly answered in the affirmative by different European investigators, but as it appears, without satisfactory proof. The conclusions to be drawn from the experiments of Dr. Folin and his assistant, Mr. Shaffer, indicate quite decidedly that the urines of the insane do not exhibit any such characteristic toxic properties.

These tests that Otto and Philip performed were largely qualitative: Was substance A present or not? However, it would take about four more decades before sensitive tests would be found for detecting metabolic products related to mental disorder.

On the other hand, Otto no doubt reasoned that with satisfactory quantitative techniques for measuring urinary constituents, particularly those substances probably derived from protein metabolism and diet, there could be a firm basis for comparing the output in health and mental disease. These quantitative techniques could also have great diagnostic value in other diseases besides insanity, and as indicators of health. The important point was that quantitative measurements must be made on human body fluids. We must be aware, however, that an adequate blood collection system would be slow in maturing, so that the centuries-old art of testing urine remained dominant simply because urine was the most readily available of body fluids.

Most important among the substances that Otto chose to measure immediately in both mental patients and in "normal" people were urea, uric acid, ammonia, phosphate, and sulfate; and in each case Otto found that to get reliable data, the analytical methods either needed improving or replacing, a project that would dominate his professional career.

Otto's mind had been stewing for two years. He had kept up with the new literature in physiological chemistry both in Chicago and Morgantown. Now, he and Phil apparently tackled the analytical problems in all of the above mentioned methods almost simultaneously. This is the only explanation of how it was possible for them to submit three articles for publication within a space of less than four months.

In Otto's early years at the McLean Hospital there were few American journals for biochemical publication. In addition, Otto was undoubtedly eager to contribute to the journal that had first given him a voice and had the ear of the world's most appropriate audience: *Hoppe-Seyler's Zeitschrift*. Moreover, he probably wanted to show his friends and "enemies" what he could do. He would have no trouble publishing because at least four of his European advisers and friends were now editors. Kossel remained the editor-in-chief, and Kutscher was now an editor along with Hammarsten and Salkowski. The American journals that could have provided an outlet for his work were the *American Journal of Insanity* and the *American Journal of Physiology*. The *Journal of Biological Chemistry* was not founded until 1905.

We must make special note of the fact that the first three papers published by Otto once he was securely established in his own well-equipped laboratory, and with his mind at ease with respect to his family and fortune, were on the determination of urea, ammonia, and uric acid in urine.

The first paper of the triad was a highly original method for determining urea in urine, reducing the time necessary for the procedure from more than eight hours to a little over an hour and a half (Part II, p. 370). The second paper presented a shortened, simplified method for determining free ammonia in urine (Part II, p. 370).

The third and last paper in 1901 from the chemistry laboratory of the McLean Hospital, authored by Otto and Phil Shaffer, was a lengthy one on uric acid. The method that Otto had published from Salkowski's laboratory in 1897 had been found wanting in accuracy by two separate investigators, Wörner and Jolles. The paper dealt with the objections of both men, which Otto admitted were somewhat valid, as well as with flaws in the Salkowski and Hopkins methods. With the modifications of the Folin procedure for uric acid offered in this paper, Otto and Phil had confirmed and extended the usefulness and quality of the method. For the moment it was the best one around (Part II, p. 371).

There is no direct information at hand as to how the Folin family passed their few months in Cambridge. They became good friends with their landlords, the Morrisons, who lived next door. Otto had written to Sweden for a girl, and "Hilma" had arrived, probably in the spring or earlier. Laura was evidently on a project of some kind involving economics because she had requested and received a list of books available on that subject from a Cambridge neighbor, J. N. Carver, who Otto mentioned was working at the McLean Hospital, but who was also on the faculty at Harvard.

Otto had kept in touch with Professor Jacques Loeb, and had asked him, probably in the spring of 1901, about the possibility of his spending part of the summer doing work at the Marine Biological Laboratory at Woods Hole. He had ideas about studying rigor mortis and wanted to test them on frog muscles. Loeb, who spent practically all of his summers in teaching and research at Woods Hole, unquestionably helped Otto in this endeavor. Both he and his younger brother, Leo, would be there. Otto had met Leo in Chicago in 1898, and they would also become good friends. By July 30, Otto was working in the lab at Woods Hole on research that would lead to later publications.

In 1902, Otto published four papers, with Phil as a coauthor on the first. These were the products of work continued after Otto returned from Woods Hole the previous summer. The first two were Otto's first presentations in English since his thesis, six years previously.

The first paper was a study of phosphate metabolism in one patient and must have excited Otto and Phil. Dr. Hoch diagnosed a patient who was available to them for study as a case of manic depressive insanity. (This diagnosis was in error. The patient had a "general paralysis." See *Am. J. Insanity* **61**, 329 (1904–5).)

> Any sufficiently characteristic alteration in the patient's metabolism was deemed important, because we can scarcely be said to have as yet the proof of the existence of any abnormal metabolism whatever which is characteristically associated with any of the mental diseases.

In fact, the results were important enough to warrant publication in a journal devoted to general physiology, *The American Journal of Physiology* (Part II, p. 372).

The next paper from Otto, which immediately followed the phosphate paper, dealt with improvements on the determination of total sulfates in urine (Part II, p. 373).

A paper submitted before Otto and his family went to Woods Hole in August 1902 was his second paper on urea determination in urine. He wished to provide data that verified the quality of the unique method that he proposed in 1901 and to change slightly some of the conditions for its use (Part II, p. 374).

Several days after Otto had submitted the above paper he received the latest copy of the *Zeitschrift*, which contained an article by Arnold and Mentzel that had evaluated the Folin method for urea analysis. Otto therefore appended an addendum to his paper. We stress again that it is important to learn about Otto's earliest analytical methods to see them evolve. They would be criticized, and he would, fortunately for clinical chemistry, react, often by improving the method or by finding an alternate approach. He would not be indifferent.

The Folins spent the summer of 1902 in Woods Hole under the auspices of the U.S. Fish Commission, which not only provided Otto with laboratories to use, but arranged other conveniences such as the housing and board they required.

Otto's work there was an attempt to disprove the idea that rigor mortis was the effect of spontaneous coagulation of certain proteins of muscle plasma after death. Kühne, working with muscle plasma, had obtained from fresh frog muscles a coagulable protein that apparently was not present in muscles in rigor. The plasma coagulated spontaneously at room temperature and at 40°, where the frog muscles go into rigor. This idea had been dominant since it was first proposed by Kühne in 1859. Otto's experiments to disprove this concept were begun with fish at Woods Hole, and later with frog muscles. As the results were not published for another year, we can assume that the work was finished after Otto had left Woods Hole and returned to McLean Hospital. Therefore, we shall discuss his paper on rigor mortis with the 1903 output.

The fourth and final paper that Otto published in 1902 was another highly original one on determining ammonia, which moved that ponderous procedure all the way into daily clinical applicability, and more important brought urea analysis one step closer in the same direction (Part II, p. 375).

For the first time, the analysis of blood was also included. Otto had found that his method for determining urea did, after all, give low results in *urine*, though not in pure solutions, because some NH_3 remained behind during distillation.

In his Annual Report for 1902, Dr. Cowles stated,

> More extended metabolism experiments were begun, and accurate analytical data are now being collected, which, it is hoped, will contribute to the question as to what classes of mental diseases are, and what are not due to disorders of metabolism. These experiments have consisted in keeping patients, for a short time, on a uniform, but acceptable diet of known nitrogen value, and carefully determining quantitatively the forms in which this nitrogen is eliminated in the urine; the purpose is to learn whether any unusually large fraction of the nitrogen so eliminated appears in forms, or in quantities, unknown to the normal nitrogen metabolism.
>
> The work here outlined is regarded by Dr. Folin as the most important part of the investigations that should be carried on, for some time to come, in the chemical laboratory. As the analytical investigations already made have helped very much to make this metabolism work more effective, so the plan for its continuation should be to broaden these experiments as much as may be feasible. In this way only can conclusions be reached concerning the important question whether any tangible relation between faulty nutrition, or other faulty metabolism, and different forms of mental disease can be established. It is clear that this work, calling for a very large number of analytical determinations, all of which must possess the greatest possible accuracy, is a large undertaking. The progress of this work will be aided by the continuous refine-

ment of methods, and the constant discoveries of new features in physiological and pathological metabolism. This line of inquiry may be hopefully pursued in the expectation that the exact study of clinical facts by such methods may reveal explanations of them that will throw some light upon the main purpose of these practical researches through which guidance for treatment is to be sought. The report is concluded with the remark that the equipment of the chemical department is practically very complete; and chemical investigations can now be carried on in the McLean Hospital under unusually advantageous conditions. The only material want that is not fully met in this department is that of files of certain periodicals; some complete sets of these are yet needed for the frequent consultation that is required of the work of many other investigators in physiology and chemistry, whose progressive activity was never greater than at the present time, making convenient references to the newer and older literature of the subject an important aid to present inquiry.

It should be mentioned, as a matter that is believed here to be of essential importance, that this department was placed under charge of an expert in pure chemistry, thoroughly trained also in physiological and pathological chemistry, in order that the full advantage might be gained of having our problems approached from that point of view.

About June of 1902, Laura was pregnant. Added to the joy of this occasion were the comforting facts that Laura and Otto lived in a comfortable home with a servant girl, surrounded by friends, and with skilled physicians and medical institutions in abundance. Not only had their landlord's sister, Dorcas Morrison, been a specially good friend, but Bertha Weiner and her family whom Laura had met while at the University of Missouri moved into a pleasant house only two doors away on Avon St.

Otto became a good friend of the Wieners, in his first intimate contact with a thoroughly Jewish but very unorthodox family. The diminutive, but sinewy Leo (1862–1939) was a proverbial dynamo of activity and a mental giant, the likes of which Otto had never seen; not even Nef or Kossel was his equal. Leo was pleasant company, but seldom available, a philologist who was a creator of that field of knowledge, a linguist of astounding erudition, with whom Otto could discuss, in Swedish if he liked, the literature of Scandinavia, yet Wiener taught Slavic languages at Harvard, and to earn a decent living, taught languages at Radcliffe and Boston University, and worked in the Boston city library reference system. He was a good mathematician and biologist; and because the Wiener's were vegetarians, he was adept at mycology (mushroom fancier) and did some light farming at his place at Foxboro. Bertha and her son, Norbert (1894–1964) were frequent visitors.

About June 6, 1902, Laura and Joanna went to visit her parents in St. Paul. Otto was lonely, but busy. For his coming visit to Woods Hole, "I have written and invited the Lommens," Otto's friends from his Berlin days (1897) and the University of Minnesota (1891). Otto was working at least one and one-half hours an evening on learning French. Though he was making abundant errors he felt he was not missing the essentials. Already, the scientific articles looked familiar.

Otto had hoped to visit Axel, but had learned that Axel's house was torn down temporarily. He was expecting to send Axel $156 as payment on his debt.

He was now planning his research on rigor mortis at Woods Hole. Lommen had written that he was coming alone and would see Otto at Woods Hole. Lommen was now at the South Dakota State University in Vermillion.

By the end of June, Laura wrote that she would come to Woods Hole, and Otto suggested that she rest a day or two first in their home in Cambridge. He would try to find a comfortable place for them. Leo Loeb, who was now in Buffalo, expected to spend some time in Woods Hole and, of course, Jacques would be there.

In July, Philip Shaffer wrote several letters to Otto from the chemistry laboratory at McLean Hospital. He was carrying on his research on methods for ammonia determination, and thanked Otto for inviting him to come to Woods Hole. He declined, however. He had looked up some literature that Otto had requested on coagulation and protein precipitation, particularly in crustaceans, and on haemocyanin in the blood of the king crab.

In 1903, Otto published three papers and had launched a major study on the composition of urine, as Dr. Cowles had stated. The first two papers dealt again with sharpening the tools for his metabolic studies: One was his third paper on urea determination and the other was on measuring the acidity of urine. The third report was a completion of his work on rigor mortis. During 1903, Otto also began his studies on creatine and creatinine determination, another lifetime pursuit, that would attract a good number of other early biochemists. As Philip Shaffer was now a full-time graduate student the position of research assistant was in the hands of Lucian A. Hill, though Phil's thesis work was also done at the McLean laboratory. Hill was Otto's assistant for one year until his place was taken in the autumn of 1904 by Christian Ostergren who remained with Otto to late in October 1907.

Otto's first paper of 1903 was a brief one. In his second paper on urea Otto had vigorously defended his method against its attackers, Arnold and Mentzel, who had disputed its accuracy and had found that uric acid, hippuric acid, and creatine partly decomposed and consequently interfered by producing NH_3. Having successfully responded to his detractors concerning the first two issues, only creatine remained to be studied; and in his third paper on urea, Otto now proved that creatine (and creatinine) did not interfere (Part II, p. 377).

Otto's second paper of 1903 was a system for determining the acidity of urine. Current methods for determining the acidity by direct titration were beset by several difficulties. Calcium salts were present and interfered with determination of the acidic monobasic phosphates. Ammonium salts made direct titration worthless. Phosphate analysis was inaccurate. And Otto pointed out that organic acids account for as much as half of the total acidity. Otto's object was to determine the total acidity of urine and the mineral acidity so that the difference between them would represent the acidity contributed by organic acids (Part II, p. 377).

Otto's work on rigor mortis also was published in 1903. As we mentioned before, Otto wished to disprove the theory that rigor mortis represented a process analogous to blood coagulation. Some evidence favored the coagulation theory: The opaqueness that usually accompanies the rigidity and the experiments of Kühne that showed that muscle plasma from muscles in rigor did not contain the mother substance of myosin fiber as did muscle plasma from fresh muscles. The coagulation of this mother substance was supposed to be the cause of the rigor. However, Otto asserted, the findings could be due to other chemical changes in frog muscles "during the 40–50 hours that one must wait for them to go into normal rigor" and he set out to prove it (Part II, p. 378).

From late 1902 to early 1904, Otto, with the help of one assistant, made the most comprehensive analyses of urinary constituents to that date. For this study seven healthy adults and 46 patients at the McLean Hospital were reported. How many others were actually used and discarded were unrecorded. This number may have been considerable because of problems in methodology and patient noncompliance.

Because of the heroic nature of this undertaking we must let Otto state his own reasons for doing it.

> ... there is still no record in the literature of a single complete analysis of one and the same twenty-four hour quantity of urine—complete in the sense of including all the more important substances that are frequently determined singly or in small groups. ... It would certainly be a point of advantage if, by means of a few such records, it would be possible to check, at least in some measure, ridiculously loose usage in medical literature of the terms "normal," "increased," "diminished" as applied to the various urinary constituents—terms which seem intended to imply that quantitative determinations have been made when such is not the case.

For metabolic experiments it was essential that the greatest possible number of constituents in urine be determined. Much remained to be learned in regard to the "laws of urinary excretion and of animal metabolism by a sufficiently complete system of analysis which would record many peculiar, unexplained variations which are sometimes exceedingly pronounced even under uniform conditions of diet and occupation."

Otto was prophetic in writing:

> Abnormal metabolic processes must unquestionably play a more or less important part among the pathological conditions that produce various forms of insanity, but it may well be that the chemical methods at present available are entirely too crude for the determination or study of these deviations from the normal.

It was possible insofar as studying metabolism in insanity to "plan such experiments comprehensive enough and extensive enough to yield all the information that the present state of chemical technique can yield along this line."

Characteristically, Otto reviewed the literature of urine analysis and found about 267 reports that dealt with individual substances or topics such as phosphates, glycosuria, urinary weight, acetonuria, sulfates, albuminous substances, creatinine, toxins, chlorides, and oxalic acid. All of the reports were fragmentary, and the conclusions were open to question. There were a few attempts at comprehensive studies of the more important constituents but they were incomplete or published without reference to the analytical methods used. But, as Otto amply demonstrated, the analyical methods were far from reliable for many constituents.

References

10-1. (a) Hurd, H.M., The new McLean Hospital. *Am. J Insanity* **52**, 477–502 (1895–96).
   (b) Cowles, E., The laboratories of the McLean Hospital for research in pathological psychology and biochemistry. *Institutional Care of the Insane in the United States and Canada.* **2**, 618–36 (1916).
   (c) Tuttle, G.T., McLean Hospital, Waverley, Mass. (1811–1914). *Institutional Care of the Insane in the United States and Canada* **2**, 599–617 (1916).
10-2. Weiner, N. *Ex-Prodigy, My Childhood and Youth.* Cambridge, Mass.: M.I.T. Press, 1953.
10-3. Shaffer, P.A., Otto Folin, 1867–1934. *Biogr. Mem. Natl. Acad. Sci. U.S.A.* **27**. 47–82 (1952).

Chapter 11. A Classical Period (1903–7)

Otto's quantitative studies of human urine appeared in two parts. The first dealt with introductory material, the analytical methods and measurements to be made, the uniform diet to be administered, and the results obtained with "normal" persons. The second paper dealt with the results on patients, discussion, and conclusions.

Because Otto was working concomitantly on modifications of the analytical methods, the procedure he used might not always have been the latest. Conversely, as he gained extensive experience with the analysis of large numbers of urines, small but important modifications were introduced that were not yet published, or details of procedures were included that clarified the steps to be followed in each method (Part II, p. 380).

In 1903 two momentous events happened in Otto's life, one blessed, one accursed. On Feb. 21, Laura gave birth to a son she named Grant (but called George). In April, Otto had an operation that permanently affected his face. For some time he had a swelling on the left parotid gland. It was a growth that had become so large as to involve the area of the facial nerve. He put himself into a surgeon's hands at the Massachusetts General Hospital and fully expected to return home that evening. When he did not return that night, and because there were no telephones, Laura rushed to the hospital via the subway train. What a shock it was for her to find him all bandaged up with his whole face shifted to the right. To remove the "tumor" the surgeon had cut the facial nerve on the left side, leaving it irrevocably paralyzed. Otto stayed in the hospital two to three days. Much later, Dr. Leo Loeb wrote that he had examined the tumor and found it benign, but Otto wasted no time in taking out some life insurance. The distortion was less noticeable because of Otto's smooth skin and his mustache. He could not fully close his left eye, so he slept with it covered. But for this unexpected occurrence, the Folin family thrived.

Otto's surgery may have been the stimulus that prompted Laura to seek part-time work in the autumn of 1903. We have no record to determine whether she actually went to work, but she applied to teach mathematics in the Cambridge Evening School (high school). For this she was recommended by several people. A friend of theirs from the neighborhood, Mr. T. N. Carver of the department of economics at Harvard, wrote of her ". . . one of the brightest and best-educated women in my acquaintance . . . quick, vivacious, sympathetic, and she expresses herself well." Leo Wiener wrote,

> . . . she took charge of some classes in the Missouri State University at Columbia, Missouri, while I was connected with that institution. Her excellent University training, her personal magnetism and enthusiasm, and her earnestness combined to make her a very successful teacher. Of her moral and sound social qualities I cannot speak too highly. . . .

Mathematics Professor N. M. Herron of the University of Chicago wrote, "she might have carried one of the best fellowships in the department of economics had she cared to do so." She also had recommendations from S. A. Farnsworth, Principal, Cleveland High School, St. Paul; and Marion Talbot, Dean of Women, University of Chicago.

We have none of Otto's correspondence with his relatives, either in the U.S. or in Sweden. In 1903 Nils Magnus Folin died in Träryd at age 78. Gertrud remained with Eva and, of course, Gusten and Ludde were nearby. The enigmatic Carl Ossian probably was somewhere in the U.S.

In 1904 Otto published four scientific papers and a note. The note (11-1) was a not-so-subtle attack on the author of another "Erwiderung," who had tried to establish priority rights over Philip Shaffer for determination of ammonia by the vacuum distillation technique, and who inexcusably had failed to mention Shaffer's work in his article.

Philip had completed the thesis for his doctorate at Harvard in December 1903. It was titled, "An Investigation of Metabolism in the Insane." He closed the thesis as follows:

> The writer is indebted to Dr. Otto Folin for his direction of the research; to Dr. August Hoch for his hearty cooperation, and for the diagnosis and classification of the cases studied; and to Dr. Edward Cowles, Superintendent of the McLean Hospital, for the excellent facilities offered for this work and for his encouragement. . . . To those gentlemen and to Professor C. L. Jackson, he takes pleasure in expressing his thanks.

Philip received his doctoral degree in 1904, avowedly the first in "physiological" chemistry from Harvard, married the former librarian at the hospital, Nan Evans, and went to his first job as a research assistant in the Loomis Laboratory of the Cornell Medical College in New York City. He earned the grand sum of $85 per month, but his young wife encouraged him to accept this position, rather than any of several more lucrative industrial offers (11-2).

Otto's crowning achievement of his first four years at McLean already had been touched on in his second paper on metabolism, and was now the subject of a full report: creatinine and creatine (Part II, p. 384). This was not merely a "method" paper, though his unique and simple procedure was in itself a major contribution to the biochemistry of that era. Rather, Otto showed beyond doubt that creatine existed in urine as well as how to prepare a pure creatinine preparation for use as a standard, and he defended the use of the Kjeldahl determination for establishing the nitrogen in the standard. He also, as mentioned before, introduced to clinical biochemistry the use of the visual colorimeter, a new analytical tool that would soon help put clinical chemistry into daily hospital practice. Otto, who was well trained in the classic gravimetric and volumetric analysis and had been soundly initiated into gasometric measurements by Stieglitz, must have quickly grasped the potential of colorimetry to his work. He had first seen a colorimeter in use when he visited a brewery in Berlin in 1897.

It should be pointed out that the first Duboscq colorimeter that Otto used in 1903 (possibly 1902) was constructed so that the "cylinders" holding the colored solutions were stationary while the prisms were raised or lowered. This caused some fluctuations in readings due to varying distances between the bottom of the prism and the mirror below. It was not until 1913 that the newer model made the prisms stationary, and the platform holding the cups movable (11-3).

As the details of the creatinine determination already have been presented previously, we will not dwell further on this topic. The method was based on the Jaffé-discovered reaction of creatinine with alkaline picric acid solution. Jaffé had reported the reaction in the *Zeitschrift* in 1886. Thanks to Otto's introduction of this reaction into clinical chemistry, it appears likely to survive a full century, though enzymatic analysis of creatinine will soon predominate.

Otto reported on the stability of the color reaction and the upper and lower limits of concentration that could be measured. Otto also introduced the use of potassium dichromate as a stable artificial standard. It would also serve later as a measure of icterus in serum. Otto provided details on the use of the colorimeter. The product formed in the Jaffé reaction was the picric acid double salt of potassium and creatinine, and Otto shortly would describe how to use this reaction as a way of obtaining pure creatinine. He would deal with creatine more thoroughly in 1906.

Creatinine determination taught Folin the distinct advantages of colorimetric over titrimetric or gravimetric analysis. Among those were speed and ease of analysis, smaller sample requirements, and often better reproducibility of results. At this time only the creatinine procedure was colorimetric. Otto would spend a significant portion of his future expanding the applications of colorimetry. This effort provided the backbone of modern clinical chemistry.

Otto's friend F. Kutscher of Marburg had published a paper casting doubt on the applicability of the Kjeldahl nitrogen determination for known simple compounds of physiological importance. For the simple nitrogen-containing compounds that Otto was interested in, Otto argued that the Kjeldahl process was a hydrolytic cleavage with H_2SO_4 as the catalyst, and that the addition of the oxidation reagent $KMnO_4$ was unnecessary. For the highly complex organic substances such as proteins or cereal products, undoubtedly a strongly oxidative process occurred. Be that as it may, Otto could show that the nitrogen content of the creatinine he purified gave the predicted theoretical percentage of 37.2 in 15 different trials with the Kjeldahl reaction, within limits of experimental error.

Otto's final paper of 1904 was primarily a discussion and not a scientific finding. It was a reasoned argument using data that he had reported two years previously, refuting a paper that purported to measure blood alkalinity and the principle on which it was based. The principle had been proposed by none other than Salkowski, and we must assume that Otto took some furtive delight in this second opportunity to deflate the ego of his old master (11-4).

The principle, which Salaskin and Lawrow had presumably confirmed, was that the addition of neutral ammonium salts to an alkaline fluid such as blood would liberate an equivalent amount of free ammonia. In Otto's 1902 paper on the determination of ammonia in urine, he had tested this idea several times on dog blood using ammonium sulfate and chloride salts along with vacuum distillation to obtain the liberated ammonia to be measured. What he found was that continued distillation produced additional ammonia. Obviously the principle was unreliable. Salaskin had ignored Otto's findings, and instead published another paper supporting the principle. Salkowski himself had not done much to support his idea, which was without either a basis in modern theoretical chemistry or proof from the laboratory.

Though Otto made his point convincingly, it was really out of character. Otto's usual approach was to offer an alternative solution to a problem, and not merely to disprove someone's work, even with Salkowski involved.

The above paper was the last until 1913 that Otto would publish in a German journal. In the U.S. the physiologists had helped establish the *Journal of Experimental Medicine* in 1896, and the American Physiological Society had founded the *American Journal of Physiology* in 1898 (11-5).

Otto's work was, of course, being read with much approval by the leading physiologists and fellow pioneers in biochemistry who had helped create these journals. Henry P. Bowditch, the head of the physiology department at Harvard, and former dean of the medical school, was a friend and adviser of Edward Cowles and had encouraged Cowles in his ideas of creating a research department at McLean. There is no question that he approved and backed Otto in his scientific efforts. And we must not forget that the first major physiological chemist in the U.S., Russell H. Chittenden of Yale, also had recommended Otto to Cowles in 1899. These men as well as Otto's old friend Jacques Loeb were among the guiding forces for the new journals.

In the Annual Report of the McLean Hospital for 1904, Dr. Tuttle recorded:

> In his work in the chemical department of the laboratory, Dr. Folin was led, by the suggestion of Prof. Henry P. Bowditch, to pursue certain studies which have been continued throughout the year, both with normal persons and with patients; and it is believed that some important facts in connection with metabolism have been discovered. These facts indicate that the nitrogen metabolism in man must be of two kinds, each qualitatively and quantitatively independent of the other . . . the results were presented in a paper read before the Boston Society of Medical Sciences on December 6, and were also in part presented in a paper read before the American Physiological Society in Philadelphia, December 28.

Otto attended the annual meeting of the American Physiological Society in Philadelphia, Dec. 27 and 28, 1904. By invitation (he was elected a member at this meeting) he presented a paper entitled "The Nitrogen of Urine; Its Distribution Among the Four Important Constituents—Urea, Ammonia, Uric Acid, Kreatinin."

When Professor Christian A. Herter, the man for whom Alfred J. Wakeman worked (Otto's friend from the Salkowski days), privately established and financed the *Journal of Biological Chemistry* in 1905 (*11-6*), he asked Otto to become a collaborating editor (*11-7*). The editors supplied much of the material that would be printed in the early issues of the journal. In 1906 the journal was transferred to the aegis of the American Society of Biological Chemists that was founded that year and in which Otto would play an important role, although Herter remained its owner. The collaborating editors of the Journal were actually its board of directors whose sole function was to conduct and perpetuate the publication (*11-8*). Otto was the only "hospital" chemist in the bunch, and probably the only one then unassociated with a university.

The year was a banner one for Otto. At a Lake Placid conference on home economics, Otto presented a discussion of his findings on human protein requirements based on his newer studies at the McLean Hospital of the urine of subjects who had been placed on restricted or uniform diets (*11-9*). Otto's detailed reports on these studies soon would follow. Otto had mentioned this new work in the second part of his long report on the composition of urine obtained from the patients at McLean. He must have been at least partly inspired in this effort by Chittenden's statement in his 1904 book *Physiological Economy in Nutrition* that he had maintained a body weight of 57 kg for over a year on a diet in which his average protein consumption was less than 35 g a day.

In the nearly total absence of protein intake, Otto observed that there was a protein catabolism representing about 20 g a day. He had succeeded in keeping several subjects on a carbohydrate–fat diet consisting exclusively of pure arrowroot starch and 300 mL of cream for periods of one week to, in one case, 17 days.

Otto found that creatinine output in the urine was constant, and probably reflected tissue catabolism, whereas urea varied considerably and peculiarly with protein intake and was likely a product of excessive protein catabolism. When protein content was high urea output could provide 90% of the total nitrogen excreted in the urine, but when it was low, as in the low-protein dietary study, the output was reduced to 50–60%. Otto coined the term "exogenous metabolism" to represent the catabolic production of waste products of protein metabolism excessive to and not involved in tissue metabolism. His report at Lake Placid did not include the term "endogenous metabolism" to represent the essential protein metabolism, independent of protein intake. This tissue metabolism was constant for each individual and could be observed when the sole source of energy provided was fat and carbohydrate.

For a generation, owing to Professor Voit's influence, the daily minimum protein requirement to maintain nitrogen equilibrium was set at about 118 g for a 70-kg man. Otto pointed out that this was far from minimal; he had found the figure to be closer to 20 g. But what was optimal protein intake remained to be determined. The minimal requirement was probably far from optimal, yet 118 g could be excessive.

In 1905 Otto published three closely interrelated papers in the *American Journal of Physiology* in a logical sequence. The first paper, previously presented in the *American Journal of Insanity*, was a somewhat clearer display of the data on the composition of 30 urines (24-h collections) on six "normal" persons who had been placed on a standard, uniform diet (*11-10*). Moreover, Otto discussed the analytical methods individually and, as expected, described the small modifications he had made to enhance their quality. No doubt one of Otto's aims in reproducing this information was to reach a much different and wider audience with what he believed was the most thorough quantitative study of the composition of 24-h urinary collections then existing. But offering this material was the proper first step for him to take before passing on to the next one of expounding principles governing the chemical composition of urine (*11-11*). Then followed Otto's stimulating theory of protein metabolism, which was introduced in his report at the Lake Placid conference and which would, without question, leave a lasting imprint in this major arena of physiological chemistry (*11-12*).

For Otto's studies on urine, he recruited five additional staff members (four physicians and himself) and two patients, who were placed on both the uniform nitrogen-rich (milk-eggs) diet and the low-nitrogen (starch-cream diet). He used Dr. Hoch in this group, but made special mention of Dr. E. Van Someren of Venice, Italy, who had been brought to the laboratory on a visit with Professor Bowditch and who remained as a guest of the hospital long enough to permit the collection of a series of consecutive 24-h urines.

Otto proposed two general principles: a) The distribution of the nitrogen in urine among urea and the other nitrogenous constituents depends on the absolute amount of total nitrogen present, and b) the distribution of the sulfur in urine among the three chief normal constituents—inorganic sulfates, ethereal sulfates, and "neutral sulfur"—depends on the absolute amount of total sulfur present.

From his studies of the composition of the nitrogen and sulfur compounds in the urine resulting from the two dietary regimes, Otto arrived at the following generalizations:

> 1. *Kreatinin.* The absolute quantity of kreatinin eliminated in urine on a meat-free diet is a constant quantity, different for different individuals, but wholly independent of quantitative changes in the total amount of nitrogen eliminated.
> 2. *Uric Acid.* When the total amount of protein metabolism is greatly reduced, the absolute quantity of uric acid is diminished, but not nearly in proportion to the diminution in the total nitrogen, and the percent of the uric acid nitrogen in terms of the total is therefore much increased.
> 3. *Ammonia.* With pronounced diminution in the protein metabolism (as shown by the total nitrogen in the urine) there is usually, but not always, and therefore not necessarily, a decrease in the absolute quantity of ammonia eliminated. A pronounced reduction of the total nitrogen is, however, always accompanied by a relative increase in the ammonia-nitrogen, provided that the food is not such as to yield an alkaline ash.
> 4. *Urea.* With every decided diminution in the quantity of total nitrogen eliminated, there is a pronounced reduction in the percent of that nitrogen

A Classical Period   175

represented by urea. When the daily total nitrogen elimination has been reduced to 3 gm. or 4 gm., about 60 percent of it only is in the form of urea.

5. *Inorganic Sulphates.* Decided diminutions in the daily elimination of total sulphur are accompanied by reductions in the percent of that sulphur present as inorganic sulphates. The reductions are as great as in the case of urea.

6. *Neutral Sulphur.* The neutral sulphur elimination is analogous to that of the kreatinin. It represents products which in the main are independent of the total amount of sulphur eliminated or of protein katabolized.

7. *Ethereal Sulphates.* The ethereal sulphates represent a form of sulphur metabolism which becomes more prominent when the food contains little or no protein.

In order that the reader may recall the extent of the changes in percentage composition of urine from normal persons shown by the experiments, a part of Table III is here reproduced (corrected for errata).

	July 13.	July 20. (low-nitrogen diet)
Volume of urine	1170 c.c.	385 c.c.
Total nitrogen	16.8 gm.	3.60 gm.
Urea-nitrogen	14.70 gm. = 87.5%	2.20 gm. = 61.1%
Ammonia-nitrogen	0.49 gm. = 2.9%	0.42 gm. = 11.7%
Uric acid-nitrogen	0.18 gm. = 1.1%	0.09 gm. = 2.5%
Kreatinin-nitrogen	0.58 gm. = 3.5%	0.60 gm. = 16.7%
Undetermined nitrogen	0.85 gm. = 5.0%	0.29 gm. = 8.0%
Total SO_3	3.64 gm.	0.76 gm.
Inorganic SO_3	3.27 gm. = 89.8%	0.46 gm. = 60.5%
Ethereal SO_3	0.19 gm. = 5.2%	0.10 gm. = 13.2%
Neutral SO_3	0.18 gm. = 5.0%	0.20 gm. = 26.3%

Those among the readers who have used urinary creatinine output as an index of adequate collection will find the origin of that practice in the statement:

The fact that the kreatinin-elimination is normally an almost perfectly constant quantity in any given person is important in connection with metabolism experiments in hospitals, or whenever one must depend on someone else for the proper collection of twenty-four-hour quantities of urine. Any considerable loss of urine is promptly shown by the kreatinin-determination which can be made in a few minutes. As an illustration of this, attention may be called to the omission in Table VIII of the urine corresponding to April 18. 600 c.c. of urine, containing 5.1 gm. nitrogen, 136 c.c. n/10 ammonia, and only 0.81 gm. kreatinin, were brought to the laboratory as the full twenty-four-hour quantity. The preceding six days showed, however, that the patient should certainly not yield less than 1 gm. kreatinin per day. An investigation of the ward was at once instituted, and it was found that the patient had been without supervision during a considerable part of the day.

Otto asked what caused variations in creatinine output among individuals. Two factors were involved—body weight and corpulency. The leaner person tended to excrete each day a greater amount of creatinine per kilogram of body weight than the more corpulent one. This factor had to be considered in speaking of abnormally high or low creatinine values in urine. Otto cited three aspects of creatinine that needed study: changes in output during disease, the effect of certain drugs on creatinine output (he had noted that salicylates increased uric acid output), and finally that hard physical exercise could affect (increase) creatinine output. Such studies remain the core of modern clinical chemistry.

Otto had some significant new points to make concerning phosphates and organic acids in relation to urine acidity. The prevailing view was that urine acidity was chiefly (60%) the result of the presence of the weak acidic diacid phosphate (monobasic) and of the weak alkaline monoacid phosphate (dibasic). Attempts to determine urinary acidity were directed to measuring the acid phosphate mixtures in the presence of the interfering alkaline earths, calcium, and magnesium salts. But Otto showed that there was no monoacid phosphate in normal acid urines. Barium sulfate precipitated by $BaCl_2$ from such urines had been mistaken for barium phosphate, while the fact that $CaCl_2$ would not produce a precipitate with monoacid phosphate was ignored.

> The phosphates in clear acid urines are all of the monobasic kind, and the acidity of such urines is generally greater than the acidity of all the phosphates, the excess being due to organic acids. . . . For all ordinary studies of the acidity of urine, the direct titrations of the total acidity and of the phosphates give the necessary information. The excess of the total acidity above that calculated from the phosphates (7 mg. P_2O_5 = 1 c.c. n/10 acid), gives the total free acids present.

Having shown that quantitative changes in protein catabolism were accompanied by changes in distribution of urinary nitrogen and sulfur, and that these variations occurred according to laws (principles, generalizations) that could be formulated with some precision, Otto now turned his mind specifically to protein metabolism.

> It is clear that the laws governing the composition of urine represent only the effects of other more fundamental laws governing the katabolism of protein in the animal organism. From the variations in the percentage composition of urine described by the above generalization, it would seem therefore that some conclusion might be drawn in regard to the nature of protein metabolism. It is my purpose in the present paper to attempt an interpretation of protein metabolism on the basis of observed variations in the percentage composition of urine.

From his study of urines of subjects on usual and low protein diets, Otto found that there were two kinds of catabolism, essentially independent and quite different. One was extremely variable, the other constant. The variable kind yielded chiefly urea and inorganic sulfate, but no creatinine and probably no neutral sulfur. The constant catabolism was represented largely by creatinine and neutral sulfur, and to a lesser extent by uric acid and ethereal sulfates. The greater the reduction in total catabolism, the more prominent

A Classical Period

were these representatives of the constant catabolism, and less prominent were the two chief representatives of the variable catabolism.

I would therefore call the protein metabolism which tends to be constant, *tissue* metabolism, or *endogenous* metabolism, and the other, the variable protein metabolism, I would call the *exogenous* or *intermediate* metabolism.

The endogenous metabolism would set a limit on the lowest level of nitrogen equilibrium attainable. This would depend on how much, if any, urea was derived from the same catabolic processes that produce creatinine. At this limit the percentage composition of urine should be practically constant. The total nitrogen at this constant composition would indicate the lowest attainable level of nitrogen equilibrium. This hypothesis needed experimental verification, Otto stated.

Prior to Otto's work there were two theories of protein catabolism prevalent, that of Voit and the more acceptable one of Pflüger. Both had tried to explain protein catabolism on the basis of two forms of protein substance, the living protoplasm and the circulating protein derived directly from food. Voit's theory was that protoplasm was in a state of physical suspension and that chemical decomposition occurred only in solution; hence the small amount of living protoplasm that died in the course of a day was first dissolved, to become part of the circulating protein derived directly from the food. In this sense, this indicated an "exogenous" state of protein catabolism that had to take place in the circulating solution. Pflüger's idea was that for food protein to be catabolized it must first be transformed into "bioplasm," becoming an integral part of the living tissue, and would then undergo oxidative decomposition.

Voit and Pflüger agreed on the generally accepted assumption that the entire catabolism of protein took place in the same tissues and by means of catabolic processes similar to those that brought about the decomposition of the non-nitrogenous food, that is, the fats and carbohydrates. In other words, protein catabolism was supposed to take place where the greatest amount of oxidation occurs, in the muscles.

Otto's theory was based very strictly on his laboratory findings and was logical, but at the same time was open to further question. He argued against, but did not rule out completely, the concept of protein catabolism as an oxidative process.

> Such splittings are more easily accomplished by hydrolytic reactions than by oxidations. For example in a $\equiv CNH_2$ or $=C=NH$ group, the nitrogen can be split off as ammonia by the simple addition of a molecule of water. There is, therefore, *a priori* no reason for assuming that the katabolism of the protein-nitrogen is brought about by the same chemical processes as those which decompose the fats and the carbohydrates.

Furthermore, Otto observed that there was strong evidence that the nitrogen catabolism represented by urea did not take place in muscle as did creatinine. The liver was a likely site because it was shown to be capable of converting ammonia into urea. Not only that, but ammonia was a product of the proteolytic action of digestion and was not recombined into protein compounds, and hence could not have passed through muscle. This had been confirmed by the recent work of Connheim that showed that the "erepsin" of the mucous membranes of the intestine split proteins and peptones into amino acids. Kutscher also had shown that trypsin was capable of splitting protein products into small amino acid molecules.

Because intermediate products (proteoses and peptones) were not detected, it was thought that digestion was followed by synthesis of protein in the mucous membranes of the intestine. This, Otto pointed out, was a teleological explanation. "The food proteins are tissue builders and the organism therefore must not waste them." But what if intake is excessive? Pflüger and Voit would explain this by the generally accepted hypothesis, dating back to Liebig, that the organism used protein by preference, even when an abundance of fat and carbohydrate was available. Protein was preferred because of its nitrogen, despite the fact that nobody believed that a diet producing 25 g of urinary nitrogen was better for the average man than one that gave 15 or 16 g.

Otto believed that hydrolytic decomposition occurred in the digestive tract as a result of enzymatic action. This decomposition was carried further in the mucous membrane of the intestines and was completed in the liver, each splitting causing urea formation. This process removed excessive nitrogen to help maintain nitrogen equilibrium, even when excessive amounts were furnished with the food. The hydrolytic removal of nitrogen from the protein produced a non-nitrogenous residue of great fuel value which "in all probability is partly converted into fats, or at least into carbohydrate, and then becomes subject to the laws governing the katabolism of these two groups of food products."

Otto believed firmly in the correctness of his view of protein metabolism.

> ... it may well be that the digestive tract and the liver are not alone sufficient to split off all the unnecessary nitrogen. Other glands may play a part, for we know that proteolytic enzymes resembling trypsin are present in such glands.

Otto did not rule out the possibility that a certain amount of oxidation was associated with protein catabolism.

Finally, Otto showed that his theory led to the consideration of several important problems in a new light. The theory explained the persistent tendency of the organism to maintain nitrogen equilibrium, even if this meant excessive production of urea. This was shown best on subjects who, after being placed on a low-nitrogen diet for a week, were returned to the nitrogen-rich diet. The nitrogen doubled after the first day and tripled after the second.

> All the living protoplasm, and on account of the habitual use of more nitrogenous food than the tissues can use as protein, the organism is ordinarily in possession of approximately the maximum amount of reserved protein in solution that it can advantageously retain. When the supply of food protein is stopped, the excess of reserve protein inside the organism is still sufficient to cause a rather large destruction of protein during the first day or two of protein starvation, and after that the protein katabolism is very small, provided suffi-

cient non-nitrogenous food is available. But even then, and for many days thereafter, the protoplasm of the tissues has still an abundant supply of dissolved protein [Otto believed it to be colloidally dispersed] and the normal activity of such tissues as the muscles is not at all impaired or diminished.

Standard diets in use were probably unnecessarily rich in protein.

Nitrogen enough to provide liberally for the endogenous metabolism and for the maintenance of a sufficient supply of reserve protein is shown to be necessary.... The loss of 25 gm. or 30 gm. of nitrogen, due to withdrawal of nitrogenous food, does not involve loss of muscle substance, 1st, because the endogenous metabolism is affected; 2^d, because it is not accompanied by loss of strength and ability to do work; 3^d, because from a teleological standpoint, it seems highly improbable that the consumption of nitrogenous material in a system consisting of living protoplasm immersed in a highly nitrogenous solution should be at the expense of the solution. The chief reservoir for nitrogenous reserve material in the human organism therefore seems to be the unorganized protein in the fluid media rather than the protoplasm of the muscle. This is why there is normally in a man a comparatively sharp limit to his capacity for storing protein, and why all excess of the nitrogen given with the food is promptly eliminated.

A low-nitrogen diet such as Otto had used in his studies could be useful, he pointed out, for treating patients with diseases in which protein was known to be harmful, such as kidney disorders and typhoid fever.

What was the effect of work on protein metabolism? Moderate or even severe muscular work did not increase the catabolism of protein. Protein catabolism, insofar as the nitrogen was concerned, was independent of the oxidation that gave rise to heat or to energy that was converted into work. The hydrolytic splitting off of ammonia, and its elimination as urea, was independent of oxidation, except to the small extent of using (for urea formation) a relatively small amount of carbon dioxide that is produced by means of oxidation. Moderately severe physical work could have an effect on protein metabolism, but it could not be demonstrated with urea, or even total nitrogen determinations. Perhaps this would be best studied by determining creatinine, uric acid, and neutral sulfur.

The Folin theory of protein catabolism would stand for almost 40 years until the advent of isotopes as a means for tracing the fate of metabolic products and the perception of the dynamic state of metabolic processes within the living organism. Otto had provided a solid interpretation of facts based on meticulously obtained data with methods of clear ingenuity. And his work had a remarkable unity centered on nitrogen, beginning with the organic nitrogen derivatives he had studied in Stieglitz's course; to his laborious thesis on the urethanes and related products; to work in Europe on mucin, uric acid, and the so-called deutero-albumoses; to urinary nitrogen compounds; and now to a theory of protein metabolism. In a comparatively short time, Otto had established himself as a leading American biochemist, an innovator, and a careful laboratory worker.

The final paper that Otto published was unusual in that Carl L. Alsberg, and not Otto, was the senior author. His collaboration with Alsberg (1877–1940), who recently had been promoted to instructor of biological chemistry at the Harvard Medical School, shows that Otto had made professional contact with others besides Bowditch at Harvard, probably long before 1904 when this project was begun. Alsberg had been studying a case of cystinuria for some time before a paper appeared by Loewi and Neuberg (L & N) that made several points with which Alsberg and Folin (A & F) had disagreed (*11-13*).

L & N had claimed that there were two forms of cystine in protein, both derived from cysteine, but that the cystinuric individual catabolized only the isomeric form of cysteine, α-thio-β-amidopropionic acid, $CH_2NH_2CHSHCOOH$, the form presumably present in cystine stones. Whereas the normal individual catabolized both forms of cystine to sulfate, the cystinuric could handle only the isomeric form; therefore, the true cysteine was not hydrolyzed or oxidized and passed unchanged into the urine. L & N further believed that the cystinuric person had a more or less complete inability to catabolize amino acids normally, though they had not found any proof of this. The urines showed only abnormal quantities of cystine. L & N also had fed pure amino acids to cystinurics. Normally these would be metabolized and the nitrogen converted to urea, but the cystinurics eliminated the monoamino acids unchanged, almost quantitatively, while the diamino acids were eliminated as the corresponding diamines. Evidently L & N used this as a basis to hypothesize that amino acids were not normally produced in the digestive tract.

A & F, while not finding fault with the paper or the data of L & N, were unable to corroborate their findings. If amino acids were not present in the digestive tract, the cystinuric individual *should* show the presence of these amino acids, just as in feeding experiments. Also, the undetermined nitrogenous residue (nitrogen not in urea, ammonia, uric acid, and creatinine) should be greater in a cystinuric than in a normal person and should disappear when the protein was nearly or completely abolished. "These expectations were based on the assumption that the general results obtained by Neuberg and Loewi with amido acids were correct, but they had not analyzed the ordinary urine of their patient with sufficient care."

Alsberg's patient was a 23-year-old man, who had been known to "pass sand" since age 11 (1893) and had considerable trouble during 1897–99. In 1903 he was admitted to the Massachusetts General Hospital, and several cystine stones had been removed from his bladder. One of his brothers had been operated on five times for bladder stones, but no other relatives were known to have had the same trouble.

Alsberg and Folin, using the low nitrogen diet, intended to repeat and extend the studies of Loewi and Neuberg in regard to the fate of pure amino acids administered to their cystinuric patient, who evidently now was brought specially for this project to the McLean Hospital. The plan of feeding was that of earlier experiments that Otto had described, as were the analytical techniques, except that the presence of cystine necessitated a different procedure for determination of total sulfur, sodium peroxide being used as the oxidizing reagent.

This method as well as Otto's latest modifications for sulfate analysis would be described in a separate paper in 1906.

When the subject was placed on a standard protein (milk–egg) diet, the urine composition deviated unmistakably from the six normals that Otto had previously studied. The neutral sulfur and cystine sulfur were at least five times the expected concentration. The patient eliminated an estimated 1 g of cystine daily. The nitrogen metabolism, on the other hand, showed a somewhat lower urea output and only about half of the expected output of ammonia. The low output of ammonia was attributed to its displacement by a diamine that had been reported in several cases of cystinuria. Creatinine and uric acid output were normal.

There was nothing to challenge L & N's generalizations until the subject was placed on a low-nitrogen diet (starch-fat). If L & N were correct, then the undetermined nitrogen should sink to the level of that of the normal person. But this did not happen. Whereas the normal individual had an average value of undetermined nitrogen of about 0.4 g, the patient remained practically stationary at 0.8 g. While on the low-nitrogen diet, the subject was given 10 g of pure asparaginic acid. The amino acid was metabolized to urea, and there was no change in the "undetermined nitrogen" fraction. In another feeding experiment, tyrosine was similarly metabolized.

In two experiments with cystine feedings of 1.226 g and 6 g, they found no change in the neutral sulfur, but an increase in inorganic sulfate.

> These figures prove beyond reasonable doubt that in our case of cystinuria pure cystin in so far as it was absorbed from the intestinal tract was not eliminated in unchanged condition, but that its sulphur gave rise to normal katabolism product, sulphuric acid.

This showed that the disease known as cystinuria was not due to an inability to catabolize normally the hydrolytic cleavage product of protein, and that L & N were wrong in their generalizations.

A & F could not reconcile the fact that the neutral sulfur fraction did not rise when pure cystine was fed, yet did so when the protein-rich diet was used. For this reason they felt that their results should be questioned.

They recommended that cystinuric individuals avoid high nitrogen-containing diets, because less cystine was produced on the lower nitrogen diet.

The authors mentioned several things in closing. It had been shown recently that only one form of cystine existed: Hair, protein, and stone cystine were the same. The feeding experiments had shown that the cystinuric patient could catabolize amino acids. Cystine was oxidized to sulfate, but peculiarly, its nitrogen was *not* converted to urea. Undetermined nitrogen rose at the expense of urea and ammonia. For some unexplained reason, their cystinuric subject, when performing moderate work, eliminated more creatinine, uric acid, and neutral sulfur. The authors wondered if their discrepancies and their differences with L & N's data perhaps could be due to different degrees of cystinuria. There were no facts in the literature to prove that assumption. Three years later the prescient A. E. Garrod (1908) would postulate correctly that cystinuria was one of several inborn errors of metabolism (11-14), but a virtual explosion of fundamental biochemical knowledge would be required to reach our current knowledge of this disease. A & F had taken a small step in the right direction.

In October 1905 Otto was in New York City, probably for two reasons: to present a 15-minute paper on his theory of protein metabolism (11-15) at a meeting of the Medical Association of the Greater City of New York and perhaps to attend a special meeting of the founding group of the *Journal of Biological Chemistry*.

Otto continued to explore the ramifications of his protein theory.

> The striking difference in the role played on the one hand by urea, and on the other by creatinin in our daily metabolism constitutes the pivotal fact upon which I have ventured to construct a somewhat new theory concerning protein metabolism.

Tissue metabolism represented the daily wear and tear of all the cells in the different tissues, while exogenous metabolism represented the specialized activities of certain definite organs such as the digestive tract and the liver. Exogenous metabolism was independent of the size of the person. It depended normally on the protein intake in the food. The tissue metabolism depended on the condition and number of cells in the individual. "A half grown boy with a good appetite may produce more urea than a 200-pound man, but his creatinin will remain small."

In view of the two forms of protein metabolism, the question of proper or most advantageous nutrition for man had to be viewed in a new light, insofar as protein consumption was concerned. Voit and Pflüger considered protein consumption as one kind. They assumed that protein metabolism was of considerable importance, so they supported as necessary the protein consumption found for strong and vigorous men. Voit originated standard diets and claimed that 118 g of protein per day was the required protein intake, and that it was dangerous to reduce it below that amount. Otto scoffed at this. Most natural foods, if taken in sufficient amount to furnish the necessary fuel value, would contain the right amount of protein for tissue metabolism. The question of optimum protein consumption would therefore seem to depend mostly, if not only, on exogenous metabolism.

About 25 g of protein per day was apparently sufficient to maintain tissue metabolism in an average-sized person. Before surplus protein oxidized in the general tissues, it was stripped of nitrogen by special hydrolytic processes in special organs that were rich in hydrolytic ferments (enzymes). Urea formation was the result of these preliminary hydrolytic processes and had no connection with the general oxidative processes that liberate energy. The carbonaceous part remaining, once nitrogen was removed from protein, was not used up immediately, but probably was converted into carbohydrates and stored as carbohydrate or fat and used as needed.

While 118 g of protein was not an essential daily requirement, it was not necessarily detrimental to consume it. Possibly less was indicated.
Protein and meat were not identical terms. Excessive meat-eating was harmful, Otto wrote, but it was not demonstrated to be due to its protein content. It was believed that meat contained deleterious waste products, such as might be produced by autolysis, but not one such product was currently known.

Otto closed by mentioning that his studies on urinary composition and protein metabolism were made on normal persons. The conclusions, if correct, now should be applied to the study and dietetic treatment of disease, and this was a promising subject for further investigation.

The 1905 Annual Report for the McLean Hospital by Dr. Tuttle boasted of the "considerable attention" attracted by Otto's work and cited current references to him in the literature by Graham Lusk, *Science*, July 1905; P. A. Levene, *Journal of the American Chemical Society*, November 1905; D. Noel Paton, *English Journal of Physiology*, September 1905; Frank Billings, *Journal of the American Medical Association*, November 1905; C. A. Pekelharing, *Proceedings of the Academy of Science, Amsterdam.*, October 1905; W. H. Howell, *Text-book of Physiology*.

In early August 1906, Otto's niece Hildur Folin (1888–1970) emigrated from Älmhult, Småland. She was the eldest of Gusten's three children, and had come to seek educational opportunity and a future just as Otto had done. She had been a bookworm at home, and had resisted pressures to get married and "settle down." Otto and Laura had agreed to take her in and to help her learn English.

Hildur was a pretty, diminutive girl, who adjusted well to her new life with Uncle Otto's family, and no doubt she tried to earn her keep around the house. She learned the new language along with baby George, supplemented by excellent instruction from both Laura and Joanna. Through Hildur, Laura now would deepen her understanding of her husband's early struggles as an immigrant, his motivation to succeed, and why he had made a votary of biochemistry. And she would learn more about her in-laws, about whom Otto had once written and who still exchanged letters in Swedish with him. Hildur would go on to Axel's and Anna Maria's near Center City, Minn., and would follow a path reminiscent of Otto's early life.

Otto's initial paper in the first volume of the *Journal of Biological Chemistry* was on his latest refinement of sulfur determinations in urine (*11-16*). Because the determination of sulfur as barium sulfate had been shown to be beset by many problems, Otto set out to improve the method before applying it to urine. He thought that published values of sulfate and sulfur in urine were "intolerably unreliable," and included many of his own data. He established as his goal refinements of the methods for inorganic sulfate, total sulfate, ethereal sulfate, and total sulfur to the point where the maximum variation in these analyses would not exceed 1%.

A recent publication by Hulett and Duschak had postulated that HCl probably split off when $BaSO_4$ precipitates were heated because of the interaction of dissociated forms of sulfuric acid and barium chloride (e.g., BaCl', $HSO_4{'}$), not because of the occlusion of $BaCl_2$. These intermediate dissociated compounds then reacted with $SO_4{''}$ and Ba'' ions to produce compounds that coprecipitated with $BaSO_4$ (e.g.,

$$\begin{matrix} BaCl \\ BaCl \end{matrix} \!\!> SO_4; \qquad \begin{matrix} HSO_4 \\ HSO_4 \end{matrix} \!\!> Ba).$$

Otto corroborated this indirectly by finding conditions under which $BaSO_4$ tended to lose HCl and H_2SO_4 when gently ignited, and considered that other coprecipitates were possible, particularly from salts of sulfuric acid, to

$$\begin{matrix} MSO_4 \\ MSO_4 \end{matrix} \!\!> Ba,$$

where M is the cation. Otto could show that this happened when neutral $BaCl_2$ solution was dropped into K_2SO_4 solution. The precipitates obtained were much heavier than predicted.

Otto then studied every step in the production of barium sulfate in pure solution and made important modifications that brought him to his goal. He found the optimal concentration of $BaCl_2$ solution to use; how to obtain "beautiful crystalline barium sulphate precipitates . . . at ordinary room temperature," rather than by heating; and that shaking was harmful. "By the use of 10 per cent. barium chloride in the heat or by use of 5 per cent. barium chloride in the cold, without shaking or stirring, the SO_4 in pure potassium sulphate solutions may be determined with great accuracy."

Otto found that salts other than the K_2SO_4 could destroy the accuracy of the determination. For testing urinary and protein sulfur, fusion mixtures were used that contained KNO_3 and Na_2CO_3. "The conclusion seems warranted that if any of these determinations are correct, it is largely a matter of good luck. Most of them must be unreliable."

The influence of other substances was tested. Then Otto presented optimal conditions for use of Gooch crucibles and the proper preparation of asbestos mats, the ignition technique, and the titration and washing of $BaSO_4$ precipitates.

Then he turned specifically to determinations in urine. He presented methods for inorganic sulfate, total sulfate, and by difference, ethereal sulfates. And as he had mentioned in the paper on cystinuria, he described the quantitative analysis of total sulfur using sodium peroxide to oxidize the organic matter and to convert the sulfur present to sulfate. He felt that he had achieved his goal of 1% accuracy. His methods were highly useful, but analytical chemists would remain dissatisfied for decades with determining sulfur as barium sulfate, and hundreds of papers on the subject would be written. Notwithstanding this, Otto's procedures would open the doorway for sulfur analysis in clinical chemistry for many years to come.

Otto's last two papers of 1906, published abroad, were on creatine and creatinine, the important one being a chapter in a special volume dedicated to Olof Hammarsten (*11-17, 11-18*).

Despite prevailing views, Otto had found great difficulties in the efficiency of converting creatine into creatinine and vice versa. "In fact, up to the present time I have been unable to find the right conditions for converting any considerable quantity of either substance." Of course, it was presumed that muscle extracts were rich in creatine, and the animal organism was demonstrably capable of converting administered creatine into creatinine that was then eliminated with the urine. Was this true?

After showing some examples of his difficulties in obtaining complete interconversion of creatine and creatinine, despite literature references to the contrary, Otto tried feeding experiments on two normal male adult subjects. Placing one on a low-nitrogen diet, he fed him a moderate quantity of creatine and found that neither creatine nor extra creatinine was eliminated in the urine. There was no change in the urea nitrogen or ammonia nitrogen. On the other hand, when he fed creatinine, almost all of it was eliminated within the next 18 h. When Otto increased the dosage of creatine administered to 5 g he could recover a small quantity in the urine, but no additional creatinine. When he fed another subject who had been on the low-nitrogen diet for eight days a large dose of creatine (6 g), he was able to recover about 1 g of creatine, but no additional creatinine. When he fed creatine to a subject on a diet rich in protein, 50% was eliminated unchanged in 24 h, but creatinine was constant. When very large quantities of meat were fed, Otto saw only a slight increase in creatinine, whereas the creatine elimination rose markedly.

Otto established the following hypothesis:

> Kreatin, in contradistinction from kreatinin, is not a waste product at all, but a food. Under the conditions of the above experiments, on a low-nitrogen diet, neither kreatin nor demonstrable quantities of kreatin-nitrogen were eliminated, because the amount of kreatin existing in the tissues had been reduced by the low-nitrogen diet. If this view were correct, then I should obtain different results by feeding kreatin together with a diet rich in protein.

The retention of creatine by the subjects fed a low-nitrogen diet clearly indicated that creatine was not a waste product, but Otto pointed out that his experiment did not prove it to be a food. Further work was necessary.

> ... it is possible that the nitrogenous substances which serve to maintain the nitrogen equilibrium in the living tissues are special products which do not easily take part in the urea forming processes. Kreatin may be one such product, and is therefore retained in the general tissues, and consequently we find the muscles rich in kreatin. When the organism is daily supplied with an abundance of protein it may then be preparing as much kreatin as is needed for the maintenance of its normal supply. The kreatin given with the food is consequently not absorbed and retained by the muscles to the same extent as when the food contains an insufficient supply of protein.

The origin and metabolism of creatine and creatinine would remain a lifelong question of interest to Otto. Cyrus Fiske and Yellapragada SubbaRow would one day show the presence of creatine phosphate in muscle, and the importance of creatine as a "food" would then be understood. It was an active metabolite, an immediate source of energy for muscular contraction, via its "high energy" phosphate bond.

In October, 1906 Otto received a letter and circular from Professor Abel. The letter read as follows:

> The enclosed circular is being sent to the gentlemen whose names appear on the title page of the *Journal of Biological Chemistry* for the purpose of obtaining their signatures, should the project therein outlined meet with their approval. These gentlemen also are requested to name other prominent workers in biological chemistry whose signatures should be secured.
>
> When the circular has received the signatures of a sufficient number of representative men, it will be sent to a selected list of chemists in various branches of biology and medicine as an invitation to membership in the new society.
>
> The responsibility of selecting this list should fall on the signers of the proposal. Will you not cooperate in this matter? Have you any suggestions to offer in respect to the method of selection? It is of course assumed that only those who are worthy of membership by virtue of acquirements and work accomplished will be nominated.
>
> The proposal has followed as the result of conversations with biological chemists in different parts of the country, and it would seem that the time has come for bringing the project to the attention of those most capable of launching it.
>
> Details, such as the form of constitution to be adopted, the question of affiliation with other scientific bodies, annual time and place of meeting, etc., will naturally be subjects for consideration at the meeting for organization.

The circular read in part:

> *A Proposal to Form an American Society of Biological Chemists*
> The undersigned have become convinced that there is need in this country for an association which shall embrace in its membership all who are interested in the biological sciences from the chemical point of view. At present such workers are affiliated with widely different organizations, and come little in contact with one another.
>
> An organization such as we now propose should bring together the best trained and most productive workers on the chemical side of the biological sciences as a whole; not only those whose special field is physiological chemistry, but also those who deal with the chemical and physico-chemical problems that are encountered in botany (and bacteriology), zoology, physiology, pharmacology, pathology and medicine. . . .
>
> It is proposed to meet for organization in New York at the time of the winter meeting of the American Association for the Advancement of Science and of the Society of Naturalists. Notice of time and place will be sent to you. Meanwhile, please state whether you will accept membership in the proposed society.

Professor Abel received more than 20 favorable replies, including not only Otto's, but practically all of the established physiological chemists. In a second letter to the group of responders on Dec. 13, 1906, Abel called for a meeting to be held on the afternoon of Dec. 26, at the Hotel Belmont in New York City to organize a National Society of Biological Chemists.

The American Society of Biological Chemists was born at the auspicious meeting, with Russel Chittenden presiding, and following a talk by Dr. Abel, the Articles of Agreement were adopted unanimously.

> We wish to draw into our society the biological chemists of all departments of biology, including those organic and physical chemists who take a lively interest in our subject. . . . Our common meeting ground should be chemistry as applied to animal or vegetable structures, living or dead, throwing light on the life processes and functions of living structures (*11-19*).

Officers and council elected at that first meeting were R. H. Chittenden, president; John J. Abel, vice-president; William J. Gies, secretary; Lafayette B. Mendel, treasurer; and Otto Folin, Walter Jones, Waldemar Koch, John Marshall, and Thomas B. Osborne, council members.

According to Superintendent Tuttle's Annual Report of 1906 for the McLean Hospital, Otto's work on protein metabolism was appreciated locally.

> In the paper entitled 'A Theory of Protein Metabolism' (page 37) the suggestion was made that the soluble starch diet, previously described, ought to prove valuable in certain diseases, especially in typhoid fever, and it was tried at the Massachusetts General Hospital under the supervision of Dr. Minot (Charles S. Minot, James Stillman Professor of Comparative Anatomy). Four typhoid fever patients were kept about three weeks on the starch solution, which was daily sent from the McLean Hospital chemical laboratory. The last three months of the year were devoted to the study of urines of these typhoid patients. It is believed that the results will prove no less interesting to physicians than to physiologists, and would seem to indicate that the ordinary milk diet in typhoid has no justification.

The year 1907 marked for Otto Folin his appointment to the faculty of the Harvard Medical School. Before the spring was over, Otto's scientific production for that year was completed, obviously caused by a break in his research activity at McLean. On Feb. 4, he submitted two short papers for publication in the *Journal of Biological Chemistry*. The first paper was a defense of a method he did not prefer to use for determination of sulfate and sulfur in urine (*11-20*), and the second was primarily a "negative" paper in which Otto sought unsuccessfully to determine whether creatine that presumably was not metabolized could have appeared in the urine as methylurea or methylamine (*11-21*).

Because the standard Messinger-Huppert method for determining acetone and diacetic acid actually gave the sum of the two, Otto noted that there was a manifest need to determine the two ketone bodies separately. In his third paper, Otto proposed a simple modification that achieved this purpose quantitatively for the first time, and then applied the method to the urines of diabetic people (Part II, p. 384).

The Messinger-Huppert method made use of the quantitative conversion of acetone to iodoform (CHI_3) with standard iodine solution in strongly alkaline solution. Under the conditions of this reaction the diacetic acid decomposed to acetone and was also determined. The excess iodine was then back-titrated with standard sodium thiosulfate and starch as indicator in acid medium. The urine used was acidified and then distilled prior to carrying out the iodoform reaction on an appropriate aliquot.

Otto's major modification was to separate the acetone from the urine by an aeration technique that used his aerometer cylinder devised for ammonia determination. The acetone in urine could be determined in about 30 minutes. Total acetone and diacetic acid were, of course, determined on the unaerated urine.

Otto also described a method for performing acetone and ammonia simultaneously, and he boasted of the latter.

> In this connection I may be permitted to state that in my opinion no other method yet devised for the determination of ammonia is so accurate for all kinds of urine as my air current method. An excellent alternative is the vacuum distillation method as described by Shaffer. . . .

Otto made a study of acetone, diacetic acid, and ammonia in the urines of seven diabetic subjects that he obtained from Dr. Elliott P. Joslin. This meant, of course, that Otto had made contact with other faculty members at the Harvard Medical School. Otto had this to say about the results that were tabulated. "The values recorded in the above are not uninteresting because they represent the first information yet obtained concerning the relative proportions of acetone and diacetic acid in diabetic urines." In this paper Otto did not deal with the medical problems of ketone bodies in the urine, but rather with the analytical aspect.

Diacetic acid was far more prevalent in diabetic urine than acetone. Otto pointed out the potential sources of error in the acetone method and included a figure of the double absorption tube used in ammonia determination, which now was sealed into a single unit to avoid using a rubber stopper, and could be purchased from Eimer and Amend in New York City.

Otto went to Washington, D.C., to attend a special meeting (actually the first) of the American Society of Biological Chemists, May 8–9, 1907. This two-day affair was deliberately held in Washington to coincide with the meetings of the Association of American Physicians, the American Physiological Society, and the Washington Section of the American Chemical Society. Only 26 members of the society attended. The programs were held in four sessions, two on May 8 and two on May 9. It was decided at this meeting that the abstracts would be printed for the first time in the proceedings of the society as presented in the *Journal of Biological Chemistry* (Vol. 3). Among the pioneer biochemists present were Phil Shaffer, Jacques Loeb, J. J. Abel, P. A. Levene, L. J. Henderson, C. A. Herter, J. A. Mandel, V. C. Vaughan, and Lafayette B. Mendell.

A Classical Period

Otto's final paper of 1907, which appeared in July, was also the last one he would publish exclusively from the McLean Hospital. It was based on a paper he had presented at a symposium on acidosis before the meeting of the Association of American Physicians on May 9. It was Otto's twenty-seventh paper from the Hospital (11-22). The paper that Otto presented was in reality a thorough discussion of his thinking about Knoop's recent theory (1905) of β-oxidation of fatty acids and its implications as a source of acid intoxication in diabetes. No experimental data were presented.

References and Notes

11-1. Folin, O., Bemerkung zu der Erwiderung von Martin Krüger (A note on Martin Kruger's response). *Hoppe-Seyler's Zeit. Physiol. Chem.* **41**, 176 (1904).
11-2. Doisy, E.A., Philip Anderson Shaffer. *Biogr. Mem. Natl. Acad. Sci. U.S.A.* **40**, 321–26 (1969).
11-3. Folin, O., and Buckman, T.E., Footnote. *J. Biol. Chem.* **17**, 483 (1914).
11-4. Folin, O., Über das von Salkowski und spater von Salaskin benutzte Prinzip der Blutalkalescenzbestimmung (On the Salkowski and later Salaskin used principle for determining blood alkalescence). *Hoppe-Seyler's Zeit. Physiol. Chem.* **43**, 18–20 (1904).
11-5. According to H. K. Beecher and M. D. Altschule, *Medicine at Harvard*, (p. 223. Hanover, N.H.: University Press of New England, 1977), Professor William T. Porter, Harvard physiologist, founded the journal in 1897 and was its sole responsible editor and publisher for 34 volumes. In 1914, he presented the journal to the American Physiological Society.
11-6. Anonymous, Christian A. Herter, M.D. *J. Biol. Chem.* **8**, 437–39 (1910).
11-7. The first volume (1905–6) was edited by J.J. Abel and C.A. Herter. The collaborators included R. H. Chittenden, Otto Folin, Willian J. Gies, Reid Hunt, Walter Jones, J.H. Kastle, Waldemar Koch, Jacques Loeb, Graham Lusk, A.B. Macallum, J. J. R. Macleod, A.P. Mathews, L.B. Mendel, F.G. Novy, W. R. Orndorff, Thomas B. Osborne, Franz Pfaff, A.E. Taylor, V.C. Vaughan, Alfred J. Wakeman, and Henry L. Wheeler.
11-8. Chittenden, R.H. *The First Twenty-five Years of the American Society of Biological Chemists*. New Haven, Conn.: American Society of Biological Chemists, 1945.
11-9. Folin, O., Protein metabolism in its relation to dietary standards. Proceedings of Lake Placid Conference on Home Economics, 1905, 8 pp.
11-10. Folin, O., Approximately complete analysis of thirty "normal" urines. *Am. J. Physiol.* **13**, 45–65 (1905). It should be here noted that Otto did not describe the details for preserving the urine specimens. In the earlier paper, Otto mentioned using thymol dissolved in chloroform as a preservative. Specimens that nevertheless showed ammoniacal decomposition were discarded. How long the specimens may have stood at room temperature, whether they were refrigerated, or other related information was not mentioned.
11-11. Folin, O., Laws governing the chemical composition of urine. *Am. J. Physiol.* **13**, 66–115 (1905).
11-12. Folin, O., A theory of protein metabolism. *Am. J. Physiol.* **13**, 117–38 (1905).
11-13. Alsberg, C., and Folin, O., Protein metabolism in cystinuria. *Am. J. Physiol.* **14** 54–72 (1905).
11-14. Garrod, A.E., Inborn errors of metabolism. *Lancet* **2**, 1, 73, 142, 214 (1908).
11-15. Folin, O., A theory of protein metabolism. *N.Y. Med. J.* (March 3, 1906).
11-16. Folin, O., On sulphate and sulphur determinations. *J. Biol. Chem.* **1**, 131–59 (1906).
11-17. Folin, O., The chemistry and biochemistry of kreatin and kreatinin. Festschrift fur Olof Hammarsten. III. pp. 1–20. Upsala: Akademiska Boktryckeriet, Edv. Berling, 1906.
11-18. Folin, O., The metabolism of kreatin and kreatinin. *Br. Med. J.* **2**, 1787 (1906).
11-19. See Reference 11-8.
11-20. Folin, O., On the reduction of barium sulphate in ordinary gravimetric determinations. *J. Biol. Chem.* **3**, 81–82 (1907).
11-21. Folin, O., On the occurrence and formation of alkyl ureas and alkyl amines. *J. Biol. Chem.* **3**, 83–86 (1907).
11-22. Folin, O., The acid intoxication theory. *J. Am. Med. Assoc.* **49**, 128–31 (1907).

Chapter 12. Harvard—Teaching and Profession (1907–12)

On Feb. 28, 1907, Professor H. P. Bowditch wrote the following letter to Dr. Arthur Tracy Cabot, a member of the Corporation of Harvard University, and formerly a lecturer on genitourinary surgery in the medical school:

My dear Cabot,
   The teaching of physiological (or biological) chemistry in the medical school has been the subject of so much earnest conversation between us that I need make no apology for returning to the question which has now assumed additional importance through the action of the Carnegie Trustees. I have good reason for believing that Boston was chosen as the home of the new Carnegie laboratory for the study of nutrition not on account of, but in spite of the proximity of our laboratory of biological chemistry. Now it seems to me rather sad to think that, just as we are trying to develop medical education on University lines, one of our departments turns out to be officered by men who are not considered likely to cooperate effectively in research work of the utmost importance in their own lines of medical science.
   I have no criticism to make on the work of the men in charge of the chemical department. I only note with regret that they yet are to win their spurs. You know I have always urged the importance of placing at the head of that department the strongest physiological chemist the world can afford.
   Now do not imagine that I am writing you this letter for the sake of indulging in vain regrets. My object is to urge action which will prevent a physiological chemist of the first rank, now in our neighborhood, from slipping through our fingers. I refer to Folin, now in Waverly, who, as I happen to know, is anxious to get a University position. I dare say you know more about him than I do but if you have any doubt about the importance of his work, I wish you would consult such men as Abel and Chittenden. There is room for him in either the department of physiology or in that of chemistry. I am sure the physiologists would welcome him and I see no good reason why the chemists should not do likewise.
   ·I should regard it as a great misfortune if Folin were to a accept a position in another university for it would seem to indicate an inability on our part to recognize a first class man when we see him or a failure to appreciate the important bearing of his work on the advancement of medical science.

Pardon the infliction of this letter. I cannot get over my interest in the school and the very gratifying notice of my appointment as Professor Emeritus which I received this morning, may serve as my excuse for tendering unsought advice.

Very sincerely yours,
H. P. Bowditch

But Bowditch was not alone in expressing the need for strengthening the department of biological chemistry and for hiring Otto in particular. Walter Bradford Cannon (1871–1945), the newly appointed George Higginson Professor of Physiology and Bowditch's successor, had also made strong a pitch on Otto's behalf as well as for a full term in biochemistry for the medical students.

The Cannons and Folins would become lifelong friends. Mrs. Cornelia Cannon, like Laura, was a native of St. Paul, where Walter had also lived. Walter had already performed pioneer work on gastrointestinal motility using radiopaque bismuth salts, which had led to studies on the effect of emotion on peristalsis, and to broader studies of the emotions, the adrenal glands, and the sympathetic nervous system (*12-1*).

From the files of Harvard President Charles William Eliot, we find that the President was authorized by the corporation to "sound" Dr. Folin, as of May 27, 1907. On May 28, he sent the following letter to Otto:

> Dear Dr. Folin:-
> I hope to have an opportunity shortly to discuss with you a project for making you an Associate Professor in our Faculty of Medicine. Unfortunately, I am obliged to keep two engagements of long standing this week at some distance from home, and shall not return until next Monday.
> Would you kindly send a note to Mr. Jerome D. Greene, Secretary to the Corporation, 5 University Hall, Cambridge, telling him when and where it would be convenient for you to see me after Monday next?

Otto wasted no time in responding. May 29, he wrote:

> Mr. Jerome D. Greene,
>   Cambridge.
> Dear Sir:-
> Replying to a note from President Eliot requesting an interview after Monday of next week will you kindly inform him that I shall be glad to meet him in Cambridge at such time and place as he may appoint.

Mr. Greene then sent Otto a note stating that the president would see him at his office on Tuesday, June 4, at 11 a.m. Obviously, Dr. Eliot was eager for the interview. As a result of the interview, the president found that Otto had definite ideas about the condition of his appointment and the pay involved, as well as the value of the work he had done. Otto wished to have a professorship. His work had been concerned with practical subjects taught in the medical school, namely nutrition, chemistry of metabolism, chemistry of the urine, and analytical methods useful in clinical work. The quality of his work had made him into an authority on these subjects. Consequently, he did not wish to be placed in a subordinate position that might interfere with his freedom.

While noting that he had been offered a salary of $4500 or $5000 at the Rockefeller Institute, Otto agreed to accept a professorship at $4000 if he were also provided an assistant. Three-fourths of this cost could be derived from the university, and a fourth from the McLean Hospital, if Otto were allowed to spend two days a week away from the school. The assistant's pay could be $700 per year.

Under this dual arrangement of his duties, Otto would have an assistant at both the school and the hospital and could expect thereby to increase his present research productiveness.

Evidently something close to the above arrangement was offered to Otto, and he sent the president the following note on June 6, 1907:

> I shall be glad to accept an Associate Professorship in Biological Chemistry in the Harvard Medical School.

Sometime during the winter of 1906–7 the Folins moved to 60 Sycamore Street in Waverley. In March of 1907, Laura went with Joanna and George to visit her parents in St. Paul. Her former landlord's sister, Dorcas Morrison, sent her a letter on March 22 that provides an inkling of how well the Folins and their children were liked.

> My dear friend,
> I thought of you many times as you were on your homeward journey. . . .
> I hope the children keep well and happy.
> You have no idea how much we miss them both. I think Joanna must have been our idol for we loved her so much. I wish I could feel that the next six years might be as pure and sweet as those past. . . .
> I shall never lose my interest in her and shall watch with interest the changes which come year by year. . . .
> George was just as nice but he was not with us as much. . . .
> The [new] family moves into the house next week. I hope they will remain as long as you, and I am very glad they have not got any children for I would not wish to have anyone take Joanna's place.

Otto's total scientific production for three straight years following his appointment to the faculty at Harvard consisted of only two papers that involved no laboratory work. He submitted his last paper containing fresh data in the spring of 1907. The next one with original material in it was tendered for publication in the spring of 1910.

Harvard created a new dimension with which Otto had to grapple immediately. He had to teach medical students—not graduate students primarily, but beginners in the field of biochemistry. He had briefly experienced teaching the women medical students in Chicago, but the students at Harvard were presumably more sophisticated, as they had met more demanding standards of admission. After 1901, the Harvard Medical School became one of the few schools requiring a bachelor's degree for admission, though this was later modified. President Eliot, with the help of a faculty of growing distinction, had transformed the medical school from one of a decidedly and deservedly inferior status—a proprietary school with part-time instructors—into a positon of equality and prominence at Harvard.

Having hired Otto, we will conjecture that Eliot told Otto that he felt that it was imperative to upgrade the basic science aspect of medicine. For biochemistry, only a purely trained chemist could do it. He had appointed Otto, an outsider to the establishment—a non-M.D. and a non-New Englander, to boot—to teach biochemistry whether or not it related to future practice of the medical students. They had to learn the basis of biochemical processes in the human body and to grasp the normal chemical functions to understand the pathological, in the same sense that knowledge of normal anatomy must precede that of morbid anatomy. He advised Otto to focus his immediate attention on establishing a strong department for teaching medical students. The graduate students, technical help, and the finances would come for his equally important research. He could afford, if he chose, to wait a while to continue his research, because just as Stieglitz had said, "Harvard is Harvard."

Otto also must have thought of his own teachers, particularly those in Europe. Hammarsten, Salkowski, and Kossel had reached their positions of eminence via academia. They had begun as teachers in medical schools, and as their research production grew the sheer force of it brought for them promotion and honor and students from everywhere. Otto now had the same opportunity as his teachers in Europe, even more so, because he was not linked to physiology, but was in an independent department, a clear and very important element in the growth of future American biochemistry. However, the teaching function in their lives never had been subordinated to research. More than their "bread and butter," it was a stimulus to their thinking, a continual outlet for updating of progress in physiological chemistry, and a potential source of graduate students. For Otto it also meant a way to introduce new advances in clinical chemistry to medicine, including his own findings; to discover the pitfalls in the application of new analytical methods; and to watch the use of physiological chemistry grow in medicine, as the students graduated and entered medical practice.

Otto's pure research position at the McLean Hospital had been, after all, limited in its potential scope. There was no ladder of success to climb there. He could not expect to create a scientific center, in the European sense, because the mission of the institution was not primarily one of education and research, but of effective treatment of mental disorders. Otto was there as the result of one man's foresight and persistence, but not because a new approach in psychiatry had emerged involving physiological chemistry. No matter how well Otto performed, the position he was in was somewhat untenable because it had no tenure and no defined goals for growth and progress. Seven years after he had started, he still had only one assistant, and his salary had risen by only $500. He had uncovered no information of direct bearing to mental patients. Undoubtedly, Otto had seriously considered leaving McLean before the Harvard offer came.

Why did Otto not sever relations at once with the McLean Hospital? It could have been done, perhaps easily, in his communication with President Eliot. Two reasons seem plausible. He was grateful to the people who had provided him an opportunity for research, Dr. Cowles and his successor Dr. George T. Tuttle, who would have wanted him to stay on, even part-time. More important, however, was the unique cooperation that Otto obtained there, as was shown in his analytical studies of the urinary constituents of patients who were put on constant diets. Support for his very demanding needs, regardless of the inconvenience, had been unparalleled. He had no prospective basis for continuing such studies at Harvard, no relationship with the hospitals affiliated with the medical school, and no access to patients for study. As a matter of fact, he would never again duplicate the ideal milieu he had for metabolic studies at McLean, notwithstanding his later contact with the teaching hospitals.

For two years prior to Otto's appointment, the continuity in the teaching of physiological and pathological chemistry to the medical students had fallen on the shoulders of two bright young instructors, Carl L. Alsberg, M.D. (1877–1940), and Lawrence J. Henderson, M.D. (1878–1942). Alsberg, whom Otto had befriended when they worked on the cystinuric patient at McLean Hospital, had joined the department as an assistant in chemistry in 1902, when the department was headed by Professor Edward S. Wood. Upon Wood's death in 1905, Alsberg was promoted to instructor and assumed the administrative tasks for the department, and Henderson was hired.

In addition to Alsberg and Henderson and two assistants from the previous year, for the 1907–8 academic year the staff would also now include Otto and his promised research assistant, Frederic C. Blanck, Ph.D.

The newly titled Department of Biological Chemistry, under the two instructors of 1906–7, offered 92 first-year, second-term medical students Biochemistry 1, with daily lectures and a two-and-one-half hour laboratory session following each lecture. Henderson discussed "the theories of chemical constitution and a survey of those classes of chemical substances which are to be found in animals and plants," and Alsberg dealt with the general principles and more important facts of chemical physiology and pathology. The laboratory course was designed to acquaint the student with important constituents of living matter and their chemical behavior, as well as routine biochemical methods of investigation. "Conferences and discussions of selected topics supplement the main work of the course."

The remaining courses, Biochemistry 2, 3, and 4, were listed as graduate courses. Biochemistry 2, on metabolism, was given by Alsberg alone in 1906–7, but Otto joined him for the 1907–8 session. Biochemistry 3, a laboratory course on the analytical techniques, also was taught by Alsberg and Otto. Biochemistry 4 was strictly the stepchild of Henderson. Entitled "Applications of Physical Chemistry to Biology," it was a lecture course given five times a week during January. This course must have convinced Otto of Henderson's outstanding gifts as a chemist.

> [It was] designed to acquaint the student with recent application of physicochemical theories and methods to Biology and medical science. The subjects to be discussed will include the theory of solution, the concentration law, catalysis, ionization, the theory of colloids, and the physico-chemical organization of the cell. The lectures will be supported by extended reading, and opportunity

for practice in physico-chemical methods will be offered. In preparation for this course an elementary acquaintance with Physical Chemistry, such as may be obtained from Chemistry 8, offered by the Division of Chemistry of the Faculty of Arts and Sciences is desirable. (12-2)

In 1907, the Benedicts, Francis Gano and Cornelia (Golay), moved to Cambridge. F. G. had come to serve as director of the new institute that Bowditch had mentioned, the Nutrition Laboratory of the Carnegie Institution of Washington, on Vila Street adjacent to the Harvard Medical School in Boston. Benedict had obtained his bachelor's (1893) and master's (1894) degrees at Harvard and his Ph.D. at Heidelberg in 1895. Mrs. Benedict and Laura Folin were former schoolmates at Vassar. The two families would become intimate lifelong friends.

Francis Gano Benedict (1870–1957), like Otto, had trained thoroughly as an organic chemist, but had switched to physiological chemistry, particularly nutrition. He had risen to the rank of professor at Wesleyan (Connecticut) University. With Atwater, he had performed nutritional investigations on their respiration calorimeter for the U.S. Department of Agriculture. In addition to being a teacher, he was a chemist at the Storrs Agricultural Experiment Station at the time Otto had met him in 1899. Benedict would perform fundamental research on basal metabolism in man, temperature regulation of the human body, the efficiency of the human body as a machine, and energy transformation.

In the spring of 1907, Laura was pregnant with her third (and last) child, Teresa, who would be born on Feb. 5, 1908, in Waverley (Belmont). Laura and the two children probably returned for a visit to her parents in St. Paul during the Christmas holidays, while Otto attended the second annual meeting of the American Society of Biological Chemists (ASBC) in Chicago, Dec. 30–Jan. 2. The Benedicts may have accompanied them, with F. G. staying with Otto, while the wives and children continued on to St. Paul, where Mrs. Benedict's parents also resided.

Thirty-four members of the ASBC attended the Chicago meeting, and 56 scientific papers were presented in four sessions. A session was held jointly with the American Physiological Society, and another with the Biological Section of the ACS and with Section C (Chemistry) of the American Association for the Advancement of Science. Otto presented a paper on "Protein Metabolism in Fasting" on Tuesday morning, Dec. 31, in the joint session with the physiologists.

The abstract of Otto's paper consisted of three sentences, and represented work not previously reported that had been done some time earlier at the McLean Hospital.

> A detailed analysis of the urine obtained during a seven-day fast for a man whose protein katabolism previously had been reduced to a minimum. The nitrogen elimination rose from day to day during the fast. The conclusion was drawn that the nitrogen elimination during the early stages of fasting can be made high or low at will and in no case do the figures obtained have any bearing on the necessary destruction (or consumption) of protein.

At the Chicago meeting of the ASBC, John J. Abel was elected president, and Otto, vice-president. William J. Gies remained as secretary, while Lafayette B. Mendell became treasurer. F. G. Benedict became a member of the nominating committee.

Otto's election to the vice-presidency, following the election to the presidency of Chittenden and Abel, two of the prime movers of the association and the journal, was a tribute not only to his scientific accomplishments that were now substantial, but to his personality. His was a pleasant synthesis of the old world and the new, a modesty and humility that was superimposed on a worldliness and culture, mingled with a keen sense of humor.

At the Washington meeting of the society in 1907, an ad-hoc committee had been formed to make recommendations on protein nomenclature to be coordinated with those of a similar committee of the American Physiological Society. The ASBC committee consisted of Folin, Gies, Koch, Osborne, and Chittenden, with the last as a chairman. The proposal of the two societies, presented in a paper entitled "Joint Recommendations of the Physiological and Biochemical Committees on Protein Nomenclature," would have importance in the teaching of biochemistry for many years to come. Recognizing that a chemical basis for nomenclature was not yet possible and that the groupings used could not be wholly satisfactory, the joint committee made three proposals, 1) the abandonment of the term "proteid," 2) the adoption of the term "protein" and its composition of α-amino acids and derivatives, and 3) the use of certain terms to designate major "groups" of proteins and their subclasses. These groups of proteins were simple proteins, conjugated proteins, and derived proteins, with several subclasses.

Philip Shaffer cited an unpublished autobiographical note that Otto wrote on April 9, 1924, in which he assessed his research experience at McLean Hospital as follows:

> When I was appointed chemist to the McLean Hospital in 1900 it became my duty to do chemical research on problems bearing on mental diseases. As the pathologist wanted all the brain material, I took to the field of metabolism.
> It was hopeless to try to find deviations from the normal in the metabolism of the insane without far more exact knowledge of the human waste products than was available. My immediate and comprehensive problem became, therefore, the chemistry of urine. I realized that by thus interpreting my duty to the hospital, I could do work of more general interest. I probably also followed my taste, for I enjoyed the mere puzzle aspect which always is present when one tries to devise a new method.
> My papers on the Laws Governing the Chemical Composition of Urine and a Theory of Protein Metabolism (1904) will probably be considered my best; but the data for those papers came easily and naturally by the help of the new methods of determination of urea, ammonia and creatinine which I had devised during the preceding three years. *(12-3)*

On Feb. 22, 1908, Otto delivered a lecture in New York City before the Harvey Society *(12-4)*. This was a unique paper in that it was the first strong plea for the medical profession, and in particular the larger hospitals, to become involved in biochemical research. Why? Because the time was ripe to apply the new field of physiological chemistry to the multiplicity of potential practical chemical problems embodied by hospital patients.

Such an application, however, required not only discrimination in selecting among the "most important, fundamental, and alluring problems," but in the investigators involved.

> ... men of every shade of experience and training have been drifting into biologic chemistry and are working on biochemical problems.... But the larger problems, the systematic working of the important fields of biochemical research call, almost without exception, for all the ingenuity, resourcefulness, and critical judgment of a trained chemist....
>
> There is now in this country an extraordinary demand for proficient physiological chemists, yet only one university [Yale] has made effective provision for meeting this demand. The universities still are turning out doctors of philosophy in chemistry, who also know considerable physics and geology, but none or practically none who are familiar with any branch of biology. And the medical schools still are graduating only physicians. They doubtless believe heartily in the importance of the medical sciences, but they are only just beginning to understand that the development of those sciences demand [sic] suitably trained specialists, specialists who can not grow up in sufficient numbers on the basis of personal initiative alone.
>
> The fact that our first physiologic chemist [R. H. Chittenden] still is in his early fifties certainly indicates that these are pioneer days in biologic chemistry....
>
> It is useless to dream that chemical problems can be solved in hospitals unless the hospitals have men and equipment for work on those problems.... During the past two or three years almost every biologic chemist in this country has had the opportunity of making a change. And there is a waiting list.... Notwithstanding the present popularity of biochemical research, notwithstanding the general confession of belief in its importance, it still remains to be seen whether hospital staffs really want it.

Otto believed from his long experience at the McLean Hospital that the time would come when the medical profession would recognize in practice as it did in theory,

> that the large city hospitals also should be centers for biochemical research ... the destructive and regenerative processes at all times to be found in large general hospitals constitute one of the most important fields for unceasing biochemical investigation, and into this field should be called the most able men to be found anywhere.... Indeed, the most direct and important aim of biochemical investigations must be the advancement of our ability to differentiate between the physiologic and the pathologic, and this can be done only when the investigator is as alert for new points or false teachings in the domain of the physiologic as in that of the pathologic....
>
> It is a mistake to think that clinical men in the hospitals can either do or direct such chemical work, [even though] they may have the most valuable material for biochemical investigations in their possession. Out of the abundance of such material they have made occasionally in the past and in the

future they will occasionally make, observations which can be worked up into distinct contributions to pathologic and physiologic chemistry. But without adequate provision for the thorough sifting and critical investigations of the observations of the clinicians, most of their impressions must remain hopeless mixtures of the correct, the probable and the impossible. . . .

I venture to predict that we shall learn more concerning the abnormal or subnormal metabolism of the sick on the basis of kreatinin and kreatin determinations alone than could be learned in another thirty years by means of nitrogen determinations of the past.

Otto cautioned that to do exact quantitative work on tissue metabolism, 24-h urines had to be obtained along with information on height, weight, muscular development, and diet.

Where should metabolic studies be focused? His own concept of dual protein metabolism suggested the need for several studies. For example, he said, patients with acute fevers required study because their protein metabolism was enormously accelerated and their urinary waste products reflected this acceleration. Already, in this connection, there were three papers on typhoid fever in preparation from three different laboratories. This work related to hospitals because it would show why some diets ought to be better than others for these patients. It would probably take years to amass what should be learned from "fever" urines alone. The neutral sulfur and the undetermined nitrogen should be given more attention. Elimination of creatine in fever needed investigation. It was not certain whether creatine in urine was pathologic, as it was not a waste product. The concept of dual protein metabolism should not be applied only to fever, but to studies of pregnancy, convalescence, progressive paralysis, gigantism, and dwarfism. Chronic diseases such as rheumatism, nephritis, atrophy of the liver, and diabetes would be included later in this diet.

He emphasized that they had to discover the general laws of animal metabolism. They needed to know first whether the waste products in the urine could reflect abnormality. This could be done on the basis of tissue metabolism "just so soon as the necessary standards for comparison have been worked out."

Otto prophetically turned his thoughts to new work of the practical clinical chemist in diagnosis and in monitoring the course of treatment and even the therapy.

> We need, however, an abundance of reliable statistical material. The normal tissue metabolism from infancy to old age must be standardized carefully by the help of suitable and easily duplicated test diets. Then we could find abnormal tissue metabolism in severely sick patients, as well as those who are merely weak and debilitated and who need "building up." We could learn whether a given treatment has produced a general and fundamental improvement or whether it has merely removed some of the symptoms. So far as I now can see, there is no reason why we should not be able to determine with all desired certainty to what extent it is possible to influence metabolism by drugs, by diets, and by different modes of living. The very stability of the tissue metabolism (as indicated by the constancy of the kreatinin elimination) against nearly all fleeting changes of diet and conditions would seem to me to

constitute the surest guarantee that this line of work, at all events, not yield a series of illusions. If well done, it should add a new chapter to the science of metabolism and of medicine.

The philosophy that Otto Folin enunciated in this paper has not paled with time.

In 1906, thanks to the persistence and prestige of Otto's benefactor Harry Pickering Bowditch, and Bowditch's friend John Collins Warren, a new medical school was built on Longwood Avenue in the outskirts of Boston, about four miles from the Cambridge campus of Harvard. Of course, President Eliot also was involved heavily in its fruition.

The new facility consisted of five white marble buildings in a quadrangle, open on the end facing Longwood Avenue. When it was opened shortly after the dedication on Sept. 25–26, there were no affiliated hospitals nearby. It was surrounded by private houses. For the clinical teaching of the medical students, use was made of the Massachusetts General Hospital and the Boston City Hospital, which were then reached via public transportation. This undoubtedly was another factor deterring Otto's immediate hopes of continuing research work on patients' metabolic problems. Within a few years, however, the situation was remedied with the opening nearby of Peter Bent Brigham Hospital and the Children's Hospital. In addition, Otto became a consultant to the Massachusetts General Hospital and established there an outstanding dynamic research biochemist, Dr. Willey Denis.

The Department of Biochemistry, located on the second floor of Building C, to the right of the central Administration Building, would be Otto's second home for the next 27 years. The buildings flanking the Administration Building were designated "laboratory buildings," and were constructed on a similar plan. Each had two parallel wings united at the front by an amphitheatre that seated 250. Building C had four floors in each wing, with a total floor space of 62,000 square feet. Besides Biochemistry, it housed the departments of Physiology, Physical Chemistry, and Experimental Surgery.

The medical faculty met at least monthly, and President Eliot, as in the past, was unfailingly present. At the first meeting in the autumn term, the president appointed the standing committees for the academic year, among which were Course of Study, Nominations, Graduate and Summer Courses, and Admission. According to the minutes of the meeting of Dec. 21, 1907, Otto was appointed to a committee to take charge of the graduate degree programs jointly with the Faculty of Arts and Science. They would be involved with instruction and examination for the degrees of Ph.D., S.D., and S.M. in medical sciences. Obviously, this appointment was of paramount importance to Otto, who now had to look to the training of graduate students for the research he eagerly anticipated.

Following his appointment to the medical school, Otto and Laura began a search for a home nearer his work. The home on Fisher Hill in Brookline that Otto bought for $12,000 was the humblest one in a rather rich Victorian neighborhood, but of huge dimensions by today's standards. The Folin family moved in during the autumn of 1908.

Prior to moving into their newly purchased home, the Folins spent at least part of the summer of 1908 in Kearsarge, N.H. The Folins' introduction to Kearsarge, a place that was to be of major importance in their lives, may have been through their friends the Cooke sisters, Elizabeth (Bess) and Joanna (Bonnie). Bess, who had entered graduate school at the University of Chicago in 1892 to work on her doctorate under Jacques Loeb, undoubtedly knew Otto and Laura during the four years she spent there before receiving her Ph.D. in 1896. In 1906, Bess was an investigator in the physiological laboratory at Harvard, but she listed her home address as Kearsarge. Bonnie, who became especially close to the Folins, was also at the University of Chicago for a time.

During this period Otto was forced to make decisions about the two instructors in his department, Alsberg and Henderson, both of whom had understandable feelings of insecurity and who had anticipated some form of promotion for their services during the time when the department was without a professor. Obviously, there was a lack of communication between Otto and the pair, so that they felt it necessary to press their complaints directly with the president.

In November 1907, Henderson had written President Eliot of his reasons for feeling insecure. His was a temporary appointment as instructor, while Alsberg's was for three years. Nearly all of Henderson's contemporaries (in medical sciences or chemistry) had been promoted to assistant professor. Henderson felt the need for a voice in the faculty, because there were affairs of the university that concerned him as a teacher, such as admission requirements to the medical school and college preparation for the study of medicine. Also, he felt out of the mainstream of the academic community. His feeling of insecurity was coupled to his low salary of $1200 per year, and the fact that even Alsberg was given a higher position, so that he believed that his freedom in teaching and research was threatened. While he thought that Folin's appointment greatly strengthened the department and he welcomed it and while he also was sure that Dr. Folin would not interfere with his work, a promotion certainly would put his mind at ease.

During the summer of 1908, Alsberg had considered an offer from the Bureau of Plant Industry in the U.S. Department of Agriculture in Washington, D.C., to serve as a chemical biologist, and he wrote to both Eliot and Otto about it. Eliot asked Otto to comment because Alsberg evidently was warning that if things were not improved in his departmental status—perhaps a promotion—then he would leave. Otto's comment was that Alsberg was by nature an impatient power-seeker. Otto had repeatedly reassured Alsberg that his position was secure and had recommended an increase in salary that Alsberg had received. But Otto did not now propose anything further to assuage him. Alsberg resigned as of Sept. 12.

Otto, who was now back in Boston, wrote to Eliot on Sept. 23 that he regretted Alsberg's leaving. It was now too late to fill his place. Otto's only choice was to divorce himself completely from the McLean Hospital so that he could attend more closely to departmental affairs, and assume that the Harvard Corporation would adjust his salary accordingly. He would proceed to move his family to Brookline. Otto had talked to Henderson and suggested that he give up teaching at the Cambridge chemistry division and at Radcliffe and devote all of his time to the new department. In view of Henderson's reluctance to do so, Otto felt that his status as a temporary instructor need not be changed.

Once the fall term began, Henderson was in a better position to press for a promotion. He informed Otto that the Tufts Medical School had offered him $2000 with an assistant professorship, and the chance to rise to a full professorship within three years. Otto relayed this information to Dr. Eliot on Oct. 20. Otto definitely did not want to lose Henderson.

Both Otto and Henderson wrote the president on Oct. 24. Otto recommended that Henderson be given a promoton to a three-year instructorship, a seat in the faculty, and a salary of $2150, so that Henderson could sever his relationship with Harvard College and Radcliffe College from which he now received income, to henceforth spend all of his time in the Department of Biological Chemistry.

President Eliot evidently followed some of Otto's recommendations, because on Nov. 2, Henderson wrote Eliot his thanks for the recent appointment that gave him a much needed sense of security. He asked for a further consideration of his salary, indicating perhaps that $2150 had not been offered. But Henderson said that if he were to continue teaching the course at Cambridge, then $2000 would be appropriate for this year.

On Nov. 9, Otto wrote the president that he did not oppose any further Henderson's giving the Chemistry 15 course at Cambridge, in view of the strong objections made by the professors of Harvard's chemistry department. Henderson could teach at Cambridge for that year.

In spite of strong arguments from Harvard as to why Henderson should continue to teach Chemistry 15 in succeeding years, Henderson acceded to Otto's wishes. He became an assistant professor in 1910, and remained in the department through the 1919–20 year. He then would rise rapidly to a full professorship in the department of physical chemistry in the laboratories of physiology.

For the 1908–9 academic year and thereafter, Otto became a teacher of freshman medical students, along with Dr. Henderson, and the third-year assistant, Otis F. Black, as well as the new one, the Canadian Walter R. Bloor, who was pursuing his doctorate and would be the first to obtain this degree from the department in 1911. There were 70 first-year medical students in the 1908–9 class, a reduction of 35 from the previous year, probably reflecting the upgraded requirements for admission to the medical school. The next year, the number would drop further to 65.

Of this period at Harvard, Bloor later wrote to Philip Shaffer (12-5):

> I came to Harvard primarily because I had heard of Folin as the new brand of chemist, the biological variety, and my training seemed to steer me into biological chemistry. Coming to Harvard Medical School, I found the department in charge of Alsberg and Henderson. They said that a new man, Folin, had just been added to the Staff, that he was upstairs, and interested in research—a term which sounded very large and out of reach of my limited background. However, Alsberg took me up and introduced me, and I was asked to come in anytime I felt like it. Alsberg got me going on preparations and Henderson tried to interest me in pH. I felt that I would like to know some-

thing about this research business, so I came to talk a good deal with Folin. Before long I was fortunately made first assistant. (*12-6*)

At the third annual meeting of the ASBC in Baltimore, Md., Dec. 28–31, 1908, Otto was elected president for the ensuing year (1909), the next meeting to take place in Boston. There were now 103 members in this fledgling group, of whom 55 attended this meeting held jointly with three other societies: the American Physiological Society, the Society of American Bacteriologists, and Section K of the AAAS.

Among those newly elected members of the society were the two young lions of clinical biochemistry, Dr. Stanley Rossiter Benedict of Syracuse University, Syracuse, N.Y., and Dr. Donald Dexter Van Slyke, Rockefeller Institute for Medical Research, New York City. Because of them the pace of work in clinical chemistry would accelerate markedly, and Otto's work would be reinforced, challenged, and extended. We shall, of course, hear much more of them later. This American trio—Folin, Benedict, Van Slyke—with Ivar Christian Bang in Sweden, would more than any others bridge the gap between biological chemistry and medicine, a fusion that flourishes today as clinical chemistry in the modern hospital. Otto Folin would lead the way.

For the 1908–9 academic year, Otto concentrated his efforts on establishing a solid series of lectures in biological chemistry, with special emphasis on metabolism. It was such an opportune moment in the new science that it was possible for men of Otto's background to encompass the entire developing field. Steeped as they were in classical organic chemistry, most of the pioneers had received an important segment of their training in Germany, now at its zenith in science. By reading the German journals and some of the French and English, all of the important literature could be covered, and Otto could adopt and modify his lectures to include the expanding knowledge. Otto also labored to perfect the laboratory exercises for both the freshman medical students and the graduates in Biochemistry 3. Otto taught both Biochemistry 2 and 3. The lectures dealt with metabolism in man and lower animals, both normal and pathological, and the laboratory dealt with quantitative methods for metabolic research. For graduate students in biochemistry, Otto introduced Biochemistry 20, Research in Biological Chemistry.

Otto's work in the department was so well appreciated in the medical school that he was promoted to a full professorship in 1909. He became the Hamilton Kuhn Professor of Biological Chemistry, and his salary was increased accordingly to $4500. It was during President Eliot's final year at Harvard.

As mentioned earlier, Otto's scientific output was in abeyance for a period of three years as he polished his didactic material and established the graduate research program. His 1909 publications consisted of two papers. The first, which was submitted in January, was a defense of his use of sodium peroxide for determining total sulfur (*12-7*).

While Otto provided a fairly good defense of the peroxide method, it still was uncharacteristic of him, because in the past he would have gone back to the laboratory for fresh verification if the situation had been opportune.

The remaining publication of 1909 was a paper that Otto had presented to the Section of Experimental Medicine of the Academy of Medicine of Cleveland on Oct. 9. It was a review of recent progress in protein metabolism and had no new data to offer (*12-8*).

Either late in 1908 or early in 1909 Otto was offered a professorship at Washington University in St. Louis, Mo., and Professor J. J. Abel was aware of it. Evidently Otto was tempted by the clinical research opportunities to be made available at the new hospitals and by the increased salary. Otto told only Dr. Cannon about this offer. Following a letter of inquiry from Abel, Otto replied that he had decided against the offer. This position eventually would become Philip Shaffer's in 1910.

Sometime prior to the spring of 1909, Otto had purchased some land, near the community of Kearsarge, N.H., on which he hoped to build a summer house. The Folins spent part of their summer of 1909 at Kearsarge, from about Aug. 1 to mid-September, when school opened in Brookline.

The two Cooke sisters evidently rented a "camp" nearby the place where the Folins boarded. They were good friends and doted over the Folin children. Bess, the physiologist, had been bitten on her left arm by a Gila monster when carrying on a research project with Leo Loeb on Heloderma venom at the University of Pennsylvania Medical School in Philadelphia. She had suffered a severe neurological reaction. Her recovery had been slow. Otto helped her get a job at the Loomis Laboratory at the Cornell Medical School. She wrote in May:

> I saw Dr. Shaffer. . . . I can't express my gratitude to you for getting me this chance. I am going to try to hold on to the place till I am eligible for a pension and I certainly hope never to make you write any more letters for me. Dr. Beebe said that I could carry on some of my own work and that prospect is delightful to me. . . . I trust you are not finding your work too heavy. I often think what a change it has been from your quiet little laboratory at Waverley.

In the 1909–10 academic year, Professor Folin's department listed seven people. In addition to Henderson and Walter Bloor, Harry W. Goodall, M.D., replaced Black as an assistant. Support for additional research came for three new positions. Chauncey J. V. Pettibone, B.S., was specifically a research assistant who would work on a doctorate. Chester J. Farmer was termed a "teaching fellow." And Archibald B. Macallum, M.B., was an Austin teaching fellow.

The freshman course was taught again by Folin and Henderson, with Macallum, Bloor, and Farmer helping in the laboratory work. What were the freshmen expected to learn? Here are 10 of the key questions that Otto expected them to answer on their examination papers:

  1. Describe the phenylhydrazine test for sugar and explain its practical importance.
  2. Compare starch and glycogen.
  3. What is butter?
  4. Write the constitutional formulas of three of the following substances: glycocol, glycerine, lactic acid, hippuric acid.
  5. Discuss the absorption of fat.
  6. Discuss ammonia as a metabolism product.
  7. Discuss nitrogen equilibrium.

8. Describe the synthetic preparation of urea.
9. Why is it possible to pump the carbonic acid out of blood?
10. State the chief characteristics of enzyme reactions.

Walter Bloor had this to report on the 1909–10 academic year.

> Folin was much interested in the medical students' laboratory and spent the whole time there when the class was in session. He did not circulate a lot but took one student or a group and worked intensively with them, doing a piece of research with the exercise they were working on. There were some distinguished students in those first groups: George Minot, W.W. Palmer, F.R. Rackemann. Joslin was around often. Henderson continued to lecture to students. He and Folin couldn't agree on urine acidity; Folin was for titration as the best measure and Henderson was for pH. Dr. Emerson, a pediatrician, often came to work with Folin on fat in feces, and was enthusiastic about cabbage juice for babies—before vitamins were thought of. Joseph Pratt worked in Christian's laboratory and came to see Folin often. Langley Porter and another clinician were working in the laboratory.

The fourth annual meeting of the ASBC was held in Boston, Dec. 28–30. Three of the four sessions of the society were held at the Harvard Medical School, the second of these jointly with the American Physiological Society. The fourth was held jointly with the Biological Section of the American Chemical Society at the Massachusetts Institute of Technology. Forty-five members attended this meeting. Sixty-four papers were presented in the four sessions.

R. H. Chittenden listed as outstanding contributions of these sessions the following five papers: "The work of the kidney as a regulator of the reaction of blood and plasma" by Lawrence J. Henderson; "The production of sugar from amino-acids" by A. I. Ringer and Graham Lusk; "The distribution of glycogenolytic ferment in the animal body" by J. J. R. Macleod; "A method for the determination of amino nitrogen and its applications" by Donald D. Van Slyke; and "The effect of inanition and of various diets upon the resistance of animals to certain poisons" by Reid Hunt.

At the business meeting of the opening session of the society with President Folin presiding, it was announced that an executive session of the society had voted that the president of the ASBC cooperate with the president of the American Physiological Society in the appointment of a joint committee of the two societies to consider the methods of examination in the medical sciences for registration in medicine in the several states.

In the coming year the great Flexner Report would appear, exposing the sad state of medical training in most of the medical schools in the U.S. Yet members of the ASBC were pushing ahead on research into nutrition, metabolism, enzymology, acid-base regulation, and many other areas of challenge with rewarding results. It should be pointed out that most of the early members of the ASBC and its leadership came from rather prestigious private and governmental institutions where the quality of education and personnel were at the highest level in the country.

In 1910, Otto finally returned to the production of scientific papers from his role as the head of a department in one of the world's leading universities.

Despite his fondness for the laboratory bench, he knew that he must now spend only a "measured" portion of his time in the laboratory. He could work along with the postdoctorate students, but the graduate students and doctoral candidates must be given their own opportunity to experiment and discover. He must advise and clarify, but rarely partake, just as Stieglitz had done. And whenever a problem caught his fancy, he could plunge in as he wished. The days of spending long hours alone in the laboratory were gone. He would devote his time now to preparing his lectures, writing papers, assisting in the laboratories for the first-year and advanced students, preparing for the graduate seminars, editing for the *Journal of Biological Chemistry*, critically reading the literature, planning the next steps in research problems, assessing the value of the work done, attending faculty and committee meetings, and occasionally making trips out-of-town to present papers.

Once Alsberg was gone, of course, Otto was responsible for the administration of the department, meaning purchasing and receiving supplies, student grading, records, and whatever it took to maintain the "nuts and bolts" of the department and to keep its people's needs satisfied.

The list of things to do seemed unending. And it should be emphasized that by spending less time at the bench, Otto's day was hardly shortened. When he was caught up in an interesting project, Otto would go at it for seven days a week. The difference now was that there had to be others involved in the laboratory work as well, and he did not work into the night, but returned home in the evening to be with his family. As Walter Bloor and all the graduate students had found, Otto's policy was the open door. He would never forget his own experiences as a graduate student at Chicago and in Europe. He was available.

Otto's first paper of 1910 was with A. H. Wentworth on a pediatric problem.

> An adequate knowledge of fat metabolism is particularly important as a prerequisite for a rational treatment of the numerous cases of digestive disorders occurring among infants and young children. Without reasonably reliable methods for determining what can properly be called fat in the stools, such knowledge is scarcely obtainable.

There were two points to bear in mind in developing an analytical scheme for fats: 1) everything extracted with organic solvents was not fat, and 2) fatty acids and soaps were practically impossible to determine separately because of the variables affecting the equilibrium between them after defecation such as the drying process, presence of calcium and magnesium, and loss of CO_2 and H_2S.

The authors would make quantitative determinations of total fat (including neutral fat concentration) and of total fatty acids. The calculated difference between the two would give the neutral fat concentration.

For the total fat determination, extraction was done with anhydrous ether acidified with HCl gas. The HCl liberated the fatty acids from the soaps. A way was found to titrate very sharply the higher molecular weight fatty acids, using sodium alcoholate as the alkali and certain oxygen-free organic solvents not previously used. The former solvents caused diffuse endpoints due to "associating" effects as disastrous as using water. These could be overcome by substituting solvents such as chloroform, carbon tetrachloride, benzene, or toluene. (Part II, p. 386).

> Here we wish only to emphasize the empirical finding that on titrating a higher fatty acid like stearic acid in hot solutions of benzol or carbon tetrachloride, theoretical figures are obtained and that the end-point, a deep purple is obtained with the addition of the last 0.05 cc. of decinormal sodium alcoholate solution.

The method was a major improvement in fat analysis, but of course needed further verification on patient samples. The touch of the master was back!

In his second paper of 1910, sent for publication in April, Otto had taken a laborious method for preparing cystine in poor yield and modified it so that it

> could be used as a laboratory exercise for the past two years by our first year medical students and at least 75 per cent of the men have obtained very satisfactory results. The isolation of the cystine is based on its insolubility in acetic acid.

Mörner's original method for obtaining cystine from acid hydrolysis products of protein was not devised with regard to the time and labor involved. Embden had improved the method by boiling horn or hair with concentrated rather than dilute acid. But all of the earlier methods had failed to take into account the fact that cystine was quickly decomposed by alkalis such as NaOH, Na_2CO_3, and NH_3 used to precipitate it from acid solution (Part II, p. 387).

Otto's third paper of 1910 was a short one based on work that the research assistant, Dr. Frederick C. Blanck, had done during the winter of 1907–8. While unexplainedly delayed, it was published then, in conjunction with the next paper to which it was related. Otto now wished to present an improved method for preparing creatinine from urine (Part II, p. 387).

In 1910 Otto's first outstanding associate in research at Harvard came to spend nearly a decade. Dr. Willey Denis (pronounced Denee) was a woman of impeccable character and educational background, but Harvard was a private school for men. How Otto managed to add Denis to his paid staff is uncertain, but she held no teaching rank, serving in a purely research role. Later, in 1912, she would be the first woman appointed to the staff (as assistant chemist) at the Massachusetts General Hospital, where she was paid to run a research and service laboratory. It is highly probable that she was the first woman to work as a hospital biochemist in the United States. She became a member of the American Society of Biological Chemists in 1911.

Willey Denis (1879–1929), a native of New Orleans and a daughter of one of its old respected families, obtained her early education in private schools and her A.B. from the local women's H. Sophie Newcomb College of Tulane University in 1899. After two years at Bryn Mawr, Pa. (1899–1901), she returned to Tulane to do graduate work in chemistry (1901–2). Her thesis, "The Reducing Bodies in Sugar Cane Juice, Syrup, and Molasses," led to a master's degree (1902). The following year she received a fellowship at Bryn Mawr (1902–3) where she took graduate courses in chemistry and geology. Once more she returned to Tulane, this time for work on a topic at the school of medicine.

In 1905, she became Nef's graduate student at the University of Chicago to work on a doctorate in chemistry, which she obtained cum laude in 1907. She minored in physiology. Her thesis, published with Nef, was "On the Behavior of Various Aldehydes, Ketones and Alcohols Toward Oxidizing Agents." She, like Otto, was mainly interested in physiological chemistry.

Dr. Denis served as an instructor in chemistry at Grinnell College, Iowa, for a term in 1907, and then as an assistant chemist in the U.S. Department of Agriculture, Bureau of Chemistry, 1907–9. She then presumably entered the school of medicine at Tulane, where she took "partial" courses in 1909–10. During this period, she evidently communicated with Otto about a research position with him (12-9).

Otto Folin may have met Dr. Denis at the U.S. Department of Agriculture's Bureau of Chemistry, when he visited its chief, Dr. Harvey W. Wiley (1844–1930), with whom he shared several interests, and with whom he served on the Council on Pharmacy and Chemistry of the American Medical Association, a group of 16 selected by the board of trustees.

Some of Folin's research projects—determining benzoic acid in cranberries and catsup (1911), hippuric acid in urine (1912), vanillin in vanilla extracts (1912), and ammonia in fertilizer (1913)—apparently were related to mutual interests with Wiley and the Bureau of Chemistry on the analysis of foods and additives. He had maintained an active interest in the chemistry of foods and drugs, stimulated in 1899 by his Chicago experiences at the Columbus Medical Laboratory.

The first paper of the Folin-Denis team was submitted for publication in September 1910. It was a note, in effect, on a very simple method for converting creatine to creatinine without the use of dilute acid (Part II, p. 387).

By 1910, or even earlier, Otto Folin had begun developing quantitative analytical methods for nonprotein nitrogen substances in blood (Part II, p. 387)—a "tooling up" period leading to his grand outpouring of papers in 1912. This strongly paralleled Otto's first two years at the McLean Hospital, when he developed quantitative procedures for urinalysis. With the arrival of Dr. Denis, this process was accelerated. The unrecorded research work performed in the period of 1910–12, therefore, represents the major initial impetus toward establishing blood chemistry in the U.S. The founding of the science of modern clinical chemistry justifiably could be attached to Otto's publications in 1912 and 1913.

As far as the records show, 1911 was a comparatively uneventful year. Otto's annual research productivity dropped abysmally low—only one modest paper on the determination of benzoic acid in, of all things, catsup (Part II, p. 388). Catsup had become a consuming interest to a departmental head in a venerable medical school.

However, the truth was that the biochemistry department now had been transformed into a major research center. The entire professional staff including the graduate students, with the unlisted Willey Denis leading the way (12-10), were absorbed in projects that would generate 20 publications in 1912.

And all this storm of activity was guided by one patient, soft-spoken catsup master, Otto Folin.

A hint of what was brewing came from the paper, "Some New Techniques for the Determination of Total Nitrogen, Ammonia and Urea in Urine," that Otto gave at the fifth annual meeting of ASBC, held in New Haven, Conn., Dec. 28–30, 1910. Four authors were listed with the title: Otto Folin, Chester Farmer, A. B. Macallum, and C. V. J. Pettibone. The paper was unique in that the procedures involved were microscale in terms of the volume of urine used, and that ammonia was determined colorimetrically (rather than titrimetrically) using Nessler's reagent. This was a historic presentation.

> The new methods described for the determination of total nitrogen, ammonia and urea depend on the use of extremely small quantities (0.1–1.0 cc.) of urine. The ammonia formed is set free by an air current instead of by distillation and is subsequently estimated colorimetrically by means of Nessler's reagent and a colorimeter. By these methods the total nitrogen is determined in twenty minutes, the urea in about twenty-five minutes and the ammonia in about fifteen minutes.

Despite this momentous advance in analytical technique for clinical chemistry, none of it was published in 1911.

Of greater importance, perhaps, was that Otto, with Willey Denis's capable collaboration, finally had begun the next essential step in dealing with protein metabolism: determination of nitrogen compounds in blood and tissues. There was only one logical way to get at the fate of protein digestion in the human organism and that was to measure its products in blood and tissues. This was also the only approach to advance knowledge of the endogenous metabolism of protein. Blood, however, was an analytical mystery. Whereas urine had been a readily available object of study for centuries, the chemistry of blood was literally an untraveled new avenue of vast potential import to biochemistry and medicine. Otto had done perhaps the most thorough quantitative analysis of urine constituents up to that time, but no one yet had defined the quantitative chemistry of blood. Although most clinically oriented biochemists were aware of this problem, Otto Folin and Willey Denis would be the first in the U.S. to plunge into this labyrinth. They soon would have plenty of company.

In the 1910–11 academic year, Otto was unusually busy in the laboratory, particularly with experiments on protein metabolism carried out with Willey Denis. For Otto to have generated 23 papers in 1912, it is obvious that he must have already been writing in the autumn of 1911. Otto prepared all of his manuscripts in long-hand. His skill and experience at writing, which now came smoothly to him; his mental organization; and his command of his material were such that he rarely had to spend much time rewriting, revising, or rephrasing.

Everyone in the department—seven people not counting Henderson who had his own worlds to conquer on the Cambridge campus—was at work on a problem of special interest to Otto, and as Bloor mentioned, there were several people who visited the laboratory and undertook investigations on Otto's advice. The projects of this period of development in the Department of Biological

Chemistry were unreservedly the brainchildren of Otto Folin. Those who could stand on their own, like Willey Denis and future members of the faculty, would do so, but at a later time.

At the scientific portion of the sixth annual meeting of the American Society of Biological Chemists (December 1911), Otto presented a paper entitled, "A New Method for the Determination of Hippuric Acid in Urine," coauthored with Fred F. Flanders, but there was no hint from this paper of the avalanche to come.

It was at this meeting that the ASBC, following a motion by Otto, took action to form a closer affiliation, a federation, with the American Physiological Society and the newly formed American Society for Pharmacology and Experimental Therapeutics, which Otto joined. Representatives of the three groups unanimously adopted the proposal to create the Federation of American Societies for Experimental Biology. The American Society of Experimental Pathology was admitted as the fourth member in 1912.

Otto's attention to his teaching matched his attention to his laboratory work. He prepared his lectures freshly and laboriously, regardless of whether they had been given the year before, so that he could incorporate and polish new material from the recent literature. In Chicago, he had learned that he could make his best presentation if he merely used a few unadorned notes. But this did not imply a lesser need to prepare. Otto wrote his lectures down conscientiously, leaving nothing to chance or expectations that he would have flashes of insight in clarifying the intricacies of biochemistry. He was determined to cover a certain number of topics in the anticipation that the medical students would gain a practical understanding of their importance, abetted by the laboratory exercises, where possible. He used the same approach to teach graduate students and conduct seminars.

Because of his interest in protein metabolism, his devotion to research, and his critical grasp of the literature, Otto's lectures reflected an enthusiasm for his subject matter that his students could not fail to grasp. As his son said, "I got the impression that my dad was just as sold on the idea of 'doing a job' of his teaching as he was of his research."

Otto had a charismatic aura, particularly when he spoke, a result perhaps of the mixed erudite and ridiculous figure that he cut. The lean, mustached man of the twisted face, who was almost always dressed in a blue or gray suit and the little bow tie, spoke gently with an unmistakable, though pleasant, Swedish accent, a fountain of information leading up to the frontiers of the science. His eyes and his mannerisms reflected a quiet fervor so that nothing he said was taken lightly, except his frequent and deliberate interjection of bits of humor that jazzed up the important aspects of his lectures.

Otto Folin was a creature of habit, who had developed a routine in his daily schedule during the academic year, a regularity that had begun in his seven years at McLean Hospital (*12-11*). During the winter months, Otto would rise early in the morning, before six if he was to lecture to the freshmen medical students, stoke the two coal furnaces, and light the hot water heater in the kitchen. He would prepare his own breakfast, read the morning *Herald*, then work on his lectures or laboratory exercises.

When he lectured to the freshmen at 2 p.m. during this period of his life, Otto generally would leave his house before his family had risen. He would catch the electric trolley car about three blocks away, and travel a mile or so down Boylston Street, turning left at Brookline Village on Brookline Avenue, where he could get off only a short distance from the medical school and walk to Building C, which housed his laboratories and office on the third floor. Of the approximately 1000 square feet of space he had for his use, most was laboratory area.

With so much research in progress, and with the unending chores of lecturing, laboratory exercises, meetings, administrative details, editing, writing, reading, as well as the unpredictable distractions of the open-door policy, Otto's days were perpetually full, a life-style on which he thrived, thanks to a blessedly robust constitution.

Otto would return home at about six each evening, eat supper with his family, and then spend what time he could with the children. When the children were in bed, he and Laura would sit down in the armchairs on each side of the big oil lamp where they could talk and read. Otto would scour the *Boston Evening Transcript*, and time permitting, the weekly magazine, the *Outlook*, and Laura's paper, *The New York Call*, for which Bonnie Cooke had worked.

Laura and Otto were content with their lives. They were a devoted couple whose rather late-in-life marriage provided them with the advantages of a realistic understanding, deep mutual appreciation free of romantic illusions. Their children were a blessing that brought them even closer together. Laura administered the home affairs capably, and she managed their finances entirely.

Outside of managing the home, Laura took pleasure in tutoring her extra-bright children, particularly in their preschool years. She remained an avid reader and maintained a lifelong interest in political affairs and economics, but she was not an activist in the modern sense, though she strongly supported the women's suffrage movement, and she often attended lectures as she had done in St. Paul. For the Folins, who were unenthusiastic Unitarians, the church ceased to be a focal point of intellectual stimulation. Laura, who had had moments of religious fervor for several years before her marriage, now avoided church affairs entirely, first perhaps out of deference to Otto's convictions, but later out of her own.

Otto's needs were unusually simple, tempered by years of austerity and rigid routine, particularly during the weekdays: carfare, cigarette money, and perhaps a bit extra for a snack at lunchtime. An extravagant month would cost him $10, including his weekends. But this simplicity was made easier because Laura ensured the satisfaction of his other personal needs. Otto was quite clothes-conscious, and she made certain that he was attired in a manner becoming to a head professor, meaning that he wore a proper suit (blue or gray) to work that was well-fitted, though not privately tailored, moderately expensive, and in keeping with the time. At work, Otto donned a laboratory coat. His bow ties were natty, but not flashy. His shoes were practical for a lab man.

Otto may have had simple tastes, but he was a highly complex man. Although seemingly suave and loose, he was often a cauldron of ideas, about to boil over. He was a man of few words because he liked to think things through before he spoke or wrote. He carried a blue notebook with him at all times when he was out of his laboratory office so he could jot things down, often waking in the middle of the night to do so. Like other creative, dedicated men, Otto functioned around-the-clock, his one-track mind working intensively when he gave it free-rein and with strict discipline on solutions to the problems at hand. His was a slow, deliberate approach to solving a dilemma.

Otto was in marriage the same considerate, gentle, kind person he had been when he was a student. He was a strict disciplinarian with his children, but it never conflicted with his love and admiration for them. His was a close-knit family who were blessed with a capacity to relate to one another. Otto put his family on a pedestal.

On Saturdays, for the most part, Otto would return home early in the afternoon. Only during sporadic periods of very "heavy" laboratory investigation, such as must have occurred in 1911, and more frequently at the McLean Hospital, did Otto not spend the remainder of the day and Sunday with his family. Otto had a weekly ritual that he faithfully followed baking the family's bread. Store-bought bread, then, as now, was no match for the homemade variety. Otto had watched very carefully, and may have helped his mother and Gertrude as they prepared bread during his visits with them in Träryd. It seems probable that Otto had studied the quality of different wheat flours in his review of foods and nutrition when he was associated with the Columbus Laboratory in Chicago. Otto baked the weekly bread supply, however, because he and the family enjoyed the bread immensely. He prepared the dough on Saturday night and did the baking early Sunday afternoon.

On Saturday afternoons during the academic year, Otto busied himself around the house with projects such as building Joanna and Teresa a dollhouse or a fire-station for Grant. Sunday was the cook's day to come and prepare the evening dinner, usually of roast beef or lamb stew, but the afternoons were often an open house for Otto's graduate students, assistants, colleagues, and family friends, young and old. The Sunday dinner, more often than not, included invited guests and friends.

During the warm weather, beginning in the early spring, Otto's one consuming interest, besides his work, became golf. He had learned this game, as we know, at the McLean Hospital and had persisted at it. It became a Saturday afternoon and early Sunday morning recreation for him, if not an obsession. Two doors from the Folins' house lived the Walkers, and Mr. Walker was not just a run-of-the-mill golf partisan. He had erected a huge canvas sheet in his large barn, so that he and Otto could practice driving golf balls. As the spring thaws came, the neighborhood thudded with the sound of golf balls punching the canvas. At home, Otto practiced putting with one of those indoor devices that all serious golfers use for this solemn purpose, and little Teresa would retrieve the balls he nudged.

Their Brookline home was in a spacious, friendly neighborhood. With warm weather the boys would go to the schoolground to play baseball, but if their numbers were too small for a game, then they would gather in Grant's frontyard, erect the backstop (a bedspring), and play there. Obviously there was not much of a front lawn left for Otto to mow, but that mattered little because there was plenty more around the house, and much space between neighboring houses. Otto's house was one of the few in the neighborhood that had no accompanying barn. But Otto had no need for one as he did not keep a horse and buggy as many did in this transition period to the automobile. The porch that skirted most of the house was used much on warm days. Otto was fond of watching the sunset from the west side of the porch, but on hot evenings the east side was preferred for its coolness.

References and Notes

12-1. (a) Garland, J., Walter Bradford Cannon, George Higginson Professor of Physiology. *Harvard Med. Alumni Bull.* **46**, 4–8 (1971).
   (b) Biographical sketch of President Cannon. In: *History of the American Physiological Society Semicentennial, 1887–1937.* pp. 94–96. Baltimore: American Physiological Society, 1938.
12-2. The Harvard Medical School, *Official Register of Harvard University*, Cambridge, Mass.: Harvard University, 1906–7, 1907–8.
12-3. Shaffer, P.A., Otto Folin. 1867–1934. *Biogr. Mem. Natl. Acad. Sci. U.S.A.* **27**, 47–82 (1952).
12-4. Folin, O. Chemical problems in hospital practice. *J. Am. Med. Assoc.* **50**, 1391–94 (1908).
12-5. Walter R. Bloor, Letter to P. A. Shaffer, June 6, 1949. Philip A. Shaffer File, Archives, School of Medicine Library, Washington University, St. Louis, Mo.
12-6. Bloor also wrote: "I seemed to be the only one around who wasn't mixed up with Alsberg & Henderson and he [Folin] told me what was going on. It seems that A & H were interested in getting an old friend of theirs Edwin Stanton Faust as professor and head of the department. Folin who was only Associate Professor at the time was to be shelved as Research Professor. But Folin wanted contact with the students and so he horned in on the laboratory work with the medical students. Also he told Eliot what was going on. . . ."
12-7. Folin, O. The determination of total sulphur in urine. *J. Am. Chem. Soc.* **31**, 284–85 (1909).
12-8. Folin, O., Ten years' progress in the field of protein metabolism. *Cleveland Med. J.* **8**, (Nov.), 1–8 (1909).
12-9. In Memoriam. Professor Willey Denis (1879–1929). *Tulane News Bull.* **9** (Mar.), 103–4 (1929).
12-10. Late in 1910 Denis had gone to New Orleans, where she published four papers from the Department of Physiology at Tulane, but she had returned to Harvard by autumn of 1911.
12-11. Information presented here is largely derived from taped interviews and written communications with his children, Dr. Teresa Rhoads and Mr. Grant Folin.

Eva Olofsdotter (Olson) Folin (1830–1915), circa 1860.

Nils Magnus Folin (1825–1903), circa 1860.

The three Folin brothers, circa 1888, on the probable occasion of Otto's high school graduation. Left to right: Alfred (1862–1933), Otto (1867–1934), and Axel (1854–1940).

Looking north on Main Street, Stillwater, MN. Picture taken in 1885 from the Main Street stairs. Note the horses, wagons, dirt street, wood sidewalks, and telegraph lines. The tallest building on the right was the opera house where Otto Folin's high school graduation ceremony was held. (*Collected by John Runk, photographer, Stillwater, MN; Historical Collection No. 218, Minnesota Historical Society, St. Paul, MN.*)

Otto Folin, 1885, at age 18 in Stillwater, MN. This is the earliest picture available of him.

The *Ariel* staff, when Otto Folin was the Managing Editor, 1891–92. Only George Sikes, Otto Folin, and Madeleine Wallin are identified, front row, middle to right.

Julius Stieglitz (1867–1937), Professor and Head of the Department of Chemistry, University of Chicago. (*Courtesy of Archives, Joseph Regenstein Library, University of Chicago, Chicago, IL.*)

Jacques Loeb (1859–1924) in 1907. (*Photographed by Gustavius Eisen; from the Rockefeller University Archives, New York, NY.*)

Professor Olof Hammarsten (1841–1932) in 1897. (*Courtesy of University Library, Uppsala University, Uppsala, Sweden.*)

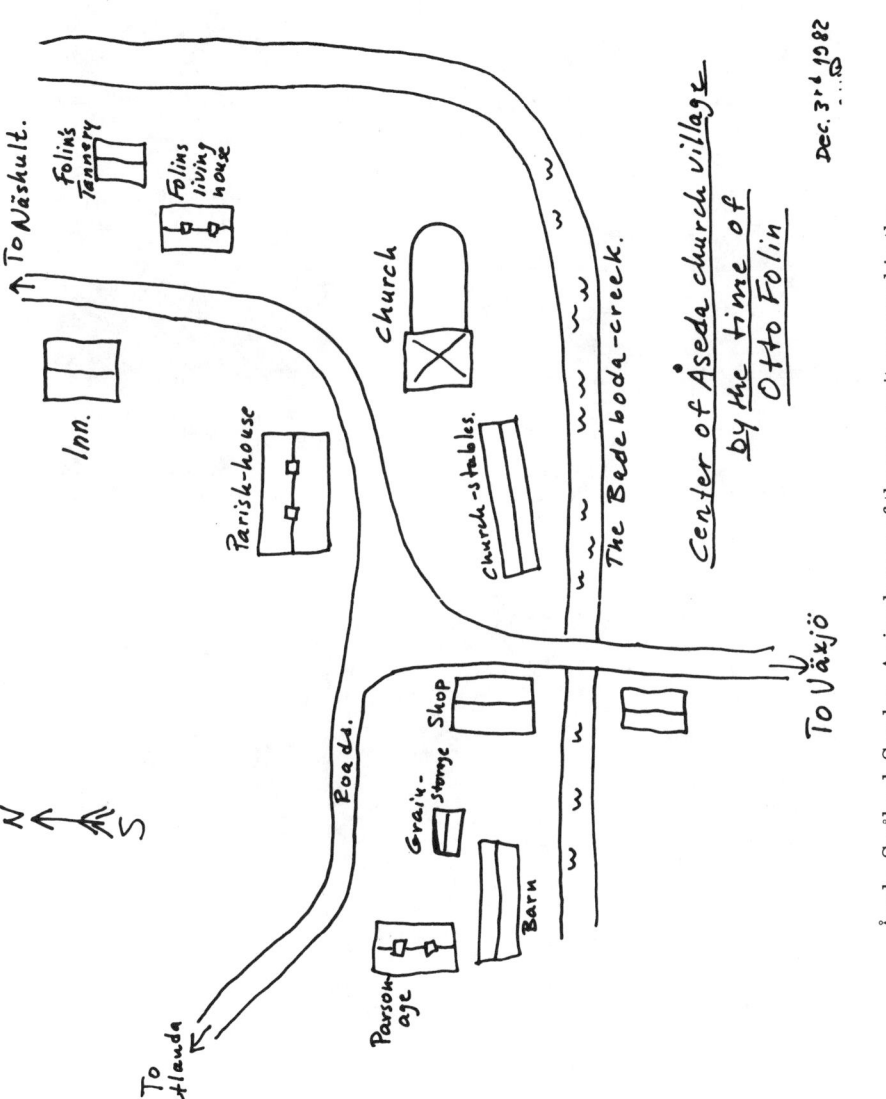

Åseda, Småland, Sweden. A simple map of the area as it appeared in the late 1860s. (*Drawn by Carl-Werner Pettersson, Brinkelid, Åseda, Sweden.*)

Otto Folin, September, 1900.

Laura Churchill Grant, December 1897.

Teresa Folin, 1912–13.

George (Grant) Folin, 1913.

Joanna Folin, 1910.

Philip Shaffer, Ph.D., 1904.

Dr. Willey Denis (1879–1929), a prodigious research worker; one of the earliest women in clinical chemistry in the United States, and probably the first. (*Courtesy of the Krewe of Proteus, New Orleans, LA.*)

Dr. Edward Cowles, Superintendent, McLean Hospital, Waverley, MA. (*Portrait by Wilbur Dean Hamilton, 1913; Courtesy of McLean Hospital Archives, Belmont, MA.*)

Otto Folin in the 1920s, lecturing on a favorite topic. Blackboard: "Explain the metabolic origin of creatine."

Laura and Otto on the mountain trail at Kearsarge, NH, circa 1920.

James B. Sumner (1887–1955) in 1917. *(Courtesy of Department of Manuscripts and University Archives, Olin Library, Cornell University, Ithaca, NY.)*

Stanley Rossiter Benedict (1881–1936). *(Courtesy of Vassar College Library, Poughkeepsie, NY.)*

Edward A. Doisy. *(Courtesy of Archives, Library, Washington University School of Medicine, St. Louis, MO.)*

Hsien Wu (1893–1959) in 1918.

Otto Folin using a Duboscq colorimeter in his office at Harvard Medical School, circa 1932.

The Department of Biochemistry, Harvard Medical School, June 1934. Front row; left to right: Dr. I. S. Danielson, Graduate Assistant; Dr. C. H. Fiske, Associate Professor; Dr. Otto Folin, Professor; Dr. M. A. Logan, Instructor; Dr. H. C. Trimble, Assistant Professor. Second row: W. W. Smith, Assistant Storekeeper; Henry B. Martin, Storekeeper; Margaret Cushman, Research Assistant; Mrs. Margaret Marvin Dupertuis, Secretary. Rear row: E. M. Stotz, Graduate Student; Dr. Yellapragada Subbarow, Teaching Fellow; N. K. Schaffer, Teaching Fellow; D. J. Mullone, Research Student; Mary Tracey, Assistant in Storeroom. (*Courtesy of Countway Library, Harvard Medical School, Boston, MA.*)

Chapter 13. Willey Denis—
The Prodigious Year (1912)

The year 1912 was filled with scientific achievement tempered by a missed opportunity, and overshadowed by a family tragedy. As we mentioned, Otto's productivity rose from the nadir of a single paper with Fred Flanders in 1911 to the zenith of 20 papers in 1912. (He wrote 23 papers in 1912, of which 20 were published that year and three in early 1913.) Willey Denis, the stout dynamo, was coauthor of no less than 11 of those papers, and was acknowledged as coauthor of one portion of a twelfth. There were three papers with Macallum; two papers with H. Lyman (the second was one of Otto's patented sharp rebukes to an attack on the first paper); two with Flanders; and one with C. J. Farmer. We shall deal with these according to the coauthors cited in the above order, rather than in chronological order.

To Willey Denis fell the grand task of determining the fate of absorbed amino acids formed in the small intestine as a result of protein digestion. As Otto had developed techniques to analyze the urinary waste products of protein digestion in his research at McLean Hospital, he now had to develop analytical methods suitable for blood. Fortunately, not only was he working on ground he had already plowed, but he now had much more capable help and resources around him, and therefore he could work almost simultaneously on new methods for evaluating blood total nitrogen, non-protein nitrogen (NPN), urea, ammonia, creatinine, and uric acid. By measuring NPN, urea, and ammonia, in particular, Otto could get at the metabolic fate of absorbed amino acids indirectly. There was no way at that time to tag an amino acid and follow its metabolic course, as Knoop had done with fatty acids, and as Schoenheimer and his associates would do much later with isotopes. Van Slyke's more direct gasometric method for measuring α-amino nitrogen in blood was also unfortunately not available when Willey Denis began to attack the problem. Otto and Willey had to arrive at answers indirectly, by determining the fraction in blood that held the amino acids, the NPN, and by analyzing the known end products of amino acid metabolism, urea and ammonia.

In their first paper on the fate of amino acids obtained from protein metabolism, Folin and Denis began what would be an extended series of experiments on nitrogenous materials injected into the intestine of cats (Part II, p. 389). We can only conjecture that Otto was unhappy with this paper, because it left the basic question of amino acid metabolism hanging in a reservoir in all body tissues without accounting for the tremendous urea production that his experiments at the McLean Hospital had shown came from the exogenous metabolism of food protein, in accordance with his now generally accepted theory of protein metabolism. However, it would not be his fortune to unravel this dilemma.

The experiments described in their second paper (Part II, p. 390) clearly showed that the large intestine was at least the most common source of ammonia in portal blood. In most of the experiments the large intestine still contained feces, and because of putrefactive bacterial action such as that of *Escherichia coli*, ammonia was produced almost constantly, accounting for the abundance of ammonia in portal blood, even in "fasting" animals. Ammonia was not produced in the intestinal wall. To the liver fell the task of disposing of the ammonia and other toxic materials derived from the large intestine. The ammonia in tissues and in experimental acidosis, as well as certain unspecified clinical applications, remained to be investigated.

Otto and Willey did not suggest how the liver handled the ammonia, nor did this paper include any determinations of urea nitrogen. No details were given for their obviously new and difficult technique for determining blood ammonia nitrogen concentration.

In their third paper, Otto and Willey returned to the puzzling problem of where urea was formed following the absorption of nitrogen compounds, particularly creatine and creatinine (Part II, p. 391). They closed the paper with some very prophetic and stimulating questions. Having proved that amino acids reach the tissues rapidly following absorption (and thereby disproving the ideas of Pflüger and Voit) where they could be used or broken down, they wrote:

> On the other hand new problems are now pressing for an answer. What is the starting impulse and the controlling factor in the urea formation? Does a given amount of absorbed nitrogen yield the same amount of urea in a given time whether it is represented by many or by only one kind of amino-acids? In other words does a certain total concentration of amino-acid nitrogen result in urea formation or do the tissues maintain a separate nitrogen equilibrium on the basis of each individual amino-acid? From a teleological standpoint it would seem almost certain that each tissue would deal individually with each kind of available amino-acids, and would tend to maintain the kind of mixture most suitable to its needs. If such is the case, each tissue maintains a certain supply of each amino-acid and the urea formation from any particular amino-acid depends, so to speak, on the "partial pressure" of that particular acid. This appears to us the simplest tentative hypothesis concerning the urea formation out of the surplus food protein.

In their next paper on protein metabolism, Otto and Willey turned to absorption from the large intestine (Part II, p. 391). Their second study had shown how readily ammonia was absorbed. Rectal feeding, by then an accepted clinical practice, already had removed all doubt that the large intestine also absorbed other substances. "Nevertheless it seemed to us worthwhile to show experimentally that the absorption from the large intestines is extensive enough to be demonstrable by means of our new methods of blood analysis." Their results, we will mention in advance, showed that absorption from the large intestine, while not as rapid as from the small intestine, was rapid enough to cause appreciable accumulation of the absorbed products in the blood, as measured primarily by increased NPN.

For their next work they studied experimental acute nephritis (Part II, p. 392), and were joined by the pathologist, H. T. Karsner. They were helped by a grant from the Rockefeller Institute for Medical Research. The rationale for this work was amply stated:

> Considerable work has been done on the subject of nitrogen metabolism in acute nephritis and somewhat variable results have been attained. The development of the colorimetric methods for nitrogen determinations in the blood has opened up new fields for work, and the object of the present study is to discover from the standpoint of blood analysis the possible retention of nitrogen in experimentally produced nephritis.

The analytical methods that Otto and Willey had been using in their studies of protein metabolism were described in a separate publication (Part II, p. 392). They pointed out that, in addition to blood, these techniques could be applied to milk, eggs, and other liquids in which minute quantities of these compounds were present.

To absorb the irritating fumes formed by sulfuric acid used in the method for determination of total nitrogen in urine, Otto and Willey devised an ingenius, simple "fume absorber" (Part II, p. 394).

The problem that Otto turned to next was one that he had left unresolved, the presence (or absence) of creatine in normal urine. A recent report by Rose in Mendel's laboratory had shown that the urine of children usually contained relatively large quantities of creatine. In adults, creatine was generally eliminated when there was excess intake in food or when there was excessive disintegration of tissue materials.

> In view of the unexpected character of the results obtained by Rose we promptly repeated the work on three normal well nourished children belonging to one of us (Folin). We had intended to continue the investigation further before publishing anything, but in view of the criticism of Rose's findings by Wolf and by McCrudden and since our results completely verify and also extend Rose's observation we have decided to publish them now. [Part II, p. 394]

The next paper reported the preparation of a uric acid reagent and a "phenol" reagent and gave the properties of these reagents from the qualitative standpoint. (Quantitative aspects would be dealt with elsewhere.) This work would have long-lasting effects in clinical chemistry, because it would serve as a basis for determining not only uric acid in blood and urine, but cystine, tryptophane, and tyrosine in protein, vanillin, epinephrine, serum proteins, sugar, phosphorus, and, of course, phenols. Tungstic acid would become a universally used protein precipitant (Part II, p. 395).

Otto had first encountered phosphotungstate and phosphomolybdate compounds as protein precipitants during his European tour, beginning in Hammarsten's laboratory, but at that time he had had no time to investigate these compounds. He and Macallum had been working on a quantitative colorimetric method for uric acid that produced a blue color with phosphotungstic acid, a reaction used universally as a qualitative test. However, phosphomolybdate was preferred, because of the poor quality of the tungstate preparations.

> As the original reagent gave a fine reaction not only with uric acid but also with a great many phenol derivatives, including tyrosine, it seemed important to discover if possible how to procure a reliable supply of this interesting substance. The research has yielded some rather unexpected results.

The two reagents were tested on a large number of organic compounds. Aliphatic substances, ketones, aldehydes, amines, and indole and its derivatives did not react. The uric acid reagent did not react with ordinary monohydric phenols except those containing an amino group in the benzene ring. The phenol reagent reacted with all oxybenzol compounds and was far more sensitive to tyrosine than Millon's reagent. The reactivity of the two reagents as well as Millon's reagent to monohydric, dihydric, and trihydric phenols and other compounds were presented in three tables. The reagents could now be made reproducibly.

Otto and Willey wasted no time in applying the phenol reagent to the determination of proteins based on their tyrosine content. There were wide discrepancies in values for the tyrosine content of proteins reported in the literature, e.g., in zein (maize) or vitelline (hen's egg) (Part II, p. 395).

They next applied the phenol reagent to the determination of vanillin (3-methoxy-4-hydroxybenzaldehyde) in vanilla extract (Part II, p. 397).

> As the Hess-Prescott method for the determination of vanillin, which in the modified form proposed by Winton and his associates is now universally used in American food laboratories is extremely laborious, although undoubtedly accurate, it has seemed worthwhile to attempt to work out a rapid colorimetric method for use in the determination.
>
> By the use of the phosphotungstic-phosphomolybdic reagent recently described we have been able to work out a colorimetric method for the determination of vanillin which gives theoretical results with purely artificial extracts containing known amounts of the substance, and gives with authentic extracts values agreeing closely with those obtained by the official method.

Otto published only one paper alone in 1912, on the determination of urea in urine based on the doctoral thesis work of C. J. V. Pettibone, whose assistance Otto acknowledged. Pettibone, Otto's second doctoral graduate at Harvard—Walter Bloor was the first in 1911, though he had gone to work in St. Louis for the man who really merited the honor, Philip Shaffer (1904)—would receive his degree in 1913, with a thesis entitled, "The Quantitative Estimation of Urea in Urine." Otto also acknowledged the contribution of Willey Denis to this paper, which was reported in four parts. (She was coauthor of the fourth portion that dealt with the determination of urea in the presence of sugar.) (See Part II, p. 397.)

Otto published three papers in 1912 with A. B. Macallum, the Austin teaching fellow. The first paper was merely a preliminary note on the use of phosphotungstic acid and alkali for uric acid determination, but was highly momentous for clinical chemistry (13-1). This work preceded and supplemented the work of Folin and Denis that we have already reported.

> The beautiful blue color which is produced when phosphotungstic acid and an alkali are added to uric acid lends itself well to quantitative work. . . . In the course of our further studies we have discovered that the color in question is given not only by uric acid but is characteristic of phenols, and that in the case of more complex aromatic compounds it is particularly, if not exclusively, those containing a hydroxyl group in the para position which give the color.
>
> It would be useless to describe the method in detail at the present time, for we have found that different samples of phosphotungstic acid (and phosphomolybdic acid) do not produce the same intensity of color. In fact the material which produces the blue color with uric acid and with phenols is probably not phosphotungstic acid. Whether it is a tungsten product at all, or some other substance as impurity, we have not yet been able to determine for lack of material. We have learned how to concentrate the active agent and to separate it from the greater part of the phosphotungstic acid, but more material and more work will be required before we shall know what it is and how to get it free from the useless, as well as expensive, phosphotungstic acid.

The next paper on this topic appeared about six months later (Part II, p. 400). With Willey Denis also studying the new reagents and their reactivity with organic compounds,

> [a] more systematic investigation soon showed that the amount of active material contained in the phosphotungstic and phosphomolybdic acid solutions was determined by the conditions under which the solutions were prepared and was particularly dependent upon the proportion of phosphoric acid used.

The new colorimetric method gave values on 20 normal urines comparable to those obtained with the Folin-Shaffer modification of Hopkin's method.

> Since the reaction for uric acid which is here used as a basis for the determination is more delicate than any other known reaction and is perfectly adaptable for quantitative work, it is clear that it ought to prove eminently suitable for the determination of such quantities of uric acid as may exist in blood. A new method for the determination of uric acid in blood, based on the principles outlined in this paper, will be described in the next number of this journal.

Otto's third paper with Macallum was a short one on determining ammonia in urine (Part II, p. 400). It presented a micromodification of Otto's original method. Results obtained by this rapid micromethod were practically identical with those obtained by the "old" air-current method.

Otto's fifth paper on protein metabolism, carried out with H. Lyman, was on absorption from the stomach of the cat (*13-2*). London, Abderhalden, and coworkers had published about 40 papers showing that there was no absorption of digestion products from the stomach, while ignoring different findings made by others. With Lyman, Otto would apply the same methods already established by Willey and him for work with absorption from the small and large intestines of the cat.

For this study, the investigators placed ligatures at the cardiac and pyloric ends of the stomach. Substances injected were glycine, alanine, Witte's peptone, creatinine, and urea. Blood samples were obtained at appropriate intervals from the femoral and carotid arteries. At the end of the experiment, the splenic and enteric tributaries of the portal vein were ligated and a sample of blood taken from the portal vein. NPN and urea were determined. There were definite to marked increases in the NPN of all blood samples drawn within about a 2-h period (glycine, alanine, urea), though less with Witte's peptone. Creatinine was not absorbed. Urea increased only following urea injection.

> The results recorded in the above experiments clearly prove that nitrogenous digestion products are absorbed from the stomach. The peculiar results reported by London and his associates are not necessarily inconsistent with ours but the interpretation drawn from them is, we believe, erroneous.

Otto had certainly mellowed since his early postdoctoral days! Nevertheless, London was not convinced, publishing an "erwiderung," to which Otto and H. Lyman replied (*13-3*). First, London had argued that the increase in NPN observed by Folin and Lyman could have been due to absorption from the intestinal tract. However, Otto and Willey had been unable to get an increase in NPN from the intestine in the short period of the experiments involved. Absorption from the stomach was more rapid. The absorption of urea from the stomach was undeniable. Second, London felt that it was strange that Witte's peptone caused a small increase in NPN, but none in creatinine. To this comment, Otto and Lyman answered that the facts spoke for themselves. The NPN rose. Third, London resorted to the "vague and all too common argument that the facts observed have anyhow no bearing on what happens normally." Because death occurs in a few days following ligature of the pylorus in dogs, London insisted that the absorption from the stomach of cats in the first few minutes after the application of such a ligature was to be explained "as a pathological rather than as a physiological phenomenon." Such argument could be advanced against almost all experiments with animals. London's work, moreover, was inconsistent with that of several others.

Otto published two papers with Flanders in 1912. Because they had worked out and published an improved method for determining benzoic acid in food products, they logically assumed that it would be simple to determine urinary hippuric acid by hydrolyzing it to benzoic acid, then extracting it out and titrating it. Although there were unforeseen difficulties to overcome and some surprises, the final outcome was reasonably satisfactory (Part II, p. 400), and the authors felt that their method was markedly superior to any other available.

Bunge and Schmiedberg's well-known method for the determination of hippuric acid in urine was published in 1876. That method is neither accurate nor convenient. It has survived evidently only because no one has succeeded in devising anything better.

The titration of acid in nonaqueous medium in their hippuric acid determination prompted the next paper by Folin and Flanders (13-4). "Viewed from the standpoint of electrolytic dissociation theory, the conditions of these titrations are unusual, and apparently not entirely in accord with the current views."

A presumed parallel between ionization and chemical reactivity and between chemical reactivity and conductivity was a fundamental concept of the theory of electrolytic dissociation.

"Our experimental evidence proves that all classes of acids may be titrated by sodium ethylate, in organic solvents. The only limitation seems to be that of solubility of the acid in the solvent employed." The authors dissolved substances in chloroform but first had to put them into absolute alcohol, and then titrate them with sodium ethylate. They were able to titrate such acids as oxalic, succinic, malic, citric, tartaric, HCl, acetic, propionic, and others dissolved in chloroform. When they tested the conductivity of these acid solutions it did not relate to their capacity to be titrated.

> We would emphasize that our titrations have been made with solutions of acid and alkali which have almost no conductivity, that the resulting salt solutions also have practically no conductivity, that the reaction is instant[aneous], and that the action of the indicator does not differ materially from that observed in aqueous solutions.

It would not be until 1923 that the Brønsted-Lowry theory of acids and bases as proton donors and acceptors and the interacting role of solvents would be proposed. The broader concept of acids and bases as electron acceptors and donors in chemical bonding as proposed by G. N. Lewis did not occur until 1938. Otto and his young colleague were not prepared to offer a solution to their dilemma.

Otto's twentieth paper of 1912, coauthored with C. J. Farmer, was a new method for determining total nitrogen in urine (Part II, p. 401). It was in keeping with Otto's maturing lifelong goals to achieve accurate, rapid, microchemical methods that used simplified apparatus, and to move toward practical clinical applications, utility in metabolic investigations, and suitability for use in student laboratory exercises. The method published in this paper had been developed in 1910 and employed in Otto's laboratory since that time.

> In principle our new method may be described as a microchemical method based on the Kjeldahl-Gunning process for decomposing nitrogenous materials and on the methods of Nessler and of Folin for the determination of ammonia.

Whereas the ordinary Kjeldahl determination used 30–100 mg of nitrogen, this method required only 1 mg without sacrificing accuracy. Urines were therefore diluted so that 1 mL contained between 0.75 and 1.5 mg of nitrogen.

Had the problem been purely a problem of total nitrogen determinations it is doubtful whether it would have been worth all the time that it has cost to develop the colorimetric procedure after it once had become clear that the color reaction seemingly could not be applied directly to the digestion mixture. As will be seen from the other analytical methods now published the total nitrogen determination was only one part of a general colorimetric scheme of analysis.

In the not-too-distant future, Otto would develop a procedure for total nitrogen that applied directly to the digestion mixture, and it involved nesslerization, thereby eliminating the need for the air-current and a receiving tube.

As mentioned previously, not all of the papers that Otto submitted for publication in 1912 were published that year. Three of them, the last sent in late December, were published in 1913. In short, the actual number of publications received by the journal editors was 23. A dam had broken loose! Of the 23 papers, 20 were in the *Journal of Biological Chemistry*.

Otto held strong convictions about the relationship of his research to the part in it played by graduate students and research assistants. Evidently when Professor Henry A. Christian enlisted the laboratory services of Willey Denis on a project, he ran afoul of Otto's displeasure. Otto wrote Christian the following letter on April 2, 1913:

> When I came to Harvard Medical School I requested and obtained a private research assistant. You will find I think that the universal custom is to regard the work of such assistants as that of the professors who have them. Their function is to increase the amount of work done by the professor just as you and most others in the school use technical assistants of various grades and for various purposes. My practice has been usually, but not always, to make my research assistants joint authors of the paper in the work for which they have taken considerable part. This is my policy with regard to Dr. Denis and she understands it.
>
> Her work is however my work just as much as the sections performed by your girl are your sections or the determinations of Dr. Benedict's 12 or 15 girls are his determinations. The only difference is that because I have only one assistant and a good one I make her a joint partner in everything—but I do not lend her or my private laboratory to other departments.
>
> The situation is different in the case of regular assistants and students. Like yourself I do not follow what you call the German plan in regard to putting my name on all the publications of the juniors in my department. Mr. Farmer published several papers last year jointly with Dr. Kendall of Rosenau's department, and while I gave advice with the work and the writing of the papers they published those papers alone.
>
> Mr. Bloor while working here on a problem assigned by me and under my supervision also published his results in his own name. And Dr. Goodall has also published papers alone embodying work which he scarcely could have done alone.
>
> I should be perfectly willing to let anyone doing joint work with me include his technical assistants as joint authors of joint papers if he should desire to do so.
>
> My opinion is that in joint work one has a right to publish the work in any form except by the consent of the other partners, and I confess to having been

angry with you for having announced on the program of the Medical Sciences Society joint work with my department without my knowledge and without any reference to my department. Nor did I regard it as friendly on your part to offer no explanation after you had heard from Dr. G. I would only speak to him since he is an independent worker and you and I have had no consultation about the work. However I am glad to have heard from you now, and my feelings in the matter have gone by.

Now as to our joint work—I shall be glad to let you have the results to do with as you see fit, since I know that it has cost much of your time.

I think that it is unfortunate and not for the good of research not to have friendly and fair cooperation between your department and mine, but unless you are willing and able to recognize my right to the work of my research assistant further joint work between you and me and my research assistant had better be discontinued as you suggest.

I recognize your desire to play fair in your proposal to refer the matter to Dr. Cannon and I am mailing to him your letter together with a copy of this.

Perhaps as early as 1909, and certainly before the spring of 1911, work began on the Folin summer house at Kearsarge under the watchful eyes of Bonnie Cooke, who now was teaching grade school in the area. By the time Joanna and George arrived shortly after mid-June, only the finishing touches remained to be done. Situated on about 50 acres of property, the house was at the edge of a forest, perhaps a mile northeast of the Kearsarge post office. Of course, the house was beautifully situated in an area dominated in every way by the mountains, forest, and wildlife.

The Folins would hike up Mt. Kearsarge several times during their summer visits, where the view was breathtaking from all directions. At the top one could look north to Mt. Washington or east into Maine, and on a clear day, see Portland, 60 miles to the southeast.

The woods were thick with white pine, spruce, hemlock, maple, oak, beech, and birch. Around the house there were several red pine trees. Across the Kearsarge trail up a gentle slope, there was a field topped by a huge boulder, dubbed Mt. Teresa by the Folins.

Joanna and George, as mentioned, were with Bonnie about mid-June of 1911. Teresa stayed with her parents in Brookline. Joanna, who soon would become 11 years old, was precocious and communicative, and her letters of that brief period give a glimpse of life at Kearsarge.

June 20, 1911

I'm awfully sorry to hear that baby [Teresa] is sick and I hope she'll be better soon. . . . This morning Bonnie and I went down to the garden and then came back and sowed grass seed until dinner, so you can see that we haven't fooled around much.

Yesterday, Bonnie had to go down to the post-office herself, so the rest of us went too. While Bonnie went in, we stayed out on the piazza. Presently a man came up and doffed his hat to Mabel [a family servant], when he came out of the post-office. He started to talk to her, and at last after a good deal of talking he took us all everyone of us up to his house where he took some baked beans out of a hole in the ground. They gave us something to eat and insisted on our taking home a small pot of beans and a jar of marmalade. We left for the post-office at 4:30 P.M. returned 8 P.M. Had a nice time. No doctor's bills yet and no prospects of any.

July 4, 1911

It's been *so* hot here. It was 95° on our usually cool back porch in the shade, this noon. . . . Mr. Whittemore, the man who is shingling the house, is awfully nice and he bought 2 boxes of torpedoes and 6 boxes of caps. . . .Night before last we heard loud hissing noises and a light rose slowly into the air in the direction of the brook. Last night I woke up about midnight and saw lights filling the sky and noises like thunder. Both of these were fireworks. . . .

I'm dripping wet with sweat and I'm so glad that it will soon be time to plunge into the "Fishes" Pool. . . .

July 10, 1911

It will be three weeks Thursday since we came and it doesn't seem so very long, though I'm crazy to have all of you up here. Does the house seem very quiet without us? . . . Do you think Grandma and the rest are coming east this summer? . . . We haven't had any rain with the exception of thundershowers since before we came. One minute Bartlett was invisible and in five more minutes the clouds had risen again. I'm sleeping on the front porch and when I woke up this morning the Ledges and Moat were half covered with those fleecy white clouds that give the appearance of snow drifts. The sun's first rays were striking the northern part of Moat, causing such wonderful light and shadow among the clouds, the like of which I have never seen. Through those banks of clouds, portions of the sky could be seen; it was that exquisite blue, that is rarely seen, even in the sky.

Timothy took George and me up Kearsarge to find some blueberries which are *not* so *very* plentiful. We found several pretty good patches, but he wasn't satisfied at all, so altogether we didn't get more than 1 quart.

Tell Baby that "Nanna" [Bonnie] is waiting patiently for her. . . . Tell Dad to expect junipers "galore." Don't any of you get sick or work too hard.

July 24, 1911

Just two weeks from tomorrow is my birthday and please be here. It's raining today, and the brook is about an inch and a half deeper. . . .Yesterday we had some beans from our own garden, and the peas are ready, the corn has begun to tassel nicely and everything looks favorable. . . .

We're going to climb Hurricane someday soon after blue berries. Already a few black berries on our land.

August 1, 1911

I'm just crazy to know whether I'm coming home or not. I want to come. . . .

We went up Hurricane after blue berries and we got a *scant* 9 quarts. It's just lovely up there. . . . The grass has begun to come up everywhere that it's planted.

We're building a dam across the bottom end of the fishes pool and in the deepest place it is about 5 inches above my knee. Before the rains and the dam it wasn't any deeper than my knee.

The kitchen and "dining room" floors are going to be painted tomorrow. Bonnie scrubbed the kitchen floors yesterday, and she is scrubbing the "dining room" floor now. . . .

Please induce grandma to come, because I want to see the whole bunch of folks at home sweet home. . . . One week from today is my birthday, remember that!

Bonnie told me the secret a few days ago and George doesn't know it yet. Bonnie is thinking of taking him up Mt. Washington the day I go so he won't feel so bad. . . . Peas, beans and apples are all cooking on the stove. The first two were picked in the garden yesterday afternoon. We had our first chicken Sunday and it was awfully good.

You'll think that it's just lovely up here. The house is not *too* small or too big, it's just right. It has been quite cold and rainy but it is warm again now.

Please write and tell us the very minute that the plans are certain because I'm dreadfully excited. . . . Tell Dad there are lots of chores for him to do. Among them, hang up the swing and cut junipers and take us up Kearsarge. . . . I transplanted a pine tree about a week ago and it's coming along at a great rate. . . . Are there any flowers in the garden [Brookline] that I planted? The candy tuft was in bloom when I left but the nasturtiums were not. (I don't know how to spell some of these words and I'm just too lazy to look them up in my dictionary). Be sure and R.S.V.P. With true love. . . .

In early 1912, Teresa caught scarlet fever, a disease with frequently devastating, even fatal, consequences before the age of antibiotics. To avoid exposing the other children—Joanna, Grant, and Ruth Anderson, a child for whom Laura was caring—Teresa was placed in the local hospital for contagious diseases. In April, when Teresa was brought home, nine-year-old George was sent to Kearsarge to live with Bonnie Cook to protect him from exposure. Joanna and Ruth remained in Brookline, because they both had had scarlet fever and there was less fear that they would contract it. Nevertheless, Joanna contracted the disease and missed the final few weeks of the ninth grade. Otto and Laura had planned to spend a good part of the summer at Kearsarge, because there was much that Otto wanted to do in the way of cleaning the shrubs away, sawing and chopping wood, mowing, and pruning trees. Before leaving for Kearsarge, however, Otto wished to attend a meeting in Boston with the Shaffers whom the Folins had invited. Thus in mid-July the children were sent on ahead to Kearsarge.

Toward the end of July, Joanna who had been left weakened by the bout with scarlet fever, came down with diphtheria. One of two local physicians Bonnie called in failed to recognize her malady, and thought it to be mumps. Diphtheria, which has a rapid onset, grows in danger the longer it is left untreated. In this era before the availability of toxoid immunization, diphtheria was fatal in almost 20% of cases untreated for five days. An antitoxin, however, was available.

When Bonnie called the Folins, Otto and Laura, accompanied by Philip and Nan Shaffer, came to Kearsarge on the next available train. Not having the diagnosis of diphtheria, Otto made no effort to bring the antitoxin. Although the exact details are vague, as the precious moments slipped away, the diagnosis was, according to Teresa's impression, shifted from mumps to diphtheria by Shaffer's judgment. Otto hastily called a specialist from Boston City Hospital who had the antitoxin and took the next train to North Conway, five hours away, and the wagon ride for the last three miles to the summer house.

The antitoxin failed. Though Joanna's throat was improved, the toxin had reached her heart, and she died on Aug. 11, one week after her twelfth birthday. She was buried in the Kearsarge cemetery. The Kearsarge physician insisted that the children remain there under quarantine, but Laura would have none of that. The Folins returned to Brookline, where Grant, but not Teresa, came down with diphtheria. No time was lost in making the diagnosis and in using the antitoxin, and he survived, though he, too, hovered on the edge of death.

The tragedy that had befallen the Folin family shattered Laura's dream of a summer retreat. For several years to come there could be no unclouded joy for her at Kearsarge. She accepted its refuge for Otto's sake because he could restore his spent vigor there. For her the mists rising up the mountainsides in the dawn were filled with shadows of despair, guilt, and loss.

In the autumn of 1912, Otto interviewed a one-armed, though vigorous appearing, young man who requested that he be permitted to enter the graduate program in biochemistry, at least to earn a master's degree. The applicant had graduated from Harvard in 1910 with a major in chemistry, and had since taught undergraduate chemistry courses in Canada, before serving as an assistant at the Worcester Polytechnic Institute.

Otto was downright skeptical. How could this handicapped student carry out a research project "singlehandedly"? Otto's intent was to train graduates in the European tradition, not as teachers alone, but as investigators carrying on independent research. Even with two hands there was difficulty enough assembling apparatus, preparing reagents, handling hot solutions, and manipulating glassware.

Unperturbed, the young man explained that as an undergraduate he already had published his first paper with Professor H. A. Torrey, "Note on Some Properties of Piperonyloin," in the *Journal of the American Chemical Society* (**32**, 1492–94 [1910]). This, of course, swayed Otto greatly, but he suggested that a man with one arm missing to above the elbow, even with such resolve, would do much better at law.

The prospective student, James Batcheller Sumner (1887–1955), however, was not to be denied, and he would become one of Otto's most celebrated students. Otto started him on the long trek that would lead him to share a Nobel prize (he received half) in 1946 with J. H. Northrop and W. M. Stanley for his crystallization of a protein-enzyme, urease (*13-5*).

References

13-1. Folin, O., and Macallum, A.B., On the blue color reaction of phosphotungstic acid (?) with uric acid and other substances. (Preliminary Paper). *J. Biol. Chem.* **11**, 265–66 (1912).

13-2. Folin, O., and Lyman, H., Protein metabolism from the standpoint of blood and tissue analysis. Fifth Paper. Absorption from the stomach. *J. Biol. Chem.* **12**, 259–64 (1912).

13-3. Folin, O., and Lyman, H., Absorption from the stomach—a reply to London. *J. Biol. Chem.* **13**, 389–91 (1912).

13-4. Folin, O., and Flanders, F.F., Is ionization, as indicated by conductivity, a necessary prerequisite for the combination of acids with bases? *J. Am. Chem. Soc.* **34**, 774–9 (1912).

13-5. Maynard, L.E., James Batcheller Sumner (1887–1955). Biogr. Mem. Natl. Acad. Sci. U.S.A. **31**, 376–98 (1958).

Chapter 14. Developing Clinical Biochemistry (1913–18)

Otto published 11 papers in 1913, of which Willey Denis coauthored seven. As mentioned before, three of these were submitted for publication in Otto's previous peak year of productivity. Now that Otto and Willey had developed their phosphotungstic acid reagent for uric acid, they immediately applied it to the determination of uric acid in blood while Macallum worked on urine (Part II, p. 402).

Blood, of course, is a much more complex body fluid than urine to analyze because of its protein content. For performing colorimetry, the proteins must be removed to prevent interferences from turbidity and color production. Whole blood contains about 20% by weight of protein, of which almost 75% is hemoglobin. Improving the method for removal of protein remained a continuing problem for Otto to explore in the years ahead.

The method they developed first was cumbersome, though a solid beginning. Removing protein was not the only problem encountered in developing a workable method of blood analysis. The need to precipitate uric acid as a silver salt introduced a number of potentially error-inducing steps, and along with the need for weighing, lengthened the analysis time. The filtrate had to be evaporated and the uric acid precipitated; then the excess H_2S had to be removed, and to ensure its removal, the product had to be was tested with lead acetate, making the method too complex for clinical application. The use of *centrifugation* and *decantation* (rather than filtration) to obtain the final uric acid solution for the color reaction was a new, and smart trick in analysis. However, isolating uric acid as the silver salt was the Ludwig-Salkowski approach that Otto had once belittled. Nevertheless, Otto and Willey again applied the technique, this time to urinary uric acid analysis (Part II, p. 402).

The Folin-Macallum procedure using an alcohol-ether extraction to remove polyphenols, though rapid, was unfortunately unsuitable for quantitative analysis of urines containing "albumin" or sugar. Otto and Willey therefore proposed a new method based on separating the uric acid as a silver salt, as in their method for blood, but without the need for protein removal at the initial step. Most important they prepared a new stable standard of uric acid in formaldehyde.

With their procedures for uric acid, NPN, and urea ready for application, Otto and Willey now sought to obtain normal values in humans, and "to measure various degrees of nitrogen retention and urea accumulation due to kidney inefficiency with much greater accuracy than had heretofore been possible" (14-1). The authors pointed out that although there had been dispute as to whether uric acid existed in blood normally, their last paper had shown that uric acid was normally present.

> All three determinations can be made without requiring more than 20 to 30 cubic centimeters of blood. We have therefore in a combination of these three determinations practically a new system of blood analysis—certainly one well worth trying out on human blood for clinical purposes.

Otto and Willey first made a general survey of the triad of substances in the blood of several animals.

> From the results obtained it would appear that the uric acid in the blood of the rabbit, sheep and horse is almost nil, and that in the case of the other animals that amount though a trifle larger is still extraordinarily small—0.2 of a milligram or less per 100 grams of blood.

Uric acid concentration, however, was much larger in birds than in mammals, while urea was much smaller. In their first examination of the uric acid values in human blood, they wrote, "It is a curious and interesting fact that human blood contains several times as much uric acid as does blood of any other mammal whose blood we have had the opportunity to examine." Others had recently pointed out that man is unique in not converting uric acid to allantoin.

Otto and Willey had tested the uric acid concentrations in the blood of 38 mental patients and obtained a range of values of 0.7–3.7 mg/100 g [42–220 μmol/kg]. But, of course, they did not consider these values "normal." On the other hand, they tested NPN and urea nitrogen in the blood of 16 healthy male adults (medical students and instructors, ages 20–45, but mostly 22–23 years old). NPN was 22–26 mg/100 g (15.7–18.6 mmol/kg) and urea nitrogen (UN) 11–13 mg/100 g (7.9–9.3 mmol/kg).

How about pathological cases? Of 63 syphilitics, only 13 had normal NPN and UN, about 7–10 of 21 insane patients had normal values for both. Of 11 patients with "chronic" nephritis none had either normal NPN or UN. The last group was also analyzed for uric acid, with the finding that there was apparently no relationship between uric acid and the UN and NPN present. "The figures recorded in tables 5 and 6 are noteworthy in that they probably represent the first analyses on record where the uric acid, the urea and the total non-protein nitrogen have been determined in the *same samples of human blood.*" Based on the values obtained on the 63 syphilitics and on the mental patients, the authors tentatively regarded the "normal" uric acid range as 1–2.5 mg/100 g (71–149 μmol/kg). They did not evaluate the concentration in the 16 healthy individuals. These values were for "whole" blood and not serum or plasma.

The urea and total non-protein nitrogen in the blood must in the main be inversely proportional to the general efficiency of the kidney since the kidney represents practically the only outlet for the nitrogenous waste products.

Twelve patients with high uric acid concentrations (gout, leukemia, lead poisoning) were also tested. The NPN and UN were practically normal in some of the cases of gout.

The blood in gout does not exceed 6 milligrams at least so far as our present experience goes. . . . In the case of uric acid it seems to be purely a matter of chance, purely a matter of insolubility, that corresponding or even smaller degrees of kidney insufficiency and slight uric acid accumulation should result in all the serious consequences involved in the development of gout.

Otto would continue to examine nitrogenous substances in the blood of patients with nephritis and with gout. He would collaborate with his medical colleagues and those in the associated hospitals of the medical school. In the next paper, Otto and Willey joined Dr. C. Frothingham of the Peter Bent Brigham Hospital, and Professor R. Fitz, of the department of Theory and Practice of Physic at the Harvard Medical School. This team set about determining the relation between NPN retention and phenolsulfonephthalein (PSP) excretion in experimental uranium nephritis produced in rabbits (14-2).

PSP had been introduced by Rowntree and Geraghty [*J. Pharmacol. Exp. Ther.* **1**, 579 (1910)].

The numerous observations recorded by Rowntree and his associates on the speed with which phenolsulphonephthalein is eliminated by normal as well as by nephritic kidneys seems to show that the elimination of this substance when injected intramuscularly is indeed a valuable index to the general efficiency of the kidney.

They wanted to discover if the rate of removal of PSP related to the rate of removal of free metabolic products and if it correlated with actual retention in blood of nitrogenous waste products. The methods that Otto and Willey had recently published made possible a continuous study of experimental nephritis in ordinary small laboratory animals, such as cats and rabbits.

The authors used rabbits induced into acute nephritis with a single dose of uranium nitrate (1.25–3 mg) given subcutaneously. Two series of experiments were done. In one series, the animals were eventually killed (under anesthesia) to examine the pathology of the kidneys. In the second series, the animals were allowed to recover. In all cases, the blood was analyzed at intervals, and PSP tests were made periodically. The amount of blood taken (2–3 mL) did not cause a drop in the hemoglobin. Each rabbit was placed in a cage over a glass funnel so that urine samples could be collected. The measured urines were routinely examined for albumin and casts, and if PSP had been administered, half of the urine would be used (at 70 min following injection) for determining PSP.

It was found that the normal rabbit had an NPN of 30 mg/100 g [21.4 mmol/kg], and a PSP excretion of 60% or more in 70 min when 6 mg of PSP was injected.

The authors found that NPN and UN accumulated gradually in the blood and returned to normal gradually with recovery, while PSP excretion responded to the nephritis by dropping rapidly to its lowest point and returning rapidly to normal with recovery of the kidneys. Both PSP and the nitrogen retention agreed well with the destruction of the kidneys demonstrated histologically.

> In general these tests parallel each other as indicators of renal function, but have this essential difference: the amount of phenolsulphonephthalein excretion shows the renal function at the moment; the amount of non-protein nitrogen and urea in the blood is rather a measure of an accumulating difference between the amounts of waste nitrogen produced in the metabolism and the amounts eliminated by the kidneys. The time element, the duration of the condition, is therefore an important factor in this test.

These results must have made the investigators eager to assess the value of these tests in human nephritis. It would not take long.

Sometime in 1913, Willey Denis was officially made a member of the paid staff at the Massachusetts General Hospital and was given a laboratory there on the third floor of the powerhouse. Otto was made a consultant. This formalized arrangement meant that it would be easier for Otto to get at the patient population and to collaborate with the research-minded professional staff. Willey Denis would not only continue on projects with Otto but would launch many of her own, alone and with other staff members. Her laboratory would be one that might be considered a transition or bridge between research and application that led to the modern clinical chemistry laboratory. She was certainly among the first of the true clinical chemists and probably the first female clinical chemist in the U.S.

Otto would also become a consultant to the Peter Bent Brigham Hospital and the Children's Hospital, and this led to joint projects that were later published.

The phosphotungstic acid reagent that Otto and Willey had formulated could detect a polyphenol compound such as epinephrine, with a sensitivity of one part epinephrine in 3 million parts of water. This was 10 times the sensitivity reported for any other color reaction. They therefore joined Professor Walter Cannon in creating a new method for quantitating the hormone, on which Cannon had done much pioneer work (Part II, p. 403). Their purpose was to measure the epinephrine in commercial and "home made" extracts of suprarenal glands.

Cannon would carry out the physiological blood pressure test on cats using the method of Elliott. "Elliott's statement that the largely denervated circulatory system will respond with almost the accuracy of a chemical balance to any submaximal dose of adrenalin we were able to confirm." Cannon's values were obtained without prior knowledge of the results determined by Otto and Willey with the chemical method, and vice versa.

The authors found that the hormone gave almost exactly three times the blue color of the same weight of uric acid.

> With uric acid, the phosphotungstic acid reagent and a high grade colorimeter available, it is now only a matter of a few minutes' work to assay a given solution of epinephrine. Using 10 cc. volumetric flasks, as little as three or four hundredths of a milligram of epinephrine, or even less, can be determined with a very satisfactory degree of accuracy.

The question was whether the method would work on extracts rather than pure solutions. After all, uric acid and other substances might be present to interfere. It was found, however, that the reagent did not react much with extracts of other tissues outside the suprarenal glands.

There was good agreement of the chemical method with the physiological method using extracts of the adrenals of sheep, a lamb, and cattle. The method was sufficiently sensitive to show that blood taken from the adrenal veins gave a stronger color reaction after stimulation of the splanchnic nerve than before, adding chemical evidence to the already established physiological evidence.

Otto and Willey were subjected to two published attacks by Abderhalden and his co-workers. The first [*Z. Physiol. Chem.* 81, 468 (1913)] denied that amino acids were absorbed from the small intestine and then distributed to different tissues as Otto and Willey had shown in their experiments with cats, but offered no laboratory data in rebuttal. Otto and Willey rejoined (*14-3*), "We recognize that our facts need verification, and anyone who will learn to use our quantitative methods can easily repeat our work." Abderhalden preferred the theory that absorbed amino acids were immediately deaminated or regenerated into protein, despite the fact that he and Lampé had qualitatively demonstrated the presence of amino acids in blood, at all times, and particularly after "a few minutes absorption of any ordinary amino-acid."

Attack (Abderhalden and Fuchs):

> The colorimetric method of Folin and Denis cannot be used to replace the determination of tyrosine by crystallization so long as it reacts with other amino acids and consequently yields values too high. [*Z. Physiol. Chem.* 88, 468 (1913)]

Reply (Folin and Denis) (*14-4*):

> The statement contained in the above sentence is given as though it embodied a conclusion derived from the experimental work described in the paper whereas in point of fact it represents only an opinion. The only observation recorded by Abderhalden and Fuchs in support of that opinion is that tryptophane and a certain obscure substance ("oxytryptophan") which, so far as we know no one but Abderhalden has yet observed, also give a slowly developing blue color with our tyrosine reagent.

Touché! Otto and Willey conceded that they could be wrong about tryptophane, because they had none available and they had perhaps falsely assumed that tryptophane could not survive the boiling mineral acid treatment used for splitting off the tyrosine. On the other hand, the excess of tyrosine indicated by the Folin-Denis method for casein (with 1.5% tryptophane) was no greater than that for zein, which had no tryptophane.

Unless Abderhalden had worked out a method for isolating tyrosine quantitatively and had published revised figures for at least some proteins, there was no proof that the Folin-Denis values were too high. Meanwhile, "we would rather insist that our tryosine figures are more nearly correct than those heretofore recorded in the literature."

No doubt Otto and the great biochemist, Emil Abderhalden, corresponded further on this privately. One outcome was that Otto wrote a chapter for Abderhalden's *Handbook of Biochemical Procedures*, published in 1913.

During the 1912–13 academic year in the department of biological chemistry, James Lucien Morris, A.M., succeeded A. B. Macallum as the Austin teaching fellow, Francis B. Kingsbury, A.B., remained as a teaching fellow, and the three assistants were Harry W. Goodall, M.D., Alfred Willson Bosworth, S.B., and Chester J. Farmer.

As part of his work on his doctoral dissertation, Morris published with Otto his findings on nitrogenous constituents in rat urine (*14-5*). Because the urinary output of the rat was only 5–50 mL a day, the microchemical methods of Folin and Denis first had to be modified somewhat if protein metabolic studies were to be carried out successfully.

The rats used in this study were kept in specially constructed cages, each resting on top of a funnel permitting urine collection into a container charged with 2 mL of 1 N HCl. The daily urines were approximate because no attempt was made to have the rats void. This was overcome by keeping each rat in the experiment for 9–15 days of urine collection so that results could be averaged. We shall not present details of the method changes. Essentially, smaller volumes of urine were used, and dilutions were decreased so that more intense colors were produced for uric acid, creatinine, total nitrogen, NH_3, and urea. Creatine was also measured in an essentially scaled-down version of the macroprocedure.

The paper was valuable for several reasons: 1) It showed how analyses of the nitrogenous constituents of rat urine could be carried out on the very limited volumes available; 2) it gave the composition of normal rat urine; 3) it reported values for the same nitrogenous constituents in rat blood (Surprisingly, the uric acid concentration was quite similar to that of the human. "The purine metabolism of rats is, therefore, like that of man and unlike that of other mammals hitherto investigated."); and 4) it showed that rat urine resembled human urine in its percentage composition of nitrogenous constituents, including uric acid output.

> It seems rather remarkable that the one animal which (excepting man) produces the most uric acid in the course of normal metabolism should lack the ferments capable of producing it. In this connection we can state that investigations conducted by J. B. Sumner in this laboratory have shown that aqueous extracts from rat livers are as capable of destroying uric acid as similar extracts

obtained from the livers of cats and sheep. The significance of 'purine ferments' as obtained from organ extracts in relation to the formation and elimination of uric acid in the course of normal metabolism is, therefore, far from clear.

Though it required a rather large volume of blood, the Folin-Denis method of analyzing uric acid in blood now made possible the study of its metabolism. Popular at the time was the uricosuric drug, 2-phenylquinolin-4-carbonic acid (atophan). "The experiments described below were undertaken for the purpose of determining whether phenylquinolin carbonic acid does or does not reduce the uric acid content of the blood" (14-6).

Seven human subjects were used in the study, five with gout, one who, though on a low nitrogen diet, still had a high blood uric acid content, and one normal control (Lyman). The patients were all kept on purine-free diets for 203 days prior to the first blood samples being taken. Blood and urine samples were collected before and after the administration of atophan. NPN and UN analyses were performed as well as uric acid. The patients were taken from Boston City and Massachusetts General Hospital.

For the five people with gout, the authors found that

> [so] far as the uric acid in the urine and blood is concerned the results recorded in the above experiments appear to us to be quite unmistakable. The administration of the phenylquinolin carbonic acid in every case led to an increase in the uric acid elimination and to a marked reduction of the uric acid in the blood. The increased output therefore represents the elimination of uric acid which had previously accumulated in the blood and the previous accumulation represents a corresponding kidney inefficiency.

In three of five of these subjects in whom the NPN and urea had not been elevated, the results could only be explained on the basis of a selective activity of the kidney. In the other two cases, the atophan reduced the uric acid along with the NPN and urea.

In the control, as with the rest, atophan caused a marked drop in blood uric acid and a rise in urinary uric acid output. Of prime importance to this study was the fact that the analytical method made it possible.

Otto's ninth publication of 1913 was the chapter in Abderhalden's handbook (14-7). He presented complete details of his analytical methods for urine and blood. Undoubtedly this chapter was a good one for Otto to write because of the large number of biochemists in Europe who may not have followed Otto's work in the *Journal of Biological Chemistry* and elsewhere. In addition it gave him the opportunity to collate his methods and point up his recent modifications. Otto was moving so very fast in methodology that this chapter may have prevented a rash of "erwiderungen."

Otto published a paper with Alfred W. Bosworth, an assistant in his department, on the use of the Folin-Macallum method for determining ammonia in fertilizer (14-8). Bosworth had used the method for determining ammonia in cheese, evidently while he was at the Bureau of Chemistry, U.S. Department of Agriculture. It had also been used on meat.

Otto's eleventh and final paper of 1913 was a lengthy "abstract" of his presentation to the Subsection of Chemical Pathology at the Seventeenth International Congress of Medicine in London, England (*14-9*). The congress had opened on Aug. 6. In conjunction with this meeting Otto also attended the Ninth International Physiological Congress in Groningen, Holland, Sept. 2–5. We have no record of what he presented there.

There is strong reason to believe that Otto had planned this European trip for some time. Several of his American colleagues were to attend the meetings, and, of course, Otto would be able to renew old acquaintances and to meet others whose work had interested him. The scientific sessions were likely in German, and Otto was at home in this language.

Of great importance was his wish to see his mother and other relatives in Sweden. Fifteen years had passed since he had last seen them, and his mother was now in her eighties.

Otto's paper in London was directed toward the physicians attending the Congress of Medicine. From the standpoint of the clinical applications of pathological chemistry, only a few substances of disease had been detected in urine. Metabolic disorders showed little. Recently, it had been found that ammonia excretion in acidosis was abnormal, a discovery that was "a brilliant contribution of biochemistry to clinical medicine." Urea, water, and chlorides had become important in the study of nephritis.

> The animal body is not capable of oxidizing the amino groups contained in nitrogenous materials taken as food, and the urea formation, as I look at it, is perhaps the most important mechanism for preserving the neutrality of the body fluids. . . . Except in the case of certain rare and unusual "inborn errors of metabolism" the urea formation is persistently normal, however much the urea elimination may vary because of normal or pathological variations in the total nitrogen metabolism.

The urea nitrogen was 50–95% of total nitrogen, depending on the relative proportion of food to tissue metabolism.

Creatinine output was practically independent of food intake, the amount eliminated depending on the size of the person. Its output was increased in fever and in certain diseases.

> It is to be noted that we are as yet entirely ignorant of the origin and significance of the creatine which is so abundant in muscles, and it is scarcely to be doubted that fundamentally important metabolism problems are connected with the muscle creatine and urinary creatinine, but there are as yet problems of normal metabolism, and it is too early to say whether or in what way light may be thrown on clinical problems by studies of these products.

The fact that muscles of man contained 0.3–0.4% of creatine and only traces of urea showed that creatine probably served some important function. Also, a metabolic disease or two might be associated with creatine. Otto persistently returned to this problem of creatine origin and creatinine production, and through his persistence, the momentous day would come when the origin and significance of creatine would be elucidated by colleagues in his department at Harvard.

Urine analysis was particularly useful in studying normal metabolism from the waste products excreted, but not accumulated in the human. For products that do accumulate in the body, however, it was better to analyze blood. "The primitive notion about 'impure blood' has a kernel of truth in it, and by chemical analysis of blood it is now possible to reveal clinically important facts."

Blood had 22–30 mg NPN/100 g (15.7–21.4 mmol/kg). About half of NPN (12–16 mg) was urea, and the remainder amino acids, various soluble materials, and traces of uric acid, creatinine, creatine, and barely perceptible ammonia. When the kidneys were not normal the total NPN in blood rose as did the proportion represented by urea. "Moreover, in middle-age adults perfectly normal kidneys are the exception rather than the rule." Otto was now pointing out that normals (or reference values) varied with age, at least insofar as NPN and urea were concerned. Of 150 syphilitics' blood examined in the previous two years in Otto's laboratory, three of four had abnormal values. Of 45 nephritics, only nine, or one in five were normal. By blood analysis, Otto announced, it was possible to determine whether a patient was excreting nitrogenous waste products normally. Only 5 mL of blood was required. Thus he was spreading the gospel of clinical chemistry.

Otto showed how, by use of high carbohydrate–low protein diets, it was possible to reduce the nitrogen retention and possibly the work of the kidneys of nephritic patients. Otto then reviewed the recent work on uric acid excretion of gouty patients given atophan.

> There is in my mind not the slightest doubt but that by means of chemical investigation of blood, clinical problems of many kinds will be elucidated to a much greater extent than has been possible by means of urine analyses, though the two must usually go hand in hand.

In the autumn, the department of biological chemistry grew to 10 members. James B. Sumner, A.M., now joined James L. Morris, A.M., and Harry W. Goodall, M.D., as an assistant; Frederick S. Hammett, M.S., became a teaching fellow; and Richard D. Bell, M.D., Fred F. Flanders, Ph.D. and Henry Lyman, M.D., were fellows; while Roger S. Hubbard, A.M., was the Austin teaching fellow. Three of the graduate students would obtain their doctorates in 1914 under Otto's direction: F. B. Kingsbury (I. A contribution to the role of bile in fat absorption. II. The determination of benzoic acid in the presence of hippuric acid in the urines of the rabbit and the dog); J. L. Morris (Protein metabolism of the rat, with special reference to tumor problems); and James B. Sumner (The formation of urea in the animal body).

We shall have more to say about Richard Dana Bell, the fellow who had joined Otto's staff in the spring of 1913 immediately following receipt of his medical degree at Harvard. During his tragically abbreviated career, this brilliant young biochemist would have the unique distinction of working with both Sumner and Doisy and, of course, with Otto Folin. A year behind Bell in the medical school was another future member of the faculty, Cyrus Fiske, on whom Otto was keeping an eye.

Otto Folin published 11 papers in 1914, all of which were submitted by the first of June. Seven of these were written with Willey Denis as a collaborator. Six papers dealt specifically with creatine and creatinine.

In his paper presented in London the previous year, Otto had discussed studies on nitrogen retention in nephritic patients in relation to diet and high blood pressure. The full report was Otto's first paper of 1914 (*14-10*).

> The chief purpose of our investigation was to determine the extent to which it is possible by means of diets to vary, particularly to reduce the waste nitrogen in blood of such nephritis and incidentally determine whether there is any relationship between the blood-pressure and the accumulation of nitrogenous waste products of the blood.

PSP tests would be included for comparison of "kidney efficiency" with nitrogen retention by blood analysis.

Twelve nephritic patients with high blood pressure were selected for study. Three diets were used: the ordinary hospital diet, a low protein–high calorie diet, and a high protein diet.

Both NPN and urea nitrogen values dropped considerably lower in the 12 patients when they were fed a low protein diet compared with the high protein diet. Uric acid values also dropped somewhat with the low protein diet. Five of the patients could take care of the protein in the usual hospital diet, without much nitrogen retention. There was no correspondence between the PSP test of kidney efficiency and the degree of waste nitrogen retention. It appeared that the PSP could be reduced by one half or more before any abnormal accumulation of waste nitrogen occurred in the blood.

> It would seem from these results as though the direct determination of the non-protein nitrogen (and urea) in the blood furnishes a more reliable guide to what might be called the protein tolerance of patients than can be obtained from any "direct" test of kidney efficiency, for of all tests yet devised for this purpose the phenolsulphonephthalein test of Rowntree and Geraghty is admittedly the best.

In March 1914, the *Journal of Biological Chemistry* received six papers from Otto. Of these, Otto was the sole author of three, his latest experiences with every aspect of creatine and creatinine analysis. The first paper dealt with the isolation and purification of home-made creatine and creatinine from urine, for which Otto had some new "twists," and presented his favorite method for preparing standard creatinine solutions (Part II, p. 403).

Now that he had devised a method to prepare a stable, pure creatinine standard solution, the artificial potassium bichromate solution no longer needed to be used for this purpose. Otto's second paper, published with the assistance of J. L. Morris, presented a way to stabilize the standard as the picrate for as long as 24 h (Part II, p. 404). A stable standard made possible multiple analyses of urines without the standard picrate having to be prepared repeatedly. Of course, stability was one advantage of the artificial bichromate solution that Otto did not mention.

The only important modification of his creatine procedure Otto declared was carried out in 1907 by Victor C. Myers [not to be confused with the distinguished German organic chemist, Victor Meyer]. Myers used higher temperatures by means of an autoclave to shorten the conversion time of creatine to creatinine from three hours to half an hour.

Using the new creatinine standard, Otto made the determination of creatinine and creatine in blood, milk, and tissue almost as simple as their determination in urine (Part II, p. 405).

With this procedure operational, Otto (with Buckman) now applied it to the determination of creatine in the muscle of several animals, including the cat, rabbit, dog, sheep, hen, and turtle.

> The results of our determinations indicate that the creatine contents of the muscles of cats, rabbits and hens vary within substantially the same limits. The variations found appear to be too large to permit the use of average figures in calculations as to the alleged relationship between the creatinine elimination and the total amount of creatine in the tissues.

In this paper, Otto also reported on the use of the modified Duboscq colorimeter, in which the prisms were in fixed position, while the platform supporting the cups was movable. The paper included two tables of data for creatine content, but the first table did not state the muscle tested, whereas the second was for heart muscle (14-11).

With Willey Denis now situated at the Massachusetts General Hospital, the opportunity for Otto and Willey to apply their latest modifications and analytical methods was enhanced considerably, particulary for blood chemistry, as the need for these quantitative tests began to be appreciated. For creatinine and creatine analysis,

> [it] therefore appeared to us worthwhile to make a series of such determinations in the blood of different animals and particularly in human blood in order to learn whether in pathological conditions there is any specific retention of creatinine similar to the retention of uric acid in gout.

If enough blood was available, then all the common NPN constituents would be determined (14-12).

Nitrogenous constituents were examined in the blood of 200 patients. In general it was found that there was no specific creatinine retention. Creatinine was removed by the kidneys "with remarkable ease and certainty" and was matched in completeness only by the removal of ammonium salts. "Except in extreme conditions of retention approaching anuria the creatinine in blood remains at the normal level." Otto and Willey, perhaps without being aware of it, were developing a concept of "normal values" for the analytes determined. They presented three tables of values obtained on patients. The first group of 12 had no nitrogen retention, and determined values [NPN, urea, ammonia nitrogen, uric acid, creatinine, and creatinine and creatine (combined)] were all lower by far than those of the 17 patients who had nitrogen retention. A group of 37 patients with mixed disorders generally had intermediate values. Of interest, but without explanation, was the range of combined creatinine and creatine concentrations in the three groups 6.0–10.0 (0.53–0.88 mmol/kg), 6.0–46.0 (0.53–4.07 mmol/kg), and 0.6–15.0 mg/100 g (0.053–1.33 mmol/kg) of blood, respectively.

Values for creatinine and for creatinine and creatine combined were given for eight animal species—cow, sheep, pig, cat, and rabbit in one group, and the fowl, hen, pigeon, and goose in another group. The birds had no creatinine, but a high creatine, whereas the other group had values similar to those of the human.

The sequence of papers would hardly have been complete without a fresh attempt to explain the origin and interrelationship of creatinine and creatine in the animal body. Such was Otto's seventh and final paper of this 1914 series with Willey Denis (14-13).

> With the observation made independently by Folin and Klercker (1906) that the administration of creatine is not accompanied by an appreciable increase in the creatinine elimination came the problem of explaining why these two substances so nearly related chemically seem to be so distinct and independent of each other in the metabolism, and what is the significance of the large accumulation of creatine in the muscles.

Otto and Willey considered two ideas concerning the relationship between the two substances. One, that of Myers and Fine, regarded the creatine in an animal as a fixed quantity. In this theory, the loss of creatine on fasting would correspond to the loss of creatinine (and creatine) in the urine.

> It does not seem entirely plausible that fasting or partially fasting animals should not form new precursors of creatine and creatinine out of other nitrogenous materials, since that is exactly what fed animals, including herbivorous animals, must be doing all the time.

Our authors were more intrigued by a suggestion made by Hofmeister in 1907 that creatine might be "in some sort of combination" with muscular protoplasm, an idea with support in the work of Urano, who showed that creatine in the living muscle of frogs was present in a non-dialyzable form. No attempt had been made to confirm this finding, and Urano's technique was rather crude.

> Creatine must, however, be a waste product or synthetic product serving some special function, or as a synthetic product it may in fact be a part of the active living protoplasm. We believe that the last named alternative represents the facts, and in support of this hypothesis we now propose to show that the so-called creatine of muscles is a post-mortem product and that there is very little creatine in living muscle.

Current analysis had shown cat or rabbit muscle to have creatine values of 450–550 mg/100 g (3.43–4.19 mmol/kg) of fresh muscle, but blood to have only traces at 8–10 mg/100 g (61–76 μmol/kg). If the origin of the creatine were in the food and not from decomposition then the tremendous gradient between tissue (at saturation) and blood had to be maintained if creatine was injected, and the creatine in muscle should not have risen appreciably. On the other hand, if there were no gradient, then creatine should have moved freely into blood and both muscle and blood would have had high values of creatine.

To test this, Otto and Willey used cats whose blood supply to the kidneys was tied off with ligatures. After taking preliminary blood and muscle samples for the determination of creatine and preformed creatinine, they introduced a fresh solution of creatine into the small intestine that had been ligatured below the stomach and above the caecum. At the end of each experiment, blood and muscle samples were again obtained for analysis.

The results showed that muscles absorbed creatine extraordinarily well from circulating blood that had also risen markedly in creatine concentration. When creatine was injected directly into the blood via the jugular vein, again the muscle creatine rose rapidly.

> These results warrant, we believe, the conclusion that the absorption of creatine from the blood by living muscles is just as rapid and extensive (if not more so) than the corresponding absorption of urea, creatinine or amino-acids. Our explanation of this phenomenon as already stated is that living muscles contain virtually no creatine, and that the creatine found on analysis is a postmortem product originally constituting a part of the living protoplasm. . . . It is conceivable of course that creatine is present in muscles in some other form of organic combination than that with the living material, but the organic compound involved is an extraordinarily unstable one since the creatine is set free when muscles (of frogs) are killed by mechanical injury so that it can then be extracted with water or saline solutions.

A footnote indicated that T. E. Buckman was investigating this subject from a quantitative standpoint.

Otto and Willey then presented arguments that the hypothesis that creatine was part of vertebrate protoplasm was useful because it plausibly explained the occurrence and behavior of creatine and creatinine metabolically: 1) When tissues die creatine is set free, whereas normally the "breakdown" (tissue metabolism) yields creatinine; 2) in unusual stress (e.g., fevers, fasting, pathological conditions) the normal conversion to creatinine is accompanied by abnormal conversion into creatine; 3) traces of creatine in urine may occur spontaneously from catabolism of creatinine, just as traces of creatinine may be formed from creatine of food; 4) hypothetical ferments (enzymes) proposed for converting creatine to creatinine are metabolically superfluous because these transformations are postmortem changes; 5) the reason why creatine and creatinine could not be traced to food sources is that creatine is synthesized only in connection with growth or renewal of protoplasm; 6) creatinine elimination is a clear cut index of total normal tissue metabolism; and 7) the liver plays no special role in creatinine formation.

Left unexplained was the high creatine in the urine of children and growing animals, as well as the alleged excretion of creatine instead of creatinine by animals in whose metabolism uric acid replaced urea as the chief nitrogenous waste product.

> The creatinine figures recorded in connection with our experiments clearly indicate that the creatinine does originate in the muscles, since the preformed creatinine small as it is, is nevertheless invariably greater than the preformed creatinine found in the blood. We have several other experiments in addition to those recorded in this paper all of which have yielded the same result.
>
> Our experiments have failed to show any creatinine formation out of the administered creatine.

A report on "emotional glycosuria" that Walter Cannon had made in 1911 and published in 1912 [Cannon, W. B., Shohl, A. T., and Wright, W. S. *Am. J. Physiol.* **29**, 280–87 (1912)] had intrigued Otto because it suggested a basis for the frequency of moderate glycosuria observed in the insane. Had this paper appeared while he was working in the laboratory of the McLean Hospital Otto would undoubtedly have rushed to test its validity on the patients there. Cannon's work had shown that when cats were excited by being restrained (without pain) or by being caged and barked at by an active dog, they exhibited glycosuria, which increased the quantitative output in 24-h urines, as well as qualitative positive tests for glycosuria. Animals with glycosuria failed to have it when adrenalectomized and subjected to prolonged periods of excitement, indicating a possible relationship between emotional glycosuria and epinephrine production.

Otto with the collaboraton of Willey Denis and W. G. Smillie now attempted to tackle the question of emotional glycosuria in man (*14-14*). They tested the urines of 192 patients at the McLean Hospital, and the urines of a group of male students and a group of female students before and after important examinations.

The tests for urine sugar were unfortunately qualitative, and probably somewhat oversensitive. Otto was not one to trust qualitative testing, but in this case the state of the art was still poor for quantitative testing of sugar in urine; and determinations of blood sugar were embryonic. The qualitative tests used were a combination of Nylander's reaction, Benedict's test, and the phenylhydrazine (osazone) reaction.

Of the 192 patients at McLean Hospital, 22 or about 12% gave unmistakable positive sugar reactions. "The great majority, but not all of those who had sugar in the urine, suffered from depression, apprehension or excitement."

Results on the two sets of students were also interesting. Of 33 male second-year Harvard medical students, six or 18% had small but unmistakable traces of sugar in their urines passed immediately after taking an (exciting) examination. Of 36 second-year women at Simmon's College, six or 17% eliminated sugar in urine passed immediately after taking an examination. All of the students had shown negative tests for glycosuria before the examination.

The results suggested, however, that "pronounced mental and emotional strain may produce temporary glycosuria in man." The frequency, however, was far too high, indicating that more specific testing for glucose was necessary.

The Folin-Denis duo also submitted three papers to the *Journal of Biological Chemistry* in June 1914. The first two took advantage of the fact that the Duboscq colorimeter could also be used as a turbidimeter to quantitate ketone bodies and albumin in urine. They recommended that the light entering the colorimeter pass through a hole cut in an ordinary dark (green) window shade. The suspension to be measured had to be uniform, and when preparing it, mixing was best done by gentle rotation or inversion (Part II, p. 406).

In their second paper involving turbidimetry, Otto and Willey accomplished three things: 1) They introduced the use of sulfosalicylic acid for quantitating urinary protein (albumin). This would of course be modified in several ways through the years. It would also be adapted eventually for measuring cerebrospinal fluid proteins, and remain in use into the 1980s. 2) They prepared a stable protein standard from serum for use with the method. 3) They devised a gravimetric method for determining urinary protein against which the turbidimetric method could be checked.

Very modestly Otto and Willey stated, "In this paper we shall describe two fairly convenient methods [for determining albumin in urine] neither of which is particularly original but both of which appear to give satisfactory results" (Part II, p. 407).

The final paper in 1914 by Otto and Willey (*14-15*) was the result of Willey's access at the Massachusetts General Hospital to a patient with multiple myeloma. This patient excreted the typical Bence Jones protein or "albumose" as it was also called because it resembled a derived protein. Though the origin of this substance was still vague, three theories were in vogue: 1) It was manufactured in a diseased area of bone marrow—that is, a "foreign protein"—and passed through the body unchanged. 2) It was produced in the intestine as a result of abnormal, unknown conditions there, and could be proved by showing a direct relationship between the amount and perhaps the kind of protein in the diet and the amount of excreted albumose. In this connection, Otto and Willey pointed out that "Witte's peptone" and other peptic digestion mixtures contain large quantities of an albumose with properties close to the Bence Jones protein. This theory, however, did not have much support. 3) Voit and Salvendi (1900) had speculated that the eliminated albumose came from body protein and not food protein. According to this idea, food protein should "spare" the body protein and reduce output of the Bence Jones albumose. Two studies performed by others did not confirm this.

Folin and Denis stated,

> Our object was primarily to study the effects of different levels of protein intake on the albumose excretion, but in order to make the metabolism record more complete we have included some more detailed analyses of the urine.

They adopted the Hopkins-Savory method for determining the Bence Jones (BJ) protein: The urine was acidified with 5% acetic acid, and the BJ protein was allowed to precipitate at 60 °C. The precipitate was washed in 50% alcohol, dried at 100 °C, cooled, and weighed.

The patient was put on a low protein–high calorie diet, then switched to a purine-free, high protein diet. He was also "fasted" for two days. Though the patient excreted considerably more BJ protein on the high protein diet than on the low protein diet, the excretion did not parallel the total nitrogen output, and the authors regarded the increased output as coincidental. The patient produced as much BJ protein when he fasted as when he ate, and there was no difference in output between a 12-h day and a 12-h night urine.

> Aside from the creatine elimination when the subjective symptoms became severe the protein metabolism in multiple myeloma appears to be normal so far as the ordinary normal nitrogenous metabolism products are concerned.

Otto and Willey conjectured that the BJ protein was produced by internal autolytic digestion because it appeared to be independent of total protein metabolism. In other words, they thought it was a product of partial digestion of a protein like other "albumoses." This idea, of course, was quite wide of the mark. The real origin and nature of myeloma proteins would not be uncovered until many decades later when more had been learned about serum proteins and the structure of immunoglobulins and their relation to monoclonal gammopathies.

As previously mentioned, Otto Folin was well known to chemists in the U.S. Department of Agriculture (USDA), not only for his published papers, but as a consultant in several legal cases and for a number of research projects. When the USDA, in early 1912, sought a suitable successor to Dr. Harvey W. Wiley, Chief of the Bureau of Chemistry, Otto was asked to recommend candidates for the position. But A. Lawrence Lowell, president of Harvard, was unaware of Otto's role in this matter, so when he received the formal request from the USDA, he consulted the Division of Chemistry in Cambridge and then telegraphed suggested names to President Taft.

When Lawrence Henderson informed Otto of Lowell's action, Otto wrote to Lowell for permission to write President Taft directly. Not only would the USDA regard the names they had received from Lowell as derived from Otto, but Otto did not believe that any of them, except one, was qualified to replace Dr. Wiley. He personally recommended a scientist already working in the USDA. The Harvard president urged Otto to write President Taft.

Perhaps as early as 1911 Otto testified in a legal battle that had ensued following the seizure by the U.S. government in 1910 of bleached and overbleached flour in Kansas City. Although Otto's testimony is not here reviewed he described the contest involved in a series of Harvard Health Talks (*14-16*).

The flour, which the government had seized, had been bleached by mixing it with nitrogen peroxide, a very poisonous gas, to produce a flour that was cosmetically whiter, but no different in nutritional value.

In the initial court case, the government's claim that the flour was adulterated was upheld under the Pure Food and Drug Act of 1906. However, on a technical error of the trial judge as to what could be interpreted under the law as "adulterated food," the decision was reversed by the U.S. Supreme Court and a new trial was ordered. In effect, the Supreme Court decision was interpreted as "virtually nullifying" the food law, thus compelling the government to prove

Developing Clinical Biochemistry

that a substance added to a food product was poisonous or injurious to health, rather than compelling the producer to prove that an added substance was harmless.

Otto contended that any substance known to be a poison must be considered injurious to health, regardless of dose size, until proved otherwise. Thus manufacturers should have to prove that their products were safe for public consumption.

> The outcome of the bleached flour litigation illustrates once more the difficulty in securing appeal-proof legislation where there is a conflict between special interests and the public welfare. A general law like the Pure Food Law, the application of which to special cases must be secured through the courts, is of necessity inadequate, because the courts cannot take cognizance of the intrinsic merits of the case. No matter how useless or preposterous in spirit and purpose a given blending of food and chemicals may be, by the time the case comes before the courts the practice represents certain "rights of property," and against such rights the intrinsic worthlessness of the practice has virtually nothing to do with the decision rendered by the court.

With far-seeing vision, he advocated effective laws, detailed and specific, that would prohibit all chemicals and drugs from being added to foods to protect the public from commercial "food chemists" and their employers. He suggested that permits could be granted to manufacturers to allow them to use new combinations of substances in foods.

Otto's relations with Henderson, the assistant professor in the department, remained cold and impersonal. Their individual temperaments clashed, and there was no true professional and certainly no personal communication between them.

Otto wrote a colleague on May 2, 1914:

> I learned from this morning's papers that Dr. Henderson has been appointed exchange professor to some western colleges for next year, and that his duties there begin with the second term of the college year.
>
> Dr. Henderson's services in this department have never been very great for the reason that he has persistently declined (sometimes directly, always in practice) to identify himself with the department. As a result of considerable pressure on my part he once agreed to sever his connection with the college work, and to spend all his time here; and on the strength of this understanding I assisted in securing his appointment first to a three year instructorship with a seat in the faculty, and about a year later to an assistant professorship. This agreement has never been lived up to, I myself realizing the futility of insisting upon it. The result has been that I have been virtually alone in a large and growing department. Could I have foreseen what would happen I never should have asked for Dr. Henderson's promotion to an assistant professorship. It has amounted simply to paying him something like fifty dollars for every lecture given by him to the first year students, and leaving me without an assistant professor and without a basis for asking for one. It has been my hope that at the end of Dr. Henderson's five year period, a year hence, he would go over entirely

to the college and give me a chance to get some one else. Now it appears that he has dropped his lecture engagements in this department for the next year as lightly as he has long since dropped the department in all other respects.

What I have written above may seem to reflect on Dr. Henderson, but I wish expressly to state that I have much respect for his ability. Like most specialists of any consequence he has simply followed the bent of his own mind, only unfortunately for my department his bent naturally has led him to Cambridge rather than to these laboratories.

As head of the department, however, I cannot longer let my respect for Dr. Henderson obscure the fact that there is urgent need for a different man as assistant professor. It is now very late in the year, but not too late, I hope, to secure a capable and useful man to take up the work next October. To secure a full time assistant professor will cost twenty-five hundred dollars.

Beginning in the autumn of 1911, Henderson had used part of his time to teach a course on the history of science to undergraduates in Harvard College. When he was sent west as an exchange professor, he was relieved of all teaching duties for the first-year medical students in biochemistry, and did not resume them upon his return. The fact that Otto was not informed of this can only reflect on Henderson's irresponsibility. Henderson's cordial relationship with President Lowell may explain condonement of his cavalier behavior.

In the autumn of 1914 Otto brought Walter Ray Bloor (1877–1966) back to Harvard as an assistant professor in the department to teach first-year medical students. Bloor had been instructing biochemistry at Washington University with Philip Shaffer since he had obtained his doctorate with Otto in 1911. Bloor would become one of the foremost biochemists in the special area of lipids and would contribute heavily to the basic and applied knowledge of lipids in blood and other tissues (*14-17*).

Bloor had this to say about his return to Otto's department:

> When I returned to Harvard (1914) there were a good many graduate students; Sumner, Doisy, McEllroy, Youngburg, Pettibone. Fiske had graduated in medicine, was then an assistant. Rappleye, then a medical student, did a piece of original work, mostly biochemical, every year in school. Richard Bell and Henry Lyman (both physicians) each paid the other his salary and were active workers in the laboratory. Bell worked with Doisy and Briggs. (Lyman was a joint author with Folin in a number of papers of clinical interest).

In 1914, Cyrus Harwell Fiske (1890–1978) whom Otto had been watching since Fiske's first year in medical school, joined the departmental staff as an assistant, following his completion of the medical course. Cyrus had graduated from the University of Minnesota in 1910, and had been a "straight A" student in the Harvard Medical School.

The assistants in the department besides Fiske were now Edward P. Phelps, S.M., and Theodore F. Zucker, S.M. The new Austin teaching fellow was Goodwin LaBaron Foster, A.B., while the previous trio, Hammett, Lyman, and Bell, remained the fellows.

While Otto drew outstanding graduate students and assembled a gifted young faculty who would hasten the development of clinical chemistry and

Developing Clinical Biochemistry 243

contribute heavily to "general" biochemistry—a monument to life and health—the young men of Europe, including its biochemical talent, were leaving their classrooms and their laboratories to become soldiers. World War I had come.

Otto's niece Hildur spent most of the year 1914 with the Folin family and part of it at Kearsarge. Bonnie Cooke had gone on a trip to Europe and was in Paris when the war erupted. She wrote to Hildur on Aug. 26:

> I have not forgotten my promise to write you. But I am writing sparingly to everyone—even my family, who are served only once a week. Each letter I write has its date entered on my account book. Otherwise I should have only a jumbled notion of what I had done in the way of writing.
>
> In spite of the horrors & the suspense, time passes quickly. It is actually nearing the time when you must all be going back to town—a whole vacation—time—which has been for me a kind of waiting to decide where to jump. I have not yet rec'd a line from America. But a ship came in yesterday. Each day brings my news from home nearer.
>
> Yesterday for the first time the Parisians looked *hurt*. Every moment I met women & young girls either crying or trying not to cry. I went very early to the English consulate (for my pass to Eng.) in order to avoid the experience of the day before, when after having followed a line for about an hour, we were all turned away. Paris was a strange spectacle at that early hour when ordinarily a great city is pouring out of its homes into its shops. The wide boulevards had more the look of Brookline in August & in the fashionable streets.
>
> . . .
>
> Oh, for a United States of Europe! Perhaps they will consider it after their chosen rulers have forced all the able-bodied men to kill each other. I shall be glad if the Hohenzollern House is wiped out forever—with its brutal medieval supporters. It would be terrible, intolerable, if the German military caste ruled all of Europe. England and France will never permit it. The black shadow of their final victory will be the heightening of Russia's power. Horrible to think of the Cossacks entering beautiful Berlin! And it may happen. When I hear from her again, I expect to learn that my dear friend in Grunewald-Berlin has lost some, if not all, of her brothers.
>
> You will have seen that the Germans have already dropped explosives in Antwerp. All night long the searchlights swirl around Paris. But what protection are the biggest guns against a Zeppelin flying 8000 feet high, carrying 12 tons of explosives!
>
> Every night when I am going to bed, I say to myself that the children—my kiddos—are still in the sunlight, playing on the lovely mountain side among the blackberries or under the hemlocks. . . .

From the time Otto began teaching the first-year medical students their laboratory course in 1907, he realized that every exercise could not be taken directly from other texts, but would have to be prepared separately and modified to include new developments in the literature, his own improvements as well as those of others. By 1909, a manual of procedures, which was revised annually, was developed for the students.

For students to learn beyond the qualitative aspects of biochemistry, and particularly for those who were upper classmen or graduate students, Otto developed supplementary quantitative analytical procedures on urine and blood. From the students' use of these methods Otto and the assistants and faculty could gain practical understanding of their pitfalls, strengths, and shortcomings. The results obtained added to the early knowledge of "normal" values, however confined they were to comparatively healthy young adult males, or to slaughterhouse-derived blood.

As the scope of basic biochemistry expanded and the repertoire of quantitative procedures grew, so did the size of the informal laboratory manual used by the students. By 1914, Otto realized that it was becoming essential that he "formalize" the material in the form of a hardcover book. Otto's reluctance to answer his mail was assuredly in part due to requests to provide updated, practical versions of his own analytical methods on blood and urine, as well as his choice among others, not only for teaching purposes, but for daily clinical use in hospital laboratories that were developing, particularly in the large medical centers in the United States. Otto, after all, had called upon the hospitals to hire trained biochemists to carry out research. Just as the Massachusetts General Hospital had hired Willey Denis, now other teaching hospitals, particularly in New York City, were following suit. A trickle of new members in the American Society of Biological Chemists now came from hospitals. Their laboratories were asked to perform quantitative tests to supplement the usual qualitative testing of urine.

During 1914 and 1915, in particular, Otto asked his faculty to provide revisions of his laboratory manual for the students that would permit him to publish formally a "hardback" version. And, of course, he had valuable assistance in this from Willey Denis, who was heavily involved with the quantitative procedures, and from Philip Shaffer. Shaffer was particularly interested in Otto's development of a manual, as he too was concerned with offering his laboratory course to medical students, and unquestionably had discussed this as well as lecture material with Otto in their annual visits at the ASBC meetings, as well as at "social" visits, and through correspondence.

In addition to Denis and Shaffer, Otto received help in his revision of the manual from Walter Bloor, C. J. Farmer, F. B. Kingsbury, F. S. Hammett, R. D. Bell, and C. H. Fiske.

The laboratory manual was first published formally in 1916, and was quite small (14-18). It consisted of 10 chapters and a supplement. The 10 chapters were heavily qualitative with a judicious mixture of quantitative testing.

The supplement was actually Otto's version of current clinical chemistry as could then be practiced in a hospital laboratory. It contained special qualitative tests on urine and quantitative procedures on blood and urine.

In 1915 Otto produced nine papers, seven with Willey Denis. Several of these papers were unrewarding in that well-conceptualized experiments did not materialize in rewarding data. A couple were "stinkeroos," and perhaps should not have been presented. Was Otto beginning to tire a bit? If so, he must have realized it. He was not one to take a "sabbatical" leave, and this time had passed anyhow. At any rate, after the ninth paper was submitted in late July, he did not submit another until one year later.

The first paper seemed to have a logical underpinning. Are fat people more prone to develop starvation acidosis than lean ones? "Clinically as well as from the standpoint of metabolism it is important to know whether persons with a pronounced tendency to obesity are less capable of drawing on their own fat deposits without developing acidosis than are persons without such a tendency" (14-19).

Two obese women volunteered to undergo the investigation: Mrs. M., 48 years old, 124 cm (4'11") tall, and weight, 108 kg (238 lb), and Mrs. B., 44 years old, 133.5 cm (5'4.5") tall, and weight, 178 kg (392 lb). Each was subjected to three fasting periods of 4–5 days each over about a month's period. Daily urinary collections were made, and analyses were made of most of the urinary constituents, especially the acetone bodies. In addition Otto and Willey collected the acetone in expired air into sodium bisulfite solution to determine whether there was any truth in the recent statements that "acidosis-breath" could be detected even upon merely entering the room of a patient with such acidosis.

No controls on lean people were necessary. Mrs. B. did not, despite her extraordinary obesity, develop the degree of acidosis that Mrs. M. did. "In view of these results we are inclined to conclude that obesity cannot be regarded as a predisposing factor in the development of acidosis."

Repeated fasts elicited a lesser response, an adaptation to starvation without the production of acetone bodies. Also, "the obese lose less body protein than others in the course of moderate periods of starvation (four to six days), and on repeating the fasts the losses of body protein become still smaller."

Detecting acetone in the room of a child or diabetic patient with acidosis was found to be impossible. The amount equivalent to the acetone expired in a 24-h day (1.2 g), when scattered on the floor of a closed closet, left no odor within an hour's passage. Finally, Otto and Willey suggested that "successive moderate periods of starvation constitute a perfectly safe, harmless, and effective method for reducing the weight of those suffering from obesity."

The second paper of 1915 proved to be one of Otto's and Willey's oddities (14-20). It was based on an observation of Mörner, who had accidentally discovered a peculiar substance in the eggs of the ordinary perch (*Perca fluviatilis* L.) to which he gave the name "perca globulin." Perca globulin according to Mörner reacted as a globulin. It was precipitated by 0.75% hydrochloric acid, by ovomucoid, and other glycoproteins. But it was best characterized by its astringent taste. Mörner had been unable to find the globulin in the sea perch, and had asked Otto to look for it in the eggs of the American perch.

Willey Denis, possibly through the grace of Jacques Loeb, had spent some of her summers at Woods Hole, and had published three papers on metabolism in fish. Hence her interest was added to Otto's in acting on Mörner's suggestion. They were able to get hold of ripe and unripe roe of yellow and pike perch from the U.S. Bureau of Fisheries. They made extracts according to Mörner's directions. Though the material *dissolved* rather than precipitated with 0.75% hydrochloric acid, the characteristic astringent taste was present, pronounced and unmistakable in all of the extracts made, though more in the unripe than in the ripe eggs. The precipitate, obtained in abundance with other concentrations of hydrochloric acid and with ovomucoid, seemed to correspond approximately to the intensity of the astringent taste.

We conclude therefore that the perca globulin discovered by Mörner in the common European perch (*Perca fluviatilis* L.) is also present in the roe of two American perches, *Perca flavescens* and *Perca sorzosceidion* (yellow perch and pike perch).

While Otto had evidently silenced his critics in Europe (even before the war), he now acquired a veritable bulldog in Stanley Rossiter Benedict. Although Benedict's lifelong interests would later become intertwined with Otto's, particularly in clinical chemical methods, Benedict at first published results critical of Otto's methods and offered his own modifications; and Otto would answer with his own sharp and valid counterattacks. As they came to realize that this was unproductive, Benedict learned to look for a sounder basis for his criticism or a shortcoming perhaps in his technique, and Otto, in turn, realized that he may have failed to consider the potential audience sufficiently or may have modified his own methods so rapidly that details were lacking that could have prevented misunderstanding and error. Otto's next paper was "a case in point," an answer to a misconception that Benedict had demonstrated in a publication earlier in the year (*14-21*). No doubt this paper helped spur Otto's desire to publish his laboratory manual.

Bock and Benedict (B & B) had criticized the Folin-Farmer method for determining nitrogen and had offered two modifications that they thought gave better, though still unsatisfactory, results [*J. Biol. Chem.* **20**, 47 (1915)].

> The criticisms of the original method I cannot accept in silence inasmuch as scores of students and many mature persons are every year learning the method in our laboratories.

Otto explained that to get satisfactory agreement between the colorimetric and Kjeldahl methods generally required about two weeks' practice. The students first used a standard ammonium sulfate solution and learned the aeration-colorimetric technique so that they obtained the values expected from the direct treatment of the standards. At first, they were apt to get low values by improper use of the aeration technique, and this could be easily shown by using a second receiving tube containing HCl, which was tested with Nessler's reagent. With practice, this error in technique was overcome.

Another problem in B & B's attempt to use the Folin-Farmer method was unquestionably due to their "blank" error of ammonia in compressed air made from "room" rather than "outdoor" air. With outdoor air, Otto had found that the acid wash (through sulfuric acid, in 2-L bottles) was superfluous, though he used it.

> As a matter of curiosity Dr. Denis and I determined a short time ago the ammonia present in three wash bottles which had been used without change in my laboratory since November 16, 1911; about three and a half years. Hundreds of aerations, many lasting for hours, must have been made through each, and back diffusion through open rubber tubes could have occurred during the entire period. Originally each bottle contained 500 cc. of water and 50 cc. of concentrated sulphuric acid. The total ammonia nitrogen found in the bottle amounted in each case to 8 to 10 mg.

If the results reported by Bock and Benedict had come from a less competent investigator than Benedict, I should have suspected dirty utensils as the source of the ammonia, because of the prevalence of ammonium salts deposits in chemical laboratories.

Otto answered B & B's criticisms one by one. He attributed the error principally to the air in Benedict's laboratory. He suggested that they not only had to check their air supply, but use their pipettes more carefully, operate the colorimeter properly, and practice the aeration technique. Implied in Otto's reply was that Benedict should come and get expertly instructed on the method in Otto's laboratory. In fact, Benedict had already visited Otto's laboratory once in 1909 to demonstrate that his modification of Otto's method for total sulfur determination in urine would work with solutions of cystine.

The giant of clinical biochemistry, Stanley R. Benedict (1884–1936) was born, raised, and educated through his undergraduate university degree in Cincinnati, Ohio. Though he had actually spent his second year of college in the medical school he chose research and teaching chemistry as his career. He was influenced strongly in this direction by Dr. J. F. Snell, who had been a student of W. O. Atwater, the eminent nutritionist, with whom Francis Gano Benedict (no relation to Stanley) was associated in Middletown, Conn.

Stanley proved to be so capable an investigator from his first year that he published nine papers in the *Journal of the American Chemical Society* and one paper in the *American Journal of Physiology* before he graduated in 1906. Only three of these papers were with Snell as coauthor, and Benedict was the senior author on all of them. The papers were primarily on detecting inorganic substances, chlorides, bromides, iodides (the halogens), cobalt, nickel, acetate, cyanide, lithium, manganese, zinc, and common acids, and a paper on the role of certain ions in rhythmic heart activity.

When the professor of chemistry fell ill during Benedict's third year, Stanley was asked to take over the teaching duties until he returned. When the professor later died, Stanley taught the course during his senior year.

Benedict went to Yale in 1906 for his doctoral work with Lafayette Mendel. During these two postgraduate years he published four papers, three on the detection or estimation of reducing sugars, and one on the influence of salts and non-electrolytes on the heart.

E. V. McCollum wrote this of Benedict:

> Stanley was highly enthusiastic about his laboratory work and his reading in scientific literature. He was fast becoming a highly educated specialist. . . . His frequent expressions of skepticism concerning statements of men or books on topics which we discussed showed that he was never superficial in his thinking, and never awed by authority. In all our conversations it was evident that he was a thinker and a critic with uncommon natural endowments. *(14-22)*

McCollum also wrote a good deal about Otto Folin's early contributions to biochemistry and his influence on Stanley Benedict.

The year 1906 marked a turning point in the history of biochemical studies. Otto Folin published in Volume 13 of the *American Journal of Physiology* three papers which immediately brought him to distinction. The first of these described a new system for the analysis of urine for urea, ammonia, creatine, creatinine, and uric acid. Methods hitherto available for quantitative estimation of these substances were either seriously unspecific, as in the case of urea, or required relatively large samples for analysis, as was the case for uric acid. His new procedures were regarded by biochemists and physiologists as so great a step in advance that Harvard University created a professorship in biochemistry for the humble chemist working in the laboratory of the McLean Hospital for Mental Diseases at Waverley, Massachusetts.

Immediately, in laboratories here and abroad, analyses of urines were made from patients with various kinds of disorders, and the more fertile-minded began to set up experiments on animals for the production of pathological states, and to study the composition of the urine by Folin's methods. Folin, himself, did little further in the way of application of his methods to the study of urines. Instead, he turned his attention to refining and improving analytical methods to a degree that would make possible the determination of the constituents of urine in small samples of blood. He was the first, apparently, to realize that it is much more important to know what the kidneys have failed to excrete, among the products of metabolic activity of the body, and which products, accordingly, accumulate to harmful concentrations in the blood, than it is to know what and how much of these have cleared the kidneys. So he reaped a second great triumph when he first made public the technics by which such analyses could be accomplished. Under his leadership investigators turned at once to the applications of his small-sample, high-accuracy methods to the study of normal and pathological problems in physiology. Folin introduced the colorimeter into biochemistry when he employed this instrument for the quantitative estimation of creatine and creatinine, using the color reaction for creatinine described by Jaffe in 1895. This test depended upon the formation of a red color when a solution of creatinine is treated with picric acid and sodium hydroxide. It reveals the presence of creatinine in a dilution of 1:200,000.

With the publication of Folin's analytical methods for determining quantitatively the principal constituents of urine, clinicians had for the first time methods applicable to the study of patients on a day-to-day analysis of the blood and urine, and the study of disorders of metabolism by the new technics spread rapidly.

For many years the question of the extent of protein digestion in the alimentary tract, the nature of the products absorbed, and the manner in which they were disposed of in the body after being absorbed, had been debated, but no data existed on which a decision between conflicting views on any of these problems could be based. Folin and his students were able to secure data on the concentration of urea, and ammonia in deproteinated samples of blood from branches of the mesenteric veins, the portal vein, the liver, systemic blood and muscle and organ extracts, the samples being taken simultaneously. The nonprotein, non-urea, non-ammonia N was correctly judged to consist largely of amino acids. These data afforded an explanation of the course of events in protein digestion and the distribution of amino acids to the tissues during and after the absorption period.

Matters stood thus when Benedict entered upon his vocation of research and teaching. His first appointment was at Syracuse University, where he remained but one year. He then took charge of physiological chemistry at Cornell University College of Medicine, in New York City. Graham Lusk, the distinguished physiologist, was responsible for Benedict's appointment. Benedict often spoke in terms of esteem of Dr. Lusk.

Benedict now entered with enthusiasm upon his career as an improver of analytical methods.

Every method which Folin described during the following years was immediately submitted to a critical study and was modified and improved in some important detail by Benedict. Methods for uric acid, creatine and creatinine, total sulfur, sugar, etc., which were devised by Folin, and which at the time of their publication were the best ones known, were, within a few months, tested by Benedict and improved in various ways. It was inevitable that the regular appearance of these follow-up critical studies and replacement of methods devised with great astuteness by Folin, should cause him some irritation. Yet Benedict's great contributions to analytical biochemistry did not in the least detract from the high eminence which was accorded to Folin by chemists, physiologists and clinicians. Folin showed his broadmindedness and tolerance through all the years of Benedict's criticism and replacement of Folin's methods by Benedict's modifications of these methods by always remaining on good terms with him. (14-22)

In 1912 Benedict became a professor of physiological chemistry at the Cornell University Medical College, and for many years supervised the research work on cancer at Memorial Hospital in New York City.

Otto and Willey published an essentially clinical paper on the diagnostic value of uric acid determinations in blood.

We realize that the determinations of uric acid in blood by our method is still probably outside the range of most clinical laboratories, but we are convinced that all "short cut" methods so far proposed are bound to lead to grossly misleading results. . . . The blood of normal persons has, however, been found to carry far more uric acid (1–2.5 mg. per 100 c.c. [59–149 µmol/L]) than was formerly suspected, and the difference in the uric acid content of such normal blood and the blood of those suffering from gout is materially smaller than early investigators realized. . . .even exact quantitative uric acid determinations are not by themselves an adequate protection against frequent mistakes in the differential diagnosis of gout and other joint diseases. (14-23)

The authors noted that there were four classes of values in blood for NPN and uric acid (UA) in relation to the patients involved, as follows:
1. UA and NPN normal. This was typical of a normal man, and found in patients with alcoholic gastritis, cystinuria, diabetes, cardiorenal disease, arteriosclerosis, insanity, and others.
2. UA normal and NPN elevated. This was found in patients with nephritis, prostatectomy, infectious arthritis, bone tuberculosis, acute rheumatic fever, mitral stenosis, and others.
3. UA elevated and NPN normal. This was typical of patients with suspected gout.
4. UA and NPN elevated. This was found in patients with uremia, arthritis deformans, weak heart, pneumonia, and other forms of secondary nephritis.

In making use of blood analysis to decide whether a given doubtful case of joint disease is gout or arthritis, it is therefore absolutely necessary to determine the nonprotein (or at least the urea) in the blood as well as the uric acid. And before the blood is drawn for such analyses it is indispensable that the patient should have been on a purine-free diet for at least two days. The level of the protein metabolism should also be ascertained by means of a nitrogen determination in the twenty-four hour urine passed during the last day of the experiment.

In this paper, Otto and Willey, for the first time, did not express uric acid concentration in milligrams per 100 g of blood but in milligrams per 100 cc, usage that would be popular in the decades to come. The distinct differences noted for uric acid values in gout and in the nephritic patients were invaluable in distinguishing the normal from the abnormal.

In their next paper, Otto and Willey modified the bed pan to make possible accurate collection of women's urine (14-24).

To solve the problem, they devised a "divided pan," which they illustrated. They used the pan for many months on the women's and children's wards of Massachusetts General Hospital with practically invariable success.

Otto had long been curious about the undetermined "organic acid" component of normal human urine. Now he and Willey would make yet another application of their tungstate-molybdate reagent, described in their germinal paper of 1912, to quantitate free and conjugated phenols, a fraction of the urine organic acids (14-25).

Older methods for determining phenols were based on the iodometric titration method, which not only required a volume of 500 mL of urine, but gave unusually low values because phenols were lost by oxidation during the evaporation (concentration) in alkaline solution. The Folin-Denis procedure required only 10–15 mL of urine, and could be completed in 10–15 min.

In the tungstate-molybdate method uric acid and traces of interfering protein were precipitated with acid silver lactate and colloidal iron. "Free" (nonconjugated) phenols were determined by treating the filtrate with the Folin-Denis reagent and saturated Na_2CO_3 solution to produce the characteristic blue color. This was then compared in the colorimeter with a standard solution of phenol.

To determine total (free and conjugated) phenols, the filtrate was treated with HCl and heated in a boiling water bath to liberate the conjugated phenols. This solution was then mixed with the color-producing reagent and Na_2CO_3 and read against the phenol standard in the colorimeter. The free phenol subtracted from the total gave the value of the conjugated phenol.

The next step, of course, was to study phenol excretion in human urine (14-26). Otto and Willey presented a long review of the origin and significance of phenols in urine. We cite the summary:

> The excretion of total phenol products in the urine appears to be much greater than is indicated by the phenol figures heretofore recorded in the literature.
> The phenols are by no means quantitatively converted into conjugated phenols. The detoxification process involved in such conjugation appears therefore to furnish only a partial protection against the toxic or deleterious effect of the phenol products formed by putrefaction in the intestinal canal.

In any given individual man or animal, the total phenol excretion tends to vary directly but not proportionally with the protein intake.

The next project of Folin and Denis was to examine kidney function from the standpoint of each kidney's output of nitrogenous products (*14-27*). For study of retention, blood chemistry was the best approach.

It is possible, however, that if the urine passed by each kidney could be collected separately and quantitatively, urine analysis might afford valuable additional information concerning the effect of disease on the specific selective function of each kidney in relation to the excretion of the different waste products. . . . In the present paper we wish to record some analyses of "single kidney" urines.

Obviously, the difficulty was great in collecting separate, quantitative, simultaneous urine samples for long periods of time. Although the patients selected were far from ideal, Otto and Willey could not be choosers of their patient sources. As it was, they had access to four patients from the genitourinary department at Massachusetts General Hospital. These patients had been operated on for malignant growth in the bladder, and were appropriate for the study because "the double ureterostomy performed by the surgeons rendered possible the separate collection of the urine from each kidney."
A table of data was provided for each of the four patients.

From these it would appear that the damage produced by moderate hydronephrosis complicated by pyelitis is not particularly uniform in character so far as the excretory function of the kidney is concerned. No broad generalization was warranted. . . .excretion of urine by the kidney can be subdivided into an unknown number of different more or less independent excretions.

Otto's last paper of 1915 was his weakest (*14-28*). Current qualitative tests for urinary sugar failed to demarcate adequately normal from pathological quantities present. "The test described in this paper is proposed as a simple and instructive method for showing the presence of such normal traces of sugar." It was not intended for use on patients, obviously, because most everyone had these traces in their urine. In fact, Otto incorporated it into his laboratory manual, and it was probably developed primarily for use by the students.
The method employed "an uncommonly sensitive alkaline copper solution" and was based on the fact that in the presence of copper "alkaline picrate solutions are not reduced by sugar." This allowed prior removal of creatinine and other substances with picric acid before applying the reduction test.
We shall not present details of this qualitative method. When tested on 100 normal persons (students) a positive reduction was obtained in every case, with the conclusion that the "amount of sugar present in normal urine is therefore probably much greater than is indicated by negative findings recorded on the basis of the clinical qualitative tests for sugar in current use."
The missing factors in this paper were a measurement of the test's sensitivity and, perhaps more, a measurement of the test's specificity. Obviously it did well

to remove creatinine interference, but what of other interfering substances? Its main defect was one that was uncharacteristic of Otto's scientific approach. The sensitivity of the test could have been measured with dilute quantitative glucose solutions and by adding graded amounts of glucose to urine. The fact that 100 out of 100 "normal" urines gave positive reactions in no way meant that the reduction was caused specifically by sugar. This weak introductory paper from Otto hardly portended the grand contributions that he, and a few of his colleagues, would soon make on the analysis of sugar in blood and urine.

On April 28–30, 1915, the Washington University Medical School held a special celebration to dedicate its new buildings. Otto, who had recently had his forty-eighth birthday, was conferred with his first honorary doctor of science degree by Acting Chancellor F. A. Hall, at exercises held in Graham Memorial Chapel on the campus. Of Otto, Hall said,

> one of America's happy importations from her sister country, Sweden. A man of great talent for research, who through contributions of fundamental importance to the technique of chemistry, has brought his science into new and fruitful relations to life. The results of his work, in their bearing upon the practice of medicine, have demonstrated anew the efficient value of the standpoint of pure science.

Among others given this degree at the ceremony were Simon Flexner, Russel H. Chittenden, William H. Howell, William H. Welch, and William T. Porter.

Of course, Otto's friend Philip Shaffer was undoubtedly involved in bringing him to St. Louis. In 1915, Philip had become dean of the medical school while serving as head of its department of biochemistry. This portended well for the future of biochemistry at Washington University.

Otto gave four special lectures in association with the dedication ceremonies, in Assembly Hall, as follows: April 26, "The Utilization of Food Protein"; April 27, "Tissue Metabolism with Special Reference to Creatinin"; April 28, "Protein Metabolism, with Special Reference to Uric Acid"; and May 1, "The Occurrence and Significance of Phenols and Phenol Derivatives in Urine."

In the autumn, Edward A. Doisy, A.B., became an assistant in Otto's department, replacing Cyrus Fiske, who had become an assistant professor of biochemistry at Western Reserve University in Cleveland. Doisy would one day also go to St. Louis to work in Shaffer's department and would perform pioneer work in endocrinology, isolating and characterizing sex hormones, and on the structure of vitamin K that would lead to his sharing a Nobel prize in medicine with Carl Dam in 1943.

Otto's mother, Eva, died in 1915 at the age of 85. Her daughter Gertrud had been with her to the end, and Gusten and his family, with the exception of Hildur, were nearby.

In 1916, Otto's laboratory manual appeared. He also published six papers and received two special honors. Otto and Willey must have written these papers before and after Otto's trip to the University of Chicago to receive his second honorary doctor of science degree on June 6 at the institution's Ninety-Ninth Convocation, which was part of the quarter-centennial celebration. The candidate was presented by Professor Julius Stieglitz, Ph.D., Sc.D., chairman of the Department of Chemistry.

For Otto the moment must have been a sentimental one and a proud one for Stieglitz, because Otto had proved to be his foremost student among many distinguished ones. Otto had found challenge, direction, and knowledge at this nascent university. And now Harper and Nef, two who had made so much of it possible for him, were gone. But he owed much to Stieglitz whose teaching and research techniques he emulated.

On April 29, Otto received the following letter:

> Sir:
> I have the honor to inform you that you were elected a member of the National Academy of Sciences at its Annual Meeting held in Washington April 17–19, 1916. The Academy thus desires to express its high appreciation of your service to science, and trusts you will signify your acceptance of this election.
> In due course a diploma signed by the officers of the Academy will be sent to you.
> Assuring you of my personal pleasure in counting you as one of our members, I am,
>
> Very respectfully yours,
>
> Arthur L. Day
> Home Secretary

Otto was delighted to accept this singular honor.

Otto's experience with the analysis of nitrogenous metabolic products had reached a new plateau. He reported his findings of his latest research, which he had done with Willey, in a pentad of papers on total nitrogen, ammonia, and urea in urine, and NPN and urea in blood.

The first paper of the five dealt with the problems of simplifying the Folin-Farmer total nitrogen procedure in urine so that direct nesslerization of NH_3 was possible without the need for separating the NH_3 first by aeration of the Kjeldahl digestion mixture (14-29). The chief obstacle had turned out to be the sulfates present that caused precipitation of the colored mercury ammonia compound formed in nesslerization. The solution was to replace most of the sulfuric acid in the digestion mixture with phosphoric acid, because phosphoric acid accelerated digestion while not interfering with Nessler's reagent. Its main drawback was that it attacked glass. This, however, was a relatively minor consideration. The silica formed could be removed by centrifugation or filtration through cotton (not paper, which absorbed the coloring matter). Direct nesslerization was a major advance in the procedure, as it steadily moved toward practical clinical application.

The ratio of $H_3PO_4:H_2SO_4$ was optimal for digestion at 3:1 and did not interfere with the accessory catalytic action of $CuSO_4$, nor did the copper interfere with the nesslerization. Other accessory catalysts were tried (ferric chloride, salts of cadmium, uranium, manganese, and mercury), but from the standpoint of nesslerization, they were less desirable.

For the first time Otto and Willey reported their findings on Nessler's mercuric iodide held in solution with potassium iodide. Its use was hampered by turbidity caused by excessive alkalinity. The turbidity could be partly prevented in several ways, such as rapid mixing, predilution, and most of all, by "radical changes in the formula for the preparation of the reagent."

The KI in the reagent had to be present in optimal concentration without causing interference with color development. The double iodide, $HgI_2 \cdot 2KI$, was easier to prepare than the ordinary Nessler's reagent and it was an excellent stock for preparing Nessler's reagent of any desired alkalinity. For direct nesslerization of the digestion mixtures described in this paper a Nessler's reagent containing 2% NaOH was recommended.

They pointed out that it was very critical to adjust the alkalinity so that the digestion mixture was neutralized and sufficient remained for an optimal reaction with Nessler's reagent. They provided a table of working formulas for various ratios of $H_3PO_4:H_2SO_4$.

The acid digestion mixture was allowed to stand overnight and filtered through an asbestos mat on a Buchner funnel to remove calcium salts that interfered with the color reaction. Nesslerization was carried out in large flasks (200–250 mL) so that, if necessary, as much as 2–3 mg of NH_3 nitrogen could be handled by dilution.

While a commercial (Kahlbaum) grade ammonium sulfate could be used as a standard (after drying it), Otto and Willey provided a procedure for preparing pure $(NH_4)_2SO_4$ by aspirating NH_3 into H_2SO_4 and crystallizing it out with alcohol, then further recrystallizing it twice from water with alcohol. The stock standard (1 mg/mL) was prepared in 0.2 N H_2SO_4 to preserve it from the growth of molds. Working standards for urine and blood analyses were prepared from the stock using the modified Ostwald pipette, previously calibrated gravimetrically with water. The modified pipette (Ostwald-Folin) was drained for 10 s against the sides of the test tube and then blown clean while dragging the tip against the sides of the tube.

The precision of color comparison could be reduced to an error of only 1% if careful attention was paid to four points in the use of the Duboscq colorimeter: verifying the zero points, keeping the optical parts free from dust, overcoming inequalities in optical properties, and recognizing that differences occur physiologically between the "observers" involved. In addition, it was important to place the colorimeter where it was best used from the side rather than from behind, away from a window, and with the observer comfortably seated. After "adjusting" the eye to read the standard in both fields of the colorimeter, no more than one reading of the unknown was made. For a long series of unknowns, the standard was reread after each two readings of the unknowns.

Many more pointers on technique were presented in this paper. A table was presented to show that results obtained for 36 urines with the Kjeldahl method were practically identical with those obtained using the colorimetric technique and contained values of nitrogen ranging between 3.14 and 15.68 g/L of urine.

In their second paper of this series, Otto and Willey applied the direct nesslerization technique to the determination of NPN in blood (14-30). This was another momentous paper for clinical chemistry because they introduced a new reagent for protein removal and overcame perhaps its last major obstacle toward making NPN determination a simple analytical tool of the clinical laboratory. Their previous use of methanol–$ZnCl_2$ as a protein precipitant was flawed. Nitrogenous lipoids and fats had been extracted by the alcohol so that the subsequent digestion step was difficult; and some of the NPN components (e.g., amino acids, creatine) could be partly retained by the protein precipitate.

> We have therefore reluctantly abandoned alcohol as a protein precipitant in connection with this determination and have endeavored to find some more serviceable reagent. . . . In the course of our study of this problem it occurred to us that m-phosphoric acid might prove serviceable and peculiarly suitable. Since we are using phosphoric acid in large part for our destructive digestion, the presence of phosphates in the filtrate would be no disadvantage. . . . The product is remarkably effective as a precipitant for the blood proteins. It is better than colloidal iron and fully as good as trichloracetic acid for this purpose.

Otto and Willey used what was called "glacial phosphoric acid" to which was ascribed the formula, $HPO_3 \cdot NaPO_3$. It provided water-clear, stable filtrates that did not foam upon boiling and that contained little carbon for digestion. It promoted digestion and was easily neutralized and in no way interfered with nesslerization.

> For the preparation of blood filtrates free from proteins and suitable for the determination of non-protein nitrogen "glacial phosphoric acid" (m-phosphoric acid) is therefore better, according to our experience, than any other reagent heretofore used for this purpose.

The only drawback found in the use of metaphosphoric acid was its instability in solution where it gradually changed to ordinary orthophosphate, losing its characteristic power to precipitate protein. Fortunately, it was inexpensive compared with other protein precipitants. "Adequate recognition of the fact that phosphate solution of unknown age must not be used is, however, absolutely indispensable for reliable results."

Glacial phosphoric acid sticks dissolved slowly but completely in four parts cold water. A stick (20–25 g) dissolved in about a half an hour when the solution was shaken occasionally. To precipitate the protein in 10 mL of blood required 5 mL of a 25% solution. If kept cool, this solution was serviceable for three days. Filtrates made from bad metaphosphoric acid were clear but foamed a great deal when the water was boiled off and could cause high results.

The use of metaphosphoric acid was a major advance in protein removal for blood and urine analysis. Its defect, instability, meant that Otto and several of his contemporaries would continue searching for other reagents.

Next they tackled the question of whether direct nesslerization could be made to work for the determination of ammonia in urine (14-31). After all, there were many organic substances there that presumably might interfere by reducing the Nessler's reagent: creatinine, the arch culprit, which could reduce the mercury in the reagent to the metallic state; glucose; uric acid; and phenols. Otto and Willey found a useful answer.

> We soon found, however, that by suitable treatment of urine with blood charcoal (Merck's, which is free from ammonia) all the ammonia of the urine goes into the filtrate, while the uric acid, the phenols, and more than 90 per cent of the creatinine are taken out by the charcoal. The reducing substances are removed so effectively by this treatment that the nesslerized filtrates remain perfectly clear for several hours.

In 1913, E. K. Marshall, Jr. (*J. Biol. Chem.* **14**, 283; *J. Biol. Chem.*, **15**, 487) had made a major improvement in clinical chemistry by introducing the use of the enzyme urease, which was derived from the soybean (Takeuchi, 1909), for the determination of urea in urine and blood. D. D. Van Slyke and Glen E. Cullen confirmed and refined the work of Marshall in 1914, using the Folin aeration technique for the ammonia determination. Later (1916) with Mateer (*J. Biol. Chem.* **25**, 297), Marshall discovered a richer and more economic source of urease in jack bean meal. This profound analytical development prompted Otto and Willey to remark:

> All methods for the determination of urea in urine dependent on purely chemical processes for the hydrolysis of the urea are destined to become antiquated, if not forgotten, because of the abundant supply of urease now available for such hydrolysis.

As they had now found an improved way to make protein-free filtrates, it was easy for them to use urease to determine urea in both blood and urine. The metaphosphoric acid could quite conveniently remove both the urease (after it had achieved its intended purpose) and the other proteins (*14-32*). Marshall had used the aeration method for removal of the ammonia liberated by urease. Now, this step could be eliminated by direct nesslerization of the filtrate.

While the last paper of 1916 by Otto and Willey was not particularly remarkable, it contained several items of interest beyond its stated purpose (*14-33*).

> Inasmuch as practically no quantitative investigations are available concerning the phenol contents of stools, and since practically nothing is known concerning the quantitative relationship between the total phenol contents of the urine and of the feces, our method and our results are not without value, at least for purposes of general information with regard to this subject.

An improved phosphotungstic–phosphomolybdic acid reagent for detecting phenols had been worked up by Dr. Richard Bell, and Otto and Willey had tested it out on "artificial feces" using a precipitation method they had developed for human feces. They obtained quantitative recovery of phenol added to a mixture of soap, neutral fat, egg albumin, and calcium phosphate.

Stools (24 h) were collected from 13 surgical convalescent male patients who had been placed on constant diets of bread, butter, eggs, and milk, as well as from a few cats (meat diet) and rabbits (rolled oats diet). An aliquot of the weighed stool (20–50 g) was then treated with alum and lead acetate, diluted to a finite volume, filtered, and then analyzed colorimetrically with Bell's modified reagent.

From a study of the urinary and intestinal phenol excretion in normal men, the conclusion is reached that normally, and in the absence of diarrhea, laxatives, or enemata, a very small fraction (from 7 to 20 per cent) of the phenols formed is eliminated by the intestine.

Only the free or unconjugated phenols were found in the stools studied.

It should be noted that Otto and Willey were much concerned about two "bugaboos" of clinical chemistry, drug interferences and adequate preservation of the sample. Salicylic acid and aspirin were obviously two of the common substances that would interfere with the phenol determination, even though most of the ingested drugs were eliminated via the kidneys. Stools could be stored for 24 h at -2 °C, but showed slight increase in values if stored at this temperature for 48 h.

Otto published only three papers in 1917. Though the United States declared war on Germany in April, the impact on Otto's scientific productivity was not immediate. His papers were completed before the war-fever struck.

F. H. McCrudden (who had served as an assistant in the department of biological chemistry from 1906–8, the "transition" period when Otto became head) and C. S. Sargent published two papers in the *Journal of Biological Chemistry* in 1916 that viciously attacked the validity of the Folin micromethod for creatinine and creatine. Had the two papers stuck to scientific exposition, Otto's response would probably have come in a more leisurely fashion. But McCrudden's and Sargent's closed with the following two incendiary paragraphs:

> From the data it is clear that in the determination of creatinine in the blood the color due to creatinine is such a small proportion of the total color that analysis gives no information whatever concerning the amount of creatinine present; the slight variations obtained in duplicates can be accounted for by slight variations in the amount of picric acid in the solution.
>
> In the light of these experiments it is clear that all that has been written hitherto concerning the physiology of creatinine and creatine needs careful revision; much of it will have to be modified, some of it—all that concerning creatinine and creatine in the blood, for example—will have to be rejected altogether.

Otto promptly assigned his second-year graduate student, Edward A. Doisy, to the problem of repeating the work of McCrudden and Sargent and preparing pure picric acid (*14-34*). In Doisy's own words:

> At the outset of my second year at Harvard, Dr. Folin directed me to study the creatine-creatinine problem. A paper critical of his methods for the determination of these substances caused some unrest. Very shortly it was found that the methods were satisfactory if purified acid was used. Dr. Folin came to see me while some purified picric acid was being dried over a steam radiator, the temperature being well below the melting point of picric acid. In spite of the

mild heat in the radiator, the evaporated water was being condensed in the chilly room and formed a small cloud over the picric acid. He ran from the laboratory shouting "Your life is your own—do what you please with it."

The American Society of Biological Chemists met at the Cornell Medical College in New York in December 1916 and a paper on creatine-creatinine was on the program. Dr. Folin said, "Doisy, I expect you to discuss it." Being somewhat inhibited, as I still am, young and inexperienced, I could not get the floor until after four others had discussed the report (14-35).

As expected, Otto and Edward obtained results entirely different from those of the attackers. In looking for a cause of those differences, they examined various samples of picric acid for their color-producing potential with creatinine. McCrudden and Sargent had found that their alkaline picrate gave about 500 times the depth of color with creatinine compared with either picric acid or the alkali alone. The Harvard pair, however, examined six different preparations, of which four were in the dry form, the fifth in the wet form in which picric acid was then sold, and the sixth was crystallized from the fifth, as given below.

According to our observations creatinine gave about 3000 times as deep a color as was given by our poorest picric acid (the wet sample), when alkali was added exactly as in the creatinine determination. Using our best, the recrystallized picric acid, creatinine gave about 12,000 times as deep a color as its sodium picrate. . . . In view of the remarkably faint color of our sodium picrate in comparison with the color recorded by McCrudden and Sargent, it occurred to us that the endeavor of American manufacturers to meet the enormous demands, created by war, for picric acid, may have resulted in the production of extremely low grades of picric acid, and that McCrudden and Sargent may have been using an unusually impure sample.

Folin and Doisy obtained samples of picric acid used by McCrudden and Sargent, that were labeled "C.P.," and found that their published observations were made on the basis of an extraordinarily impure reagent, making their conclusions irrelevant.

Folin and Doisy proposed a test for the purity of picric acid and a procedure for purifying picric acid, and then presented a simple modification of their original procedure for creatinine to permit analysis of as little as 2 mL of blood (Part II, p. 408).

Folin and Doisy presented one table that compared results on three urines with the new method using purified picric acid with results using McCrudden's material. Values with the latter were much too high. A second table gave creatine-creatinine determinations on the urine of 65 subjects (normal and abnormal) by the new and old methods. The results were practically identical and also indicated the frequency of the appearance of creatine in human urine in health and disease. The war had affected the quality of commercial grade picric acid, and thanks to McCrudden and Sargent's inept paper, Folin and Doisy had solved the problem.

The war created a second problem for Otto. For determining ammonia in urine by direct nesslerization, Otto and Willey had used blood charcoal imported from Merck in Germany to remove interfering substances (primarily creatinine, uric acid, and phenols). Otto and Richard D. Bell, having tried unsuccessfully to find or prepare a suitable animal charcoal with properties equal to Merck's, looked to other sources for a suitable alternative. This resulted in their introduction to biochemistry of a much superior substance, borrowed from its use as a water softener, the zeolite permutit (14-36). Permutit was a synthetic mineral, an aluminum silicate, discovered by Gans in 1905. The American manufacturer was able to prepare a product suitable for ammonia determination.

Permutit was rather unique. It absorbed NH_3, and as a clean, moderately fine, insoluble powder settled like sea sand from water in a few seconds, thereby making it possible to separate rapidly the absorbed NH_3 by mere decantation.

Folin and Bell studied the properties of permutit. Under suitable conditions the "reagent" absorbed NH_3 quantitatively by an exchange process. The term "ion exchange" was then not yet used. A gram of the powder took up 13 mg of NH_3 nitrogen within a few minutes, but the presence of other salts (cations) could markedly influence the uptake. The amount of salt in urine presented a problem only if too much NH_3 was present. The problem was overcome by diluting the sample. Two grams of permutit would handle 1 mg of NH_3 nitrogen in the presence of 20–25 mg of NaCl.

Permutit was effective only when it contained water, and in neutral or weakly acid solutions. Its serviceability depended on the fact that in the presence of NaOH, ammonia was set free quantitatively. The longer the NH_3 stood with the permutit the longer it took to be nesslerized; hence our authors recommended that the consumer find out for himself the optimum duration for color development. They obtained 95% of the theoretical yield after 2–3 min standing, and practically 100% in 10–15 min or less. If the exchanged NH_3 was left overnight in the permutit, its liberation took longer, and a deficit of 2–3% usually remained. The zeolite was rechargeable and reusable, but in practice, the time involved in its recharging made the process uneconomical.

Folin and Bell tested their method for ammonia determination using permutit on 17 urines and obtained results quite comparable with the macro–aeration-titration method. They noted that tap water could be used to dilute the final Nessler's color, if it were rendered NH_3-free by shaking or aerating it with a few grams of permutit.

Otto's last paper of 1917 was a published lecture (Third Mellon Lecture) he had given on May 18 to the Society for Biological Research at the University of Pittsburgh (14-37). The paper was particularly valuable because of its historical perspective.

Part of the recent progress in biochemistry was attributable to the development of new analytical techniques, Otto wrote. "In the last few years there have been very great changes in the methods used for the analysis of urine, while the advancement in the field of blood analysis has been ever more remarkable."

The main characteristic of the successful modern analytic method was speed, now called "turnaround time." Obviously, a physician could hardly be expected to encourage applications to clinical problems if he had to wait four days to find out what the urinary ammonia value was. That was the time it once took to obtain values for urea, NH_3, and uric acid.

Today any analytic method in urine analysis which cannot be finished within less than two hours stands in need of further revision. Most of the common determinations in urine and blood will become from fifteen to thirty minutes for single determinations, and also several determinations will be made with little extra expenditure of time.

Once a method had become speedy and simple, biochemists would be quick to make use of it for their investigations. Clinicians, too, would use the method, not only for research, but in their practices. A large and increasing number of American physicians possessed their own laboratories. But this posed some problems about what we now term "quality control," or "quality assurance."

There is a legitimate and important use for technical assistants in the innumerable little laboratories springing up in connection with private and with hospital practice of medicine, but it seems to me extremely important that this use should conform to some reasonably honest standard of responsibility. . . . It is unfortunate that the laboratory expert and the competent clinician cannot be united in one person. It is unfortunate that methods must be simple and easy, as well as quick, before they can find any very widespread and sound application within the medical profession. The modern methods are wonders in quickness in comparison with the old ones, and they are not very complicated; but no man knows them who does not know how to check upon his own results by working against theoretical figures.

Changes in technology for urine testing had reduced sample requirements from a full 24-h urine to about 10 mL to analyze for the more common nitrogenous constituents. Otto showed from recent studies by Willey Denis on creatine excretion in adults that a full day's urine was not necessary. Traces of creatine, she had found, were not uniformly excreted from hour to hour, but were apparently confined to a short period after each meal.

Because emphasis in nutrition had shifted away from concerns about nitrogen equilibrium, there was no need to collect 24-h samples. Otto proposed the use of standardized 3-h samples to increase enormously the possibility of doing quantitative metabolic work on private patients, "office" patients, and outpatients in hospitals and to reduce the loss of sample and difficulty concomitant with collecting 24-h samples.

Of course, a 3-h sample would require a new set of normal values. "I hope shortly to furnish such figures for the first three hour morning period and for a second period representing the effect of a standard test meal." Unfortunately, these figures were never published.

Another major development in technology, Otto pointed out, was the "tendency to make the methods microchemical." Colorimetric procedures had made this possible for both urine and blood. Not long before, it had taken 100–200 mL of blood to perform a qualitative test for uric acid that gave positive results only if the blood was unusually rich in uric acid. Now the analysis could be quantitatively performed on 10 mL, or even as little as 5 mL, regardless of the concentration. This made possible the elucidation of the question of protein absorption in the animal body, and the fate of the waste products. "To any one who has actually followed this process, as Dr. Denis and I followed it in cats, there can be no doubt as to the essential features of protein metabolism, and the ability of muscular tissues to form urea." (Otto was correct about the use of the new technology, especially for protein digestion, but would be proved wrong, as we mentioned before, about the site of synthesis of urea.) The new technology made possible the study of the speed with which equilibrium is established for ordinary soluble products between blood and tissues.

Micromethods by colorimetry were now available for determining a growing number of substances in blood: NPN, urea, NH_3, uric acid, creatine, creatinine; Bloor's methods for lipoids, including fat, lecithin, and cholesterol; Marriott's nephelometric methods for acetone bodies; Benedict's method for blood sugar; and Lyman's very recent method for calcium. The colorimeter could also be used as a nephelometer. ". . . the guiding principle is to overcome the lack of concentration and the limited supply of blood by the application of corresponding intense and sensitive reactions. The amount of blood used for each determination in the series mentioned above varied from a fraction of 1 cc. up to 10 cc." Five milliters or less was most practical. For working with "drops of blood" (ultramicrochemistry), Otto had found the techniques impractical because of the need to resort to a complicated system of weighing instead of measuring (pipetting) the blood (Bang's technique). But Otto did not rule out the possibility of working with samples that small. Modern analysis was incomplete, and only about six to seven years old. Also Otto did not fail to mention that "I have made no reference to other recent useful analytical procedures, such as Van Slyke's methods for amino-acid nitrogen and for chlorides."

Otto dwelled at length on the subject of normal values in blood using NPN and urea as examples. For medical students and non-hospitalized subjects, NPN values at the upper limit were at 28–30 mg % (20–21.4 mmol/L), with urea nitrogen half this value. For hospitalized patients, the NPN was as often found to be between 30 and 40 mg % (21.4–28.6 mmol/L), as below 30 mg % (21.4 mmol/L). Was this due to kidney dysfunction? While the intake of food protein did not affect the levels of NPN and urea concentration in healthy subjects, it did affect levels in people with damaged kidneys. Would lowering their NPN and urea through restricted protein intake help them in any way? Why did some nephritic patients tolerate high blood values of urea nitrogen and others go into "uremic coma" without accompanying high values, as in toxic pregnancy?

With the help of Mr. Foster of the biochemistry department and Dr. Newell at the Boston Lying-In Hospital, Otto had determined the NPN and urea in the blood of 100 pregnant women. Because they were hospitalized patients he had expected to find rather high values. But without exception, urea nitrogen was found to be below the lower limit of normal, 11–12 mg % (7.9–8.6 mmol/L). Their urea nitrogen was 5–9 mg % (3.6–6.4 mmol/L), and a few ran as high as 9 mg % (6.4 mmol/L). What is more, whereas normal urea nitrogen represented 50% of the NPN, for pregnant women it represented 20–35%. This phenomenon was as yet unexplained.

Blood ammonia was theoretically interesting, but the amount present was so small that it was difficult to ascribe clinical significance to it. It could be responsible for uremic and diabetic coma. Creatinine determinations were useful for detecting gross errors of urine collection. It was one of the last nitrogenous substances to accumulate in kidney insufficiency. In the absence of a suitable commercial source of creatinine, determination should be done using potassium bichromate as a standard. Urine diluted with strong HCl was useful as a stable standard.

New findings on uric acid metabolism had been reported by S. R. Benedict. While most dogs excreted allantoin in urine, the Dalmatian dog, like the human, excreted uric acid. Beef blood contained extraordinary quantities of "latent" uric acid. Some kinds of human blood carried large amounts of uric acid while the NPN and urea were normal; others carried normal amounts of uric acid and extremely large values of NPN and urea. The former condition occurred in gout, the latter in nephritis. The high uric acid concentration was apparently not due to overproduction, but arose from the excretory power of the kidneys. Uric acid was about 20 times as great in concentration in glandular organs (e.g., spleen, liver) as in muscle or blood.

Other metabolic products not excreted as waste products had to be considered. Bloor's findings on the normal relationship between lipoids should lead to a detailed knowledge of the intermediate processes in fat metabolism. The practical analytical methods involved should attract the clinician's interest for studying disorders such as malnutrition in infants, obesity, lipemia, and diabetes.

Microchemical methods had made possible the study of phenols. Only urine had been studied so far. Phenols were chiefly derived from intestinal putrefaction. A suitable method for blood phenol analysis remained to be found.

In closing, Otto stated:

> It will seem to you that I have discussed little else than analytic methods—a subject which cannot be made very interesting outside of the laboratory. I am convinced, however, that both the biochemist and the clinician must pay more and more attention to this least interesting but most important aspect of research. . . . It is now only by means of finer and ever finer technic that progress can be made toward the solution of the many metabolism problems which must be solved by us and those who follow us, in order to secure an increasingly better basis for clinical, experimental, and above all, preventive medicine.

American entry into World War I evidently had a profound effect on the staffing of Harvard's Department of Biological Chemistry. The medical school catalogue for the 1917–18 year listed only four names in the department: Folin, Bloor, Henderson, and the new assistant, Guy E. Youngburg, S.M. Gone to military service were Foster, Doisy, Bell, and Lyman. As Henderson was listed for neither lecture nor laboratory duties, it must be assumed that he made no direct contributions to the teaching within the department, nor did he make any before his transfer to physical chemistry in 1920.

But the war was not a total loss for the department of biological chemistry. In 1917, a young Chinese doctoral candidate Hsien Wu (1893–1959) joined the department. He had come from the nearby Massachusetts Institute of Technology, where he had served during the previous year as a graduate assistant in organic chemsitry. Although he had enrolled at MIT in "naval architecture," two years later he switched to chemistry with a minor in biology and obtained a bachelor's degree in 1916. As Otto had once reasoned, Wu also discovered that he did not wish to confine his training and future career to pure chemistry, but rather wished to deal with living processes that the new field of biochemistry offered. The field now encompassed nutrition, metabolism, clinical chemistry, enzymology, and much of physiology. Of course, with Otto, the clinical aspects predominated.

Although several of Otto's graduate students had remarkably productive careers following graduate school, none during his student days approached the experimental depth of Wu.

Otto published only two papers in 1918, submitted together on Jan. 25, to the *Journal of Biological Chemistry*. Both papers dealt with sugar analysis. The first one (with W. S. McEllroy) made use of a modified Benedict's reagent for the qualitative and quantitative measurement of sugar (reducing) in urine (Part II, p. 408). To maintain the copper hydroxide in solution, phosphates replaced citrate (or tartrate and glycerol in other reagents). The advantages of using the phosphate reagent were presumably its cheaper cost, its lack of reducing power (upon heating), and its tendency to regulate the degree of alkalinity at a lower level of hydroxyl ion concentration than was obtained by carbonate alone, hence a given amount of reducing sugar would reduce more copper. Our authors found that the reagent worked well; but it was difficult to recommend one superior combination of phosphates, because several useful ones were possible.

Otto and Willey Denis made use of this quantitative titration method, and added a colorimetric technique for determining lactose in milk (*14-38*). In view of the fact that albumin did not interfere with the titration method on urine, this method was successfully applied to the direct determination of lactose in milk without prior removal of protein. Milk fat also did not interfere with the procedure. Because the lactose content of milk was fairly constant a prior dilution could be made with certainty of its range of values (in contrast with urine). The authors recommended a 1 + 3 dilution for cow's milk and a 1 + 4 dilution for human mother's milk. The titration procedure was essentially the same as the one for urine using the special burette. About 3 mL of diluted milk was used and the composition of the dry mixture was slightly altered.

Although a colorimetric method for milk sugar that used picric acid to precipitate the protein as well as reduction of alkaline picrate for the quantitative assay had already appeared in the literature [Dehn, W., and Hartman, F.A. *J. Am. Chem. Soc.* **36**, 404 (1914)], Otto and Willey thought their method was simpler and as accurate. Excellent correlation was obtained in parallel determinations of lactose in milk by the colorimetric and titration methods, using 10 samples of cow's milk and eight samples of mother's milk.

Otto spent the summer of 1918 teaching chemistry in a training camp for nurses established in behalf of the war effort. It was held at Vassar College in Poughkeepsie, N.Y. He was probably contacted about the program late in 1917 or early in 1918 by Professor Herbert E. Mills, its dean. It is also likely that Laura first heard about the proposed course through alumnae communications and helped enlist her husband's services.

By the autumn of 1918, Walter Bloor had gone on to the University of California and the next stage of his eminent career, and Cyrus Fiske had returned as assistant professor, to remain until his retirement. Foster, Doisy, Bell, and Lyman were listed as absent on war service. Henderson stayed on the payroll of the department. There were no assistants or teaching fellows.

From an unpublished biographical sketch of Otto Folin, prepared by Laura, we note:

> The University of Lund, Sweden, gave him an honorary degree in 1918, and shortly thereafter tendered him a professorship. This honor, I believe, touched his feelings more deeply than any other which he received. He felt, however, that it would be unwise to accept it, for he thought that he and I could not easily adjust ourselves to the University life of Sweden.

The invitation from the University of Lund came to Otto while he was in Kearsarge during the summer. A hand-penciled draft of his reply reads as follows:

> I have been in the woods on a vacation for weeks, which accounts for the delay in replying to your communication concerning the 250 year celebration of Lund's University and the honorary degree which the University has offered to bestow upon me on that occasion.
>
> I deeply appreciate the honor of having been thought of in connection with that celebration. A degree from the University of Lund would be particularly gratifying to me who though a loyal American still cherishes a deep attachment for my native land.
>
> Present conditions make a journey to Sweden on my part out of the question.
> I do not know whether or not the University of Lund grants honorary degrees in absentia; so I am doubtful as to whether, since I can not be present, it is desired that I should send the data concerning my life asked for in your communication. I am, however, enclosing the same. Also, should the degree be conferred in absentia, I should like to have and to keep as an heirloom the doctor's ring.
>
> In view of the delay in replying to your much esteemed communication I shall send a cable reply on my return to the city next week.
>
> Most respectfully,

Otto Folin

Otto was given the degree in September, and the ring with it.

References

14-1. Folin, O., and Denis, W., Protein metabolism from the standpoint of blood and tissue analysis. Sixth paper. On uric acid, urea and total non-protein nitrogen in human blood. *J. Biol. Chem.* **14**, 29–42 (1913).

14-2. Frothingham, C., Jr., Fitz, R., Folin, O., and Denis, W., The relation between non-protein nitrogen retention and phenolsulphonephthalein excretion in experimental uranium nephritis. *Arch. Intern. Med.* **12**, 245–58 (1913).

14-3. Folin, O., and Denis, W., On the absorption of nitrogenous products—a reply to Abderhalden and Lampe. *J. Biol. Chem.* **14**, 453–55 (1913).

14-4. Folin, O., and Denis, W., On the tyrosine content of proteins—a reply to Abderhalden and Fuchs. *J. Biol. Chem.* **14**, 457–58 (1913).

14-5. Folin, O., and Morris, J.L., The normal protein metabolism of the rat. *J. Biol. Chem.* **14**, 509–15 (1913).

14-6. Folin, O., and Lyman, H., On the influence of phenylquinolin carbonic acid (atophan) on the uric acid elimination. *J. Pharmacol. Exp. Ther.* **4**, 539–46 (1913).

14-7. Folin, O., Einige für Blut- und Harnanalyse bestimmte Schnellmethoden (Several rapid methods for blood and urine analysis). In: *Handbuch der Biochemischen Arbeitsmethoden.* pp. 715–26 E. Abderhalden, Ed. Berlin: Urban & Schwarzenberg, 1913.

14-8. Folin, O., and Bosworth, A.W., The application of Folin's method for the determination of ammonia to fertilizer. *J. Ind. Eng. Chem.* **5**, 1–2 (1913).

14-9. Folin, O., The clinical applications of pathological chemistry. *Lancet* **2** (Aug. 16), 468–70 (1913).

14-10. Folin, O., Denis, W., and Seymour, M., The nonprotein nitrogenous constituents of the blood in chronic vascular nephritis (arteriosclerosis) as influenced by the level of protein metabolism. *Arch. Intern. Med.* **13**, 224–34 (1914).

14-11. Folin, O., and Buckman, T.E., On the creatine content of muscle. *J. Biol. Chem.* **17**, 483–86 (1914).

14-12. Folin, O., and Denis, W., On the creatinine and creatine content of blood. *J. Biol. Chem.* **17**, 487–91 (1914).

14-13. Folin, O., and Denis, W., Protein metabolism from the standpoint of blood and tissue analysis. Seventh Paper. An interpretation of creatine and creatinine in relation to animal metabolism. *J. Biol. Chem.* **17**, 493–502 (1914).

14-14. Folin, O., Denis, W., and Smillie, W.G., Some observations on "emotional glycosuria" in man. *J. Biol. Chem.* **17**, 519–20 (1914).

14-15. Folin, O., and Denis, W., Metabolism in Bence Jones proteinuria. *J. Biol. Chem.* **18**, 277–83 (1914).

14-16. Folin, Otto. *Preservatives and Other Chemicals in Foods: Their Use and Abuse.* Cambridge, Mass.: Harvard University Press, 1914. 60 pp.

14-17. (a) Sperry, W. N., Walter R. Bloor (1877–1966). *Clin. Chem.* **12**, 897–99 (1966).

(b) Bloor, Walter Ray. *Who Was Who in America* 4, 96 (1961–68). Chicago: Who's Who, Inc.

14-18. Folin, Otto. *Laboratory Manual of Biological Chemistry, With Supplement.* New York and London: D. Appleton and Co., 1916. 188 pp. (Note: It was printed front page only.)

14-19. Folin, O., and Denis, W., On starvation and obesity, with special reference to acidosis. *J. Biol. Chem.* **21**, 183–92 (1915).

14-20. Folin, O., and Denis, W., Note on perca globulin. *J. Biol. Chem.* **21**, 193–94 (1915).

14-21. Folin, O., Note in defense of the Folin-Farmer method for the determination of nitrogen. *J. Biol. Chem.* **21**, 195–99 (1915).

14-22. McCollum, E.V., Stanley Rossiter Benedict, 1884–1936. *Biogr. Mem. Natl. Acad. Sci. U.S.A.* **27**, 155–77 (1952).

14-23. Folin, O., and Denis, W., The diagnostic value of uric acid determinations in blood. *Arch. Intern. Med.* **16**, 33–37 (1915).

14-24. Folin, O., and Denis, W., An apparatus for the quantitative collection of urine from women. *Arch. Intern. Med.* **16**, 195–96 (1915).

14-25. Folin, O., and Denis, W., A colorimetric method for the determination of phenols (and phenol derivatives) in urine. *J. Biol. Chem.* **22**, 305–8 (1915).

14-26. Folin, O., and Denis, W., The excretion of free and conjugated phenols and phenol derivatives. *J. Biol. Chem.* **22**, 309–20 (1915).

14-27. Folin, O., and Denis, W., Some observations on the selective activity of the human kidney. *J. Biol. Chem.* **22**, 321–26 (1915).
14-28. Folin, O., A qualitative (reduction) test for sugar in normal human urine. *J. Biol. Chem.* **22**, 327–29 (1915).
14-29. Folin, O., and Denis, W., Nitrogen determinations by direct nesslerization. I. Total nitrogen in urine. *J. Biol. Chem.* **26**, 473–89 (1916).
14-30. Folin, O., and Denis, W., Nitrogen determination by direct nesslerization. II. Non-protein nitrogen in blood. *J. Biol. Chem.* **26**, 491–96 (1916).
14-31. Folin, O., and Denis, W., Nitrogen determination by direct nesslerization. III. Ammonia in urine. *J. Biol. Chem.* **26**, 497–99 (1916).
14-32. Folin, O., and Denis, W., Nitrogen determination by direct nesslerization. IV. Urea in urine. *J. Biol. Chem.* **26**, 501–3 (1916); V. Urea in blood, *J. Biol. Chem.*, **26**, 505–6.
14-33. Folin, O., and Denis, W., The relative excretion of phenols by the kidneys and by the intestine. *J. Biol. Chem.* **26**, 507–13 (1916).
14-34. Folin, O., and Doisy, E.A., Impure picric acid as a source of error in creatine and creatinine determinations. *J. Biol. Chem.* **28**, 349–56 (1917).
14-35. Doisy, E.A., An autobiography. *Ann. Rev. Biochem.* **45**, 1–9 (1976).
14-36. Folin, O., and Bell, R.D., Application of a new reagent for the separation of ammonia in urine. I. The colorimetric determination of ammonia in urine. *J. Biol. Chem.* **29**, 329–35 (1917).
14-37. Folin, O., Recent biochemical investigations on blood and urine; their bearing on clinical and experimental medicine. *J. Am. Med. Assoc.* **69**, 1209–14 (1917).
14-38. Folin, O., and Denis, W., The determination of lactose in milk. *J. Biol. Chem.* **33**, 521–34 (1918).

Chapter 15. Hsien Wu—A Major Leap Forward (1919–21)

The year 1919 was a banner year for Otto in that it provided a solution to a long-standing problem that confronted the future of blood chemistry, namely how to remove proteins so that colorimetry could be carried out on soluble substances cleanly and quantitatively. Otto had introduced picric acid and metaphosphoric acid for this purpose. He had tried alcohol precipitation, and had noted the potential usefulness of phosphotungstic and phosphomolybdic acids. Trichloroacetic acid was also well known for this purpose, but Otto had little experience with it.

To Hsien Wu Otto assigned the problem of studying the various protein-precipitating agents. Although metaphosphoric acid was an excellent reagent, it had proved disappointing because of its instability and impurities. There was no question that Otto, and no doubt a good many other biochemists, thirsted for the "universal" precipitating reagent that would allow separation and analysis of the major constituents of blood. In actuality, as we shall see, the discovery of tungstic acid as a protein precipitant opened the doors of clinical chemistry and, once and for all, moved the quantitative testing of body fluids out of research and students' laboratories into the hospital routine. It irreversibly simplified the approach to the analysis of blood, and permanently changed the role of the chemical laboratory in diagnosis of disease and in monitoring therapy and health.

Of course, Wu was "new to the game" when he came to Otto's laboratory from MIT in 1917. It was a tribute to his acuity, laboratory prowess, and his sound preparation in science that he was able to absorb his graduate coursework and the language of biochemistry while carrying on this monumental work. It was his good fortune to have come to the master teacher, who could give him the sharp challenge in research that he needed and whose history was one of achievement and dedication, joined with enthusiasm, deference, humor, and friendship.

There was more—much more. The discovery of tungstic acid came early during Wu's two doctoral years. He found time to scrutinize each of Otto's favorite nitrogenous metabolites, and he made very significant modifications in the analysis for each. But Wu's greatest ingenuity, and certainly the culmination of his own effort, was his original contribution of a practical method for the determination of blood sugar, about which Otto could only exclaim, "This is worth a second Ph.D.!"

The 1919 paper of Folin and Wu is a classic of clinical biochemistry and we shall dwell on it at length.

> The main purpose of the research recorded in this paper has been to combine a number of different analytical procedures into a compact system of blood analysis, the starting point for which should be a protein-free blood filtrate suitable for the largest possible number of different determinations. It need scarcely be pointed out what a convenience and advantage it would be if one could take the whole of a given sample of blood and at once prepare from it a protein-free blood filtrate suitable for the determination of all or nearly all the water-soluble constituents, non-protein nitrogen, urea, creatinine, creatine, uric acid, and sugar.
> In this paper we deal chiefly with a semi microchemical scale of work representing only a moderate reduction of the quantities ordinarily taken for colorimetric work with the 60 mm. Duboscq colorimeter.

Uric acid determination was a major obstacle in developing a blood analysis scheme. Folin and Wu had presented a modification of the old Folin-Denis-Benedict method, but this required 25 mL of blood. They now brought uric acid determination to a new level of practicality. First they solved the pressing problem of preparing a stable standard; then they established a method requiring the filtrate from only 2 mL of blood. Another feature of the paper was a new colorimetric method for determining blood sugar (Part II, p. 410).

The second Folin-Wu paper of 1919 was on uric acid determination in urine (15-1). Although Otto had been using silver lactate instead of ammoniacal silver solution to precipitate uric acid, he had not published it because it did not seem worthwhile until a permanent uric acid standard was available. Having achieved this in the previous paper the new procedure was now offered. The reagents used were those from the method on blood.

Besides the two with Wu, Otto published five more papers in 1919. His last paper with Willey Denis was published that year. Willey had become a clinical research biochemist in her own right. In addition to those with Otto, she had published papers on blood cholesterol; creatine in muscle and its diurnal excretion; ammonia excretion; lead analysis in urine, feces, and tissues; NPN constituents in human milk; feeding effects on blood calcium; magnesium in blood; sulfates in blood; and other related subjects. She had functioned entirely in a hospital environment just as Otto had at the McLean. Now she was long overdue for a truly independent teaching and research opportunity. And this came to her with an appointment as assistant professor of physiological chemistry at Tulane University in her hometown of New Orleans. By 1925, she would rise to be full professor and head of the department, independent within the school of medicine. She was probably the first woman to be chairman of a major basic science department in a major medical school. Her productive research work continued uninterrupted until her death in 1929 at age 49.

Willey's last paper with Otto (and A. S. Minot) was a follow-up of the Folin-Denis paper on measuring lactose in milk. The method was applied to the milk of animals, particularly small ones because of the microanalysis involved. For completeness, values of fat and protein were also measured (*15-2*).

Ten animal milks were examined and data provided for protein, fat, and lactose, mainly for human, cow, rabbit, and cat, with fewer samples from guinea pig, pig, goat, sheep, dog, and horse. The protein in human milk was the lowest of any animal studied, 1.39 g/dL (87 women). The lactose in human milk was 7.06 g/dL, much higher than that of cow (4.54) and rabbit (1.8) milk. Useful data were provided for the milk of all animals tested.

Otto published a short paper with G. E. Youngburg on determining urea in urine by direct nesslerization (*15-3*). In the Folin-Denis method blood charcoal was used, but not required if the urease was purified by treatment with permutit. This was accomplished by shaking fine jack bean meal with acid-washed permutit and dilute alcohol, and then filtering. The filtrate contained the urease. One milliliter was added to 1 mL of diluted urine (usually 1 + 9) in a test tube, then incubated in warm water (40–55 °C) for 5 min, or at room temperature for 15 min. A drop of phosphate buffer was added, preferably at the beginning of the digestion period. Following the incubation, the contents of the tube were transferred to a 200-mL volumetric flask, 150 mL of water and 20 mL of Nessler's solution were added, the solution was diluted to volume, and a color comparison was made with a standard (1 mg NH_3 nitrogen) prepared and treated simultaneously with Nessler's solution in another 200-mL flask.

The method was very effective, even with urines rich in albumin. It had been used for two seasons by the medical students, but not previously published, held up in anticipation of a larger study that evidently did not take place.

Otto published a paper (with E. C. Peck) that revised the salt mixture used in the "quantitative" urinary sugar determination and added a little sodium carbonate to the copper sulfate solution (*15-4*). The revised salt mixture was used to reduce variations introduced in making up the dry mixtures, and the sodium carbonate was used to prevent reaction between the copper sulfate and the thiocyanate. Details were presented for interpreting the significance of the time of appearance of the heavy cuprous thiocyanate precipitate in terms of the amount of dextrose present in the sample.

Another paper, with L. E. Wright, was, of all things, "a macro-Kjeldahl method for the determination of nitrogen in urine which requires very little equipment and by which a urinary nitrogen determination can easily be finished in 20 to 25 minutes" (*15-5*).

This rapid method used the acid digestion mixture of the micromethod. The procedure was done on 5 mL of undiluted urine in a 300-mL Kjeldahl flask (Pyrex), using 5 mL of the digestion fluid. The contents were boiled for 4 min with a microburner, cooled, alkalinized with NaOH, and distilled into standard (0.1 N) acid via an air-cooled connecting tube that dipped below the surface of the acid in the receiving flask. The receiver fluid became heated, but there was no loss of acid. The acid was then titrated in the cooled solution using alizarin red indicator. Distillation required only 5 min. The method was intended for urine only. When the urine contained sugar, fuming sulfuric replaced the sulfuric acid in the digestion mixture, or it could be added to the existing mixture.

Otto's last paper of 1919 was a note expressing skepticism about a paper that had recently described a method for blood ammonia determination that used metaphosphoric acid to precipitate proteins and that extracted the NH_3 from the filtrate with permutit (*15-6*). Folin and Denis had tried this many times and abandoned the idea because of the errors involved. When a procedure for purifying metaphosphoric acid was carried out and a purified permutit of certain mesh size was used, there was still insufficient color produced with Nessler's reagent by the NH_3 present in blood to obtain valid results.

We note in passing that the second edition of Otto's laboratory manual appeared in 1919.

The ASBC met twice in 1919, the first time (13th annual meeting) in Baltimore, April 24–26, and Otto probably presented both papers cited with his name in the proceedings: "A New Qualitative and Quantitative Color Reaction for Amino-Acids" (Otto Folin and H. Wu), and "A Convenient Permanent Urease Preparation" (Otto Folin). At this meeting and again in December in Cincinnati, S. R. Benedict was elected president, with D. D. Van Slyke, vice-president.

The eminent clinical biochemist Victor C. Myers had this recollection of the Cincinnati meeting of 1919 in a letter to Philip Shaffer (Jan. 14, 1936):

> I recall very vividly the verbal tilts between Benedict and Folin. These were amusing and enjoyable, chiefly for the reason that Folin could always see their amusing side. At the Cincinnati meeting in 1919, Benedict picked the first Folin-Wu blood sugar method to pieces, as you will recall. After Benedict's paper Folin got up and said "Benedict is a past master in finding flaws in perfectly good methods and if what Benedict says is so I am perfectly willing to accept it," and sat down. I whispered to Benedict, "You seem to have Folin licked." He replied, "I'll bet Folin can't get back to Boston fast enough to go over the method again." Folin gave a Harvey Lecture in February (1920) and had ironed out practically all of Benedict's criticisms. Although his lecture was supposedly on another topic he could not resist coming back to this blood sugar method on several occasions. Benedict frequently got quite excited over his discussions with Folin but Folin always took them very calmly and treated Benedict somewhat as a son.

For the 1919–20 academic year, the department returned to full strength. Hsien Wu, Ph.D., was now an assistant; Bell and Lyman were back as research fellows; and Walter C. Russell, S.B., was the Austin teaching fellow.

Wu's thesis for his doctorate was entitled, "A system of blood analysis with special reference to uric acid." Its preface stated:

> This thesis contains three parts, each dealing with a separate phase of the determination of uric acid with special reference to blood. The object of the research is single, but for greater clearness the results are presented under separate headings. I wish to acknowledge my indebtedness to Prof. Otto Folin under whose guidance this research was undertaken for his helpful advice and kindly attention.

Otto's only scientific paper of 1920 was published with Wu. It put the "final" touches on the blood sugar method (*15-7*). What Victor Myers had said about Benedict's criticism of the Folin-Wu method was partly true in that it caused

Otto to do some accelerated thinking about the problem of reoxidation and to offer his amazing solution. Wu had been working on improvements in the molybdate reagent, but the idea of the constricted blood sugar tube may very possibly have occurred to Otto at the Cincinnati meeting, and not earlier. Certainly after his return on New Year's Eve he and Hsien must have worked feverishly to complete the studies of the reoxidation of cuprous oxide in relation to tube size; and not only on the design, but on the arrangements for manufacturing and marketing the "sugar" tube. Furthermore, the paper itself had to be written and mailed.

We note that the Folin-Wu paper was received for publication on Jan. 26, 1920, meaning that there were perhaps 24 days to complete the final work after Otto's return to Brookline on New Year's Eve.

In the first Folin-Wu sugar method, the sugar was oxidized by a weakly alkaline copper tartrate solution and the cuprous copper formed was estimated colorimetrically with the help of the phenol reagent of Folin-Denis (phosphotungstate–phosphomolybdate). While interference from creatinine and uric acid was eliminated, a new though less pronounced error cropped up: interference from "phenols." Carrying out the reaction in acid solution would have prevented the phenol error and also would have eliminated the "blank" error due to the alkaline copper tartrate. Unfortunately, the pronounced yellow color of the phenol reagent made color comparison difficult.

This impasse forced the ingenious pair into a remarkable resolution of the problem. They replaced the Folin-Denis reagent with another that reacted with cuprous copper in acid solution to produce a blue color, yet gave no color with phenol. This historical "phosphomolybdate" reagent was prepared as follows:

> Transfer to a liter beaker 35 gm. of molybdic acid and 5 gm. of sodium tungstate. Add 200 cc. of 10 per cent sodium hydroxide and 200 cc. of water. Boil vigorously for 20 to 40 minutes so as to remove nearly the whole of the ammonia present in the molybdic acid. Cool, dilute to about 350 cc., and add 125 cc. of concentrated (85 per cent) phosphoric acid. Dilute to 500 cc.

The sodium tungstate was added in this reagent to match that in the filtrate and to "modify somewhat the shade of the blue obtained in the reaction."

The alkaline copper solution remained unchanged. The authors recommended the use of three standard sugar solutions: 1) a stock 1 g/dL solution of dextrose or invert sugar; 2) a 10 mg/dL solution of sugar made by diluting the stock 1 + 99 with water; 3) a 20 mg/dL solution of sugar made by diluting the stock 1 + 49 with water. Xylene or toluene was added to preserve the stock and dilute standards. While the stock would keep indefinitely, the dilute standards were stable no longer than one month.

> It is a well known fact that the cuprous compounds produced by sugar in alkaline copper solutions show a marked tendency to be reoxidized to the cupric condition when exposed to air. Most of us are all too familiar with that fact in connection with ordinary sugar titrations done by overcautious students. That such reoxidation must occur to some extent in our colorimetric blood sugar determination is also undeniable. Without having made any direct experiments on the extent of such reoxidations we had satisfied ourselves that

they do not contribute any material error in our sugar determinations. Check experiments with sugar solutions 50 per cent apart and heated 4, 6, and 8 minutes had given proportionate values. And in actual blood sugar determinations the values obtained were not changed by varying the heating time from 4 to 8 minutes. We had therefore no occasion to fear that material analytical errors could creep in because of the reoxidation.

At the last annual meeting of the American Association of Biological Chemists (Cincinnati, 1919), Benedict condemned our blood sugar method on the ground of excessive, inevitable, and uncontrollable reoxidations of cuprous oxide. He cited shaking experiments by means of which more than 60 per cent were made to disappear. He also asserted that reoxidations are much more extensive in blood filtrates than in pure sugar solutions and therefore insisted that the blood sugar values obtained by our method must be too low. Vigorous shaking, as it happens, is also disastrous to the reaction between reducing sugar and alkaline picrates [Benedict's method]. Losses of 40 per cent can be secured by shaking (in 25 cc. flasks), and if agitation by an air current is substituted for shaking, nearly the whole of the sugar represented in a blood sugar determination by Benedict's method is lost. We cite these observations not as a criticism of Benedict's method but merely to show the grossly misleading character of shaking experiments.

It must be admitted nevertheless that we made something of an error in depending exclusively on indirect evidence on so important a point as the losses of cuprous oxide, and we gladly give Benedict credit for having compelled us to reexamine our method with reference to the effect of reoxidation. We have verified our earlier findings that analytical errors do not occur because of such reoxidation, but this is because all our test-tubes in which the oxidation of sugar takes place are substantially the same diameter, 17 to 18 mm. on the inside. The oxidation in such tubes, when kept in a somewhat slanting position, may amount to as much as 20 per cent of the cuprous oxide formed, yet correct sugar values are obtained.

Folin and Wu verified again that error due to reoxidation did not occur in their method, because they used tubes of inside diameter (ID) of 17–18 mm in a vertical position during the procedure. Nevertheless, they studied the reoxidation problem in terms of test-tube diameters. When the ID was above 15 mm, loss occurred and became astoundingly large at 20 mm. It was possibly caused by currents because the loss was not proportional to surface, nor did it relate to the amount of cuprous oxide present. The loss was about the same over a wide range of sugar concentrations either in pure solution or in filtrate. In short, under uniform conditions, the loss was constant regardless of the concentration of sugar, affecting standards and "unknowns" equally.

What should the tube size be? Folin and Wu designed the special blood sugar test tube. "The essential point to be observed in connection with it is, of course, that the surface of the alkaline mixture of sugar and copper shall reach the constricted part." The tube was readily made commercially available by two manufacturers—the Emil Greiner Co. and Arthur H. Thomas Co.

The blood sugar method was now carried out as follows:

> Transfer 2 cc. of the tungstic acid blood filtrate to a blood sugar test-tube, and to two other similar test-tubes (graduated at 25 cc.); add 2 cc. of standard sugar solution containing respectively 0.2 and 0.4 mg. of dextrose. To each tube add 2 cc. of the alkaline copper solution. The surface of the mixtures must now have

reached the constricted part of the tube. If the bulb of the tube is too large for the volume (4 cc.) a little, but not more than 0.5 cc., of a diluted (1:1) alkaline copper solution may be added. If this does not suffice to bring the contents to the narrow part, the tube should be discarded. Test-tubes having so small a capacity that 4 cc. fills them above the neck should also be discarded. Transfer the tubes to a boiling water bath and heat for 6 minutes. Then transfer them to a cold water bath and let cool, without shaking, for 2 to 3 minutes. Add to each test-tube 2 cc. of the molybdate phosphate solution. The cuprous oxide dissolves rather slowly if the amount is large but the whole, up to the amount given by 0.8 mg. of dextrose, dissolves usually within 2 minutes. When the cuprous oxide is dissolved, dilute the resulting blue solutions to the 25 cc. mark, insert a rubber stopper, and mix. It is essential that adequate attention be given to this mixing because the greater part of the blue color is formed in the bulb of the tube.

The two standards given representing 0.2 and 0.4 mg. of glucose are adequate for practically all cases. They cover the range from about 70 to nearly 400 mg. of glucose per 100 cc. of blood.

Cooling of the alkaline cuprous oxide suspension was not required if only one or two determinations were made. It was essential that the standards and unknowns be heated the same length of time and be at substantially the same temperature when the molybdate was added. Maximum color developed faster in hot solutions. If the procedure was performed uniformly the color comparison could be made at the end of 5 min or 1 h.

Of course, Folin and Wu tried the method out on a number of blood samples. Results were slightly lower than the "old" method, but were not very significantly different. This method would soon become universally used and evaluated.

About a month after he sent the blood sugar paper to the *Journal of Biological Chemistry*, Otto gave his second lecture to the Harvey Society in New York City, although it did not have the dramatic impact of his first lecture in 1908. Then, he had summed up his own experience at the McLean Hospital and issued a call for the larger hospitals to hire trained biochemists as investigators. He could foresee the coming role of the clinical chemist. There was no blood chemistry then; and urine analysis had offered limited prospect for opening the laboratories up to the needs of the clinicians. What a change the past decade had brought! The field of biochemistry had grown steadily across the nation as the medical schools tried to correct their sad deficiencies noted in the Flexner report of 1910. Thanks to the pioneer work of Otto and a few others, the early biochemists in large numbers had been doing intensive studies in blood chemistry. Otto summarized progress in blood analysis (*15-8*):

> Feigl [Johannes] is practically the first one in Germany to enter seriously the field of blood chemistry. The subject in its modern development is almost wholly American, although the late Professor Bang in Sweden was intensively active and developed a very comprehensive system of blood analysis which has been used to some extent in Europe.

Otto praised Bang's work. His methods were strictly microchemical, based on the use of two to three drops of blood taken from a finger or an ear and collected on a weighed filter paper. The amount of blood was then determined by weight.

After appropriate treatment of the paper with various coagulating reagents, the constituent to be determined, such as sugar, non-protein nitrogen, or urea is extracted and determined by a suitable process devised by Bang.

These methods represented good work, but would probably not generally survive because they required excessive care and specialization. They were hopelessly unsuitable for regular use in hospitals, but would not be forgotten because they would be needed for special investigations where two to three drops of blood only were available for testing. Bang had published a book in 1916, and various reports in the *Biochemische Zeitschrift* during the war. Bang's work contrasted sharply with the few blood analyses recorded in the earlier literature in which 20–50 mL were needed for a single NPN determination.

In the U.S., blood chemistry was currently a combination of volumetric analysis with colorimetry and nephelometry. The analytical balance was used solely for preparing reagents. "This combination will certainly survive." Preparing extracts on small, readily measurable volumes of blood was extraordinarily simple and the possibilities of applying quantitative color reactions was limited only "by the ingenuity of the investigator and the number of blood substances present." Only a beginning had been made, and it was clinically practical as shown by the fact that clinicians promptly adapted the new analytic methods to their diagnostic studies on patients. These applications were fine, if clinicians would only stop publishing unreliable information!

"Systems" of blood analysis were being developed in the U.S. exemplified by the Folin-Wu approach, Bloor's methods of blood lipid analysis, and a forthcoming system for measuring inorganic elements. The Folin-Wu system measured water-soluble nitrogenous substances—total NPN, urea, uric acid, creatinine, and creatine—and sugar. "To all these we are now prepared to add the amino-acid nitrogen (and chlorides)." The clinician could use the Folin-Wu methods as described except for one or two changes. The urea method was changed by replacing the alcoholic extract of jack-bean meal with a strip of filter paper impregnated with strong urease solution as follows:

Shake 30 grams of jack-bean powder with 10 grams of permutit and 20 c.c. of 16 per cent. alcohol for 10 to 15 minutes. Pour on one or two filters and as soon as the filtration is substantially finished pour the filtrates into a clean flat-bottomed dish. Draw strips of filter paper through the solution and hang them up to dry, just as is done in the preparation of litmus paper. They dry very quickly, and, once dry, the urease seems to keep just as well as it does in the original jack-bean powder. Half a square inch of such paper is enough for each blood urea determination.

A system of blood analysis was extraordinarily advantageous in the saving of time and material, including blood. Individual isolated methods could not compete with a system. But, of course, a system was difficult to work out. If the system collapsed at some important point the loss was severe and not easily repaired. This was why Otto had been disturbed by Benedict's attack. More than blood sugar was at stake. The whole Folin-Wu system was threatened.

Why? Unmentioned was the psychological impact of the indictment: If blood sugar was wrong, the system was probably all wrong! Otto then discussed the reoxidation phenomenon and how it was corrected.

Otto's lecture to the Harvey Society included a portion not incorporated into the published article—a bit of raillery. He had invited Stanley Benedict to sit in the front row opposite the speaker's rostrum. In discussing the problem of reoxidation, he stressed the need to follow directions exactly for boiling the alkaline copper-filtrate mixture. For those unable to follow directions, such as some first-year Harvard medical students (and others), he had devised a tube with a constriction in it to minimize the surface of the solution exposed to air. Then Otto withdrew from his jacket pocket a Folin-Wu sugar tube, and while holding it aloft for all to see, commented that with this tube the students now obtained accurate results, and perhaps Professor Benedict would also find it useful. Otto then good-naturedly presented it to Stanley before he continued the lecture (*15-9*).

Otto mentioned the perplexing problem of non-glucose interfering substances, particularly the high values obtained in patients with NPN values above 150. He had not yet studied this with the newly modified Folin-Wu sugar method. Other questions related to sugar analysis needed answering. Did sugar move in and out of red blood cells freely or was it confined to plasma? This question also had to be answered as well for all waste (nitrogenous) products. How does one prevent glycolysis? Denis had found that one drop of 40% formaldehyde preserved 5 mL of blood for 48 h, and that the formaldehyde did not interfere with copper reaction.

There was need to extend blood analysis. Only part of the NPN constituents were known in protein-free filtrates (PFF). Phenols were present, but other unknown substances reacted with the Folin-Denis reagent. Amino acids were present and Van Slyke's method had given much information about it. But the method was too complex and required too much blood for it to be used extensively for making systematic analysis. Otto stated that he and Wu would soon be publishing a paper in the *Journal of Biological Chemistry* on "other" nitrogenous substances in PFF, and hinted that it would include amino acid nitrogen, and possibly polypeptides or peptones. The undetermined nitrogen in blood was not merely amino acids. For example, in the blood of the bird, 40–50% of the NPN was unaccounted for. "But what we are striving for at present is to reduce the *undetermined* nitrogen in our blood filtrates to the smallest fraction that is practically attainable."

Studying the nitrogenous substances was not merely a contribution to the problem of nephritis and the efficiency of the kidneys. Prolonged statistical study was needed. Feigl had suggested from 750 complete blood analyses on seven age groups of man that "with advancing years there may be gradual decrease of kidney efficiency without involving clinical nephritis"—a suggestion that Otto and Willey had made earlier. Otto was not convinced by the figures, but was obviously urging that such studies be continued.

Were the kidneys selective in excretion of nitrogenous waste products? Was damage of this selective system involved in the removal of uric acid in gout? The kidneys were selective with respect to NH_3 excretion. There was no accumulation of NH_3 in nephritis or in diabetes accompanied by pronounced acidosis. Concentrations of uric acid and NH_3 differed greatly in blood yet they could be about the same in urine. The existence of blood NH_3 was debatable although the problem might be one of analysis. At that time, the only abnormally high value that had been reported for NH_3 was in epilepsy before the onset of convulsions. The value needed corroboration, however.

> In closing, I would again remark that I have only intended to give a fleeting sketch of what we "blood chemists" do. It is only in the realm of politics that it is proper and laudable to "point with pride" to past achievements. It is less than ten years since this line of work was begun, yet there is scarcely a reputable hospital in America where chemical blood analysis is not now recognized as an indispensable aid in the diagnosis and treatment of patients.

The Association of American Medical Colleges at its annual meeting in March 1919 instructed its Committee on Medical Pedagogy to arrange for the preparation of monographs on medical pedagogy to be presented at the annual meeting in 1920, and to appoint subcommittees for this purpose. Eight subcommittees were formed, and Otto was chosen as chairman of the subcommittee on biological chemistry, with P. A. Shaffer and A. P. Mathews (University of Cincinnati) as members (*15-10*).

A conference was held in Chicago, March 1–3, 1920. The program was jointly sponsored by three organizations: the Association of American Medical Colleges; the Council on Medical Education, AMA; and the Federation of State Medical Boards of the U.S. Otto presented the report on biological chemistry, having had one day to travel to Chicago from his Harvey Society lecture.

Physiological chemistry as part of the medical school curriculum must be considered an advanced subject, the report said. As a minimum the student needed a background of at least a quarter of a college year (i.e., one full course for an academic year) in inorganic chemistry followed by one eighth of a year in organic chemistry, both accompanied by laboratory work. A course in physical chemistry was not practical to require, though certain of its concepts would have to be taught in the medical school physiological chemistry, for example, the law of mass action. Training in analytical chemistry was desirable but this, too, would be partly covered in physiological chemistry.

These minimum requirements were not being met by the medical schools, the report continued. Some Class A medical schools offered a course in organic chemistry. This was better than having the student take a short course two or more years before he entered medical school. However, the time given to organic chemistry should not, as had been the practice, be at the cost of that allotted to physiological chemistry. Because the medical school curriculum was crowded, organic chemistry was best offered as an entrance requirement. The course in biochemistry (as in physiology) should be for a fourth of a college year.

Some provisional recommendations on the scope of the course in biological chemistry were 75–80 lectures and conferences, one hour each, and an equal number of laboratory sessions, two and a half hours long, except when written and practical examinations were given. The 75–80 lectures were used as follows: 1) two weeks in the laboratory to study standard solutions, the balance, and volumetric analysis with lectures on solutions, indicators and selected topics in physical chemistry such as mass action law, reversible reactions, and osmosis; 2) the third week in the lab to cover quantitative nitrogen determinations on pure nitrogenous compounds such as ammonium sulfate, urea, and uric acid. The lab work would allow the student to learn the quality of his work through quantitative recovery of expected values. The accompanying lectures could cover the topics of catalysis, leading to the consideration of enzymes. The first three weeks, which would cover 45 to 60 hours, would be the most exacting part of the course, training the student in precision and in paying attention to details.

Forty to 50 lectures and corresponding laboratory periods should then be devoted to the chemistry and metabolism of fats, carbohydrates, and proteins. Urinalysis could be given with protein metabolism. "In connection with the lectures, emphasis must be laid on the fundamental principles (biological, physical, and chemical) rather than on clinical illustrations and applications." A few lectures should be devoted to fuel requirements, energy metabolism, oxygen consumption, respiratory quotient, acid-base and neutrality regulation; and a few special lectures should be devoted to secretions of the body, particularly on milk, accompanied by quantitative laboratory work.

> The elementary first course in physiological chemistry, coming as it generally does during the second half of the first year, or first half of the second year, is not a course in pathologic chemistry, yet it includes the principles and methods on which the later applications to clinical materials are based.

Another problem the committee addressed was that of teachers. "Medical schools and the American Medical Association have been making commendable efforts to supply everything—except the teachers." The members of this committee (Folin, Shaffer, Mathews)—none holding an M.D. degree—found no fault with the recommendation of the Council on Medical Education (AMA) "to the effect that nonmedical men should be selected as teachers in medical schools only when medical men of equal special capacity are not available." However, the men desired were not available in biochemistry, nor would they be in the immediate future. As it was, the stricter regulations on granting the M.D. degree were strangling the laboratory departments such as physiology and physiological chemistry insofar as special students and assistants were concerned. "The difficulty is greatest in the field of biochemistry, because the biochemist must also have a respectable command of the various branches of pure chemistry, branches which are not taught in medical schools." There was urgent need for some constructive recommendations as to how the "great and growing demand for well trained and productive biochemists in medical schools and hospitals may be met." The older current professors of biochemistry in the U.S. had received at least part of their training in Europe. Schools in the U.S. had never been "self-supporting" in this subject.

Otto further expressed his views on staffing of a biochemistry department, particularly at Harvard, in the following letter, written in response to Abraham Flexner's request for such information (*15-11*):

Dr. Abraham Flexner
General Education Board
61 Broadway
New York, N.Y.

My dear Dr. Flexner,

Besides myself this department has one full time assistant professor and usually four teaching fellows and assistants. The latter are as a rule Ph.D. candidates in Biochemistry except that occasionally I get a man to remain a year after he has got his Ph.D. degree. At times I have had a research assistant giving all his time to research. I have one diener and one woman helper and from a third to one half time stenographer (at times none at all). And finally about a quarter of the time of a janitor. I have no technician and no other help except that a crew of scrub women come in and wash floors once a week. We have one, sometimes, two, "research fellows" without pay. These render no services of any kind: they are well to do men who simply like to have a place in which to work. I shall probably soon get another full time instructor or assistant professor.

We teach from 70 to 120 medical students and the amount of help obtained has not varied with the number of students. Before the war we usually had from 5 to 7 Ph.D. students and usually 1 or 2 foreigners, and, in the course of a year, perhaps 10 physicians remaining on the average, one month. Then we have usually 2 or 3 fourth year men each year coming in for one month's elective work. We give a six weeks' summer course, when we have a miscellaneous group of 10 to 16 students, but give no formal lectures at that time. I should perhaps add that I have never given any appreciable amount of time to the Medical School as a whole—to any of the numerous committees engaged in administrative work. I attend to the administration of my own department including the bookkeeping and the ordering of supplies, except local purchases attended to by the diener. (The latter is the most important member of my staff next to the assistant professor.) The teaching has been distributed very nearly equally between the assistant professor and myself; but of late rather more has been relegated to the assistant professor.

(I venture to suggest that liberal space and equipment including literature is more important than a large number of instructors and assistants for the teaching of an elementary course to medical students.) These students are intelligent and keen and will largely help themselves and each other if the conditions are right. Ph.D. candidates make better assistants than M.D.s yet none can really take the place of the experienced professor in the laboratory, partly because no one else takes quite the same interest in the course or gets the same attention from the students.

(From the standpoint of research and advanced teaching two or three professors, each capable of taking on one or two research students, is, in my opinion, far more fruitful than one professor with a large retinue of assistants and technicians; for the professor and students will then really work together, and, in the course of one or two years, the student will thoroughly understand the ideals and the points of strength and of weakness of his teacher. A good *departmental* library is indispensable.)

I am aware that the suggestions contained in the last two paragraphs were not called for in your letter.

Sincerely,

Otto Folin

P.S. If additional information is wanted please let me know. F.

Sometime in 1920 the Mayo Clinic offered Otto and Walter Cannon jobs at three times their current salary. For Otto, who was then earning $5000 per year, this staggering amount must have been ironic, if not tempting. Harvard was not paying leading salaries to its professors. Other heads of biochemistry departments were being paid a thousand more than Otto. Undoubtedly, he talked the matter over with Laura, and then with Cannon, who wrote of it, "we did not feel that we were making a sacrifice by refusing the increased pay; we chose to remain where we were assured complete liberty in following our personal research programs" (15-12). Otto had built a solid academic department and a center of biochemical teaching and research activity. Perhaps he could have accomplished as much elsewhere. But Harvard was Harvard. Would he have drawn outstanding students, assistants, and fellows in, for example, Minnesota or Missouri?

Otto was appointed "Chemist to the Hospital" in an advisory capacity to the McLean Hospital in 1920. The chemical and psychological laboratories had closed during the war. In 1921, a resident chemist was again hired.

In the autumn, Otto succeeded in enlarging the department of biological chemistry. Richard Bell now joined Cyrus Fiske as an assistant professor. Lyman remained as a research fellow. Joseph M. Looney, M.D., was an assistant, and Floyd De Eds, A.B., A.M., was a teaching fellow.

Otto produced not a single paper in 1921! One publication bearing Otto's name was explained in a footnote as follows:

> By agreement between Mr. Whitehorn and myself, this paper is published as Supplement II of the "System of blood analysis" devised by Folin and Wu— Professor Otto Folin. (15-13)

This paper will not be described here in great detail, because Otto was ostensibly not a part of it directly though the work was carried out in his department, and his "encouragement" was acknowledged.

The well-known "Whitehorn" procedure applied the Volhard method (from the Folin laboratory manual) for determining chloride in urine to the Folin-Wu filtrate. Chloride was precipitated as silver chloride from a standard excess solution of silver nitrate. The excess silver nitrate was then back-titrated with standard sulfocyanide to a salmon-red color using ferric ammonium alum as an indicator. Chloride concentration was readily calculated from the titer and the volumes used in the procedure.

"At Dr. Folin's suggestion, purification of tungstates containing added chlorides has been accomplished by recrystallization with alcohol." Chloride could be determined accurately in tungstic acid filtrates of blood, plasma, and urine.

Sometime late in 1920 the research fellow Hilding Berglund (1887–1962) came from Sweden to work with Otto. He had been educated at the University of Uppsala and then at Stockholm University Medical School where he obtained an M.D. degree in 1916 and an S.D. in 1920. He had become an assistant professor of medicine before deciding to obtain postdoctoral research experience in clinical biochemistry, not in Germany, but in the new mecca developing in the U.S. And who else would he choose but a compatriot, the eminent Professor Otto Folin?

Berglund spent the remainder of 1920 and a good part of 1921 studying the urinary output and changes in the sugar concentration of blood following intake of various carbohydrates. Later in the year he completed a study on amino acid intake in relation to urea formation.

In the autumn, Hilding Berglund became a departmental research assistant, and Mark R. Everett, D.Sc., was a teaching fellow, bringing the staff number in biochemistry to eight, a gain of two over the previous year. Cyrus Fiske was now studying the inorganic constituents in urine and published four papers in the *Journal of Biological Chemistry* in 1921 dealing with the determination of inorganic phosphate, as well as inorganic sulfate, total sulfate, and total sulfur in urine, observations on the "alkaline tide" after meals, and inorganic phosphate excretion.

At the 16th annual meeting of the ASBC in New Haven, Conn., Dec. 28–30, Otto presented a paper (coauthored with Hilding Berglund) on the "Normal Sugar Excretion in Relation to Carbohydrate Intake and Blood Sugar Fluctuations." Otto was again serving on the council of the society. For 1922, Donald D. Van Slyke was elected president, Philip A. Shaffer, vice-president, Walter R. Bloor, treasurer, and Victor C. Myers, secretary. With Stanley R. Benedict also on the council, the society was now governed by a group of extraordinarily dedicated, clinically oriented biochemists. They were lodestars guiding biochemistry into the medical school curriculum, the men who were establishing, through their research and students, clinical chemistry as a service to hospitals and as a health science. While the AMA had, as Otto's subcommittee noted, recommended that physicians serve as educators in biochemistry, the main source for this discipline was rooted among those who had obtained doctorates in chemical sciences.

References

15-1. Folin, O., and Wu, H., A revised colorimetric method for determination of uric acid in urine. *J. Biol. Chem.* **38**, 459–60 (1919).

15-2. Folin, O., Denis, W., and Minot, A.S., Lactose, fat, and protein in milk of various animals. *J. Biol. Chem.* **37**, 349–52 (1919).

15-3. Folin, O., and Youngburg, C.E., Note on the determination of urea in urine by direct nesslerization. *J. Biol. Chem.* **38**, 111–12 (1919).

15-4. Folin, O., and Peck, E.C., A revision of the copper phosphate method for the titration of sugar. *J. Biol. Chem.* **38**, 287–91 (1919).

15-5. Folin, O., and Wright, L.E., A simplified macro-Kjeldahl method for urine. *J. Biol. Chem.* **38**, 461–68 (1919).

15-6. Folin, O., Determination of ammonia in blood. *J. Biol. Chem.* **39**, 259–60 (1919).
15-7. Folin, O., and Wu, H., A system of blood analysis. Supplement I. A simplified and improved method for determination of sugar. *J. Biol. Chem* **41** 367–74 (1920).
15-8. Folin, O., Blood analysis and its applications. The Harvey Society Lectures, 1919–1920. pp. 109–20. Philadelphia, Pa.: J.B. Lippincott Co.
15-9. Hastings, A.B. *Ann. Clin. Lab. Sci.* **4**, 217–18 (1974).
15-10. Folin, O., Shaffer, P.A., and Mathews, A.P., "Report on Teaching of Biochemistry." Proceedings of the Thirtieth Annual Meeting, Association of American Medical Colleges, Chicago, Mar. 1–3, 1920, pp. 107–117. Abstracts of the proceedings were published in the *J. Am. Med. Assoc.* **74**, 825 (1920).
15-11. Letters are from the Rockefeller Archive Center, North Tarrytown, N.Y. (General Education Board, Box 700, Folder 7211).
15-12. Cannon, W.B. *The Way of an Investigator.* p. 208. New York: Hafner Publishing, 1965.
15-13. Whitehorn, J.C., A system of blood analysis. Supplement II. Simplified method for the determination of chlorides in blood or plasma. *J. Biol. Chem.* **40**, 449–60 (1921).

Chapter 16. Metabolic Studies: MLIC (1922–26)

In 1922, Otto published nine papers and the third edition of his laboratory manual. Three of the papers were based on Berglund's work.

The first was a short but important paper on determining sugar in urine quantitatively (*16-1*). It was financially supported by a grant from the Swedish Society for Medical Research. This paper was important for reasons other than its original purpose. It introduced the use of Lloyd's alkaloidal reagent (concentrated fuller's earth; kaolin) to adsorb creatinine, creatine, uric acid, and most of the coloring matter from urine. However, it did not adsorb sugar. This is noteworthy here because the same reagent would be applied later in the year to remove added creatinine from blood samples by none other than Stanley Benedict and his distaff colleague, J. A. Behre [*J. Biol. Chem.* **52**, 11–33 (1922)]; and it would also be used quantitatively for this purpose by O. H. Gaebler a few years later. Their use of Lloyd's reagent, however, led them to the false assumption that there was no creatinine in normal blood.

A second lasting contribution of this short paper was that Otto and Hilding found that dilute and concentrated solutions of glucose could be preserved almost indefinitely at room temperature in 0.3% (w/v) benzoic acid solution.

The procedure for performing urinary sugar determination was quite simple, and would be important for the next paper with Berglund. Five milliliters of 0.1 N H_2SO_4 and 10 mL of water were added to 5 mL of urine. Then 1.5 g of Lloyd's reagent was added, shaken gently for 2 min, and filtered. Two milliliters of the filtrate was used as in the Folin-Wu blood sugar method.

For "total" sugar, hydrolysis was performed with 10% HCl. The solution was heated in a boiling water bath for 75 min, then cooled and neutralized. The "Lloyd's reagent" was especially prepared and supplied by a chemist in Cincinnati, Ohio, J. U. Lloyd.

The second paper from the Swedish pair was the longest to date for Otto, and it dealt with the "rise and fall of sugar in blood and urine following the intake of glucose and other carbohydrates" (*16-2*). They wanted to determine whether there was a relationship between the concentration of sugar in blood and its elimination. Because there were also carbohydrates bound to other substances in urine and blood, sugar was determined before and after hydrolysis. The sugar in red blood cells and plasma was also studied.

For obtaining plasma and removal of its proteins, the procedure used was of historical significance. Blood was collected via a needle in a cubital vein and placed into carefully paraffined centrifuge tubes. No oxalate was added at this stage. Four minutes after the blood collection was completed, the tubes were rapidly centrifuged to separate plasma. No clotting occurred. This process took less than 8 min. To prevent subsequent clotting, 15 mg of potassium oxalate was added per 10 mL of plasma (as for whole blood). To precipitate the proteins in plasma, only half the customary volume of 10% tungstate and $\frac{2}{3}$ N H_2SO_4 was used.

The fate of ingested carbohydrates was unknown. In this pristine period of medical and biochemical history, much was being made of the terms "renal threshold," "alimentary glycosuria," and "glycuresis," the last a term Benedict had introduced to express the increase of sugar elimination above the control periods. The kidney's tubular reabsorptive capacity for glucose had not yet been discovered, so the concept of a threshold was universally accepted. In view of the discoveries to come within the next few decades the questions about the metabolic fate of the sugars seem simple and perhaps superficial. A certain amount could be metabolized and converted into glycogen. Another amount could, as Otto and Hilding believed, be stored in tissue, not in the form of glycogen but as free glucose just as it was apparently stored in the red blood cell. A portion could be bound to non-carbohydrate substances, or be converted to a di- or polysaccharide. Therefore, hydrolysis should tell how much as well as which sugar was involved. Then, urinalysis could elicit the amount of sugar excreted, and whether the excretion of one sugar was influenced by the ingestion of another.

Otto and Hilding performed all of their testing of sugars on humans, with Hilding leading the way. Following ingestion of the sugars, blood and urine samples were taken at selected intervals. "Fasting" samples were also taken. As mentioned the sugar was determined in each sample, before as well as after hydrolysis. The reducing sugar content of the red blood cells was obtained by calculation from values in whole blood, in plasma, and of the hematocrit. Sugar in the urine was expressed in terms of amount excreted per hour.

The carbohydrates tested (mostly on voluntary medical students) were glucose, maltose, dextrin, starch, fructose, galactose, and lactose. By hindsight it all seems so simple! Otto and Hilding had reviewed the literature, and except for the fate of glucose, there appeared to be little to inform them about the other carbohydrates. None of the carbohydrates raised the reducing sugar level in blood to the extent of glucose, which could presumably rise to the renal threshold values.

There was a "glycuresis" after ordinary meals attributable to "unusable" carbohydrates in the diet other than glucose, maltose, dextrin, or starch. This included pentoses and substances in grains, vegetables, and fruits, as well as decomposition products from cooking, canning, and baking—a motley variety of di- and polysaccharides. Apparently, protein or fat ingestion did not directly affect the blood sugar concentration. Hypoglycemia could occur, however, with abundant fat intake, but this was attributed to a decreased *need* for sugar transport from tissue to tissue, because they were already loaded with sugar.

Berglund conducted some interesting experiments on himself. He swallowed 200 g of glucose in a liter of water and observed the expected rise in sugar in whole blood, plasma, and corpuscles. Six hours after he ate a protein-fat meal there was no glycuresis. He drank a solution of 200 g of maltose in 600 mL of water and found that glucose rose slightly in plasma though not in whole blood and dropped in the corpuscles, with no glycuresis until he ate his dinner about 6 h later. On another day, he drank a solution of 200 g of dextrin in 800 mL of water but could find little change in anything except the glycuresis portion and the hydrolytic fraction in urine, which stayed elevated for several days. This delayed excretion was not due to delayed absorption, which Hilding proved by taking another 200 g dose of dextrin followed 2 h later by 45 g of caster oil! "The bowel was thus thoroughly cleaned out in the course of about 5 hours." The hydrolytic fraction remained elevated into the second day. This fraction, however, was probably due to some "denatured, indigestible, and unusable carbohydrate," not to dextrin. To prove it the indomitable Swede swallowed 175 g of raw potato starch in 600 mL of water and found no increase afterwards in the hydrolytic fraction. But this was not considered conclusive, so he took 200 g of a dextrin-starch mixture (partially digested starch), with similar results.

Berglund's bouts with fructose ingestion were at first somewhat disturbing. Upon taking 200 g of fructose there was a slight rise in plasma sugar and a glycuresis lasting several hours not due to fructose. This rise was attributed to impurities in the fructose. To prove it, 200 g of pure levulose dissolved in 300 mL of water was autoclaved at 145 °C for 1 h. The levulose was destroyed to the extent that it lost its sweet taste. Hilding took a small sample and observed no unpleasant symptoms of any kind beyond a "negligible" diarrhea. The following day he took the equivalent of 150 g of the solution.

> In about 1 hour a most violent diarrhea began and lasted for several hours. Considerable nausea was also experienced, but no headache or other intoxication symptoms. We deem it important to call attention to the vigorous laxative effects of decomposed levulose solutions. Maple sugar, molasses, and certain candies doubtless owe their laxative effects to similar decomposition products of sugar. . . . Because of the unexpected and dramatic outcome of this experiment we unfortunately neglected to pursue the original purpose for which the experiment was made as long as we should have done.

The urine samples they did manage to obtain "between runs" showed marked glycuresis. Fearlessly, Berglund took 200 g of another manufacturer's fructose that, in powder form, looked white enough, but in solution was much darker than similar solutions of the first manufacturer's fructose. "The diarrhea which followed the taking of this product was almost as intense as that obtained from the levulose which had been decomposed in the autoclave."

The investigators wondered at the fate of the fructose ingested and concluded that it raised blood sugar slightly and, like glucose, was absorbed by tissues. Its absence from urine was attributable to a renal threshold for fructose. Fructose might prove useful in treating diabetes because it did not increase the blood sugar level to any extent, and might be better utilized. The fructose would have to be given in small doses at frequent intervals to avoid excessive use of fructose in the tissues.

Berglund next took galactose in separate doses of 10, 30, 30, and 100 g, and determined the glycuresis for up to 8 h later. This was repeated for lactose on doses of 10, 30, 45, 50, and 200 g. Galactose was assimilated "very imperfectly" in that large amounts, as much as 10 g from the 100 g dose, were excreted. With the 100-g dose, despite the loss in the urine, the blood sugar rose slightly, about equal to that when fructose was taken. Its slow rate of conversion to glycogen and dextrose might make galactose useful to the diabetic. It apparently had no renal threshold. Lactose behaved similarly insofar as raising the blood sugar, and it gave rise to lactosuria and galactosuria. Of interest was the fact that when Hilding took a mixture of 100 g of galactose and 100 g of glucose, practically all of the galactose was retained. This fact was particularly important in explaining why milk-sugar was advantageous for the young. The excretion of excess galactose was prevented by glucose, an advantage because galactose was needed for the building of nerve tissue. Less lactose and some pure glucose or maltose might be better than lactose alone for infants "whose rate of growth is subnormal or whose urine contains sugar." Galactose rather than milk could be useful to nursing mothers whose milk was low in lactose.

It must be mentioned that Berglund tested the effects of protein and fat on himself. In one trial, he took 1000 g of raw egg white, and in another, 135 g of gelatin in 900 mL of water (no other food taken all day). In another trial he took 200 g of olive oil and 250 mL of water. Otto's contribution apparently was the testing of his "fasting" blood sugar and at intervals after one "mixed" lunch.

Otto and Hilding believed that the red blood cell water had a higher concentration of sugar than plasma and that part of this was converted into a polysaccharide other than glycogen. On hydrolysis their hypothesis was substantiated. Of course, they were not aware of the sugar phosphates that were present to account for the effects of hydrolysis. It would later be shown by others that the glucose in the red blood cell could be accounted for solely on the basis of its water content, and that there was no elevation in its concentration compared to glucose in the water of plasma (*16-3*).

This paper was picked to shreds by Stanley Benedict [Benedict, S.R., and Osterberg, E., Sugar elimination after the subcutaneous injection of glucose in the dog. Including a discussion of the paper on observations on carbohydrates by Folin and Berglund. *J. Biol. Chem.* **55**, 769–94 (1923)]. His paper was a point-by-point acid commentary on most of the interpretations made by the Harvard pair, some of it valid, much of it merely unscientific counterargument, and mainly immoderate and unnecessary. Otto's third paper of 1922 was on the determination of amino acid nitrogen (AA-N) in blood (and urine) (*16-4*). Although the work had begun in 1919, Wu was unable to complete it before he left, and Otto had to do much more with it before submitting it for publication. Otto not only wished to determine the AA-N in blood and urine, and in milk, gastric contents, protein hydrolysates, meat extracts, food, and medicinal preparations, but he wished to clarify further the contents of the undetermined nitrogen and possible polypeptides in the Folin-Wu filtrate.

In examining the literature for a color reaction characteristic of amino acids, Otto found an old observation (Wurster) that o-quinone gave a color reaction with AA and proteins. Though o-quinone itself proved unsuitable, it spurred Otto into a study of o-quinone derivatives. The one that proved most suitable was already in his stock of chemicals, β-naphthoquinonesulfonic acid (BNSA), which had long been known to give a bright red precipitate with aniline. The fact that it gave a highly chromatic and decidedly stable color with amino acids had escaped previous discovery. The reaction with aniline derivatives was known to occur as follows:

$$\text{(BNSA)} + RNH_2 \rightarrow \text{product} + SO_2 + H_2O$$

Among all of the nitrogen products in the Folin-Wu filtrate the only known interference present was NH_3, which was readily removable. BNSA did not react with indole, but would form a blue color with it in strong alkali (a fact that would be used later), whereas it gave a red color with AA.

Otto established the conditions for the method. Glycine was used as the standard. The alkalinity was adjusted with sodium carbonate. Otto provided a detailed description of how to prepare BNSA from β-naphthol, a grand project for an old, experienced organic chemist. The PFF (either 5 mL, or preferably, 10 mL) was allowed to react with freshly prepared BNSA in the dark for 19–30 h. Excess reagent was removed with sodium thiosulfate, which did not affect the colored complex with amino acids in the presence of previously added acetate-acetic acid (buffer) solution, which not only enhanced the color but prevented turbidity from sulfur that could be generated by the thiosulfate.

Otto noted the shortcomings of the method. It reacted with only 16% of the arginine, and results were excessively high for proline and tryptophane. Obviously a procedure requiring 19–30 h standing was not to his liking. [This important method was later simplified and extended by Otto's last graduate student, I. S. Danielson (16-5), and more recently by Frame, Russell, and Wilhelmi (16-6). It became a highly useful procedure on 0.2 mL blood requiring less than an hour to complete.]

It must be noted that Otto had one other important reason for introducing this method. He was still eager to determine the metabolic fate of amino acids, particularly in the liver, and whether it related to urea formation.

Otto immediately applied his new method to urine (16-7). Ammonia was first removed with permutit. Because the concentration of amino acids and ammonia varied greatly in urine, it was not always easy to select the proper amount for analysis on the first trial. Excretion of amino acids, however, was substantially independent of volume (4–12 mg/h), so one could almost always get suitable amounts for the color reaction. Once the NH_3-free urine or a dilution of it was prepared, the procedure followed was the same as for blood filtrates, with some changes in the volumes used.

Now Otto turned again to the question of amino acids and urea formation, this time with Hilding Berglund (*16-8*). This work was done in conjunction with that on carbohydrates.

Previous work by Folin and Denis had shown that deamination and protein regeneration were not features of amino acid absorption; and there was experimental proof that urea formation was not localized in the liver. Van Slyke had confirmed the first part, but maintained that urea formation occurred especially in liver. But Otto and Hilding pointed out, as had Fiske and Sumner, that Van Slyke's data could be interpreted differently. The extraordinary disappearance of amino acids from liver was not caused by urea formation but due to the enormous supply of blood in liver, which initially accounted for its rich amino acid content, and from which the muscles and other tissues abstracted much of the amino acid content.

Berglund's feeding experiments with carbohydrates were now extended to testing the relation of amino acid concentration to urea in blood. When he ate a mixed diet that was not excessively rich in protein, there was a slow rise in the values of the blood amino acids, which accompanied an expected increase in the amino acid content of muscles. When he ate 135 g of Knox gelatin, an easily digested protein, the plasma amino acid nitrogen doubled in 2 h from 5.5 to 11 mg % (3.9 to 7.9 mmol/L), while the urea nitrogen sank from 17.3 to 14.2 mg % (12.4 to 10.1 mmol/L). Then, in the next 1¼ h, the AA-N sank to 7.2 mg % (5.1 mmol/L) and the urea nitrogen rose to 20.2 mg % (14.4 mmol/L). This trend continued for several hours. The urine showed similar data. Had the liver been involved, our authors maintained, the urea nitrogen should have risen rather than fallen in blood in response to the increased amino acid concentration. The rise in urea nitrogen occurred when the AA-N fell, indicating that the tissues were absorbing the amino acids as urea synthesis occurred. What further proof was needed? "We therefore refrain from referring to the mass of other evidence showing that the urea production continues when the liver is virtually eliminated, damaged, or destroyed." However, the work of Bollman, Mann, and Magath [*Am. J. Physiol.* **69**, 371 (1924)] established more convincingly that the liver truly was the main site of urea formation. They removed the liver from dogs and showed that amino acids in blood rose steadily while the urea nitrogen declined.

Folin and Berglund, despite their incorrect reasoning about urea synthesis in the liver, correctly affirmed that amino acid excretion could serve as a test of liver function. When Hilding ingested 0.25 g of glycine, they noted that the amino acid nitrogen rose significantly in 20 min in blood and plasma and that the urinary nitrogen rose steadily for the next 4–5 h. This was not tested, however, on patients with known liver disease.

When Hilding ingested glycine, he lost glycine in the urine. When a subject was given a mixture of amino acids (predigested casein), the blood AA-N concentration rose significantly in 45 min. The AA-N in urine rose also. This increase in blood, plasma, and urine was confirmed in two other subjects given predigested casein in one case and casein in suspension in the other. Evidently there was no threshold for the amino acids.

In contrast to a recent assertion that the red blood cell was free of NPN constituents, Otto and Hilding correctly believed that their data proved "beyond possibility of doubt that for the nitrogen constituents human blood corpuscles are as permeable as we have previously shown them to be permeable for glucose." The higher concentration of urea in erythrocytes than in plasma following a heavy intake of protein they interpreted wrongly as meaning that the corpuscles synthesized urea.

Otto had many inquiries about a source for sodium tungstate. The manufacturer he had cited had ceased making it. Now he passed on his recent experiences with tungstate in a brief note (*16-9*). He had encountered a tungstate that was not excessively alkaline, yet gave erroneous figures in blood analyses. It was found to contain "acid tungstates" that produced clear filtrates with blood yet gave low total nitrogen values and more or less completely removed uric acid and creatinine. These acid tungstates were poorly soluble, and Otto provided a method for their rescue by converting them to the simple sodium salt via titration with sodium hydroxide.

Otto and J. M. Looney, the departmental assistant, published an important paper on determining tyrosine, tryptophane, and cystine separately in proteins. The colorimetric method for tyrosine proposed by Folin and Denis in 1912 gave erroneously high results because tryptophane and its decomposition products produced the same blue color as tryosine with the phenol reagent. This investigation was undertaken to remove the tryptophane error, but it soon "became evident that the color due to tryptophane and its decomposition products is quantitative for the amount of tryptophane present." Therefore, tryptophane and tyrosine could be measured separately.

T. Zucker, an assistant a few years previously, had observed that cystine in the presence of sodium bisulfite gave an intense blue color with the *uric acid reagent*. The reaction was evidently due, as Heffler had shown, to reduction of cystine to cysteine by the sulfite. Looney's task was to make this quantitative (Part II, p. 418).

Hsien Wu applied an improved Folin-Denis phenol reagent to determining plasma proteins based on their tyrosine content [*J. Biol. Chem.* **51**, 33–39 (1922)]. He published two later modifications with S. M. Ling [*Chinese J. Physiol.* **1**, 161–68 (1927); **2**, 390–402 (1928)].

Otto presented a review of the "state of the art" of the knowledge of the NPN in blood (*16-10*). As this was the main arena of his own scientific work it provided an insight into his personal views of the progress that had been made, often catalyzed by his own efforts.

Otto's final paper of 1922 was an update and standardization of his method for determining uric acid in blood (*16-11*).

It had been reported that uric acid added to sheep blood could not be quantitatively recovered, a fact that Otto confirmed. The solution, however, was not to heat the precipitated blood as proposed but to mix thoroughly on addition of acid and when possible to make a 1:20 dilution of blood rather than a 1:10; and to use plasma rather than whole blood. Otto, however, was not able to overcome an 8–10% loss of uric acid added to sheep blood.

Because of the major improvement in the uric acid procedure (introduced by Benedict), the use of cyanide to complex with the silver and to intensify the color reaction with the uric acid reagent, Otto was forced to find a better way to preserve his standard. The sulfite in the standard hindered the intensifying effect of cyanide. So Otto returned to the preservative he had used in 1913, formaldehyde. To avoid the formation of addition products of uric acid with formaldehyde, Otto found the best concentration to lie between 25 and 50 mL of formaldehyde per liter, a concentration that prevented both "chemical" and bacterial decomposition.

The standard (1 mg/mL) was prepared by dissolving exactly 1 g of uric acid in water heated to 50 °C and containing about 0.45–0.5 g of dissolved lithium carbonate. This was cooled, diluted to about 500 mL, and then 25 mL of 40% formaldehyde was added and mixed. The solution was acidified with 3 mL of glacial acetic acid and then diluted to 1 L. It was stored in full, small bottles in a dark place at room temperature and diluted as needed (usually 1:250 or 1:10).

The uric acid reagent had been used practically without modification since Folin and Denis introduced it in 1912. Wu had made a thorough study of it in 1920. [Wu, H., Contribution to the chemistry of phosphomoldybdic acids, phosphotungstic acids and allied substances. *J. Biol. Chem.* **43**, 189 (1920)]. Others had tried to improve the reagent, but Otto found none better from the standpoint of accuracy and convenience.

Although Benedict had incorporated arsenic acid in the uric acid reagent so that it could be applied directly to the filtrate without isolating uric acid with silver lactate, Otto had not yet tried it. If it worked it would be superior to the Folin-Denis reagent. However, Benedict had used other conditions that Otto found immediately useful. NaCN could be used as the alkali, eliminating the need for the carbonate. More color was produced this way. Also, the application of heat (water bath) intensified the color. With these conditions it was now possible to determine uric acid in 5 mL of filtrate (0.5 mL of blood).

Details for obtaining better results with the silver lactate procedure were also given in that paper, but this method was on its way to perdition. Otto discussed his experiences in detail with these new modifications, reagents, and interferences. Benedict's work had brought major improvements.

The third edition of Otto's *Laboratory Manual of Biological Chemistry* (With Supplement) was completed in 1922, though it appeared in 1923. The page, type, and margin sizes were about the same so that its growth since 1916 from 188 to 301 pages (printed one side only) was genuine. New methods in the quantitative section (supplement) included phosphates, sulfates, total base, sugar and amino acids in urine, and chlorides and amino acids in blood. Cyrus Fiske had made important additions to the discussion of solutions and the use of indicators.

In the autumn, Harry C. Trimble, Ph.D., replaced Joseph Looney as an assistant in the department, and would also become a close friend of the Folin family. Sidney W. Bliss, B.S., replaced Floyd De Eds as a teaching fellow. Whether Richard Bell, the brilliant young assistant professor, was actually helpful to the department from this time on is doubtful. The absence of publications from him after 1921 indicated that he was already in the throes of the prolonged fatal (but unrecorded) malady from which he died in 1925.

In 1922 Folin was appointed to a committee of the National Research Council's Division of Chemistry and Chemical Technology. This six-man group (P. A.

Levene, chairman; J. J. Abel; H. Dakin; O. Folin; W. A. Jacobs; and T. B. Osborne) was the American committee on nomenclature of biological chemistry. It represented the U.S. in the commission on nomenclature in biological chemistry of the International Union of Pure and Applied Chemistry (IUPAC).

Otto served until 1925 when the entire committee resigned because the international body took "certain actions" not in accord with the committee's views. A new American committee was subsequently formed, but prior members were evidently uninvolved.

In 1921, Dr. Augustus S. Knight, medical director of the Metropolitan Life Insurance Co. (MLIC), had offered Otto Folin the directorship of its Home Office laboratory in New York City. Almost certainly they had met to discuss this possibility, and the idea that a new laboratory would be built emerged from this discussion. But Otto had declined the offer for the same reasons that he had passed up the offer from the Mayo Clinic.

The Home Office of MLIC had established its first chemical laboratory in 1905. Most of the work was apparently urinalysis, with emphasis on qualitative testing. The importance of urinalysis in the granting of life insurance policies had been emphasized in a detailed report that Director Knight made in his president's address at the 1921 annual meeting of Association of Life Insurance Medical Directors of America (ALIMDA) (*16-12*). The report, which covered the decade of 1905–15, included an investigation of the mortality experience on the 6904 applicants rejected by MLIC because of the presence of albumin, albumin and casts, or sugar.

At the same meeting, Elliot P. Joslin had given a thorough discussion of "Diabetes and Life Insurance." An increasing number of diabetics were being discovered by life insurance companies.

Joslin had spoken at length about the laboratory tests on urine and blood for the diagnosis of diabetes. He recommended the Benedict test on urine, but considered it a conservative test tending to give false positives. Newer quantitative tests on urine were coming into vogue: in particular the recent tests of Benedict and of Berglund. Benedict removed interfering substances from urine with mercuric nitrate, whereas Berglund used Lloyd's reagent.

> All of us, medical insurance directors, and practitioners, should take cognizance of these new tests, because they may be of great advantage, not only in diagnosing the case of diabetes, but in discovering the potential diabetic. The advantage of these tests is that they do not in the least annoy the patient.

Joslin had mixed feelings about the use of blood sugar determinations in detecting diabetes. They were less satisfactory than urine tests because they were more complicated and far more prone to error. He had already learned of serious mistakes made in reputable laboratories, and others with experience similar to his own had confirmed this.

> The tests are in their infancy; they are not yet sufficiently simplified. Excellent technicians simultaneously examining the blood will obtain somewhat diverse results even if the same method is employed, and if different methods are followed the divergence is still greater. Fortunately there are several excellent tests: the Folin test, and the Benedict test with its various

modifications, and in Europe Bang's method is extensively employed. The Benedict test gives the highest results for blood sugar, next comes the Folin test, and the Bang method gives the lowest results.

After Otto's meeting with Augustus Knight sometime in 1921, what eventually emerged in 1923 was that Otto agreed to serve as his laboratory director. This was a commitment without reservations, though it hardly impaired Otto's duties, purpose, and loyalty to Harvard, as we shall see. Knight literally asked Otto to write his own terms of service, and by doing so could not have demonstrated a higher esteem for him. He would pay Otto at least twice the annual salary of his professorship.

Otto probably asked for several costly conditions that would ensure the success of his directorship and of the future laboratory, and Knight met each of them: 1) that Stanley R. Benedict be asked to serve as a joint director, 2) that a new laboratory be built for research and service, according to Folin's design, and 3) that he be allowed to hire a chemist to be in charge of the research laboratory. There was no implication that Otto and Stanley would have to spend any regularly scheduled time in the New York laboratory, except as they saw fit.

Knight's objective was obviously to carry out research that would upgrade the quality of chemical laboratory testing used by his company for applicants and physical examinations. Perhaps it had been further prompted by Joslin's lecture, or even from direct communication with him. Following Knight's action, the entire industry would shortly follow suit in upgrading its clinical laboratories.

On April 24, 1923, Haley Fiske, president of MLIC wrote the following company report:

> At a meeting of the Board of Directors held this day, OTTO FOLIN, Ph.D. and STANLEY R. BENEDICT, Ph.D. were appointed Directors of Laboratory, taking effect on April 1st and May 1st respectively; and FRANCIS B. KINGSBURY, Ph.D. was appointed Chemist-in-Charge of the Laboratory, taking effect July 15, 1923.
>
> Doctors BENEDICT and FOLIN are Biochemists of great international reputation, the former having been Professor of Chemistry at Cornell University Medical College since 1913 and the latter Professor of Biological Chemistry at Harvard Medical School since 1907. They will continue their Professorships and their work at these Universities and the time and efforts given by them to this Company will be largely in the way of Direction and Consultation with reference to research and routine chemical problems of our Home Office Laboratory. They have already devised two new methods for making quantitative tests for sugar and albumin that we are using successfully in our Laboratories and that will give us quantitative results that can later be studied accurately and profitably by the Statistical Division. These men, with Dr. KINGSBURY as the Resident Chemist, will doubtless work out many problems of research to the advantage of the Company, of the policyholders and of all who are connected with the business of life insurance.
>
> Dr. FRANCIS B. KINGBURY was graduated from Harvard University in 1909 with distinction in Chemistry, obtained his degrees there of A.M. in 1912 and Ph.D. in Biochemistry in 1914, was Instructor in Physiologic Chemistry at

the Medical School of the University of Minnesota from 1913 to 1917 and since his return from Military Service in 1919 has been at the University of Minnesota in his capacities of Assistant Professor and Associate Professor. He has already demonstrated signal ability in published results of his research works on biochemical problems and he certainly comes to this Company well equipped for the duties of his office.

Certainly Otto had contacted Benedict in 1922 concerning his becoming a codirector of the laboratory. The two chemists discussed writing a paper together covering the analysis of sugar and albumin in urine that would become the company's standard of performance in its laboratories. Benedict dealt with problems of sugar analysis and Otto with albumin. The paper—Otto's only publication for 1923—would serve as the theme of their discussion at the ALIMDA meeting in the autumn. They undoubtedly discussed in detail the research problems that Kingsbury would initiate. Though Benedict was an individualist, and a seeming rival and continual critic of Otto's, he was actually an admirer and friend (*16-13*). They could hardly fail to touch upon a constant theme of their lives, the non-protein nitrogenous constituents of blood and urine, their most recent research, particularly on uric acid. Otto was now heavily involved with Berglund, Derick, and others on a long study of uric acid metabolism in several animals and in man.

The paper that Otto and Stanley wrote in 1923 for the Metropolitan Life Insurance Co. was entitled "The Application of Quantitative Chemical Methods in Examinations for Life Insurance." The paper was noteworthy for the insurance industry (and anywhere else appropriate), because it instituted quantitative chemical testing of sugar and albumin in urine on an objective basis, replacing the divergent personal interpretation that was so important in qualitative testing.

Specific directions were provided for testing under uniform conditions, with prescribed glassware and standards. Sugar was tested by Benedict's method using reduction of picric acid in alkaline medium to picramic acid, and results were compared with those obtained with pure glucose solutions converted to picramic acid standards. A stock picramic acid solution was stable for a long time and the diluted standards for at least one month. The dilutions of the stock were made so that they visually checked against the glucose standards carried through the procedure. The method used was adapted from Benedict's article in the *Journal of Biological Chemistry* [**48**, 51 (1921)].

The procedure on urine was simple, requiring 1 mL to be mixed with 3 mL of 0.2% picric acid solution and 5 drops of 5% acetone. This mixture was heated in a boiling water bath for 12–15 min, cooled, and diluted to 25 mL, then mixed and compared with the standards. The reaction was valid for sugar concentrations to 0.5%. Greater urinary concentrations required predilution.

For albumin determination, the sulfosalicylic acid test of Folin-Denis (1914) was described in detail. The standard was sheep blood serum diluted 1 + 7 with 15% (w/v) NaCl solution. This diluted serum, considered 1% (1 g/dL) and presumably stable for months, was diluted further for use in the procedure. One milliliter was diluted to 25 mL with 2% (w/v) sulfosalicylic acid and allowed to stand for 10 min; the turbidity was then compared with the standard to obtain the protein concentration.

Otto and Stanley recommended that if a urine had a specific gravity of 1.015 or less, then 0.2% of sugar or less could be accepted as normal; and if it had a specific gravity of 1.025 or more, then sugar up to 0.3% could be accepted as normal, in the absence of other information indicating diabetes. For specific gravities between 1.015 and 1.025, sugar concentrations up to 0.25% could be accepted as normal, barring any reason to suspect diabetes. (These permissible "sugar" concentrations reflect the poor state of the analysis at this time).

When the applicant's urine showed more glucose than indicated for a normal finding, then 100 g of glucose or an ordinary meal should be given and another urine sample tested about 2 h later, unless the sugar value was already so high as to preclude further examination.

The phenylhydrazine test for sugar could also be used for testing urines that showed excessive sugar, as could quantitative copper titrations, but no procedures were provided.

Otto and Stanley cautioned that the tests recommended could not prevent fraud, nor could they deter rejection of subjects with renal glycosuria, or those who excrete sugar without having a disease. For such cases, as well as for untreated mild diabetics, blood sugar determinations should be done, with the upper limit (fasting) considered as 110 mg/dL (6.11 mmol/L).

The two laboratory directors did not recommend any other test for albumin. Amounts of albumin exceeding 0.1 g% were to be considered distinctly pathological, whereas quantities below 0.05 g% were considered the same as the former interpretation of "slight trace."

By Oct. 3, 1923, Kingsbury was in New York City at his new job and by the end of the year, the new laboratory at MLIC was fully operating. Kingsbury was at work on a number of projects that will be revealed shortly.

At the 34th annual meeting of ALIMDA, which took place in Hartford, Conn., on Oct. 18–19, Otto and Benedict gave detailed oral reports on their brief history as directors of the Home Office Chemistry Laboratory and their findings and recommendations for a future course of action for the insurance industry.

Benedict indicated that he and Otto had found that "laboratory conditions were very unsatisfactory." Was this criticism valid from the practical standpoint of insurance or merely academic? They considered the laboratory at MLIC a leader in the industry. Yet some very fundamental problems had not been tackled. First came the question of preserving urine. A large percentage of samples brought to the laboratory were decomposed, yet these samples were being analyzed for sugar, albumin, casts, etc., and the results reported. Boric acid was being used as a urine preservative though it was useless for that purpose. Finding an answer to this needed immediate attention because neither he nor Otto had one to recommend.

Methods employed in the laboratories and the interpretation of results varied widely. A slight trace of albumin recorded in one laboratory was considered "no albumin" in another. Considering such borderline cases and their interpretations, what good were statistical inferences about them? Yet such inferences were being made and reported. Running and interpreting urinary sugar was just as bad or worse. The test procedure involved Fehling's solution. It was heated, and then a small, unmeasured amount of urine was added ("poured" in). The tube then stood, before results were read, for 2 h in one laboratory and 24 h in another!

Benedict then discussed the problems of analyzing urine sugar qualitatively and quantitatively. All urines contained some sugar along with interfering substances that could either inhibit the reaction or react directly with Fehling's solution. Without considering which sugars were present (other than glucose), a relatively simple procedure could be recommended. He then demonstrated the picric acid procedure using two samples of urine, one which had given a positive Fehling's reaction and the other a negative reaction in the laboratory at the MLIC, yet both contained the same amount of sugar. "The big thing has been the elimination of the reaction of interfering substances." The acetone in the reagent eliminated the interference caused by creatinine.

If everyone followed this procedure exactly, and uniformly used the same reagents and glassware, then all should get identical results; and consistency would be achieved.

Otto then spoke. A general finding about the laboratories of life insurance companies was that their chief workers were not laboratory men very long. These "chiefs," no matter how expert, ended up passing their laboratory work on to others who had little training and experience. The expertise (if any) was lost. No progress was possible. The atmosphere was wrong, sterile.

At the laboratory of the MLIC, a committee (unspecified) had recommended that changes be made in the chemical tests used to make them quantitative. Benedict discussed the sugar method. Now Otto would discuss the methods for urinary albumin determination.

Otto then explained how it was practically impossible to interpret the nitric acid test quantitatively (Heller's ring test for "albumin"), since its width varied with the contact of the nitric acid layer and the urine. The reaction was also possibly misinterpreted because of the presence of interfering materials such as potassium iodide, resinous material, and uric acid.

Otto then explained the use of the quantitative determination adopted by the committee, the sulfosalicylic acid method that employed the use of albumin standards for comparison.

A discussion of the two papers was then held. Dr. Exton said, "I feel sure that their association with life insurance is going to help get things done which I have been trying to do ever since I came into life insurance from clinical medicine." Exton had kept up with the literature, and had been using the Lloyd's alkaloidal reagent that Otto had recommended to remove interfering substances, along with Benedict's copper test, which he preferred to the picric acid procedure. Exton did not like Otto's albumin test because it was not a clinical test and unsuitable for use by the examiners but was intended to accompany the qualitative test. The standards used were not really good standards. He had himself obtained a set of solid turbidity standards suitable for rough work.

Dr. Balch, who was in charge of one of the two MLIC laboratories, denied that he had not tried to stay abreast of the progress in urinalysis. He had resigned from the first laboratory because he had been denied the opportunity to modernize it. Under his direction, things were changed, against much opposition. He had adopted the Benedict-Osterberg method for quantitative sugar only two days after it was published, and after he had acquired a colorimeter. He explained in effect that there had been a conflict within the company concerning the need to change the laboratory approach and that he had been on the side of modernization.

Benedict and Folin closed the discussion. Otto responded to Balch and to Exton. Constructive research work was needed by the insurance companies. It was not so much a question of modernization as it was the need to solve specific problems by research from within the industry.

The sulfosalicylic acid test was every bit as sensitive as the nitric acid test; and the method was certainly simple enough to be used by examiners in the field, rather than the crude nitric acid test that involved so much guessing. The problem that the examiners had, Otto stated, was one that they had not yet given full consideration.

In the autumn, the Harvard Medical School announcement of courses showed only one change in the department of biological chemistry. Hilding Berglund joined the medical staff at the Peter Bent Brigham Hospital, and Clifford L. Derick, M.D., was now listed in his place as assistant in medicine and research fellow. Richard D. Bell was officially on leave of absence; this meant that Cyrus Fiske was now carrying the heaviest teaching load of the department.

The Harvard Medical School under President Lowell in 1917 had decided, as an emergency war measure, to allow qualified women to work in doctorate programs in medical sciences. The candidates were enrolled at Radcliffe, and the Ph.D. degree at Radcliffe was countersigned by the Harvard president. The program continued after the war, though without any blessings from the Harvard administration. Otto Folin offered three courses, first published in the Radcliffe Catalogue 1921–22. These were general biological chemistry, the regular course for first-year students of medicine; advanced biological chemistry, a two-month full-time course requiring three other chemistry courses at Harvard College as prerequisites; and research in biological chemistry. In the first seven years that these courses were given, only a handful of Radcliffe students took them.

Early in 1923, Olive Watkins, advised by her faculty adviser at Vassar to contact Otto Folin concerning her desire to get graduate training in biochemistry, wrote him a long letter explaining why she preferred to be a biochemist rather than a physician like her father. Research work was an emotionally less demanding occupation. She ended the letter by asking, "Do you or do you not take women students in your biochemistry department?" Otto returned the last page of the letter. Under Olive's signature he had written, "Sometimes we do and sometimes we do not."

During the spring vacation, Olive went to see Otto in his laboratory.

> He listened to me and said nothing, as usual; and finally he said, "Vell, I tink we can find a place." . . . Years later I asked Folin why he decided to find a place for me and he said, ". . . you seemed like a very determined young lady, you looked strong—and you had on sensible shoes."

She also reminded him of his mother.

Olive Watkins enrolled in the graduate school at Radcliffe that autumn, and began fulfilling requirements for the doctorate, which she obtained in 1928. To avoid strain on her father's finances she worked her first two years as an assistant in the chemistry department at Wellesley, where she stayed in the faculty house and commuted to town by train. Though anatomy was a required course for physiology there was no institution in Boston that would allow her in the dissecting room.

Graduate students in first-year physiology courses were expected to score 90 or better and pass an oral examination. Olive had learned that on the basis of the oral exam she (and her friend, Hazel Hunt) would not receive an A grade. On her way to tell Otto about it she met Hilding Berglund,

> [a] great, big blustering person and he had on a big rubber apron that he always wore and a cigar in his mouth, and as he was talking in front of the library there, he said, "I am going to give Dr. Cannon a little hell." Well, Dr. Cannon was a mild man but was also a man of immediate action. Instead of answering Dr. Berglund, he picked up the telephone and reported that Hazel Hunt and Olive Watkins were going to get B grades, not A grades in physiology. And of course Otto was very upset about Dr. Berglund because he was sure that Dr. Cannon wouldn't have done that if Hilding hadn't gotten him so mad (*16-14*).

We shall hear more from Olive Watkins later. Hazel Hunt, a Vassar graduate, wrote on July 10, 1981:

> In the early '20s it was my great fortune to be able to take my graduate work under Prof. Folin at the Harvard Medical School. He was an excellent teacher and made the science of chemistry so easy for us to understand.
> Soon after I began my studies at the Medical School, Prof. Folin sent me to meet Dr. Joslin who was in touch with the Canadian young men, Banting and Best, who were working on insulin. An agreement was made with Dr. Joslin that I would establish a laboratory for him when the original work on insulin had been completed. When this occurred, insulin was placed exclusively in Dr. Joslin's care for one year.
> When I left the Medical School to go to the Deaconess, it was also agreed that I should keep in touch with Prof. Folin and do the laboratory work on many of the experiments he was conducting.
> This was many years ago and my memory hasn't been too clear but I shall always be grateful for my association with such an exceptional teacher.

In 1924, Otto published three papers. The first deserves mention only because it was his only one in the British *Biochemical Journal* (*16-15*). It was a note in answer to a publication in that journal maintaining that the Folin nesslerization method for determining nitrogen as ammonia "fizzled" due to the turbidity of the product. Otto replied that turbidity was the chief obstacle he had overcome in developing Nessler's reagent. His naive critic had not bothered to read the directions for its use, so Otto kindly gave him a few pointers, and suggested that he refer to the literature that he was citing.

The second paper of 1924 was the longest of Otto's career—101 pages, in collaboration with Berglund and Clifford Derick. The work was a detailed investigation into the metabolic fate of uric acid in animals, man, and in patients with gout (Part II, p. 418). Derick (1894–1972), a Canadian who had obtained his medical degree from McGill University in 1918, was a National Research Council Fellow in medicine. Berglund and Derick were on the staff at the Peter Bent Brigham Hospital where some of the work on the humans involved in this project was carried out, later with the added laboratory assistance of Miss Hazel Hunt.

Much of what was reported in this paper was substantiated by later work. Two factors were found to account for the so-called "destruction" of uric acid, the "miscible pool" (equilibrium with body water) in which injected uric acid promptly participates and which normally amounts to about a gram a day in males, and the removal of uric acid via the gastrointestinal tract wherein bacteria convert it (uricolysis) into urea, NH_3, allantoin, and respiratory CO_2 (16-16). A small amount is metabolized to allantoin by leukocytes. The "absorbing" power of the kidneys relates to the tubular reabsorptive capacity of the human kidney. Among other factors causing hyperuricemia in primary gout, excessive production of purines and renal retention stand out.

Otto's third and final paper of 1924 was based on work carried out in collaboration with Trimble on improving the quality and preparation of the uric acid reagent (16-17).

Harry Clyde Trimble (1889–1962) was already well established in an academic career as an associate professor of chemistry at the University of North Dakota when he decided to become a biochemist, a move that Otto would admittedly have encouraged because Trimble had obtained his doctorate in 1918 with Julius Stieglitz on "The Effect of Electric Currents in the Attempt to Induce Molecular Rearrangements." The encouragement that Trimble received from Otto was the result of a "chance" meeting. Trimble came to Harvard as a National Research Council Fellow in 1923. In 1924, he was appointed an assistant professor, responsible for instruction of first-year dental students in the department of biological chemistry. He was an excellent teacher, modest, gentle, unassuming (16-18).

In this paper, a successful effort was made to remove molybdenum from the sodium tungstate used in the preparation of the uric acid reagent. Molybdenum was a contaminant of tungsten ores and caused the uric acid reagent to become more sensitive to reducing agents, particularly to phenols. All of the commercial products, including C.P. grades, contained Mb. "One particular bad sample had on the container the trademark 'Folin's purity' in addition to the customary C.P."

The authors provided a qualitative test for Mb in tungstate based on the reddish color that it formed with potassium xanthate, which was extractable into chloroform. They provided a method for preparing the potassium xanthate from KOH and CS_2 in ethanol.

They described a process for removing Mb from tungstate. In the presence of phosphoric acid, Mb could be precipitated as sulfide with H_2S. Although some contamination from Mb did not hurt the usual analysis of uric acid on human blood and urine, heavy contamination could cause errors equivalent to 1 mg/100 mL.

Also included in the paper was the preparation of the uric acid reagent from sodium tungstate, sodium paratungstate, or phosphotungstic acid. All gave substantially identical results using 1 mL of reagent, 2 mL of 0.5% NaCN, and 5 mL of PFF, up to a uric acid content of 8 mg %, provided the cyanide did not produce a blank. The reagents produced a white turbidity with the filtrate in the cold, but this subsequently disappeared on heating on the water bath and did not return.

Kingsbury kept Otto informed in detail of his research progress at MLIC. He and his assistant had made some temporary bichromate-nitroprusside color standards in the hope of creating solutions permanent enough to replace the standard picramic acid solutions used for sugar determination. If successful these could be applied later to creatinine methods. It was urgent that a preservative for urine be found. At least 3% of the urines received for analysis by the laboratory were decomposed, and this percentage would increase in warmer weather. Formaldehyde had so far proved to be the best preservative. The problem was how to dispense it, and Kingsbury was trying to overcome this by using hexamethylenetetramine, which in an acid urine generated formaldehyde. This could be dispensed along with sodium acid phosphate as a tablet. Sodium benzoate worked fairly well in this way with acid phosphate. Sodium fluoride preserved urine well but interfered with the examination of the urinary sediment.

Otto was pleased with Kingsbury's progress, but advised him not to push too hard administratively until he had turned out some good publications.

By the end of June, Kingsbury reported his progress to Dr. Knight on the urinary preservative, glucose tolerance tests, and on permanent albumin and sugar standards. Most promising at that moment was the preparation of a permanent albumin standard.

A committee from four insurance companies was devising a suitable albumin method, and in the hope of achieving this before the October ALIMDA meeting, Kingsbury asked that a particular chemist-physician he had in mind be hired. Knight approved this, but suggested that Kingsbury discuss it with Folin and Benedict. On July 24, Kingsbury wrote Otto that he wanted to go over the work that he and Dr. C. P. Clark (of another insurance company) had completed on the albumin standards, and hoped that Otto could spend the night of July 29 with them in Stamford. Undoubtedly this meeting took place.

As mentioned previously, Harry Trimble became an assistant professor in the department of biological chemistry in the autumn term; Harry H. Powers, A.B., became a teaching fellow; Elliott T. Adams, S.B., replaced Mark Everett as the Austin teaching fellow. There were two new research fellows, Lloyd H. Newman, M.D., and Tsang Gi Ni, M.D., Sc.D. Richard D. Bell remained on leave of absence.

Otto and Laura Folin continued to spend time at Kearsarge during the summer. In August 1924, for the first time since they had initiated this annual retreat, neither of the children was present. Teresa had finished the Brookline high school (at age 16) and then gone to Europe in May with a friend of Laura's and her aunt. Grant (George), now 21, was working as a junior freight solicitor and freight office clerk for the Northern Pacific Railway.

Kearsarge was a refuge for Otto, a place to appreciate and grapple with nature, to recharge his marrow; and that was why he and Laura had been drawn to this isolated wilderness, not merely because it reminded him of his boyhood life in Småland. There was a tranquility here, a place to ruminate and to recall the past, to muse at the stars, and to be the delighted amateur geologist and botanist. He brought no reading matter with him except the profound and growing works of the incomparable Thorstein Veblen. On a rainy day, Otto would sit for hours in the doorway of the woodshed and ponder, perhaps on Veblen's essays, but certainly on the direction and progress of his own life's work. It was a time for tiring physical exertion at woodcutting and household chores, walking, hiking the trails, of seemingly non-mental pursuits; yet there was also much leisure for indulgence in man's finest luxury, thinking, undisciplined like the winds in the surrounding forest. What a delight it was not to have to pick up his pen!

The shirt-sleeved informality of Kearsarge was a welcome interlude from the sedate life of Brookline and the medical school. Otto's relationship with his colleagues was never, in those days of decorous behavior and address, other than straight-laced. One dressed conservatively. Otto was not seen in public without jacket and tie (except when golfing); and when traveling in cold weather, never without hat and coat. At home, he relaxed in his shirtsleeves, but would don a jacket and tie should a visitor arrive. He never used first names in speaking to or about colleagues. This right was reserved only for family and a few intimate friends. The speech and attire of that era were probably a matter of respect and evolution in human relations; and not derived from the "stuffed-shirt" syndrome of position and eminence.

Wherever they went, the Folins had a circle of friends. They had slowly made acquaintances with some of the permanent citizens of Kearsarge, and with the "people of summer" such as themselves. To make a friend of Laura, in particular, was to make a lifelong bond. Otto's circle was drawn mainly from the people he worked with in Building C as adviser or as colleague and from his golfing companions. Now there was a growing number of his graduates performing outstanding work elsewhere, and he saw most of them at least annually at the meetings of the federation: Shaffer, Bloor, Pettibone, Kingsbury, Morris, Sumner, Hammett, Wu, Doisy, Foster, and most recently, Everett.

Laura had her educated friends with whom she went to political and often, for those days, radical meetings at Ford and Faneuil Halls. Sometimes one would spend the night with them. Otto rather liked them and never objected to their presence. Most of her friends had maids at home, and consequently had plenty of time on their hands.

Laura always had a lively interest in world and domestic affairs. Twelve years before, Laura had dragged Grant, who was nine or so, to a suffragette parade in downtown Boston so he could carry and wave the "Votes for Women" placard. She and Otto would attend a lecture of some sort about twice a month. She was an avid reader of the *New Republic* since its founding in 1914. "She was ready to discuss any topic of the day at length." She always had opinions and voiced them, even at times untactfully.

Laura stayed in touch with the friends of her youth: from St. Paul, Marion Craig Wentworth and Ida Lusk Holman, who was a founder of the League of Women Voters; from Vassar days, Cornelia (Golay) Benedict and the Sands sisters, Adelaide and Georgiana (who had married Leo Loeb); from the University of Chicago, Madeleine Sikes and the two doctorates who had settled in the

Boston area, Eleanor Hammond and Cora Scofield (also a former Vassar classmate). In the Boston area, she had many other friends: Bertha Wiener; the Cooke sisters, Beth and Bonnie (mostly at Kearsarge); Nora Holman (no relation to Ida); Miss Grout; and Miss Sarah Perkins, who had taken Teresa off on the lark in Europe. There were, of course, friendly ties to the families of Otto's faculty colleagues, the Cannons, Joslins, Fiskes, and others.

Both Grant and Teresa graduated from high school at age 16. Grant completed two years at Harvard, but then had strong desires to make his own living. He would definitely not follow in Otto's footsteps. After all, Laura believed that Otto was a great man, and had repeatedly told this to the children from the time they could first grasp its significance. Grant, though a fine student and an amateur telegrapher, was frankly not interested in pure science and teaching, and this, of course, disappointed Otto greatly. He who had worked so hard to acquire an education beginning as an immigrant boy, could hardly understand how his own son, who had displayed so much aptitude, could give up this golden opportunity. Laura was more sympathetic. From her experiences in her own family, she knew that one could not "force" the desire for education on a youngster. In their Brookline and academic environment, the peer pressure was potent for obtaining a college degree and entering a profession. But Grant was certainly encouraged to think for himself; and he chose to try working for his living without a degree. The student life was too austere for a young man who had discovered stimulating interests—girls, automobiles, travel, and the practical business world. Besides, he had had one codfish supper too many at home! Laura had managed the Folin finances very carefully, and the essential needs of the children were taken care of. But there was never any "real" spending money for a young man. Otto's windfall from the MLIC came too late for Grant's college years, though it would help him later.

During their stay in Kearsarge in 1924, Otto must surely have examined his scientific field and the direction of his career. Biochemistry was taking several interesting turns into specialties and subspecialties. With more than 250 biochemists at work in the United States, the expansion in research was almost breathtaking. The medical schools had stiffened their requirements for admission to the point where most of them were requiring a degree. Biochemistry departments were now part of every major medical school. In addition, new centers were emerging in the agricultural, nutritional, and biological areas.

Chittenden had expressed little expectation that the science of biochemistry could be advanced, except in a limited way, in the medical schools. There, he believed, a specialized field of pathological chemistry could be developed, with special reference to the techniques of clinical chemistry. The broad concepts and development of biochemistry could not be channeled into the narrower interests of medicine, and there was much apprehension, when in 1923, the Sheffield Laboratory of Physiological Chemistry, after 49 years of independent pioneering and growth, had been incorporated into the medical school (*16-19*).

But Otto had already stated that there was little need to fear the potential narrow orientation of the medical school. He did not see any need to use the term "pathological chemistry" for medical school biological chemistry. The quality of the research would depend on the merit of the faculty, and not on the direct applications to medicine.

Otto's own research interest as he reached the last decade of his life had certainly narrowed toward clinical chemistry. The need for the application of soundly developed analytical methods to measure human metabolic function in health and disease was not only stronger than ever, but biochemists had made fruitful progress with them already. What would Banting and Best have accomplished with insulin during the summer of 1921 had they been unable to use the Myers-Bailey modification of the Lewis-Benedict method, and the reference Shaffer-Hartmann procedure for blood sugar determination? How quickly a good analytical method was taken for granted!

Shortly after their return from Kearsarge in early September, Kate called Laura to come to St. Paul. Their father was very ill. Mr. G. J. Grant died on Sept. 19, at the age of 83. Otto, in a letter of Sept. 26, encouraged Laura to stay in St. Paul helping Kate, until all the details were taken care of. Teresa was now at Vassar. Their finances were in good shape. He would take care of the home front.

Otto had no publications in 1925 except the fourth edition of his laboratory manual. He made several trips to New York to confer with Kingsbury, Benedict, and Knight. In a memo to Dr. Knight, Kingsbury wrote on April 6, "In a conference with Dr. Folin on April 2 nearly all the aspects of the work in the laboratory were taken up." Otto had advised that applicants for large insurance policies who had shown possible carbohydrate disturbance, come to the laboratory for a blood sugar analysis. The applicant was to fast for 14 h prior to having his blood drawn, rather than be given a test meal.

Evidently Benedict had expressed the view that applicants with pentosuria presented no risk in connection with life insurance, but Otto had pointed out that the literature showed that pentosuria occurred in morphine and cocaine addicts. Kingsbury was studying this question now.

> In connection with the use of the new preservative I reported to Dr. Folin that of the first 103 specimens received containing this preservative [hexamethylenetetramine + acid phosphate], only one had been received decomposed. The usual percentage of decomposed specimens at this time of the year with boric acid as a preservative is from 6½ to 7½ per cent.

Dr. Folin had suggested that he come to the laboratory for a conference with Knight and Benedict within a few weeks, "at which several questions could be taken up which could not be taken up at our conference." No doubt these related to establishing new standards for interpreting the test results for the granting of insurance. Meanwhile, Kingsbury was having success in the preparation of stable standards for the determination of urinary albumin. The formaldehyde that Otto and Willey Denis had recommended for preserving blood had now proved useful for urine, as well. The reagent used for generating formaldehyde (hexamethylenetetramine) would prove useful in the preparation of the stable albumin standards.

In the autumn, three staff members of the department of biochemistry departed, Newman, Bliss, and Tsang Gi Ni, but there were five new faces, making the number in the department a record 11. There was the heroic Yellapragada SubbaRow, M.B., research fellow, of whom much would be heard;

a second research fellow, Charles S. Woodall, M.D., who was also a Fellow of the National Research Council; and the three teaching fellows, Joseph DeFrates, B.S., Clarence A Morrell, B.A., and Embree Rose, M.A.

At the annual meeting of the ASBC, Otto was appointed chairman of the editorial committee for the *Journal of Biological Chemistry* (term 1926–31). Directors of the journal included S. R. Benedict and D. D. Van Slyke.

In December, Richard Dana Bell died at the Deaconess Hospital, one month before his 39th birthday, after a prolonged illness. He had been on a leave of absence since 1923. Bell, of nearby Somerville, was a devotee of biochemistry and of Otto Folin. Once he had obtained his bachelor's (1908) and medical (1913) degrees, he had foregone a hospital internship to join the department of biological chemistry as a research fellow without pay. This was probably voluntary because Bell was financially well off. Bell was considered almost painfully shy, friendly, exceptionally loyal to his friends, and generous. Following his military service, Bell returned as a research fellow for a year and was then promoted directly to assistant professor in 1920 (*16-20*). His devotion to Otto was demonstrated in the bequest of $100,000 (at least $600,000 in terms of the 1983 dollar) that he left, a 50-year endowment.

> The annual income from said sum shall be expended for the benefit of said Department of Biological Chemistry under the direction of the ranking professor in charge of teaching the regular course of Biological Chemistry to the Medical Students in the Harvard Medical School. . . .
> 1. Departmental library (books, salaries of librarians), up to one third of the income.
> 2. Fellowships, not over one half of the income.
> 3. Loans up to a total of $500 annually to students, junior members, and workers in the department. At the discretion of the ranking professor, a portion could be used to defray expenses incurred by junior faculty members attending scientific meetings.
> 4. Unexpended income could be used as a contingency fund by the ranking professor for the promotion of research in biological chemistry (supplies, salaries, equipment, but not maintenance, repair, alteration, and construction, of buildings).
> 5. Unexpended income of any one year could be expended in the following years.
>
> In general it is to be understood that in creating the foregoing bequest it is my purpose to benefit the Department of Biological Chemistry alone, and this purpose will be defeated if because of such bequest the Department fails to receive the full annual appropriation which it would otherwise normally receive (*16-21*).

It is noteworthy that in the bequest's provision on fellowships, Bell specifically stated that

> at the discretion of said ranking professor these fellowships may be awarded to women, and until such time as women become eligible for fellowships in the Graduate School for Arts and Sciences of Harvard University, such women applicants may be nominated by agreement between the Dean of Radcliffe College and said ranking professor.

Kingsbury's efforts at MLIC to prepare and evaluate stable albumin standards came to fruition late in 1925 with the publication of a paper that appeared the following year [Kingsbury, F.B., Clark, C.P., Williams, G., and Post, A.L., The determination of albumin in urine. *J. Lab. Clin. Med.* **11**, 981–89 (1926)]. Otto was intimately involved with this paper, but probably felt it to be in Kingsbury's interest not to include his name. The paper was a milestone for the life insurance industry because it provided an objective means for rapidly interpreting the turbidity obtained in using the Folin-Denis sulfosalicylic acid (turbidity) test for urinary protein (albumin). Kingsbury worked on the method with the two female assistants at MLIC, G. Williams and A. L. Post, and with Dr. C. P. Clark of Mutual Benefit Life Insurance Co., Newark, N.J. The paper described how to seal fixed, graded amounts of turbidity into test tubes.

This turbidity, which was created by mixing hexamethylenetetramine (urotropin) with hydrazin sulfate, was found accidentally. The product formed (formazin) was suspended in 10% gelatin. A series of these "standards" was prepared and matched to give turbidities equivalent to that produced from sulfosalicylic acid reacted with appropriately diluted protein standards made by diluting sheep serum (as devised by Folin and Denis). A "kit" of sealed tubes was produced containing 10, 20, 30, 40, 50, and 100 mg protein/dL placed in alternate holes of test-tube rack, so that the tube (of equal size) from the patient's test could be placed between the standards to obtain the nearest match visually. To the back of the rack a black line on a white strip was affixed to facilitate sharper reading of turbidity. The test for protein was somewhat modified from the original Folin-Denis method. One volume of urine was pipetted into a test tube and three volumes of 3% (w/v) sulfosalicylic acid were added and mixed, then the resulting turbidity was compared against those of the fixed standards. The turbidity standards were stable for six months, and were shown to be resistant to sunlight and to room temperature (40–110 °F; 4.4–43.3 °C). Otto designed a special elongated Mohr pipette graduated at 1.0 mL and 3.5 mL to allow pipetting of urine for both the sugar (1.0 mL) and albumin (2.5 mL) tests. The test tubes (12–14 × 125 mm) were graduated at 10 mL so that after the urine was pipetted into the tube, the sulfosalicylic acid could be added to the mark.

The "Kingsbury-Clark" method was put into use for a year by the MLIC, and for several months by three other insurance companies, prior to its publication. It became not only the standard for the industry but for urinalysis universally, and remains to this day as a rapid semiquantitative method.

Otto published two papers in 1926 on the determination of sugar in urine and blood, the second with Andrea Svedberg (*16-22, 16-23*). In the first paper Otto agreed with an idea of Benedict's that a quantitative method giving lower sugar values in urine should be the most accurate for blood sugar determination, i.e., it should have better specificity than a method more susceptible to non-sugar reducing substances. But Otto could not resist tweaking Benedict's nose, knowing full well that Benedict would soon respond in kind. After all, both were editors of the *Journal of Biological Chemistry*. Otto believed that Benedict's method for urine sugar was superior because it employed citrate rather than tartrate to obtain a stable copper reagent, whereas Benedict thought the essential feature of the reagent was that it used carbonate instead of hydroxide to make it alkaline. Citrate also caused less reduction of copper to take place with reducing sugar, a favorable decrease in sensitivity compared to tartrate. Benedict proposed that his citrate reagent be used for determining blood sugar.

Otto's purpose was to reduce to a minimum the action of the non-glucose-reducing substances in urine and blood. This could be achieved significantly by lowering the alkalinity of the copper tartrate reagent with a combination of sodium carbonate and bicarbonate rather than sodium carbonate alone. Otto also modified the acid molybdate reagent to produce more color from a given amount of cuprous oxide and to increase its acidity so that it could be used with almost any alkaline copper reagent, including Benedict's. These reagents made the sugar results very sensitive to small variation in acidity, requiring titration of urine or protein-free filtrates to near neutrality prior to sugar determination and making the method clinically impractical.

In checking Benedict's method on blood, Otto found that, contrary to Benedict's report, blood sugar values were often higher or the same, but not, as Benedict claimed, lower than values obtained with the Folin-Wu method. Otto found that the higher values were due to the deterioration of the bisulfite in Benedict's copper reagent. Once this problem was corrected, Benedict was right. The new "Folin" method gave values on normal blood lower than those from the Folin-Wu method, and about the same as those of the Benedict method. On urine it gave results strikingly less than the Folin-Berglund procedure, but similar to those of the Benedict method. In diabetic blood there was an unusually high concentration of non-glucose-reducing substances that needed further investigation.

By modifying the copper and molybdate reagents, Otto had obtained a quantitative method for measuring sugar in blood and urine that was apparently more specific for glucose than the Folin-Wu and the Folin-Berglund procedures. But he had omitted one important element: recovery experiments. Benedict would point out this omission among many things in his response to Otto's paper [Benedict, S.R., The estimation of sugar in blood and normal urine. *J. Biol. Chem.* **68**, 759–67 (1926)].

In his paper, Stanley denied every objection but one that Otto had raised. Benedict believed that the tartrate reagent was more sensitive than citrate, but less satisfactory because it was less stable and specific. His use of the tungstic acid reagent for color development with cuprous oxide was entirely justified because it was not interfered with by uric acid, and if the method were used as directed (a slap at Otto) there was no problem with undissolved cuprous oxide. Moreover, the tungstate reagent was less susceptible to the reducing effects of foreign compounds. Stanley did acknowledge that Otto was correct about the instability of the bisulfite used, and he therefore (presumably) abandoned it. He then pointed out that Otto's new bicarbonate-containing copper reagent, while more specific, was now too weak to neutralize minute amounts of acid, and therefore exact neutralization of filtrate would be essential for its use. He highly recommended the Folin-Wu tubes for sugar analysis, but defended his own recommendation on using an ordinary graduated test tube layered with benzene to prevent reoxidation of cuprous oxide. Rather than use benzoic acid as a preservative for glucose standards, Benedict added a few drops of toluene. He then pointed out that when Lloyd's reagent was added to standard glucose solution as in the Folin-Berglund procedure for sugar in urine, there was considerable loss of glucose with the new Folin method, perhaps caused by the effect of dissolving some of the Lloyd's reagent. "Obviously the new Folin procedure is not adapted to the determination of sugar in normal urine." Stanley's arguments were weakened by one central deficiency: he supplied no data to prove them. He modified his own alkaline copper solution and tungstic acid color reagent, yet provided no study of its merit with which to convince his audience. The answers, as Otto had characteristically shown, were at the laboratory bench.

The exchanges between Otto and Stanley were certainly an important factor in the development of sound analytical methods for clinical chemistry. No matter how ingenious and novel the approach was in developing a new procedure, it was always open to improvement. One investigator could rarely supply a "complete" analysis of anything in human body fluids, though huge strides could be made. This was certainly true of blood and urine determination. The Folin-Wu method was a major advance in the field of clinical chemistry, yet it had shortcomings that only further research brought out.

One of the shortcomings of the Folin-Wu procedure was its lack of specificity, as Benedict had stated, and Otto attacked this important point with Andrea Svedberg in his second paper of 1926. Andrea Andreen Svedberg (1888–1972) was Otto's second postdoctoral research fellow from Sweden (Berglund, after two years with Otto and two on the staff as assistant professor and associate in medicine at the Peter Bent Brigham Hospital, had gone in 1925 to a chair in medicine at the University of Minnesota).

Andrea Svedberg was a graduate of Uppsala University (1908) and had completed her clinical training as a physician in Stockholm in 1919. She was married to Theodor Svedberg (the Nobel Laureate) from 1908 to 1915 and was the mother of two children. Prior to coming to Otto's laboratory in 1926, she had worked as a physician in the department of medicine at the Sabbatsbergs Hospital in Stockholm (1921–23) and in clinical chemistry at a private laboratory (1923–25), and one also at the Board of Health of Stockholm.

Before tackling the question of specificity of sugar analysis, Otto and Andrea first resolved the problem of recovery experiments in the urine sugar method. Not only Benedict but Kingsbury had pointed out this flaw in using the new Folin reagents. As Benedict had reasoned, the problem was apparently that some Lloyd's reagent was dissolved, thereby setting free large amounts of calcium salts that interfered with the new copper reagent. Otto and Andrea attributed this to the acid used, and found that the loss could be reduced to 5–7% by using oxalic acid with Lloyd's reagent. This loss could be completely eliminated by following the Lloyd's reagent treatment with a further shaking with permutit. Once this was done, recovery of 40 mg% of glucose added to urine was quantitative.

Otto and Andrea found that the new Folin copper reagent tended to lose CO_2, and thereby became too alkaline, hence it had to be stored in small, well-filled, tightly corked bottles and used no longer than about a week. Six precautions for properly carrying out the procedure were provided. The advantage of the new method was primarily that it was more specific than the Folin-Wu procedure, and allowed Otto and Andrea to examine in detail the specificity of the Folin-Wu method and the fraction containing the non-glucose-reducing substances.

First of all, however, Otto showed that the recovery of glucose added to sheep blood in concentrations of 50–400 mg was quantitative with the new reagents.

With diabetic blood, they found that as much as 15–25% difference existed between values of sugar obtained by the Folin-Wu and the new Folin method. This non-glucose reducing fraction was now further studied by an elegant yeast fermentation procedure carried out on the protein-free filtrates of blood. A cake of fresh Fleischmann's yeast (3.25 g) was added to 20 mL of filtrate, then incubated for 7–8 min at 37–40 °C. The yeast was then removed with a small teaspoonful (1.75 g) of kaolin. Results were read on the colorimeter by comparison with a low glucose standard (2.5 mg/dL) under reduced light, with "an astonishing degree of accuracy." The Folin-Wu and the Folin methods were now applied to blood filtrates before and after fermentation. The results were quite revealing.

The "non-fermentable" or "rest reduction" value of the Folin and Folin-Wu methods were (with blank correction) 5–6 and 15 mg%, respectively, and these were constant values regardless of blood sugar concentration. The 5–6 mg% in the Folin procedure could be entirely accounted for from the nitrogenous products in the filtrate. Otto and Andrea proved this with an alkaline copper reduction technique that measured only the reducing properties of the nitrogenous products (creatinine, uric acid, etc., probably obtained by adsorption on Lloyd's reagent), but not glucose. There were no "unknown" non-fermentable products in the filtrates from normal and diabetic blood.

The fermentable sugar was greater in filtrates determined by the Folin-Wu than the Folin method. This was termed the "non-glucose sugar." What was its origin and nature? Otto and Andrea believed it was derived from the intermediary metabolism of carbohydrates. It was present no less abundantly in plasma than in the red blood cell. Hydrolysis experiments had ruled out maltose as a possibility. On administering insulin in small doses to normal persons, "In five of the six subjects investigated, the fall of the non-glucose sugar was practically just as sharp and decisive as the drop of the total sugar." But in three cases the drop in non-glucose sugar was even more dramatic. Total sugar fell 60–65%, whereas in three cases the non-glucose sugar fall was from 15 to 6–7 mg%, from 19 to 3 mg%, and from 14 to 4 mg%. Adrenalin was also tested, but the results were not clear cut. The question remained: Does orally administered glucose get converted at all into the non-glucose sugar? No evidence for this was found.

Though the Folin method did not become a standard clinical method, as mentioned, it provided a means by which Otto could, for the first time, examine the specificity of the Folin-Wu procedure, and take a scientific look at the non-glucose-reducing substances. Except in nephritic blood the nitrogenous substances were of little significance as interferences with the method. The urine sugar method was now much improved, but Benedict's procedures would rule the roost, here. The use of Lloyd's reagent and permutit remained useful for examining the reducing substances in normal human urine.

On June 17, 1926, Kingsbury wrote that he hoped to see Otto in New York before he went on vacation near the end of the month. He had had some success with trials on a new sugar reagent, and Benedict had approved his continuation with it. On Aug. 2, Otto wrote Kingsbury that he was expecting to go to Kearsarge on Aug. 4. "As we leave our daughter behind, taking a summer course in Harvard for two weeks, I cannot very well leave Kearsarge next week, because our place is a little too lonely for Mrs. Folin alone. Teresa will be back with us on Aug. 14th." Otto mentioned that he was looking over his manuscript for another sugar paper before sending it to the journal (Folin-Svedberg paper), and that it dealt in part with the determination of sugar in normal urine.

> If in the course of your own work you find something which is sound yet gives low values don't discard it, for the sugar of normal urine *is* materially lower than the values obtained by the Folin-Berglund process. You are presumably adjusting your methods to the results of the picrate method, and that is well and good, but sooner or later, cognizance will have to be taken of the fact that the sugar in urine is really much lower.... The sugar in urine is still a rich field and no one is in a better position than yourself for getting something good out of it.

On Aug. 4, Kingsbury wrote that he expected to see Otto on the 18th, which would be satisfactory with Dr. Knight, unless he heard to the contrary. Kingsbury mentioned the low values for sugar in urine that he had encountered, and hoped to discuss this further with Otto at their next meeting. He lamented the slow pace of his research, due in part to the illness of one of his two research assistants and to their being busy furnishing standards and preservative tablets to other companies. Parke-Davis had begun providing the tablets on an experimental basis. His work on a colorimetric assay of hippuric acid was at a standstill because of the lack of help.

The Kingsbury-Clark paper had appeared, and Kingsbury hoped to order 3000 reprints [later changed to 8000] for distribution. Sumner had visited and he had shown him the laboratory, and introduced him to Dr. Knight. Sumner "had done very well with the urease problem."

Otto and Stanley attended the ALIMDA meeting in October (37th annual meeting). A detailed report was presented by Dr. William Muhlberg, medical director of the Union Central Life Insurance Co., "A Practical Survey of the Newer Chemical Urinary Tests." He noted, in advance, that 48% of the insurance companies were using the new procedures and that he hoped to convince the remaining 52% to join in as well. "I have found that the tests are very readily made and require not much more technical skill than the older tests, and since the greater part of the difficulty in connection with the preparation for doing this work has been now solved by the manufacture of standards for both albumin and sugar tests, I think if you care to adopt the method as practiced by our company, you will find very little difficulty in training your laboratory technicians in doing this work." He mentioned work of his company on permanent sugar standards and their availability, and praised Dr. Kingsbury's tubes. His laboratory was testing urine preservatives. Dr. Muhlberg, in preparing his paper, had sent a questionnaire on laboratory tests to the members of ALIMDA. There were 73 who responded. As mentioned before, 48% were using the new quantitative procedures for albumin and sugar, another 30% were contemplating doing so, and 22% were satisfied with the old methods. These tests were run by the home office laboratories, and the companies were opposed, for the present, to instructing the field examiners in the use of the tests. About 60% were using either Urofix tablets (oxyquinoline alum) or formaldehyde as preservatives for urine, whereas another 27% were still using the useless boric acid.

> We are really greatly indebted to the research studies of the experts who have been working in the laboratories of the larger companies. They have not only modernized the methods and improved our technique, but have jolted us out of a self-complacency that might eventually have meant stagnation.

The paper then gave details of the Kingsbury-Clark albumin method. Clark had devised an excellent "albuminometer" that could be purchased, and Exton (Prudential) had obtained a manufacturer for a complete kit for his own version of the sulfosalicylic acid test. Then the Benedict method for urine sugar was described and directions provided for preparing permanent artificial standards, which could also be purchased. Muhlberg's group had found that formaldehyde was the best of a large number of preservatives studied. He recommended the use of Kingsbury's tablets, or the tablets made by Sharpe and Dohme consisting of hexamethyleneamine and salicylic acid, and finally Urofix, mentioned above.

Others who spoke on the use of the new methods, standards, and preservatives were Clark, Kingsbury, Exton, and eventually Folin and Benedict. Kingsbury had modified his preservative tablet, urotropin-acid phosphate, so that it now also contained sodium benzoate, benzoic acid, sodium bicarbonate, and mercuric oxide. This allowed 2% decomposition of urines in the cooler months and 3.5% in the hottest months of the year, whereas Urofix allowed 5–10%, depending on the time of the year. Obviously, the major insurance companies in the U.S. were now actively improving and standardizing their quantitative urinalysis practices, and Otto and Stanley were catalytic forces in the process.

In the autumn of 1926, changes in the department of biological chemistry at Harvard were the following: Vintilo Ciocalteau, M.D., became a research fellow, replacing Charles Woodall. E. Harry Lundin became a teaching fellow. Joseph S. DeFrates became the Austin teaching fellow replacing the departed Elliott Adams. The number in the department remained at 11.

While the fruit of Otto's lifelong work at this moment in his life was beginning to reach into an ever-increasing number of universities, federal and state laboratories, private clinically oriented laboratories, and hospitals, his former students, fellows, and assistants of the past three decades were also making their mark. A superficial glimpse at the topics of some of their publications in

Metabolic Studies

the *Journal of Biological Chemistry* for 1922–27 shows that most of these had direct impact on clinical chemistry.

Sidney Bliss published on the site of ammonia formation and the role of vomiting in ammonia elimination.

W. R. Bloor wrote on the determination of fatty acids, cholesterol, distribution, and form of combination of unsaturated fatty acids in blood plasma; plasma lipoids in experimental diabetes; and the distribution of unsaturated fatty acids in tissues.

F. De Eds had a paper on a simple microvessel with electrode for measuring the hydrogen ion concentration of blood and other body fluids.

Willey Denis continued her massive research output at Tulane on a variety of topics; phosphates in blood plasma; renal excretion of inorganic salts, serum inorganic constituents in nephritis and in experimental nephritis, NPN in the blood of marine fish, glycolysis in diabetic and non-diabetic blood, effect of temperature on protein intake, determination of total sulfates in serum, distribution of injected sulfates in tissues, effect of excessive calcium ingestion on calcium content of tissues with and without application of ultraviolet light, and determination of non-protein sulfur compounds in blood.

Edward A. Doisy reported on buffer systems of blood; the relation between arterial and venous blood, particularly affecting plasma chloride; properties and purification of insulin; preparation and properties of an ovarian hormone; and lactic acid in diabetes.

Mark Everett published two papers on the estimation and properties of the total sugar of blood and urine.

Cyrus H. Fiske worked on the estimation of total base in urine, the hydrolysis of amides in the animal body, ammonia and fixed base excretion after administration of acid, the colorimetric determination of phosphorus, the fate of acid in the body, analysis of potassium, nitrogen metabolism in the chick embryo, and the "inorganic phosphate" of muscle.

G. L. Foster compared blood sugar values in venous and finger blood and made an interpretation of blood sugar following glucose ingestion, separated hexone bases from protein hydrolysates, studied the effects of posterior pituitary ablation in the rat, prepared a respiration apparatus for small animals, and studied fat and glycogen in the tissues of experimentally induced obesity in rats.

Frederick S. Hammett performed research on determination of magnesium; creatinine and creatine in muscle tissue extract; creatine formation; blood constitutents, bone ash, water content, and the refractive index of thyroparathyroidectomized and parathyroidectomized rats; chemical composition of and changes in bone with age; and growth and the role of thyroid and parathyroid. He also published a chemical study of thymus involution.

Francis B. Kingsbury reported on the benzoate test for renal function, variations in the sugar content of human urine, and the effect of glucose on the condensation of formaldehyde and its use in determining urinary sugar.

Joseph M. Looney investigated the determination of cystine and cases of cystinuria; the relation between undetermined nitrogen in blood and its toxicity to *Lupinus albus* seedlings; the effect of parathyroid hormone on control of muscular rigidity; the estimation of tyrosine, tryptophane, and cystine in protein; the blood changes in acute mercuric chloride poisoning; and the preparation and use of colloidal carbon solutions.

J. Lucien Morris reported three studies on the chemical properties of saliva and on saliva as an index of glandular activity.

Philip A. Shaffer worked on antiketogenesis in diabetes, ether anesthesia, the ketolytic reaction, coupled reactions, and insulin.

James B. Sumner dealt with the use of dinitrosalicylic acid in determining titratable alkali in blood and sugar in diabetic urine, and with isolation, crystallization, purification, and properties of urease.

Hsien Wu reported on a new colorimetric method for plasma proteins, separate analyses of corpuscles and plasma, and on studies on heat denaturation of proteins, and the Donnan equilibrium and osmotic pressure relationship between cells and serum.

Guy E. Youngburg wrote on pentose metabolism, a study of analysis of furfural; a method for determining pentoses and pentosans; a method for organic phosphorus determination in urine, and on the effect of temperature on metabolism, particularly of protein.

The records are unclear on the thesis advisers for several of the doctorates during the period 1922–27, but it seems likely that Cyrus Fiske was the main guiding hand rather than Otto for K-H. Lin (1924), M. R. Everett (1924), E. T. Adams (1926), E. H. Lundin (1927), and H. H. Powers (1927) and several others that followed. As Otto's sixtieth birthday neared, his productivity had declined, and his research interests had narrowed considerably to what appeared as an effort to refine his earlier contributions rather than to forge ahead on new frontiers. Clinical chemistry was solidly launched, but would need continued support from minds such as his, and he chose to consolidate and extend his established interests in clinical chemistry. From this would come useful, important work, but the great fires of his youth were beginning to dampen.

Yellapragada SubbaRow (SubbaRao) (1896–1948), who had become a research fellow in the department in 1925, was an Indian physician who had spent a year with Prof. Richard Strong in the school of tropical research at Harvard. He was from Madras State in India, where he had developed an almost messianic fervor to conquer tropical diseases such as tropical sprue and cholera that killed thousands of his countrymen including his own brother.

After obtaining his medical and master of science degrees from Madras University in 1921, followed by a year there as a demonstrator in physiology, he spent a year at the school of tropical medicine at London University. There he met Professor Strong, who later said of him, "He plied me with questions I couldn't answer. I never encountered such a probing mind, such a conviction of destiny. His zeal bordered on the fanatical. . . . I suggested he come to Harvard and pursue his quest."

But unfortunately for the historian, SubbaRow was also a legendary, mysterious, and quasi-mystical figure. One day, as director of research at Lederle Laboratories, he would lead team efforts into the synthesis of folic acid; diethylcarbamazine (Hetrazan), for treating filariasis, ascariasis, and other related diseases; aminopterin; and chlortetracycline (Aureomycin).

After a year with Strong, SubbaRow concluded that he needed to learn biochemistry. In 1924, he went to see Otto about becoming a postdoctorate student. Otto agreed to have him on a one-year "probationary" basis. This meant that SubbaRow could expect no financial support, a situation he already endured. He had arrived in the U.S. with $25 in his pocket. To survive at Harvard, he had tended furnaces, worked as a hospital orderly, collected stray cats for the laboratory, and later was a "night librarian" in Building C. He subsisted on less than $50 a month, and was said to have lived at times on baked beans, shredded wheat, and milk (16-24). He was off the probationary status in three months, and Otto assigned him to the aegis of Cyrus Fiske, a pure blessing for both of them.

In the next few years, SubbaRow would become an adept chemist and a master of research. Fiske immediately had "Subb" join him in his momentous work on phosphorus compounds in muscle extracts. For this they needed a more refined method than that of Bell and Doisy for the analysis of phosphorus. Within a year, the pair had worked out a colorimetric method that stands as a classic and remains in use today in clinical chemistry [*J. Biol. Chem.* **66**, 375–400 (1925)]. They would go on, as we shall note, to characterize some of the most fundamental phosphorus compounds in nature: phosphocreatine, adenosine triphosphate, and α-glycerophosphate.

The method for analysis of phosphorus was intimately related to Otto. As mentioned previously, a logical progression had developed from the Folin-Denis introduction in 1912 of the phosphomolybdates and tungstates into modern biochemistry. Wu, in the course of developing tungstic acid for blood protein precipitation and phosphomolydate for sugars, had made an extensive study of the chemistry of the molybdates and tungstates [*J. Biol. Chem.* **43**, 189–200 (1920)]. Bell and Doisy had made use of phosphomolybdate reduction with hydroquinone in their phosphorus method [*J. Biol. Chem.* **44**, 55 (1920)]. The 1,2,4-aminonaphtholsulfonic acid used by Fiske and SubbaRow as the reducing agent to develop the blue color of phosphomolybdate was a compound that Otto had prepared for his studies of amino acid analysis in 1922.

In May 1926, Otto was elected to the National Board of Medical Examiners (then 15 members), and in 1930 became chairman of chemistry of its examination committee. The latter met at least once a year to determine the composition, administration, requirements, and the components of the examinations. Otto gave freely of his time and advice.

References and Notes

16-1. Folin, O., and Berglund, H., A colorimetric method for the determination of sugars in normal human urine. *J. Biol. Chem.* **51**, 209–11 (1922).

16-2. Folin, O., and Berglund, H., Some new observations and interpretations with reference to transportation, retention, and excretion of carbohydrates. *J. Biol. Chem.* **51**, 213–73 (1922).

16-3. (a) Somogyi, M., Distribution of blood sugar between corpuscles and plasma in diabetic and in alimentary hyperglycemia. *Arch. Intern. Med.* **42**, 931–38 (1928).

(b) Somogyi M., The distribution of blood sugar in normal human blood. *J. Biol. Chem.* **78**, 117–27 (1928).

(c) Meites, S., Preservation, distribution, and assay of glucose in blood, with special reference to the newborn. *Clin. Chem.* **25**, 531–34 (1979).

16-4. Folin, O., A system of blood analysis. Supplement III. A new colorimetric method for the determination of the amino-acid nitrogen in blood (with the assistance of H. Wu). *J. Biol. Chem.* **51**, 377–91 (1922).
16-5. Danielson, I.S., Amino acid nitrogen in blood and its determination. *J. Biol. Chem.* **101**, 505–22 (1933).
16-6. (a) Frame, E.G., Russell, J.A., and Wilhelmi, A.E., The colorimetric estimation of amino acid nitrogen in blood. *J. Biol Chem.* **149**, 255–70 (1943).
(b) Russell, J.A., Note on the colorimetric determination of amino acid nitrogen. *J. Biol. Chem.* **156**, 467–68 (1944).
16-7. Folin, O., A colorimetric determination of the amino-acid nitrogen in normal urine. *J. Biol. Chem.* **51**, 393–94 (1922).
16-8. Folin, O., and Berglund, H., The retention and distribution of amino-acids with especial reference to the urea formation. *J. Biol. Chem.* **51**, 395–418 (1922).
16-9. Folin, O., Note on the necessity of checking up the quality of sodium tungstate used in the system of blood analysis. *J. Biol. Chem.* **51**, 419–20 (1922).
16-10. Folin, O., Non-protein nitrogen of blood in health and disease. *Physiol. Rev.* **2**, 460–78 (1922).
16-11. Folin, O., A system of blood analysis. Supplement IV. A revision of the method for determining uric acid. *J. Biol. Chem.* **54**, 153–70 (1922).
16-12. Folin, O., Proceedings, 32th Annual Meeting. *Assoc. Life Insurance Med. Directors Am.* **8**, 6–28 (1922).
16-13. According to daughter Teresa, the Benedicts visited at Kearsarge at least twice, and the Folins made one visit to the Benedict summer house. Benedict later wrote of often playing chess with Otto in the summer.
16-14. (a) Olive Watkins Smith. Tape interview, Dec 15, 1981.
(b) Smith, O.W., The men in my life. *Radcliffe Quart.* **67**, 25–27 (1981).
16-15. Folin, O., LXIV. Nesslerization and avoidance of turbidity in nesslerized solutions. *Biochem. J.* **18**, 460–61 (1924).
16-16. Wyngaarden, J.B., and Kelley, W.N., Gout. In: *The Metabolic Basis of Inherited Disease.* 5th ed. pp. 1075–77. J.B. Stanbury, et al., Eds. New York: McGraw-Hill Book Co., 1983.
16-17. Folin, O., and Trimble, H., A system of blood analysis. Supplement V. Improvements in the quality and method of preparing the uric acid reagent. *J. Biol. Chem.* **60**, 473–79 (1924).
16-18. Ball, E.G., Harry Clyde Trimble, 1889–1962. *Harvard Med. Alumni Bull.* **37**, 2–3 (1962).
16-19. Chittenden, R.H. *The Development of Physiological Chemistry in the United States.* American Chemical Society Monograph Series. New York: Chemical Catalog Co., 1930.
16-20. (a) Richard Dana Bell. Harvard College Class of 1908, 15th and 25th Anniversary Reports. Cambridge, Mass.: Harvard University.
(b) Bell. *Boston Med. Surg. J.* **193**, 1178 (1925).
16-21. Richard Dana Bell Bequest, 1926 ($133,958.59). *Endowment Funds of Harvard University, June 30, 1947.* pp. 250–52. Cambridge, Mass.: Harvard Medical School, 1948.
16-22. Folin, O., The determination of sugar in blood and in normal urine. *J. Biol. Chem.* **67**, 357–70 (1926).
16-23. Folin, O., and Svedberg, A., The sugar in urine and blood. *J. Biol Chem..* **70**, 405–26 (1926).
16-24. Dr. Yellapragada SubbaRow (1896–1948). *Lederle Chevron* 5–9 (Nov. 1948).

Chapter 17. Modifications, Micromethods, Unlaked Blood (1927–31)

Otto published two papers in 1927. The first one, with V. Ciocalteau, put the master touch on the determination of tyrosine and tryptophane in protein, and prepared it for its eventual application to serum protein analysis (*17-1*). The method that Folin and Looney had published in 1922 applied to a limited range of protein concentration and was bothered by turbidity. Also, the tryptophane did not develop the maximum color expected from it on a molar basis.

After modifying the phenol reagent, primarily by adding lithium sulfate to it, Folin and Ciocalteau were not only able to eliminate turbidity attributed to the precipitation of sodium salts, but to obtain a blue color reaction with tyrosine that was linear with concentration over an extended range. In fact, they boasted: "The color reaction obtainable from the phenol reagent and pure tyrosine is one of the most perfect reactions to be found in the field of modern colorimetry, when the color reaction is developed under suitable conditions." Furthermore, tryptophane, on a molar basis, now gave all of the color reaction that could be expected from it.

Because tryptophane was too expensive and impure as obtained "on the market," tyrosine could be used as the standard for tryptophane analysis. They provided a procedure for the recrystallization of tyrosine.

In this paper, alkaline hydrolysis of protein was carried out for 18–20 h using NaOH rather than $Ba(OH)_2$. Folin and Ciocalteau discovered that Millon's reagent reacted at room temperature with tyrosine if the protein hydrolysate was pretreated with mercuric sulfate in strong sulfuric acid prior to a short heating of the mixture in a water bath and the addition of sodium nitrite. The brilliant red color that formed became the basis for the new tyrosine determination in place of the phenol reagent.

The mercuric sulfate used for tyrosine determination (above) quantitatively precipitated out the tryptophane. After the precipitate was washed, the tryptophane could then be quantitatively measured with the new phenol reagent by either of two equally valid methods. Pure tyrosine served as the standard for the measurement. In one method, the mercuric sulfate, which tended to precipitate in the alkaline medium of the reaction with the phenol reagent, was removed as the sulfide by treatment with H_2S. In the second method, the mercuric sulfate was kept in solution by the addition of sodium cyanide during the color development with the phenol reagent.

The authors presented some new values obtained for the tyrosine and tryptophane content of casein, egg albumin, edestin, gliadin, and zein. Because the mercuric sulfate available commercially had impurities in it (primarily mercurous sulfate and possibly reducing salts such as ferrous sulfate that reacted with the phenol reagent), Folin and Ciocalteau provided a procedure for purifying mercuric sulfate.

The tyrosine-phenol reagent reaction became a key feature of Greenberg's widely used determination of serum proteins [*J. Biol. Chem.* **82**, 545–50 (1929); *J. Lab. Clin. Med.* **21**, 431–35 (1936)].

Otto's second paper of 1927 (with Trimble and Newman) was an effort to determine the immediate fate of glucose injected intravenously into nephrectomized guinea pigs and dogs (*17-2*). Otto's work with Berglund had postulated that the tissues take up and hold sugar as the initial mechanism to prevent its excessive accumulation in blood. Was there any basis for this hypothesis? The Coris of Washington University, St. Louis, had shown that 90% of absorbed sugar was metabolized or stored as glycogen in 4-h experiments. But this did not explain the fate of sugar before oxidation or storage. Hagedorn had shown that arterial blood gave up much sugar to tissues. Differences of up to 30 mg occurred between arterial and venous blood within 30 min of glucose injection that could not be attributed to glycogen formation.

Folin, Trimble, and Newman could not find the answers by injecting glucose into nephrectomized dogs. Within 5 min of injecting 2 g of glucose per kg of weight, all but 15–30% left the blood, but only a small part of the remaining 70–85% could be traced. After 30 min the blood sugar was still very high, but the muscle concentration was only moderately elevated, indicating perhaps that there was no free and unregulated diffusion of glucose from blood into muscle. If it was not due to glycogen formation then where had the glucose gone? The only answer was to do whole body analysis. The dog was too big for that, so the authors used guinea pigs. Injected nephrectomized guinea pigs could be dipped into boiling water to help remove the hair, and then passed through an ordinary meat grinding machine to produce a "uniform hash."

The trio first established that they could quantitatively recover unmetabolized sucrose 30 min after its injection into guinea pigs. The sucrose was measured in protein-free filtrates made from the extracted hash by analyzing the sugar present before and after acid hydrolysis using the Folin-Wu procedure. The average recovery from eight guinea pigs was a remarkable 94%.

Sucrose did not present an analytical problem because there was none originally present in the tissues, nor were any other potential disaccharides present to cause a blank value. Of course, glucose was another story. Fortunately, the investigators found that by using three guinea pig (uninjected) controls for each set of three injected, the average sugar values of the controls could be satisfactorily used as a base-line value.

The first trial showed that there was no loss of glucose 30 min after injection, nor was there any evidence of glycogen formation, and this meant that there was also no evidence for any unknown transformation into non-carbohydrate material. The second trial gave the same results.

Then the "hash" method was tried on two puppies. The recovery of free glucose in 30 min was 75 and 79% of the injected glucose with no tangible evidence that there was transformation into non-carbohydrate materials. The guinea pig work had suggested that some tissues took up more sugars than others, and certainly more than the skeletal muscles. Now it was time to return to the dogs and "regional tissue analysis" to account for the large losses observed. Would tissues other than muscle show high concentrations of glucose?

In 30 min following glucose injection, it was shown that skin and other tissues accumulated sugar. All the glandular organs as well as the intestines and stomach took up much more sugar than the skeletal muscles, with the skin leading the way. In fact, the skin had values comparable to the blood. This was not to be interpreted, however, as a special capacity for holding and conserving carbohydrate, but as Bell had shown a few years previously with NaCl, it was a passive diffusion phenomenon, a response to the high level in blood. There was no special absorption or metabolic process involved. In modern terminology, they had found what Otto had observed but could not explain for the disappearance of uric acid, the "miscible pool."

The glycogen content of skin did not change in 30 min following injection, while the change in blood and skin sugar were parallel, lending weight to the idea that the movement of sugar was by a diffusion or an osmotic process. But the diffusion process did not seem to apply to muscle where the sugar concentration was comparatively low, and the glycogen-forming power was great.

The question was what happened to glucose 3–4 h after injection. The blood sugar had returned from an initial value of 600–700 mg/dL (33.3–38.9 mmol/L) to about normal. The glycogen found in muscle and liver could account for 50% of the injected glucose, showing that glycogen formation was a slow process and that the initial distribution of free sugar injected was entirely different from the subsequent distribution of the glycogen. Again the investigators expressed surprise that the muscle sugar was rather low compared with other tissues after glucose injection, and that there was no evidence that non-carbohydrate materials were being formed as a consequence.

At the 38th meeting of ALIMDA held at the Hotel Biltmore in New York City on Oct. 27–28, Otto presented a paper entitled "Some New Observations on the Distribution of Sugar Within the Animal Body" [*Assoc. Life Insurance Med. Directors Am.* **14**, 428–35 (1927)].

Otto rehashed briefly the work just described of Folin, Trimble, and Newman on the fate of injected glucose in the animal body within the first 30 min. He placed more emphasis on the implications of high sugar concentration in skin. The discovery of skin as a temporary reservoir for surplus sugar could be of practical importance. In diabetes, the skin should be "loaded with sugar." Perhaps the frequent infections in diabetics leading to gangrene and amputations could be a consequence of chronic high concentration of sugar in the skin. Skin specialists ascribed some skin eruptions to faulty diets. Could the excessive eating of sweets by young people cause much of their skin eruption?

In his conclusion, Otto noted that the skin acted as an overflow receptacle not only for sugar, but for all diffusible substances in blood. Uric acid, for example, was taken up like glucose.

> ... and I suppose that the purple colored skin which is produced by prolonged internal use of silver nitrate may be taken to indicate that in this case also the skin has received more than its share of the circulating chemical.

For the 1927–28 academic year, the department of biological chemistry lost four of its 11 people with no replacement. Gone were the four fellows, Embree Rose, Harry Powers, E. H. Lundin, and V. Ciocalteau.

Otto introduced a startling new blood sugar method in his sole paper of 1928 (17-3). It was followed by a supplemental note submitted at the end of the year (17-4). In this latest creation, sugar was oxidized with alkaline potassium ferricyanide, and the ferrocyanide produced was measured colorimetrically as Prussian blue. Gone were the alkaline copper reagent and phosphomolybdate solution of the Folin-Wu procedure.

The new method was developed for several reasons. There was a strong need for a micromethod for sugar analysis on blood collected by skin (capillary) puncture. The Folin-Wu filtrate was not readily "scaled down" for use with small volumes; the filtrate was originally intended for application as a system of multiple analysis of the NPN constituents. Nevertheless, the blood sugar test was the most frequently requested of laboratory chemistry tests, and most often it was requested alone. Venous puncture was still required because the microvolumes obtained from skin puncture could not be readily made into protein-free filtrate because of the need for unavailable special glassware and the inaccuracies involved. The amount of oxalate anticoagulant used in the collecting vials was intended for large volumes. If too little blood was collected, the excess oxalate interfered with the action of tungstic acid in protein precipitation.

The new microprocedure actually had only one "micro" step, and that was the collection of 0.1 mL of blood. For this step, Otto recommended that a pipette be used that could be filled without introduction of air bubbles. Commercial pipettes available at the time had a tendency to collect air bubbles, so Otto proposed that the pipettes be home-made. Eimer and Amend, however, had agreed to produce them. In this era, before the U.S. National Bureau of Standards had set any specifications on pipettes, Otto's contribution was significant. He recommended using a pipette that would fill by capillary action. This was possible in a glass tube having an internal diameter of 1.0–1.7 mm. Capillary action could fill a pipette with this diameter about 5.0–8.0 cm from the tip. Otto described how to calibrate this "automatic" *to contain* pipette by use of mercury weighing. This permitted rinsing of the blood sample (in the case of blood sugar, into 10 mL of tungstic acid) so that the retention in the pipette, after rinsing, was negligible, in contrast to the *to deliver* pipette that could not be rinsed.

Otto described his blood collecting system. The area of the skin to be punctured was first washed with soap and hot water. The puncture was made with a spring lancet. The pipette was filled to the mark. Excess blood was wiped from the tip (outside) with a piece of soft filter paper or cloth. Excess blood in the pipette was removed by slowly drawing it, as though it were a fountain pen, along a glass surface. If too much was removed it could be recovered by merely moving the pipette to a more horizontal position.

The blood in the pipette was then directly transferred to 10 mL of tungstic acid (dilute) in a centrifuge tube, and rinsed. A major step and source of contamination was removed: the typical receptacle with anticoagulant for use with venous puncture. Another usual step, filtration, was then avoided: After mixing (with the pipette or a glass rod) the tube was centrifuged for 3–5 min, and the 9 mL of water-clear supernate obtained was adequate for duplicate analysis in the method (requiring 4 mL per test).

The procedure was quite simple. Details including reagents will be covered in the description of the 1929 modification by Folin and Malmros. The 4 mL of supernate was transferred to a test tube graduated at 25 mL (not a Folin-Wu sugar tube). Then 1 mL of ferricyanide solution and 1 mL of cyanide-carbonate solution were added. The tube was heated for 8 min in a boiling water bath, and then cooled. Finally, 3 mL of acid ferric iron solution was added. This solution stood for 5 min developing its blue color and was then read in a colorimeter opposite a similarly treated glucose standard.

When compared with the Folin-Wu procedure on the same blood samples, the new method gave unmistakably lower values. Results were also generally lower on blood samples obtained by skin puncture than those obtained on venous blood by the Folin-Wu method. The Prussian blue values were lower because the method measured true reducing sugar more specifically.

When determinations with the new method were made on blood collected from the same person simultaneously by skin and venous punctures, three of five cases showed higher values in the skin-puncture samples. This finding tended to support what would be shown later: "Capillary" blood reflects arterial blood so that it shows much higher postprandial blood sugar values, values not approached by venous blood until several hours later. Of course, in the fasting state, there was a slight arteriovenous difference, the arterial (capillary) blood being always a bit higher.

Miss Anna L. Post of MLIC had been testing Otto's new method, and had uncovered a problem that Otto corrected (along with others) in the published supplementary note (17-4). Tungstic acid preserved with toluene was unstable when exposed to sunlight. Otto confirmed this and recommended that no toluene be used, and the tungstic acid be stored in brown bottles in the dark. Sunlight was a danger for other reagents as well, particularly the ferricyanide. Otto reviewed each reagent and recommended storage conditions, as well as giving their "shelf life."

The dilute tungstic acid used in the procedure had been troublesome because it was often made inaccurately, so that there was either too much acid present that hindered the reaction with alkaline ferricyanide or too little, so that the proteins were incompletely removed with consequent turbidity. The answer was to make the tungstic acid reagent accurately as described.

The Prussian blue reaction produced a colloidal suspension that was stabilized with gum arabic. Unfortunately, gum arabic quality was not uniform, so Otto proposed that gum ghatti replace it. Otto gave a procedure for purifying gum ghatti with potassium permanganate and incorporating it into the iron reagent. Gum ghatti was five times as effective as gum arabic and was stable in solution for at least two months. The Prussian blue obtained in the sugar reduction would now remain for weeks in a clear, uniform suspension.

The new micromethod was obviously a blessing for pediatric patients and for adults on whom venipuncture was difficult. It would find application as well to sugar studies on small animals. Otto did not intend replacing the Folin-Wu method, but to supplement it when the need arose. His venture into microchemistry was a solid demonstration of what could be done in that area, a truly pioneering effort that would not be fully appreciated in clinical chemistry for another three decades. His particular contributions would be used but their source forgotten. Several of the new principles he had stumbled onto would prove useful as a basis for ultramicrochemistry in blood analysis: 1) the use of *to contain* mercury-calibrated micropipettes, 2) the minimization of the number of transfers, 3) centrifugation to replace filtration, and 4) a new chemical reaction that allowed ample color measurement for microvolumes of blood.

Olive Watkins received her doctorate degree in 1928 at Radcliffe. Her research thesis was entitled "The Occurrence of Lactose in the Urine with Especial Reference to Pregnancy and Lactation." She would be Otto's only female doctoral graduate student, though there were two who obtained master's degrees via Radcliffe, Marjorie Henderson in 1925 and Katherine Jones in 1929. It had not been easy for Olive, as she had shuttled between Radcliffe, MIT, Wellesley, and the Harvard Medical School. Not long before being awarded her degree, Olive was told that she did not have enough "resident" courses.

> But my dear Prof. Folin came to my rescue. He called up Dean Chase at Harvard, saying "Why are you being so mean to my first woman Ph.D.?" When the dean looked me up and stated flatly that without enough resident credits the degree could not be granted, Dr. Folin made up a course for me—one for special students, he said, not listed in the catalog. "It is a full-year course," he added. "Please record it for her, with an A grade." This got me by—but not for free. It was followed by a fat bill for Biochemistry 201.

Otto had obtained a small amount of financial assistance for Olive during her first five years. She never had a title, and her monthly check merely said, "Wages." More than likely, the money for this was derived from the Bell endowment. Following her degree, Dr. Watkins became a research assistant in biochemistry, but could not be listed officially on the staff of the department. In Otto's remaining years, Olive, who was a native of nearby Worcester, would become a close friend of the Folins.

Olive cherished her memories of Otto Folin, "this humble genius," and the following "Advice to a Young Ph.D." that she received from him in 1928:

> *Ideas.* Ideas are a dime a dozen. Anyone engaged in research has more ideas than he can possibly follow up on in one lifetime. What really counts are consistently careful work, an alert and open mind, and the patience of Job.

The Literature. You must know the literature. Not just the current, but the background in your field. That is very important! (I don't have to. I have Fiske.)

Publications. Do not be in a hurry to publish a new finding. If someone else gets it out ahead of you, you will be confirmed before you publish. That is a good thing. You will be sure to have something to add, and will write a better paper.

Discussions. Never talk through your hat. It is better to pretend that you don't know than to pretend that you do. You learn more that way.

Five "Ins" for the Investigator. Interest, Intelligence, Ingenuity, Integrity, Indefatigability.

Teresa Folin had obtained her bachelor's degree with honors at Vassar. In September she entered the Boston University School of Medicine and lived at home.

In 1928 Grant had become the Detroit sales representative for the Union Twist Drill Co., Athol, Mass., after having served for a year in Great Britain in that capacity. Prior to that he had worked as an apprentice at the parent company in Athol (1925) where he had learned to operate the machine tools, and in 1926 had been the representative in New York. In 1931 he became the sales manager.

In July 1928, George Sikes died at age 60, and Madeleine wrote Laura and Otto about it. He had been active in public service to the end. In 1923 he had become secretary of the Police Pension Board of Chicago and had helped clean up its scandalous mismanagement by a prior regime. He later also became secretary of the Chicago Pension Commission, a group appointed by Mayor Dever to make a study of changes needed in pension legislation. Prior to that he had served on the Chicago Bureau of Public Efficiency which helped to prepare reports on the organization and operation of the various local governing agencies of Chicago and Cook County. He had also been engaged to assist the Taxpayer's Association of California in the preparation of a report on the "City and County Consolidation for Los Angeles."

Following receipt of Madeleine's letter, Laura wrote on July 26:

> We received your letter this morning. Otto seemed to feel very badly indeed. I know how deeply he respected your dear husband, and of course he realizes what you have lost in losing him. I, too, had a sense of shock which persisted a considerable time. . . . If my words seem incoherent, forgive me. I am deeply moved, and wish I could see you. You will understand the feeling. . . .
>
> Please give our deepest sympathy to your children.

At the Metropolitan on Sept. 1, Norman Robert Blatherwick, Ph.D. (1887–1961), replaced Francis Kingsbury, with the title of Director, Biochemical Laboratory. Before obtaining his degree, Blatherwick had worked as an assistant chemist at the Montefiore Hospital in New York City, and after his doctorate from Yale, he had been director since 1920 of the chemical laboratory of the Potter Metabolic Clinic, Santa Barbara, Calif., where he had conducted studies in a search for an insulin substitute and in carbohydrate metabolism (*17-5*).

Otto and Dr. Blatherwick got on very well together, the latter evidently being thankful to have Otto and Stanley to lean on in his new position. He would be an aggressive, capable research leader and tackle basic metabolic problems as well as the practical needs of the company.

In December, Otto exchanged letters with Dr. Blatherwick. The first was about the manuscript of the supplementary note on the new ferricyanide method. Otto enclosed a copy of the part that mentioned the work of Miss Post at the Metropolitan laboratory on the light-sensitive tungstic acid, and hoped that he had adequately credited both her and MLIC. Blatherwick replied that the credit was adequate. He mentioned that fresh tungstic acid would not decolorize dilute permanganate. Even more important, he was trying to adapt Otto's new sugar method to field use. The problem was first how to collect and mail samples for analysis.

> Our plan is to absorb the blood on paper, dry it, and send it to the Home Office for analysis. We are using sodium fluoride for a preservative. The preliminary work looks encouraging. We have had samples which have retained their original values for one month.

Otto replied (Dec. 24) that his own idea was to collect a couple of drops of blood without preservative into an accurately preweighed centrifuge tube that would be corked tightly and then heated for 2–3 min in boiling water. The tube could then be weighed later in the laboratory. He had tried it and found that he could extract plenty of sugar from such tubes by stirring with tungstic acid. "By supplying a suitable wire holder, any doctor could heat the blood in a tea kettle of boiling water before shipping it, and from the number of the tube, the laboratory would get the weight of the blood." This eliminated simply the need for the field man to be accurate.

For Blatherwick's idea, Otto recommended the use of Whatman No. 41 filter paper. It was almost free from soluble reducing material and did not require preliminary washing.

For the 1928–29 academic year, the department of biological chemistry again included 11 members. H. Lyman and C. A. Morrell were gone. Samuel B. Nadler, A.B. was a teaching fellow. There were now four new research fellows, Jayme A. Cavalcanti, M.D., Stephen J. Maddock, M.D., Agustin D. Marenzi, D.Biochem, and Juan M. Munoz, M.D. A fourth member of the teaching faculty, Milan A. Logan, Ph.D., was now added as an instructor. Milan had obtained his doctorate in 1928 under the guidance of Cyrus Fiske.

Otto worked with at least three postdoctoral research fellows in 1928–29, two under the auspices of the Division of Education of the Rockefeller Foundation, Agustin D. Marenzi of Buenos Aires and Haqvin Malmros of Lund. The third, Andrea Svedberg, returned for a two-year sojourn, having gone back to Sweden after her stay of 1926.

Otto published eight papers in 1929, the first three on his own, that related to blood sugar. We have already presented his supplementary paper on the ferricyanide (Prussian blue) method, labeled henceforth the "Folin micromethod," to distinguish it from the two copper-reduction methods, the "Folin-Wu" and the "Folin" methods.

The second paper of 1929 was Otto's response to a paper by Michael Somogyi and H. D. Kramer [*J. Biol. Chem.* **80**, 733–42 (1928)] that denied the finding of Folin and Svedberg that there was probably present in blood some non-glucose reducing sugar (*17-6*). Recall that Otto and Andrea had found more fermentable sugar by the Folin-Wu than by the Folin method. Otto pointed out that even if Somogyi and Kramer were correct in their denial, their data contained nothing in proof. They used three analytical methods; the Folin-Wu and two titration methods. The lowest blood sugar values obtained before fermentation were with the Folin-Wu method, indicating the non-specificity of all three methods. Therefore, they could hardly make valid statements about the non-glucose reducing substances.

Somogyi and Kramer had also modified the Folin-Wu alkaline copper reagent by substituting Rochelle salt for the tartaric acid; and they concluded that the special sugar tubes were unnecessary to prevent reoxidation of cuprous oxide. Otto pointed out that the Folin-Wu tartrate reagent would keep indefinitely and produced little autoreduction (blank) with heat. This was accomplished by keeping the tartrate (and copper) concentration as low as possible, contrary to the increased amount used by Somogyi and Kramer. Keeping the blank color minimal was essential for measuring low concentrations of sugar, and Somogyi and Kramer had made no studies of the blank. This constricted sugar tube helped reduce losses that could be as great as 25% if tubes of 15–16 mm in diameter were used.

Otto stated euphemistically that Somogyi and Kramer, who had worked extensively with titration methods for sugar analysis, were lacking experience in colorimetry, and consequently did not get consistent data with the Folin-Wu method (nor with the Hagedorn-Jensen procedure).

In his next paper Otto provided an updating of the Folin and Folin-Wu methods for blood sugar determination (*17-7*). At low concentrations of sugar (50 mg/dL) a loss of proportionality (color) was now corrected by diluting the 6 mL in the Folin method (or 4 mL in the Folin-Wu) to 25 mL in the sugar tubes with dilute phosphomolybdate rather than with water. The dilution did not help as much in the Folin-Wu method when old reagents were used.

Otto discussed the basis for his alkaline copper reagent. Several factors had to be considered. A balance of tartrate concentration was needed between its inhibition of the reducing power of sugars and its tendency to autoreduction. But the optimal concentration to achieve this balance had to be somewhat tempered to get proportionality (of color vs. concentration range). The more alkaline the reagent, the faster it deteriorated, and the greater the blank. The deterioration could be somewhat delayed by refrigeration.

Otto reviewed points in the Folin method that caused problems to the user. Storage in tightly stoppered bottles was essential to prevent a rise in alkalinity. The 10-min heating period was chosen to give proportionality, but this could be extended to 12–15 min (no more) for slight improvement in proportionality and 20% more color. Details were presented for preparation and storage of the reagents, including the dilute phosphomolybdate.

Otto had been using the Bausch and Lomb model of the Duboscq colorimeter for many years. Because its cups were enclosed in a metal frame, the acid molybdate reagent tended to corrode the metal, and there was seepage of solution between the glass and the metal. This seepage could be prevented by periodically replacing the rubber ring at the bottom of the cup. The space between the cup and the glass at the top of the cup could be kept sealed with stopcock grease, and the rim could be protected from molybdate by daily rubbing it with beeswax. In modern parlance, Otto was describing a preventive maintenance program for the colorimeter.

Otto was now convinced that saturated or nearly saturated benzoic acid was a reliable preservative for stock glucose standard solutions. He preferred to make fresh "working" standards, but they could be preserved with a few drops of toluene or formalin.

The glucose methods were meant to be used in the range of 50–200 mg % (2.78–11.10 mmol/L). Deviations that occurred at the two ends of the range were constant, so that corrections could be made from a table arrived at empirically.

Otto and Agustin Marenzi converted the Folin-Ciocalteau method for determining tyrosine and tryptophane in 1–2 g of protein to a micromethod requiring only 0.1 g (17-8). In doing so, several revisions of the procedure were made, including a change in the system of protein hydrolysis.

But first of all there was a need to defend the Folin-Ciocalteau tyrosine method against criticism that it (along with others) had received from Hanke. In Hanke's method for tyrosine, published in 1922, he removed histidine with silver oxide, and in doing so, also removed a substance that he labeled an "X factor," the presence of which, he claimed, proved the superiority of his method. It had turned out that this mysterious substance was tyrosine, so that all of Hanke's work and his claims were false. What is more, he had abandoned the procedure and was now using a modified Folin-Ciocalteau method. But Hanke had not tried recovery experiments with his modified method. When Otto and Agustin tested it, the recovery of tyrosine varied, but was quite poor at lower concentrations.

But Hanke did raise one useful point. Cystine, or its residue in alkaline hydrolysates of protein, could precipitate with the mercuric sulfate used in the Folin-Ciocalteau procedure and cause turbidity. Folin and Marenzi studied this possibility and found that cystine did not interfere. The tyrosine value remained the same whether it was determined from acid or alkaline hydrolysates.

For the hydrolysis of protein in the micromethod, Folin and Marenzi replaced Kjeldahl flasks with test tubes to which air-condensers were attached. Hydrolysis of 0.1 g of protein was carried out in 2 mL of 20% NaOH, in a steam or in a water bath for 12–18 h. The hydrolysate was then acidified, diluted to 25 mL, shaken with some kaolin, and filtered. The filtrate (20 mL) was then used in the determination and scaled down from the Folin-Ciocalteau procedure for tyrosine and tryptophane, as previously given.

Although tyrosine and tryptophane could not be determined in casein with the new method, probably because it required a higher temperature of hydrolysis, the other proteins tested gave reliable results, including egg albumin, cottonseed globulin, gliadin, edestin, hemoglobin, and zein.

Folin and Marenzi then improved the Folin-Looney method for cystine analysis in proteins (*17-9*). Cystine was unique among amino acids in that it reacted with the uric acid reagent, but if the reagent were contaminated with phenol reagent (molybdate), then some tyrosine would be included with the cystine values. There was also too much blank in this method, and the rate of reaction was not optimal because cystine did not react as rapidly with the reagent as did uric acid.

In the new modification, a purer uric acid reagent was used (to be described in the next Folin and Marenzi paper) that eliminated the phenol reaction with tyrosine. The step requiring the conversion of cystine to cysteine with sodium sulfite was speeded up remarkably by adding this substance before rather than after the carbonate. This change not only reduced the amount of sulfite used, but made the blank negligible. The proportionality range was improved by making the final dilution in dilute sodium sulfite instead of in water, thereby eliminating a bleaching of the blue color. There were several other minor refinements of the Folin-Looney method that in toto constituted major improvements.

This was not a micromethod. The acid hydrolysis of protein was carried out for 18–20 h on 1–5 g of protein in a Kjeldahl flask equipped with a vertical condenser. The hydrolysate was treated with kaolin to decolorize it, filtered, and diluted to a standard volume. An aliquot was then treated as in the Folin-Looney procedure, with the prior addition of the sulfite; then the carbonate, lithium sulfate, and the uric acid reagent were added.

Values for cystine now obtained on a number of proteins were higher than those previously reported (except for gliadin), but they were considered more valid, in view of the improved procedure.

The uric acid reagent had to be free of phenol reagent, as mentioned, so that tyrosine would not interfere. For this to happen, the contaminating molybdate had to be removed from the sodium tungstate used. In 1924 Folin and Trimble had made an attempt to eliminate the molybdate. Wu had shown (1920) that the reactive material in the uric acid reagent was what he labeled phospho-18-tungstic acid, whereas ordinary phosphotungstic acid consisted almost entirely of inactive phospho-24-tungstic acid. Manufacturers had not provided a better uric acid reagent. In fact, except for Merck's best "reagent" tungstate, the present brands apparently contained more molybdate than ever.

In the next paper Folin and Marenzi published the revised Folin-Trimble method for purifying sodium tungstate (*17-10*). They made five observations on which they based their modifications.

- Sodium molybdate did not combine with H_3PO_4, but sodium tungstate did at room temperature.
- After adding H_3PO_4 to sodium tungstate, the uncombined molybdate reacted fairly rapidly with H_2S.
- Most of the molybdenum was converted to insoluble sulfide by H_2S and could be removed by filtration.
- Part of the molybdate was converted to a soluble sulfur compound with H_2S, and more soluble sulfide formed with longer treatment.

- The sulfomolybdate was *very* soluble in alcohol and could be quantitatively removed by a single 5-min extraction with alcohol.

On the basis of these findings they provided a method for converting 100 g of sodium tungstate completely free of phenol reagent to 1 L of uric acid reagent in about 2 h.

> The reagent which we obtain by the process described above is better than any uric acid reagent we have ever had before in this laboratory, not only for cystine determinations, but also for uric acid determinations. It is very active, yet possesses the highest obtainable degree of specificity, hence gives the minimum of blanks, with sulfite, cyanide, etc., and does not give a trace of color with tyrosine. In addition, it shows less tendency to give disturbing precipitates than do our other uric acid reagents.

Otto's micromethod for blood sugar, published the year before, now received further study and modification with Malmros (*17-11*). The result was one of Otto's finest papers, another hallmark in clinical chemistry. In this paper, the range of blood glucose concentrations that could be measured was greatly extended to meet the needs of diabetic clinics. It was made possible primarily by extending the application of colorimetry through the introduction of a filter.

The original micromethod was limited in its applicability to blood sugar concentrations of 50–200 mg/dL (2.78–11.10 mmol/L). For the diabetic, a preferable upper range was at least 600 mg/dL (33.3 mmol/L), and a rapid (or stat) analysis was needed.

The main problem to solve was how to measure reliably the Prussian blue color in the presence of the yellow ferricyanide background. The answer that Folin and Malmros found was to make and use a yellow light screen for the colorimeter. This was accomplished by staining filter paper a deep yellow with picric acid and preserving it with paraffin, then pasting it over the lamp window of the colorimeter, and finally testing to make sure that two fields in the colorimeter cups were equal with the filter in place.

> The success which we have obtained with the picric acid light filter leads us to believe that by the help of suitable light filters (in the form of colored paper or in the form of solutions) the scope of practical colorimetry may be greatly enlarged.

With the modified method the range of the determination of blood sugar was extended: 25–400 mg % (1.39–22.2 mmol/L) in a single determination, and approximate values up to 700–800 mg % (38.9–44.4 mmol/L).

> When such light filters are used, the method becomes rather more photometric than colorimetric, for the blue color is mostly replaced by light absorption, and the end-point in the colorimeter adjustment is more or less like one observes in a polariscope. The two fields remain different in two respects, namely intensity of light and quality of color until the point of equality is reached.

> This comparison seemed easier to make and surer than the usual color comparison where the quality of color was uniform, and only the intensity

varied. Once the filters were introduced, the stage was set for the introduction within the next decade of the photocolorimeter, and then the spectrophotometer into clinical biochemistry.

Now that the three Folin methods had been improved, it was time to take another look at the non-glucose reducing substances in blood. For this, not only would these methods be used, but Otto, with Malmros, would examine the most popular sugar method in Europe, the Hagedorn-Jensen procedure (17-12). The latter method, Otto originally thought, might help solve the non-glucose question, but early work showed that it had no new information to offer on that score and was more laborious to perform than other colorimetric methods.

Folin and Malmros improved the Hagedorn-Jensen method. In this analysis the protein in 0.1 mL of whole blood was precipitated in a gelatinous suspension of zinc hydroxide. The filtrate derived was heated with alkaline potassium ferricyanide, and the reduced ferricyanide was determined by adding an iodide solution and acetic acid, then titrating the iodine set free with standard thiosulfate, using starch as an indicator. For the recovery of glucose quantitatively, it was essential, they found, that the cotton plug used in the filtration be of the right size, or significant losses occurred.

The venous blood sugar of 28 subjects (including diabetics) was measured by the three Folin methods and the Hagedorn-Jensen procedure. The two ferricyanide methods gave nearly identical results. The few instances of differences between them, attributed to the oxalate or fluoride in the blood, would therefore probably not happen with capillary blood collected without additives. The Folin-Wu values were the highest and the Folin values the lowest of the four methods.

Folin and Malmros now repeated the work of Folin and Svedberg on the fermentable sugar in the blood of another 28 subjects, mostly diabetic. They now used a washed yeast preparation and a fermentation period of 10 min.

In 14 of the 28 samples, results by the two copper methods were essentially the same. Differences in the other 14, however, were outside the limit of experimental error. Our investigators made certain of this by doing recovery experiments.

> . . . we never obtained a difference of over 2 mg. percent in the triplicate analyses of nine different samples of blood whose sugar content ranged from 78 mg. percent [4.33 mmol/L] to 332 mg. percent [18.3 mmol/L].

The difference between the two copper methods (Folin-Wu vs. Folin) in fermented filtrates was (in those 14 cases) at least several milligrams and, in one case, almost 10% of the total fermentable sugar. There was no question that some non-glucose reducing sugar was present in these bloods and that the work of Folin and Svedberg was confirmed.

It is now known that fructose, mannose, galactose, and pentoses are absorbed into the blood in traces by the portal system, depending on the amount of intake. Lactose has also been reported in the blood of some lactating women. As Folin and Malmros were analyzing primarily random samples obtained from perhaps a good number of non-fasting subjects, their results were both valid and understandable.

Otto kept Blatherwick at the Metropolitan laboratory informed of his progress in sugar analysis. He had begun to collect data on the "normal" glucose tolerance test using a 50-g dose, with blood samples taken before the glucose was given, and at 30, 60, 90, and 120 min after. The subjects were primarily medical students. On March 24, he wrote about the picric acid filter and how it greatly extended the range of the colorimeter to meet the needs of the diabetic in a single determination. He sent Blatherwick a few of the filters.

Blatherwick was working on a method for blood collection that the field examiners could use, and was trying both the "spot" collection and the pre-weighed container ideas. He was using the Folin (1928) micromethod for the sugar determination. On April 10, Otto wrote:

> . . . we have by dint of considerable effort, secured a remarkable improvement in the micro method along the lines indicated in my other letter, by the use of light filters. . . . The method has become very much easier to carry out by those who are not expert, and I think that you would do well to incorporate the new form in your application to mail order blood. . . . If you are skeptical about this, as you well might be, and if you would like to see the application, I could come to New York and show it to you, for it is now past the experimental stage. . . .

After further correspondence from Blatherwick, Otto planned to arrive on Friday morning, April 19, and recommended:

> Your place ought in part to serve as a teaching laboratory, and it would not be bad if something of that sort could grow out of your filter paper adaptation, since by that process you might be called upon to check up both their standard solutions and their results on blood. That would be very useful work.

In a letter of May 13, Otto reported to Blatherwick that he had completed 22 glucose tolerance tests on first-year medical students, primarily to get information for use in insurance examinations. They could make a joint report of their findings at the autumn meeting of the medical directors, or if Blatherwick preferred, he could report it himself, and would certainly be welcome to the data.

By June 26, Blatherwick had agreed to a request by the president of ALIMDA that he join in the discussion of the laboratory papers at the autumn meeting. In addition, two papers from the Metropolitan would be given, "Blood Sugar Curves After the Ingestion of 50 Grams of Glucose," by Otto Folin and N. R. Blatherwick, and a "A Blood Sugar Method for Field Use," by N. R. Blatherwick. Otto would present the first paper. Blatherwick reported to Otto that he had two people working on the blood collection problem. The pre-weighed bottle system was probably "out," and there had been problems with the "spot" collection, too.

Otto wrote on July 1 that he had now had results of tolerance tests on 50 normal young people. He was scheduled to speak before the medical alumni in Minneapolis in the autumn and hoped that there would be no conflict with the

ALIMDA meeting. If so, Blatherwick would have to give both papers. If necessary, Otto could come to New York any time before the meeting to talk things over, except between Aug. 13 and Oct. 1. Part of that time Otto had set aside for the 13th International Physiological Congress, which was to take place in Boston the week of Aug. 19-23.

In letters of July 10 and Aug. 13, Blatherwick included results of 31 glucose tolerance tests he had obtained, mostly on laboratory-associated people of all ages. Otto would prepare the written report for the Proceedings of ALIMDA (*17-13*). Otto had glucose tolerance reports on 50 subjects, mostly students, and another six subjects on whom he had done multiple tolerance tests to determine whether warming the glucose solution to be administered would improve its palatability, as well as the effect of "fasting."

At the 13th International Physiological Congress, Otto Folin and H. Malmros presented a demonstration entitled "Improved Micro Method for the Determination of Blood Sugar." Malmros, on his own, had a demonstration of "Lundgren's Angle Centrifuge." More than 1700 people attended this meeting, and certainly the host group of physiologists, pharmacologists, and biochemists in the Boston area must have worked hard to ensure the success of the meeting and make foreign guests comfortable. Among those biochemists attending the congress were a number who would greatly influence clinical chemistry in the U.S., particularly in the methodology: W. R. Bloor, M. Bodansky, A. B. Hastings, P. B. Hawk, L. J. Henderson, V. C. Myers, Joseph H. Roe, J. Sendroy, Harry Sobotka, W. M. Sperry, and H. Wu.

In his talk at the ALIMDA meeting held in Newark, N.J., Oct. 24-25, Otto discussed the implications of his and Blatherwick's findings with the glucose tolerance test. For testing normal people, a 50-g dose of glucose was preferable, perhaps, to a 100-g dose, because the blood sugar rose about as high, and gave typical results. It was not necessary to have the subjects tested in a fasting (overnight) condition. Of the 87 subjects reported, most had eaten breakfast 2-3 h before the glucose was given. In short, a 2-h fast was sufficient.

With the 50-g dose, Otto reasoned, glucose concentration would be apt to return to the normal level faster. Though highest values would be reached in about 30 min following the glucose intake, the maximum height was not of any significance (within limits). The important point to stress was whether the sugar value "will come down to the normal level at the end of a certain definite period. In this case, we chose two hours as the period at which the blood sugar level ought to be back to the normal." This final sugar value would perhaps be no more than 120-125 mg % (6.66-6.94 mmol/L). The highest figure that their data showed at the end of 2 h was 119 mg % (6.61 mmol/L).

Otto had advocated at the previous year's meeting the use of cane sugar (sucrose) rather than glucose for tolerance tests because sucrose was purer and tastier; and again he made a pitch for it. Glucose, in large doses, had caused nausea in some subjects. And warming it to 50-60 °F, or hotter, was definitely no help.

Otto cautioned that among people he had tested, there were four or five who showed urinary sugar at the end of 2 h. These results were simply due to renal glycosuria, but were subject to misinterpretation, if only urine sugars were determined. The blood sugars were normal. There was a time when glucose was given and only the urine was tested to rule out diabetes.

Blatherwick added commentary. The method used was the Folin micromethod. Most of Otto's subjects were comparatively young (male) medical students, whereas his own varied in both age and sex. The blood sugar at the 90-min interval was lower in 30–40% than at the end of 120 min, for which he offered no explanation.

In the department of biological chemistry for the 1929–30 academic year, Cyrus Fiske was promoted to associate professor. Three research fellows left: Cavalcanti, Marenzi, and Munoz. There were three new teaching fellows: Irvin S. Danielson, M.S., George H. Hitchings, S.M., and Clarence A. Morrell, A.M. Steven Maddock, who remained as research fellow, was also listed now as a Fellow of the National Research Council.

Nineteen thirty saw the last of Otto's unbroken productive output insofar as publications were concerned. He remained the heart of his department: the "Chief"; the friend and adviser; the skilled lecturer, full of humor, the quasilegendary figure; the apt target of medical students' jokes. But his youthful vigor was waning. His blood pressure had risen alarmingly, and his "waterworks" had slowed down, as he later wrote. Aging had become obvious, and his declining health was an important factor of his ebbing years.

Otto's life was full of many interrelated laboratory themes, of which the foremost was the analysis of uric acid. Of all the blood and urine substances he had worked on, uric acid had received the most attention, beginning in Salkowski's laboratory in the summer of 1897. Despite all of the progress he had made, even with Benedict's significant contributions, the determination was beset with severe inaccuracy. Besides the recurring problem of preparing a molybdate-free uric acid reagent from sodium tungstate, it had two disturbing factors: 1) poor recovery of uric acid whether it was added to the Folin-Wu filtrate or to the original blood sample, and 2) a "blank" of non-uric acid substances that reacted with the molybdate-free uric acid reagent. Both factors, Otto found, were due to substances released when red blood cells were laked (disintegrated) in the preparation of the Folin-Wu filtrate. The poor recovery was related to inhibitors released from the cells. In short, there were disturbances of the color reaction produced from both a positive and a negative direction. Could the red blood cell also be the nemesis for the other analytes determined in blood?

To prove that laked red blood cells were anathema for uric acid analysis, Otto ingeniously devised a method for preparing protein-free filtrate without disintegrating the cells (17-14). By cutting the strength of the tungstate and sulfuric acid used in the Folin-Wu precipitation of blood proteins and by adding sodium sulfate, Otto could make a water-clear filtrate of blood in which the red blood cells, though in a probable hypertonic medium, remained intact, while the plasma proteins were precipitated.

Now, when he tried recovery experiments with uric acid by adding known amounts to the unlaked blood filtrate, or to the blood before the protein precipitation, he was able to eliminate the "blank" almost completely, and to obtain quantitative recovery of added uric acid. He invited the reader to repeat his observations and provided detailed experiments for carrying this out.

But Otto was apologetic. All of the Folin-Wu procedures had been repeatedly modified since 1920 except one, the preparation of the filtrate. Now, he himself was toppling this last bastion of consistency. In doing so, however, Otto had reached a major crossroad. There were two avenues open for avoiding problems from erythrocytes: either use unlaked blood, or change to using plasma or serum. The strongest argument for unlaked blood was that the present "system" of analysis could be retained totally. One filtrate was being used universally to analyze a number of substances. The change in the precipitating agent would be relatively minor—in view of the benefits gained—not only in uric acid analysis, but for sugar, NPN, and amino acid analysis (as we shall see), and probably others. Plasma, on the other hand, represented only about half of the total blood sample collected so that methods would have to be modified to work with smaller volumes. The "normals" for plasma were entirely different, so that these would need fresh study.

One must conjecture that these barriers would hardly have prevented Otto from developing the system for plasma analysis a few years earlier. It was a drastic step, but he would probably have taken it. However, as a consultant to a major life insurance company, he must have been well aware that the medical examiner in the field had no access to a centrifuge to separate the red blood cells from the plasma or serum. And opening such a Pandora's box in methodology might have brought much condemnation from the hospital laboratories. Furthermore, using plasma meant that Otto would have had to start over in many respects, so at least for the moment, he chose not to do so. Meanwhile, unlaked blood had much to commend it, and Otto now provided the proof.

In his second paper of 1930, Otto made major improvements in the uric acid method, not only because the filtrate from unlaked blood had removed interferences in the analysis, but because there were better reagents to use (17-15). For this study, Otto acknowledged the help of Andrea Svedberg and a candidate for her master's degree at Radcliffe, Katharine Jones.

To avoid turbidity, to maximize the blue color reaction, and to promote stability, the sodium cyanide used had to be pure. Lithium salts and urea helped prevent turbidity. Refrigeration preserved the sodium cyanide–urea solution for many months. Slow formation of carbonate weakened the cyanide reagent. A stock uric acid standard, stable for at least five years, was made by adding to uric acid, lithium carbonate, 40% formalin, and sulfuric acid. Working standards were prepared fresh from the stock and were stable for many days. Of course, the new uric acid reagent, free of phenol reagent, was to be used.

The procedure remained simple. It was carried out on 5 mL of protein-free filtrate to which 5 mL of the cyanide-urea and 1 mL of uric acid reagent were added. After standing 4 min, it was heated for 2 min in a boiling water bath, diluted to 25 mL, then read against a standard in the colorimeter.

When Otto compared the uric acid values of unlaked and laked-blood filtrates from 10 patients, he found that the Folin-Wu (laked) filtrate gave deceptive values. When the uric acid was elevated, the values were too low because of poor recovery of uric acid, and when the values were low, they were nevertheless too high because the "blank" was a part of the value found. Otto showed that the recovery of uric acid added to unlaked blood was 96–104%, whereas for the laked blood, the recovery was 64–71%. The unlaked blood needed much more testing, however. Otto could only conclude, now that he had unmasked the fallacy in the Folin-Wu filtrate,

> Whether or not much valuable information has been hidden by this deceptive character of the uric acid determinations, it is impossible to say, but it is certainly a fact that as yet there is not available any true picture of the fluctuations of uric acid in human blood.

Andrea Svedberg published three papers with Otto dealing with the nonprotein nitrogen constituents of blood. In the first paper, she worked on an improved method for distilling the ammonia liberated by urease in the urea determination into Nessler's solution (17-16). The earlier distillation method was unpopular, compared to the aeration technique, because of three factors: bumping, foaming, and back-suction. All three of these obstacles were now overcome. A new, very simple microdistillation (test tube) apparatus was devised and the reagents used were improved.

Usual antibumping materials used had proved futile, so the remedy provided was an old one, "a fine glass tube closed at one end and open at the other," which would be put into the reaction tube (Pyrex or Jena test tube, 30 mL) for the distillation. Foaming was prevented during distillation by the addition of two drops of toluene-diluted fuel oil used in oil-burning furnaces (nothing more refined). Back-suction was prevented by increasing the capacity of the delivery tube (a bent 5-mL pipette). Use was made of a strip of urease paper (1 cm × 2.5 cm), a phosphate buffer, borax solution, and Nessler's reagent. Gum ghatti was added to stabilize the color produced in Nessler's solution by NH_3.

Folin and Svedberg next used the unlaked blood filtrate to establish and update micromethods for NPN, urea, uric acid, and sugar (17-17).

> There is a perfectly legitimate place for micromethods in so far as they can be made reasonably accurate. Such methods are not only indispensable for studies on small animals, but they should also prove important to life insurance examiners, and even in regular hospital laboratories they are often useful, because of the limited amount of blood that is sometimes available.

Obviously the needs of the pediatric patient were not yet a mentionable consideration. In these methods, the goal was to use about 0.1–0.2 mL of blood per test, perhaps an order of magnitude greater than the true need established more than a generation later. Yet this was a major step in the right direction, whatever the motivation. (Andrea, who as a physician, had worked in clinical laboratories in Stockholm, must certainly have felt the need for microchemistry among many of the patients, particularly the elderly, the obese, and others with veins difficult to puncture, not to mention children. She would take back to Sweden with her some highly practical methods, a step beyond the valued procedures in the fourth edition of Folin's laboratory manual, and those of Hawk and Bergeim.) In this paper the methods were presented individually, not as a system derived from a single protein-free blood filtrate.

For the NPN determination, 0.2 mL of blood was used, but it was possible, if necessary, to use 0.1 mL. An unlaked protein-free filtrate was prepared by adding the blood to 4 mL of the sulfate-tungstate solution; and after a 15-min stand, 1 mL of dilute sulfuric acid was added. After centrifugation, 4 mL of supernate was digested as in the macro-NPN procedure, and the final digest diluted and nesslerized. Values found on 12 samples tended to be slightly higher than those of the macromethod. The urea determination was that described in the preceding paper.

Uric acid was also performed as Otto had already published. But the authors could not forgo mentioning that the volume of blood required for the determination was dramatically diminished. In 1907, Brugsch and Schittenhelm and others had used 150–300 mL of blood for one analysis, and evidently were able to get significant values of uric acid with as little as 75 mL. In 1913, Folin and Denis, with their highly original phosphotungstate reagent, succeeded in determining uric acid on 15–25 mL of blood. Now it could be done on 0.2 mL of unlaked blood, using 4 mL of filtrate. The micromethod agreed well with the macromethod in the normal range of uric acid values, but was a bit less accurate at higher values.

For sugar analysis on unlaked blood with the Folin-Malmros technique, some new modifications were essential. The method was not suitable for urine because of its excessive interferences. Two simple methods could be used. The first was to prepare a filtrate as in the NPN procedure with 0.1 mL of blood and to use 2 mL of filtrate for the sugar analysis. Using 0.1 mL of blood ensured the removal of traces of protein. After the 8-min boil in a water bath, gum ghatti was added to stabilize the final color. The second method combined the two protein-precipitating solutions into one reagent (20 g Na_2SO_4, 10 mL of 10% sodium tungstate, and 10 mL of ⅔ N H_2SO_4 in 250-mL volume). In this method, 0.1 mL of blood was added to 2 mL of the tungstic acid–sulfate solution. After centrifugation, 1 mL of the supernate was reacted with 1 mL of ferricyanide. Of course, for both methods, the picrate light filter was used. Both methods gave values virtually identical with the Folin-Wu method on unlaked blood protein-free filtrates.

In their final paper together, Otto and Andrea opened with a sanguine condemnation of the red blood cell (17-18). Otto, in particular, was seeing red.

> ... the conclusion seems inescapable that for nearly every kind of metabolism study, it must be obscuring and misleading to include in the analyses of blood as a transportation system, an unknown mixture of residues obtained from the destroyed cellular elements which have no direct connection with the circulating waste products or food products.

However, the use of unlaked blood for analysis would cause a break in the continuity of laboratory data because of the switch from values obtained with the contaminated laked blood. By using unlaked blood, however, there would be some immediate advantages. For example, methods such as blood sugar would not require so much modification. With unlaked blood, the Folin-Wu analysis now gave answers identical with the Folin modification so the latter became superfluous. The non-sugar reducing substances were eliminated in the unlaked blood.

The purpose of this paper was to compare laked and unlaked blood. But a more important issue was attacked: the differences between unlaked blood and plasma.

What was the distribution of diffusible products between plasma and corpuscles (red blood cells)? Ege and Roche [*Skand. Arch. Physiol.* **59**, 75 (1930)] had recently published that sugar was distributed equally in the water of cells and plasma on the basis of a (water content) ratio of 80:100, cells to plasma. From the hematocrit and the sugar values in plasma and whole blood (unlaked), Otto and Andrea calculated the sugar values in the cells, compared them with those in plasma, and found them to be in a ratio of 59:100 for 15 normal subjects (fasting) as compared with 69:100 for 10 diabetics. The lower value for the normals could have been due to the assumed hematocrit of 42% for all of the mostly male individuals, obviously too low. Later work by others would show that the distribution was indeed a matter of water content.

Unlaked (whole) blood gave values in the 15 normal subjects (above) about 17% below the plasma values. Laked blood gave values 11–17 mg % above the unlaked blood, and this was almost completely attributable to the presence of non-sugar reducing substances. There was hardly an excuse left to use laked blood for sugar determination.

If the sugar in the unlaked corpuscles was determined after extracting it directly with salt solution, how close would these values come to the calculated values from unlaked blood and plasma? For seven samples tested, the calculated values fell between 91 and 98% of the determined values, a solid vindication for the use of unlaked blood in sugar analysis. Unlaked blood, then, provided true values of "whole blood" sugar.

From determinations on the unlaked blood and plasma of 10 subjects, Otto and Andrea were willing to concede that urea was distributed according to the water content of corpuscles and plasma in the ratio of 80:100. This was not surprising; Folin and Denis had shown in 1912 how readily urea distributed between blood and tissues in general.

It was expected that creatinine should be distributed in plasma and cells according to the water content. But the analytical error was too large to permit this conclusion. The values found, however, for the ratio of cells to plasma in 10 subjects averaged 80:100, but the range was 58–90%. However, use of unlaked blood was a vast improvement over laked blood because it eliminated interferences in corpuscles that previously caused laked blood values to be much larger than those in plasma. The creatinine values from the unlaked blood of the 10 students tested averaged 1.09 mg/dL (96 μmol/L), and that of plasma, 1.22 mg/dL (107.8 μmol/L).

Otto had already demonstrated the superiority of unlaked to laked blood for uric acid analysis. In fact, uric acid was the reason that prompted him to look for an alternative to laked blood to get rid of interferences from the red blood cells. The work of Folin, Berglund, and Derick had shown that the distribution of uric acid in tissues was small. The average ratio of distribution now found between corpuscles and plasma was 22:100 in 10 subjects tested.

Earlier analyses of amino acids (and creatinine) had shown higher values in whole blood than in plasma. Because Van Slyke and the team of Folin and Denis had shown that values in tissues were greater than those in blood after amino acids were administered, it seemed reasonable that whole blood (and therefore, the red blood cells) should have higher values of amino acids than plasma. But the results of the analysis of unlaked blood and plasma showed that, as in the creatinine determination, the higher values previously reported for (laked) whole blood were misleading. The amino acid content of plasma was much greater than that of corpuscles. Values of the corpuscles of eight subjects ranged from 14 to 50% of the plasma values. Obviously the laked cells released gross interferences, now held back by the use of the unlaked cells.

Previous work had already shown that the NPN of unlaked blood was about 10 mg % lower than laked blood; this difference was found to be due to what had previously been termed the "undetermined" or "rest" nitrogen of the NPN determination. From comprehensive work on 19 students, it was found that 100% of the NPN in unlaked blood was accounted for by the sum of the nitrogen of urea, amino acid, uric acid, and creatinine. There was no rest nitrogen, in contrast with laked blood, which showed about 7.6 mg %. The rest nitrogen in unlaked blood was absent in these subjects before breakfast or 2.5 h after breakfast, whereas it was at 8.0 mg % in the laked blood. With unlaked blood, the NPN for these normal subjects varied between 13.8 and 20.8 mg % (9.9–14.9 mmol/L), whereas for laked blood it was between 25.6 and 42.0 mg % (18.2–30.0 mmol/L). For unlaked blood, the nitrogen was 68% of total nitrogen, whereas it was only about 40% for laked blood.

Incidentally, it should be noted that Folin and Svedberg tested the effect of eating on the NPN values. This effect remains a problem because one cannot always "fast" the patient or subject whose blood is to be analyzed. At 2.5 h after breakfast Otto and Andrea found that the NPN of their 19 subjects had not changed. The amino acid and the uric acid tended to be somewhat higher than the pre-breakfast concentrations, whereas urea was somewhat lower. And, of course, Otto had already shown that normal individuals' blood sugars could be tested reliably about 2–3 h after eating breakfast when the glucose tolerance values should have returned to the near-fasting state.

Could nitrogen retention be better detected with unlaked than with laked blood? Otto and Andrea tested this by determining the NPN and urea nitrogen in the unlaked and laked blood of seven patients with mild or moderate retention from the Peter Bent Brigham Hospital. For three of these patients who had unmistakable nitrogen retention (NPN > 40) (>28.6 mmol/L) it made no difference which of the two blood preparations was used. But in the other four, the NPN of unlaked blood was higher than that of the highest average normal previously studied. Unlaked blood values were also higher for urea nitrogen, but laked blood values were also high in the four patients. In short, both types of blood were effective with urea nitrogen, but unlaked blood was better for NPN.

With this paper the clear superiority of using unlaked blood over laked blood was established. For NPN, creatinine, uric acid, amino acids, and sugar there was an unquestionable release of interfering substances from the laked red blood cell. Only in the case of urea determinations were comparable values obtained. Otto was more than justified in believing that the simple tungstic acid filtrate was no longer the ideal protein-precipitating agent for whole blood.

Among the students who provided the blood samples for Otto and Andrea's last study on laked and unlaked blood was Bradford Cannon, the son of Walter Cannon, who wrote in February 1983:

> My closest association with Folin was as a medical student in 1930. We enjoyed the lectures and particularly his visiting us in the laboratory when we were learning various chemical tests. We were each asked to collect a 24-hour urine specimen. One of these was on the lectern in one of his lectures. In order to test the sugar in the urine he dipped his finger in his mouth for the sweet taste. Of course one had to notice that he put his index finger in the urine and then tasted another finger. We got even with him, however, because we put some white of egg in the urine specimen of one of our classmates. When the albumen was discovered on testing the urine he was immediately rushed to the Brigham Hospital for all sorts of tests before the hoax was finally revealed.

In the autumn, Y. SubbaRow replaced J. S. De Frates as the Austin teaching fellow in the department of biological chemistry. The remaining two research fellows were gone. Norwood K. Schaffer, M.S., replaced S. B. Nadler as a teaching fellow. C. A. Morrell was gone, having received his doctorate. In 1930 SubbaRow also received his doctorate, his thesis entitled "Phospho-creatine." We shall hear more of this epic work a bit later.

At the Metropolitan laboratory, Dr. Blatherwick was still working on the blood collection problem and remained hopeful of sampling dried blood spots for the micromethod for sugar determination. Otto had written him of his new findings on unlaked blood and had suggested in a letter of Feb. 15 that the dried blood spot could be extracted with the sulfate-tungstate reagent, and then the dilute acid could be added to precipitate the proteins. Following up on Otto's preliminary study on eight subjects, Blatherwick was making an extensive study of sucrose tolerance tests, and would report his findings at the fall meeting (41st) of ALIMDA.

At that meeting Blatherwick spoke just ahead of Otto. He had run the sucrose tests on 27 subjects on whom the glucose tolerance had been done the year before so that the data could be compared. While sucrose was advantageous because of taste and purity and the tolerance curve (blood level vs. time) was about the same as that of glucose, three of the 27 had flat curves. When one of these was repeated with glucose, the expected increase was obtained. Also, when the sucrose was given again, and blood samples taken at shorter intervals thereafter, there was no increase, indicating either slow absorption or lack of inversion from intestinal sucrase activity. Failure to get an increase in blood in three cases, Blatherwick felt, was reason enough to drop the idea of using sucrose.

Otto discussed Blatherwick's paper next, before giving his own. He pointed out that it was a matter of interpretation as to whether cane sugar could replace glucose in the tolerance tests. He stressed, as he had done the year before, that the important feature of such tests was that the sugar values should drop to the expected range in 2 h. The height that the sugar reached was of secondary importance, in the absence of diabetes.

Otto continued:

> However, that is not what I stepped up here to talk about and no discussion was called for. What I rather wish to do this afternoon is to talk, as the Chairman had indicated, about another topic, and this time I have written a very brief manuscript on that subject. But before reading this—I am a poor reader anyhow—may I make just a few informal remarks?
>
> It is now exactly twenty years within a few months, I think, since I first seriously entered the field of blood analysis and I have been in that field more or less continuously ever since. Our first papers on the subject were published in 1912 and our first analyses, extensive analyses, of albumin [in] blood were published in 1913. In that paper we first established the normal level of non-protein nitrogen and urea in blood and thereafter made a survey of different types of clinical material, and I remember well the impression that those results have made because of the tremendous differences which we found between the hospital material and the normal materials which we had obtained in our own laboratories from students and assistants. In the latter we have taken nitrogen figures varying only between 22 and 26, whereas, for example, when we took the outside material—we happened first of all to take syphilitics—we found out of 663 only 13 which came down to the normal figures which we had obtained on persons which we had every reason to believe were normal.
>
> Then we examined the insane because to include them in the whole would represent the average population, and there we found that only about 50 per cent of them came down to the normal figures.
>
> Then we took in the nephritics and there we found that in every single case the results obtained were entirely different from our normal values.
>
> I have never forgotten that and it was my intention at that time to keep on making studies of the blood and accumulating data under different conditions, but unfortunately for reasons which it is not necessary to dwell upon here, our new methods, while they seemed very good in comparison with those we used before—there had been something of the sort in Germany and a little something had been published before in this country—our methods seemed to be superior when we began to use those methods, but then as one goes on one takes things for granted. We ourselves saw that there was room for improvement and so, as a matter of fact, the trend of the investigations has taken a different turn and during by far the greater part of these twenty years we have been busy with improving the technique.
>
> A second series of methods was published in 1916. Another system of analysis which has been used for the past ten years came out in 1919. In the meantime Benedict's original method for the blood sugar came in and that also became incorporated in systems of blood analysis, and so we have perhaps given a little too much time to the technique to have accumulated as much information that might otherwise have come. But on the whole, I believe that

the loss has not been great because of that fact. So many others have entered into that kind of work.

And now, if you please, I shall read you a very short paper.

Otto's paper was on "The Determination of Non-Protein Nitrogen in Blood and Its Relation to Nephritis" [*Assoc. Life Insurance Med. Directors Am.* **17**, 319–22 (1930)]. He mentioned that he had been working on micromethods for determining nitrogenous products, methods as easy and simple as that of sugar analysis. He believed that chemical analysis of blood must gradually take a larger part in the examination of life insurance clients. The major stumbling block was the "business risk involved in collecting 5 or 10 cc. of blood from the applicants." If it were not for this risk, Otto felt, blood analysis by life insurance companies would probably have become as common a procedure as it was in outpatient departments of hospitals. On the other hand, merely taking a few drops of blood for a microtest would be far less risky. As a result, Dr. Knight and Otto had "conspired this summer to put into your hands reprints of the paper describing the new micro methods." Of these methods, Otto stressed the one on NPN.

Otto felt that there were more people in the general population with unsuspected kidney damage and high levels of NPN than there were diabetics. Certainly in his audience, Otto stressed, there were more people with high NPN than with high sugar. With 0.2 mL of blood, he mentioned, important information on potential kidney disease could be obtained. This would not be a "routine" chemical analysis, but intended to provide more certain information for doubtful cases where the physical and urine examinations were at the "border-line of acceptance." These cases, as well as elderly people applying for large amounts of insurance, would be the material for the NPN test. This was more practical than carrying out kidney function tests such as the PSP because the latter took too much time.

It had not yet been shown to what extent the NPN must rise before kidney damage is revealed. On the other hand, it was believed that considerable excretory power could be lost before the waste products were not removed from the blood. This needed further study. At any rate, an increase in the NPN would be very significant. Otto closed with a review of his findings with the micromethod on the normal NPN in medical students. During the morning (8 to 12) unless breakfast was eaten, the NPN should not exceed 25 mg % (17.8 mmol/L) in anyone. The students' upper limit was 21 mg % (15.0 mmol/L). Heavy protein consumption could raise it to 30 mg % (21.4 mmol/L) but otherwise any person with 25 mg % (17.8 mmol/L) was open to suspicion of abnormal kidneys.

While the paper evoked a mostly favorable response, its importance was that it helped push the medical examiners and laboratories of insurance companies into using a wider repertoire of chemical tests on their clients; and micromethods such as Otto provided offered a practical means for doing so.

Otto's only publication of 1931 was written early in the year, a sketch on the work of P. A. Levene (*17-19*). Otto had been asked to write a tribute honoring Levene's receipt of the Willard Gibbs Medal for 1931.

Otto discussed the work Levene had begun with Jacobs in 1909 on the purification of yeast nucleic acid. The work cleared up the structure of the two known mononucleotides, inosinic acid and guanylic acid, and proved that a nucleotide was a combination of phosphoric acid and sugar-purine (or pyrimidine). They proved that the pentose obtained from their nucleotide was D-ribose, not previously known to occur in nature, and that it was attached to purine. Levene formulated the hypothesis that nucleic acids were polynucleotides and that yeast nucleic acid was a tetranucleotide, which he proved. It wasn't until 20 years later that Levene identified deoxyribose in animal nucleic acid. In 1931, Levene was at the height of his activity and had already published four papers. (He would eventually publish about 400 papers and make important contributions to many other areas of biochemistry: lipids, lecithins, sphingomyelins, cerebrosides, and more.)

The American Society of Biological Chemists commemorated its 25th birthday at its annual meeting in Montreal, April 8–11, 1931. There were 175 members present, almost half of the recorded membership of 370, composed of "active workers in American universities, colleges, agricultural experiment stations, medical schools, hospitals, and research institutions." A special tribute was paid to the founders, and particularly to its first president, R. H. Chittenden. Thanks for service to the society during its first quarter century were tendered to Henry B. Dakin, Alfred N. Richards, Otto Folin, Donald D. Van Slyke, and Stanley R. Benedict.

Also presented was a progress report of the *Journal of Biological Chemistry*. Ninety-three volumes had appeared since its beginning in 1905. The journal was self-supporting and its reserve fund untapped, thanks to efficient management and steady growth. D. D. Van Slyke had been the managing editor from 1914 to 1925, and since then, Stanley R. Benedict. Otto had resigned as chairman of the editorial committee and P. A. Shaffer became the new chairman.

There seems to be little doubt that the ASBC, with its prestigious journal leading the way, had been a potent force in the development of biological chemistry in the U.S., both as a science and profession. The clinically oriented biochemists were among the foremost contributors to this solidly established new discipline.

Evidently, Otto was hospitalized for about two weeks in June 1931. Earlier in the year he had suggested a research problem on allantoin for a Dr. Larson who apparently was now working for or sponsored by Dr. Blatherwick. On July 6, Otto wrote Blatherwick:

> Your note arrived the day after I left the hospital. I am still a little shaky, but by Saturday of next week, the date preferred by Dr. Larson, I should be able to give him all the time he wants and I shall be very glad to see him.
>
> I cannot go down to the school so please have him come to the house—any time A.M. will be all right.

On Sept. 28, Otto wrote Blatherwick concerning two matters.

> It gave me more or less of a thrill to see your blood sugar outfit, for it shows that it is only a question of a little more time until a beginning will have been

really made in extending the field work to blood analysis. I congratulate you on the neat way in which you have solved the blood collecting problem.

This method would be formally presented the following year. Otto also wrote because he was disturbed by the paper of Exton and Rose that he had been called on to discuss at the coming ALIMDA meeting. He had the galley proof. Exton referred to the successful application of sugar tolerance tests to applicants in the field.

> ... as though the collection and preservation of the blood represented no special problem worth mentioning ... publishing a long series of blood sugar determinations which seem to be utterly meaningless, and which will tend to throw such determinations into disrepute.

Exton and Rose had proposed a two-dose oral glucose tolerance test at the October meeting of ALIMDA. Two 50-g doses were administered 30 min apart, and blood and urine samples were collected at 0, just before the 30-min dose, and at 60 min. They had tried the test on a number of diabetic and non-diabetic subjects. Otto was unimpressed with the preliminary results. The test made use of the peak values to which sugar rose at 30 and 60 min, and of course, Otto insisted that such values were too varied to be relied on.

The Exton-Rose test did receive some vogue in the insurance business, but as with tolerance tests in general, there were too many factors and other diseases that made setting criteria for normals versus diabetics almost impossible. On the other hand, the use of the blood sugar determination or the glucose tolerance test by the insurance industry, as a follow up on glycosuria, considerably sharpened its capacity to screen applicants for diabetes.

The "Rose" involved above with the Exton-Rose test was Dr. Anton R. Rose, who had joined the laboratory at the Prudential Insurance Co. in 1924. He and Otto had been good acquaintances for years, with strong mutual interests. A few years later, Rose wrote to Berglund:

> My Chief [Exton] and Folin were not friendly—their spirits clashed. ... Why did Folin mean more to me than anyone else in our field except Mary S. Rose and L. B. Mendel? Folin came from Småland, as did both of my grandfathers; Folin grew up in Chisago County, as did all my uncles, one of whom—Alfred Bloom—knew him well; Folin went to High School in Stillwater—12 miles south of my home; Folin graduated from Minn. Univ. as did also I; in Chicago Folin boarded with a married fellow chemist student—Sam Swartz—this was my wife's uncle; in Chicago Folin earned some of his cash by working in F. B. Turck's laboratory; when F.B.T. moved to New York it was to my lot to organize his chemical laboratory; when I attended my first Biochemical Soc. meeting, I was welcomed by Otto Folin, and I shall never forget it; when I got mixed up in an Insurance laboratory there was a conference between Prudential and Metropolitan and there I sat next to Otto Folin!

The medical school schedule of courses for 1931–32 indicated that Carl J. Klemme, S.M., had replaced S. Nadler as a teaching fellow.

On Nov. 14, 1931, Otto wrote the following to Philip Shaffer:

Dear Shaffer,
Your letter made me feel guilty and if you don't mind my using a pencil I will write you a little right away. Whether I write to you or not does not matter so far as attitude of sentiment is concerned for among my colleagues you always stand as my oldest and best friend.
The fact is that I had done practically no work since the end of May until just about two weeks ago when I went at it with my old time vigor—and as a result I have, as usual, begun to let mail take care of itself.
One unanswered letter is from Doisy. . . .
The important point about me is that I have got older and toward the end of the last school year my water works ceased entirely to function. Under the circumstances there was only one thoroughgoing thing to do and I took the chance. The removal of the prostrate seems to be a risky proposition and I have noted, since May, the departure of at least half a dozen well known people following a similar operation, but mine came off very well indeed. And except for the fact that I get tired more easily than formerly I am all right and very happy to be through with it.
The rest of the Family are well. Teresa is in her fourth year of medicine (at Johns Hopkins) and Grant ("George") is married and is doing well enough.
You are evidently working hard and effectively and in a very difficult field. Your last paper seemed particularly fine and clearcut. With that kind of work I suppose that your Ph.D. men must get good training just as Fiske's men do here.
With respect to the JBC and Ed. Committee I have gradually come to feel that the present arrangement has serious shortcomings since the journal is the property of a national society. The managing editor has too much power. The time will come I think when something like a system of regional editors for at least a preliminary acceptance or rejection of manuscripts will be demanded rather than have everything clustered in New York. My only reason for not feeling more strongly now about the matter is the fear that at least in some sections the caliber of work submitted and accepted would become poorer.

However, Otto's health was not fully restored. He had begun to have circulatory problems, particularly to his feet, and the aches were reminiscent of the time when he had worked at the boom in Stillwater.

References

17-1. Folin, O., and Ciocalteau, V., On tyrosine and tryptophane determinations in proteins. *J. Biol. Chem.* **73**, 627–50 (1927).
17-2. Folin, O., Trimble, H.C., and Newman, L.H., The distribution and recovery of glucose injected into animals. *J. Biol. Chem.* **75**, 263–81 (1927).
17-3. Folin, O., A new blood sugar method. *J. Biol. Chem.* **77**, 421–30 (1928).
17-4. Folin, O., Supplementary note on the new ferricyanide method for blood sugar. *J. Biol. Chem.* **81**, 231–38 (1929).
17-5. Norman R. Blatherwick, Ph.D. *Metropolitan Life Insurance Home Office Bull.* **56**(13) (Jan. 25, 1961).
17-6. Folin, O., The nature of blood sugar. *J. Biol. Chem.* **81**, 377–79 (1929).
17-7. Folin, O., Two revised copper methods for blood sugar determination. *J. Biol. Chem.* **82**, 83–93 (1929).

17-8. Folin, O., and Marenzi, A.D., Tyrosine and tryptophane determinations in one-tenth gram of protein. *J. Biol. Chem.* **83**, 89–102 (1929).

17-9. Folin, O., and Marenzi, A.D., An improved colorimetric method for the determination of cystine in proteins. *J. Biol. Chem.* **83**, 103–8 (1929).

17-10. Folin, O., and Marenzi, A.D., The preparation of uric acid reagent completely free from phenol reagent. *J. Biol. Chem.* **83**, 109–13 (1929).

17-11. Folin, O., and Malmros, H., An improved form of Folin's micro method for blood sugar determinations. *J. Biol. Chem.* **83**, 115–20 (1929).

17-12. Folin, O., and Malmros, H., Blood sugar and fermentable blood sugar as determined by different methods. *J. Biol. Chem.* **83**, 121–27 (1929).

17-13. Folin, O., and Blatherwick, N.R., Blood sugar curves after the ingestion of 50 grams of glucose. *Assoc. Life Insurance Med. Directors Am.* **16**, 155–63 (1929).

17-14. Folin, O., Unlaked blood as a basis for blood analysis. *J. Biol. Chem.* **86**, 173–78 (1930).

17-15. Folin, O., An improved method for the determination of uric acid in blood. *J. Biol. Chem.* **86**, 179–87 (1930).

17-16. Folin, O., and Svedberg, A., An improved distillation method for the determination of urea in blood. *J. Biol. Chem.* **88**, 77–83 (1930).

17-17. Folin, O., and Svedberg, A., Micro methods for the determination of non-protein nitrogen, urea, uric acid, and sugar in unlaked blood. *J. Biol. Chem.* **88**, 85–96 (1930).

17-18. Folin, O., and Svedberg, A., Diffusible non-protein constituents of blood and their distribution between plasma and corpuscles. *J. Biol. Chem.* **88**, 715–28 (1930).

17-19. Folin, O., The scientific work of P. A. Levene. *Chem. Bull., Chicago Section, Am. Chem. Soc.* **18**, 99 (Apr. 1931).

Chapter 18. The Final Years (1932–34)

Otto published two papers in 1932. Because the micromethod for blood sugar had been modified in several important details since its introduction in 1928, he now provided the latest, and for him, final updating of the procedure (*18-1*). The blood sugar values obtained by this analysis on unlaked blood were close to true glucose values. As previously mentioned, the non-sugar reducing substances of red blood cell origin were largely eliminated.

The second paper showed that Otto had lost none of his touch as the master chemist. The determination of blood NH_3 nitrogen was a bit of unfinished business (*18-2*). Although Folin and Denis had introduced the aeration-nesslerization technique in 1912, Otto had apparently lost interest in blood NH_3 nitrogen when Nash and Benedict showed that the kidneys accounted for the ammonium salts in urine. No disease had been associated with NH_3, the technique was particularly subject to error, and the normal values were indefinite. But the challenge remained. Who could tell whether it would prove to be important someday?

It was time for Otto to correct an "error" that he and Willey had made. They had abandoned the use of the special ammonia-absorption tube, and therefore this made impossible the analysis of the trace quantities present in whole blood.

Otto redesigned the special absorption tube so that it would fit into a test tube with an inside diameter of 20–22 mm, and provided a figure with its dimensions. Otto then studied the optimal aeration pressure for recovery of NH_3. Although 7–8 L/min was the maximum speed of air current for complete recovery of 0.05 mg NH_3 nitrogen, Otto felt that for the smaller quantities in blood it was best to play it safe at 6–7 L/min. Otto recommended a receiver tube 260 mm long with an outside diameter of about 25 mm and graduated at 25 and 50 mL.

In general, the method consisted of passing compressed air, acid-washed in a large bottle to remove NH_3, into a stoppered 300-mL Erlenmeyer flask containing 10 mL of blood to which an alkali solution (oxalate-carbonate) was added. The air current carried the liberated NH_3 through a tube in the stopper into the absorption tube that was dipped into the Nessler's reagent.

Because the blood foaming problem was great, Otto got around it: "Instead of violently agitating the blood in a test-tube, by means of a rapid air current, I now sweep it out by letting the air current play over the blood in an Erlenmeyer flask." There were no air bubbles or spattering with this technique. The air was admitted via a tube whose point was 1–2 cm long, with an inside diameter of 1.5 mm, and which was kept above the blood surface about 2–3 mm.

After testing various alkalis to liberate the NH_3, Otto found that the best was a mixture of potassium oxalate and potassium carbonate.

To ensure the clarity of the Nessler's product, Otto added 2 drops of gum ghatti to the absorbing solution. The aeration time finally chosen was 40–45 min.

For blood NH_3 nitrogen determination the reagents all had to be NH_3-free. Otto provided a method for removing NH_3 from the oxalate-carbonate mixture.

Otto was able to get quantitative recovery of 0.05 mg of NH_3 nitrogen added to water and to blood. But it was easier to get good recovery than it was to determine the original NH_3 in blood because "These experiments have shown that the escape of ammonia never comes to a sharp and definite end." Otto showed that NH_3 was continuously generated in four 30-min consecutive aeration periods on eight human blood specimens. This was the basis for selecting the 40–45 min period in the determination.

The method gave blood NH_3 nitrogen values that would still find acceptance today. The range of values found (in the first 30-min period) was 60–116 μg/dL (43–83 μmol/L).

Otto did not discuss the collection and preservation of the blood specimen for this determination. Obviously his results were based on rapid analysis of the freshly drawn blood, because blood contains a potent glutaminase that liberates NH_3 from glutamine in unpreserved blood on standing, and the values may rise to 800–1000 μg/dL (571–714 μmol/L). That was why Otto could not find a favorable time period for the aeration technique. Obviously, the shorter the period, the better.

Since his prostatectomy, Otto's health had been "spotty." His vigorous mind had to grasp the fact that his body had lapses of strength. At his best he had slowed down physically, but he was able at least to play golf. At home, a hired man came to remove the ashes, tend to the yard, and do other chores. At Kearsarge, others came to clear the road and trim the trees. No longer could Otto get his "pole" saw ready to cut the dead overhead branches from the evergreen trees. Otto would have attacks of "intermittent claudication" so that he could not walk far. Climbing Mt. Kearsarge required frequent stops if it could be done at all.

In March 1932, Otto was made a member of the "Kaiserlich Deutsche Akademie der Naturforscher." He was sent a formal announcement of it and an imposing document signed by "der Präsident der Akademie, Emil Abderhalden."

The time was ripe for a perhaps final visit to Sweden. Otto had not been to Sweden for 19 years, Laura had never had the opportunity to meet the Folins or partake of their heritage, and Teresa was just finishing her last year of medical school and could accompany them before she began her internship. Accordingly, plans were made for the trip to take place in July. Grant could not come. Not only was he now married, but he had made Otto and Laura grandparents of the only grandchild (son) that Otto would see.

On June 29, Otto and Blatherwick met in New York, probably to discuss the results of their successfully launched field method for collecting blood. Highly encouraging data had been obtained on the mailed samples collected for the glucose tolerance tests. Blatherwick would report these findings at the October meeting of ALIMDA.

Less than three weeks later, Otto was again in New York City, this time with Laura and Teresa, where they boarded the *S.S. Drottingholm* en route to Sweden, Norway, and Denmark for, as the passport stated, "pleasure."

It was a moment of memory, joy, and sentiment. Fifty years had elapsed since Otto had left his family in Åseda. Then, his heart had raced at the thoughts of the grand adventure opening before him in America. Now if it raced, it was with a sense of fulfillment. He had done all that he was able to do. He had achieved more than he had had a right to expect. Could he not afford a little complacency? And what a delight it was to have Laura and Teresa with him! How could he have waited so long? He would see his Swedish past and the present through their eyes as well as his own. What pleasures, indeed, this six-week trip would bring!

When they landed in Gothenburg, a Buick automobile awaited them (arranged by Andrea Svedberg), and Teresa was the chauffeur. They visited Gertrud in Träryd in her neat house with flower beds made lovely through her personal care. They visited Gusten in Almhult. Otto had maintained contact with Gusten during the intervening years, and Hildur had also kept Otto and Laura informed of family events in the homeland.

They drove to Åseda, of course. Much, like the Folin tannery, had vanished, but the old church tower was still there and the house remained. They met a few of Otto's boyhood companions, now gray and hardly recognizable.

In Stockholm, the Folins visited Andrea Svedberg and possibly the Berglunds, as well as Malmros. Andrea was in private practice, but was also a physician for a Stockholm High School for Girls, and at the Stockholm Board of Health clinical laboratory. She also held a teaching appointment in physiology at the Women's Teachers College. Andrea showed them the sights of Stockholm.

The Folin family continued their trip by driving from Stockholm into the fjords of Norway. Otto could not walk much, and had to stop frequently to rest, nor was Laura more mobile, but it was another cherished moment of the trip. They then spent two days in Copenhagen.

At the 43rd ALIMDA meeting, Oct. 27–28, at the Waldorf-Astoria Hotel, Blatherwick unveiled the results of his study on glucose tolerance tests using the proposed "field method" for blood collection. The field examiners were provided with a printed data sheet to accompany each test, with instructions on administering the glucose and on what to record. Also included were a detailed set of directions and figures for drawing blood via skin puncture. Accompanying the paperwork was a kit containing four pipettes (0.1 mL), each containing a measured amount of preservative and specially banded for each timed specimen; rubber bands to seal the pipettes; and a spring lancet.

On the whole, the first trial on 339 cases had been heartening. Another 201 applicants had refused the test. About 47% were collected properly, but some of these were not correctly timed. An abnormal curve was found in 93 instances, and in 55 cases, a normal curve accompanied a significantly high urine sugar. Both types of results were highly important from the insurance standpoint.

In a small series of tests carried out in February, mailed versus unmailed samples showed little difference. Three different blood samples with widely different glucose values were mailed in six separate boxes in a two-week period. A table showed the percentage change in 88 fresh blood samples kept for various intervals of time from 1 h to 12 days. Losses were slight, ranging from 0 to a high of 7.5%, but generally at 2–3%, regardless of the time of standing.

About 14% of the examiners who tried to collect the blood specimens failed because they did not follow the directions. This problem could be readily remedied by appointing one examiner, or perhaps one laboratory, to collect all of the glucose tolerance tests.

Otto was called upon to discuss Blatherwick's paper. To the audience of some 180 people, he said:

> I have not written any comments on this paper. It is so simple and so clear to me, at least, that it does not seem to call for any except that I would say that it is evidently a very clear and careful and probably very important piece of work tending to meet a very serious problem that has, of course, been with us now for a number of years, namely, the problem of how the average physician is going to get the benefit for his patients of the undoubtedly useful analytical technique which is now available in the field of blood analysis. This innovation of Dr. Blatherwick's seems to me a good one and at least one line of attack on this important problem.
>
> I might say in this connection that I ran into another field of blood analysis during this Summer during a little trip which I took into Sweden. This has to do with the determination of alcohol in blood. The whole country there is very much devoted to the use of alcohol in various forms, and there is no objection to it. The people get all the beer and all the wine they want, and in addition to that, up to 4½ quarts of 96% alcohol per month, which seems to be adequate for most of them. Also there is no objection to the amount that they take within the prescribed limits, nor is there any objection to their driving an automobile or a motorcycle while under the influence of all that liquor. But yet, if a person who has imbibed too much runs into an accident, the question comes in whether he was intoxicated or not, and that is the difference between them and us. That fact not only greatly increases the presumption of his guilt, but it also brings with it a very much more severe punishment than if he had been sober. Now the point I come to is as to how they determine whether the man was sober or not. They do that by means of blood analysis. The various people involved in such an accident at once are taken to the police station, or perhaps to the hospital, and the first thing which is done is that a sample of finger blood is taken from him. That small sample is sent to the central office in Stockholm.

There the alcohol is determined, and on the basis of the findings, the man is either sober or drunk. The test is so sensitive that if one takes one glass of 3% beer, it will show that there was alcohol in the blood. Therefore, the determination has to be quantitative and a definite limit has been set which shows some intoxication, and, of course, the test being quantitative, it also shows varying degrees of that state.

The point I am coming to is that if they can consider it worth while to have a mail order business in quantitative blood analysis for the sake of proving whether a man is drunk or sober, I think we ought to be able to use it for the more serious purposes of life insurance, and I hope also for the more widespread need of the medical profession in general. I think that this line of attack of sending blood through the mail certainly represents one important attack on that problem.

I am fairly well in touch with the situation and have seen for a number of years the keenness with which doctors receive every simplified method for the making of such determinations, and it is a question of how they are going to get the benefit of the results. Many of them are, of course, trying to do the work themselves, but one is a little bit afraid of their results, with their meager office facilities, their lack of chemical experience and, above all, their lack of time, and one is a little afraid, of course, of just how the results might be some time. In quantitative work, it must not be forgotten that poor results, very poor results, are worse than none at all.

Just a few weeks ago here this Summer, a new experiment was being made in Massachusetts in that a new colorimeter, which is very inexpensive, has been turned out and with it a set of all the needed standard solutions. Two hundred and fifty thousand were made at once before any attempt was made to sell them and all of them were sold within a short time, so that the supply is now exhausted, showing how widespread is the hope and the need of the medical profession for getting hold of these simple devices to help them in their study of cases.

There is one aspect in Dr. Blatherwick's paper that perhaps it may be worth while to call to your especial attention, and that is the times when the blood samples are taken—one sample before the giving of the sugar, another thirty minutes later, and the final sample two hours after the taking of the sugar. These seem to me to represent the very best that can be done on the basis of 50 grams of glucose. I mention these facts because I had occasion a short time ago to see the results of a questionnaire sent out by Dr. _____, from which I learned that the practices were very far from uniform among the different life insurance companies. It seems to me that it would be better if you settled down to some more uniform definite conditions, and I think that these conditions mentioned by Dr. Blatherwick are as good as any that you can make, and I say this although personally, I don't use glucose anymore. I still prefer very much to use cane sugar. We have again tested our cane sugar and find that 75 grams are quite equivalent to 50 grams of glucose, and will give results similar to it in every way. I have done this on normal persons and Dr. Joslin has, at my request, done it on patients, and his conclusions coincide with mine that 75 grams of cane sugar will give substantially the same results as 50 grams of glucose. That is, however, just a small minor point.

I think that the conditions laid down by Dr. Blatherwick for the making of sugar tolerance tests are so good that you are not likely to improve upon them, and I suggest that you adopt them.

Otto recommended that two pipettes be collected at the 2-h interval in the tolerance tests. While he did not personally put much store in the other sugar values, he felt that because Joslin and others wished to interpret the height to which the sugar concentration rose, the 30- and 60-min samples were to be continued. But to reduce failures, the 2-h samples were critical, whereas the others were not. Otto had not given up on replacing glucose with cane sugar, and this was being further tested on his students, with continued success, and in Joslin's laboratory.

In the autumn of 1932 the only change listed in the department of biological chemistry at the Harvard Medical School was the departure of Carl J. Klemme. The same group would be listed for the 1933–34 academic year: Folin, Fiske, Trimble, and Logan, faculty; and the teaching fellows, Danielson, Hitchings, Schaffer, and SubbaRow. In 1933 under Otto's supervision, I. S. Danielson completed his doctorate on amino acid nitrogen in blood and its determination.

Otto had two published papers in 1933, but only the second was scientific. The first was a special article for the alumni journal that described the research activities of the department of biological chemistry (18-3). The ratio of printed space that Otto allocated to the work of the current faculty, Trimble, Folin, and Fiske, was about 1:2:6, with a mention of Logan.

Otto thought that the department was best known for its blood analysis, particularly the Folin-Wu system of 1919. This method had been changed recently by converting from laked to unlaked blood filtrates to make the analyses more precise and to sharpen the differentiation between the normal and abnormal. Micromethods were introduced, particularly for blood sugar, and these had proved welcome. The development of a dependable method was often enough a modest accomplishment compared to the more startling discoveries in other fields of biochemistry.

"The refinements and improvements in technique which have been to some extent the final objectives in Dr. Folin's work, are but means to an end in the work of Dr. Fiske and his co-workers." Fiske had made several fundamental studies of organic phosphate substances including the characterization of phosphocreatine, adenosine triphosphate, and glycerophosphoric acid. Recently the work had concentrated partly on adenylic acid and its effect on blood supply and heart rate. A dozen new substances had been separated from other tissues; "[they] are being purified and analyzed with a view to establishing their identity as rapidly as facilities permit."

Fiske and Logan's method for determining minute amounts of calcium had found use in blood analysis. Trimble had worked on various metabolic problems. He found that injected sugar in animals was more freely distributed in the skin than in the muscles; and he measured the rate of absorption of sugar from the small intestine. He showed that the level to which the blood sugar rose in the first 30 min had no significance. At present he was studying metabolism after the complete removal of the liver, and the ensuing cause of death. His findings suggested leads for the treatment of partial liver destruction, e.g., in acute yellow atrophy of the liver.

Otto paid tribute to his three colleagues, but the part played by SubbaRow, who remained so long as a research fellow, was sadly unmentioned. Although

The Final Years

organic phosphorus was Fiske's "show," their classic pentad of papers was a joint effort, and SubbaRow's brilliance alloyed with Fiske's brought the work to fruition (*18-4*), just as Otto's great year in 1912 was made possible because Willey Denis could implement and carry out so much research.

The purity of the phosphotungstate used in the uric acid reagent persisted as a problem that had not been solved in Otto's two previous papers on the reagent. Otto, with the assistance of Margaret Cushman, now found that while the last traces of molybdate could not be removed, if the phosphoric acid concentration used to make the phosphotungstate was reduced, the troubling phosphomolybdate (phenol reagent) would not form. The preparation of the reagent was, therefore, revised, and along with it the three Folin methods for uric acid determination (*18-5*).

Otto pointed out that the new purification procedure was ineffective if more than trace amounts of molybdenum were present. He found that his previously favored Merck "reagent sodium tungstate," which was now made "according to Dr. Folin," was unusually rich in molybdate. Thus he would modify the uric acid reagent a "final" time in his 1934 paper.

The urea-cyanide reagent was modified to ensure 100% recovery of uric acid, which was not always possible with the reagent proposed in 1930. Lithium salt was now omitted to prevent turbidity, but lithium oxalate replaced disodium phosphate for removing calcium hydroxide. The procedure for preparation of the reagent was the following:

> Transfer 75 gm. of Merck's Blue Label sodium cyanide to a 2 liter beaker, add 700 cc. of water, and stir until the cyanide is completely dissolved. Add 300 gm. of urea and stir. Then add 4 to 5 gm. of calcium oxide and stir for about 10 minutes. Filter, at once if necessary for immediate use, but preferably not until the next day. Add to the filtrate about 2 gm. of powdered lithium oxalate, shake occasionally for 10 to 15 minutes, and filter.

Otto provided a way to prepare the lithium oxalate from lithium carbonate and oxalic acid.

There were now three Folin methods for uric acid determination, and they were updated and simplified. Two were "direct" colorimetric procedures, a macro- and a micromethod, carried out without a preliminary separation of uric acid. The third was "indirect," because uric acid was separated as the silver complex, prior to its determination. The procedures were remarkably easy, despite their use of the lethal cyanide salt.

The range of proportionality was now greatly extended in the macromethod thanks to the improved (phenol-free) reagent that removed the blank. The heating step was eliminated. The procedure briefly was as follows: Five milliliters of unlaked blood filtrate was pipetted into a test tube graduated at 25 mL. Standards were treated similarly. With a cylinder, 10 mL of urea-cyanide solution was added and mixed. Four milliliters of uric acid reagent, "double the regular concentration," was added. The mixture was allowed to stand 20 min, though 40 min gave more color. The solution was diluted to volume, mixed, and compared on the colorimeter with the standards.

The micromethod was equally easy. The blood filtrate was prepared by adding 0.2 mL of blood to 4 mL of sulfate-tungstate. After standing 15 min, 1 mL of sulfuric acid was added. The precipitate was then centrifuged. The sulfate-tungstate contained 20 g of anhydrous sodium sulfate and 3 g of sodium tungstate per liter. The acid was made by diluting 12 mL of ⅔ N H_2SO_4 to 100 mL. The procedure was otherwise the same as the macromethod except that 4 mL of filtrate was used.

Although the "indirect" method for determining uric acid had fallen into disuse since 1922, Otto had retained it for "check" purposes. The indirect approach served as a "reference" method against which others would be evaluated. However, the "direct" methods carried out on Folin-Wu blood filtrates had proved inadequate because of poor recovery until Otto had corrected the problem by using unlaked blood and the purer uric acid reagent. Nevertheless he still felt the need for the alternative procedure, and he now presented his final version of it, leaving an unbroken thread that dated back 36 years to his summer in Salkowski's laboratory and to the Salkowksi method. His own modification of the Hopkins method had passed into limbo.

The indirect method of 1922 had not been completely quantitative. Recovery of uric acid from aqueous solution was only 90–95%. Otto now found that the lower yield was due to the "depressing effects of the dissolved silver on the color reaction." There were three ways to remedy this: by cutting down the chloride used, by using a more effective cyanide solution, or by extracting the silver precipitate with acid chloride.

In the revised procedure, 5 mL of unlaked blood filtrate was treated with 2 mL of an acid-silver solution (containing silver nitrate, lactic acid, and sodium carbonate). This was immediately centrifuged, and the precipitate was treated in one of two ways, either extracted with 1 mL of a 10% solution of NaCl in 0.1 N HCl and washed with 4 mL of water, or it was dissolved in 10 mL of the urea-cyanide solution and washed with 5 mL of water. Otherwise the procedure followed the direct macromethod.

Although recovery experiments were carried out on a large number of blood samples, to save space only 10 were reported. The values before and after the addition of 5 mg of uric acid to blood were obtained with the two alternate indirect methods and the direct macro- and micromethods. The agreement and recovery were outstanding.

Otto could certainly crow about his latest findings on blood, but urine was another matter. Though there was little call for uric acid analysis of urine, it remained a challenge to him. As he had long ago found in Berlin, uric acid could not be determined reliably in urine containing bile. And there were other interferences present, particularly reducing substances. The question was which of the methods was best to use.

He prepared a uric acid–free urine with Lloyd's reagent. Then he tested the direct and indirect methods from the standpoint of their recovery, and the "blank" reaction with the uric acid–free urine. The direct methods gave values 5–10% (or more) too high, whereas the indirect method gave 100% recovery. "We are therefore inclined to consider this a standard method, the first really standardized method for the determination of uric acid in urine."

Although details will not be presented here, the method was designed to meet the needs of the hospital laboratories already using the Folin methods on blood, so that duplication of reagents and equipment was minimal. A direct method was also included. No data were provided on any urinary studies, other than the recovery experiment.

Sometime before the academic term ended in June 1933, Otto was inducted into the Nu Sigma Nu student medical fraternity, a fact of importance primarily because Otto was asked to address the membership at the final dinner meeting of the year. This short talk, reproduced here verbatim, illustrates Otto's wit and timeliness.

> I have not attended any meeting of just this sort before, and have not been entirely clear as to its purport or significance. But I take it that it is a sort of double celebration—a meeting of welcome between you seniors and the alumni on the one hand and also something of a farewell meeting with some of us professors on the other. In former times such a transition on your part would probably have been referred to as passing from the study of medicine as a science to the study and application of it as an art. It is true, however, that all your professional associates hereafter will be practical physicians whereas some of us in the school, including particularly myself, know extremely little about medicine even as a science and nothing at all about it as an art. Personally, I therefore cannot give you the few last words of counsel and advice that might be expected and usually would be obtained from a professor when he speaks for the last time to a body of departing graduates. My own one sided personal experience does not enable me to give you a single professional word of advice as to what you should do or what you should not do in the future. That is too bad for I know that you could love and cherish and heed any well considered remarks of that sort that might be made to you at this time.
>
> With regard to important matters of this kind we professors now must simply pass you on to your older and more experienced brother alumni, including your immediate superiors in the various hospitals where you must begin your work. But I must take this opportunity to call your particular attention to the distinguished president of your alumni association, Dr. Quinby. Some of you may not know him so very well as yet. He is a little taciturn and not so easy to get at, but he is an artist all right—an artist at getting results and also at getting personal satisfaction out of his work. He can teach you a lot about medicine as an art.
>
> I had a rather prolonged occasion to observe his ways and methods a couple of years ago, and since I cannot give you any professional advice I don't know now that I can do any better on this occasion than to tell you a little something about my experience with Dr. Quinby [urologist].
>
> I went to him at just about this time of the year and I said, "Dr. Quinby, I am a sorely troubled man and I believe that you are the only one around here who can help me. But I already have been through the hands of one wonderful surgeon [1903] and you see what he did to me and I don't want to be disfigured that way again." Well, he looked me over in his own intimate sort of way and finally said, "I think that I can do a satisfactory job on you." And in the course of time he did do a magnificent job; I might say a real work of art.
>
> Because of being an old colleague of his, he was probably a little extra thoughtful and a little considerate. He even arranged it so that I should be able to follow what he was doing, if I wanted to and I tried, but many of the details

proved rather baffling—not to say murky, and I regret that I cannot describe them.

Then came one of his master strokes. He provided me with a perfectly lovely nurse. This helped tremendously to make the succeeding fortnight pleasant for me, and as I discovered by and by, also proved equally cheering to Dr. Quinby. In fact, I don't believe that he begrudged at all the times he spent in my room at the Brigham. My friends, real, worthwhile success in life is very largely a matter of being able to enjoy one's work. I found that out long ago, and in those days at the Brigham, I decided Dr. Quinby had come to have the same opinion.

Perhaps I have already said enough or even too much about Dr. Quinby and his work, but a person who has had a narrow escape of any kind rather likes to talk about it. And there is one other little point which I have not yet made perfectly clear. It seems to me that modern surgery in the hands of a really competent exponent can now be compared with so-called invisible mending, at least so it proved in my case. All my fears of another awful disfiguration proved quite groundless. In this case the scar is really invisible to the naked eye. I had quite a time convincing some of my golfing pals of that fact. One chaffs and bluffs so much on the links that it is difficult to tell when he is serious and when he isn't. Having noticed that my friends were casting surreptitious glances at me in the dressing rooms and being quite sure of my facts, I finally allowed one of them to look me over. He did so very carefully, and finally discovered a faint white line (which he had probably made himself a moment before) and concluded that that was the scar, and then admitted that it was practically invisible.

And now why have I told you all this? You may not know it or recognize it after I tell you but I have been expounding and illustrating a text, a message of truth, which I want to leave with you, a truth which I verily believe has at all times been of great practical importance in the careers of all kinds of physicians. That message is simply this: dead men tell no tales, but living men do. And I hope that soon living men and women will begin to tell good, decent human tales about every one of you.

During 1933, Otto worked on the revision and updating of the fifth and final edition of his laboratory manual that would appear the following year. In the preface he acknowledged his indebtedness to Drs. Logan, Danielson, and Fiske for "many improved details and also for some of the major revisions," as well as to Professor Bloor (now at the University of Rochester) for the methods for determining fats and cholesterol in blood. Among the new features were three micromethods for determining sugar, NPN, and uric acid, and a revised method for creatinine determination in blood that would be published posthumously.

The book was now almost double the size of the first edition of 1916 (184 versus 94 printed pages). The supplement on quantitative urinalysis had grown from 17 to 28 pages, whereas the supplement on quantitative blood analysis had expanded from 14 to 54 pages. The actual constituents of blood analyzed, however, had increased by only three, from 10 to 13, but now included were alternate as well as micromethods for several analytes. Details for reagent preparation and for carrying out the determinations were much enlarged because of recent modifications and longer experience with the methods. The text, while particularly useful in the student biochemistry laboratory at

Harvard, was now overshadowed by several others for general use in the teaching of biochemistry and for clinical chemistry practice, including Hawk's *Practical Physiological Chemistry*, in its 10th edition in 1931, and Peters and Van Slyke's classical *Quantitative Clinical Chemistry*, the "methods" volume published in 1932, following the "interpretations" volume of 1931.

On Dec. 20, 1933, the following letter was sent to a large number of Otto's colleagues, associates, and friends, past and present:

> Next spring Dr. Otto Folin will have completed his twenty-fifth year as Hamilton Kuhn Professor of Biological Chemistry in the Harvard Medical School. In anticipation of this event the undersigned committee has undertaken to secure a portrait of Dr. Folin for presentation to Harvard University.
>
> Contributions of from five to twenty-five dollars are requested for this purpose, and may be sent to Dr. C. H. Fiske, Harvard Medical School, Boston.
> WALTER B. CANNON
>   Chairman
> CYRUS H. FISKE
> REGINALD FITZ
> REID HUNT
> HARRY C. TRIMBLE

The last three papers that Otto published represent an ironic, though fitting, end to his lifelong research, almost as though a cycle had been completed. He had made his first intimate contacts with patient biochemical problems at the Pathological Institute in the Charity Hospital of Berlin, and had there discovered in uric acid analysis the direction for his career. Now 37 years later, he provided an almost triumphant final touch to that analysis in two papers on how to free sodium tungstate from the interference of molybdate. And it seems particularly appropriate that the last paper he published should be in German in the *Hoppe-Seyler's Zeitschrift*, on a topic that perhaps more than any other had brought him his first "fame," the venerable creatinine method. Was it not the editors of that pioneering journal, particularly Hammarsten, Salkowski, Kossel, and Kutscher, who had provided him the forum he needed to report his findings?

The two papers on the purification of sodium tungstate brought an end to a long-standing frustration of Otto's (*18-6, 18-7*).

> For many years I have labored, off and on, with the problem of preparing a uric acid reagent "completely free from phenol reagent," and each time I succeeded and concluded the problem was solved, only to find at a later date, with different chemicals, that the problem had not been solved.

That the problem had again not been solved became apparent almost immediately after the 1933 paper was published! As Wu had indicated, the interfering chromophoric phosphomolybdate (phenol reagent) had to be removed, once and for all, from the two chromophoric phosphotungstates in the uric acid reagent.

Otto now found the solution to the problem lay not in trying to remove traces of molybdate in sodium tungstate as the insoluble sulfide, but rather in converting the molybdate into highly colored sulfomolybdate that, in contrast to tungstate, was soluble in alcohol.

> This is presumably my last paper on the preparation of the uric acid reagent, and I hope that the method for the determination of uric acid developed last year is also final.

One mystery about the uric acid method remained. Only 10% of the color produced by the reduction of phosphotungstate came from the uric acid itself. The remaining 90% was due to the presence of cyanide, and this enhancing effect was a phenomenon specific to uric acid, and was subject to great variations

> according to the quality of the uric acid reagent, the quality of the cyanide, and other modifying factors, such as the degree of alkalinity, the presence of carbonate, of amino compounds, of some phenol products, etc. . . . Because of these obscure factors, the colorimetric method for the determination of uric acid represents probably the most complex reaction that we have in the whole field of practical colorimetry.

The second paper published in English probably at Marenzi's instigation, differed from the first in only one detail. It included the method for purifying the potassium xanthate used in, as well as the procedure for, detecting the presence of molybdate.

Otto's final paper was on creatinine and written for two purposes: to discuss a recently proposed modification of the alkaline picrate method by Lieb and Zacherl and to offer his own latest version, available only in the last edition of his laboratory manual, and only in English (18-8). Lieb and Zacherl, in their modification, had, according to Otto's view, been cavalier in using creatinine standards of questionable purity, and in their method of purifying picric acid. Rather than use a colorimeter, Lieb and Zacherl had measured the creatinine–alkaline picrate in a Pulfrich photometer, and employed tables of creatinine values corresponding to the readings obtained. The creatinine was obtained commercially, and the sole criterion of its purity was a nitrogen determination of dubious value. Otto recommended that creatinine zinc chloride be used for standardization. Its purification had been described in 1914, and there were numerous studies of it in the American literature.

While Lieb and Zacherl emphasized the necessity of using pure picric acid, the procedure they employed to purify it was one that Benedict had tried and later abandoned: crystallization from benzene. Otto now supplied a method for its purification from acetone, along with treatment with animal or bone charcoal.

Otto also pointed out the unsuitability of using laked blood (Folin-Wu) for creatinine determination. The red blood cell contained other substances besides creatinine that reacted with the alkaline picrate, so that values reported were too high. Otto then provided his final version for determining creatinine and creatine in blood.

He remarked that though the reaction of alkaline picrate with blood filtrate was perhaps not entirely due to creatinine, the determination was nevertheless important for use in renal function tests, particularly when creatinine was administered. Of course, Otto's persistence was vindicated. Not only is creatinine present in blood, but it serves today as one of the best markers of kidney disease.

Early in April, Otto attended the meetings of the Federation of American Societies for Experimental Biology in New York City, but did not stay for the entire session, as he felt a cold coming on or feared worse. Shaffer wrote him on April 12, "It was the best part of the New York meetings to have the few chats with you and to see you so well. I was sorry you left early."

Hildur Folin came to visit her aunt and uncle for a few weeks in the summer of 1934 while she took a summer course at Harvard University. Hildur was still a grade school teacher in Omaha, where she had first taught exceptional children, and was now teaching the handicapped children. Having noted how much Otto had aged and his poor health, she wrote:

> One time during this visit, while talking about his early life to me, he said, 'Life has been good to me, and I have enjoyed it, but now I am through. My work is finished or almost finished.' Then he laughed and added, 'And even my portrait is finished.'

Evidently in July on into early August, the portrait artist, Mr. Emil Pollak-Ottendorff, had completed his sittings with Otto. Otto had made plans to go to Kearsarge after the artist was through.

The coming academic year was to be officially Otto's last. In October, he would be 67, the age of retirement. Plans for the 1934–35 term were completed. Obviously reflecting a refreshing, long-overdue change in the Harvard administration, a woman, Mary D. Baker, A.B., had been appointed a research associate and formally listed as a departmental member. Elmer H. Stotz, S.B., who was Otto's last graduate student to receive the doctorate (1936), was now a teaching fellow. Gone were G. H. Hitchings and N. K. Schaffer. Irvin S. Danielson and Yellapragada SubbaRow remained. With the faculty of Folin, Fiske, Trimble, and Logan, the department numbered eight in Otto's final year. In view of his pending retirement, Otto had probably expressed, at least orally, his strong recommendation that Cyrus Fiske become the next chairman of the department.

Laura and Otto drove to Kearsarge in spite of Otto's failing health. How long they were there and how long they waited after Otto's genitourinary problems flared up is uncertain. Otto had had recurring problems since his prostatectomy, but he took a gamble that the climate, serenity, and beauty of Kearsarge would be restorative. Besides, Kearsarge beckoned, a habit that could not be readily dropped. Otto was also a realist, and the options for these visits were fast playing out.

By the last of August, Otto's genitourinary system had become badly infected, and the Folins returned immediately to Brookline. Otto was admitted at once to the Peter Bent Brigham Hospital on Sept. 1 under Dr. Quinby's care. He had cystitis and epididymitis. This was still the era before antibiotics and parenteral therapy. "Forced" fluids were prescribed and, no doubt, catheterization. Otto spent 10 days in the hospital and was sent home much improved to continue on the oral fluid treatment. He was free of pain, but the copious fluid intake did not permit him to visit his laboratory.

The chief had to take it easy, so his friends called upon him both at the hospital and at home. Harry Trimble was particularly helpful in briefing Otto on the departmental affairs and was Otto's most intimate friend. Olive Watkins Smith came by with her infant child.

Otto had time on his hands for reading and reflection. He was quite proud of both of his children. Teresa had finished her internship at the New Haven Hospital, and on July 1 had begun training in Chicago as a pediatrician. Grant had, in 1933, founded the Ace Drill Company in Detroit. Though only 30 years old, he was exceptionally well prepared for this new venture, having successfully mastered every aspect of the business in the eight years he had worked with the Union Twist Drill Co. Despite the Great Depression, his business thrived.

Laura's health had improved, especially following surgery for an enlarged ovarian cyst in 1927. Their financial status was adequate. They had lost some money in the stock market crash of 1929, but continued income from the consultancy and the steady salary had made this inconsequential.

On Sept. 27, Walter Cannon wrote Philip Shaffer:

> I am glad to be able to tell you that the portrait which has been made of Dr. Folin is highly satisfactory. It is what might be called a "speaking likeness." We are now thinking of some appropriate way of presenting the portrait to the University. Tentatively we have selected the plan of arranging for dinner on November 16, at Vanderbilt Hall. At the dinner we wish to have someone speak concerning Dr. Folin's contributions to chemistry and someone else concerning his contributions to the methods used in clinical medicine. We thought that nobody could speak with better authority and with more personal interest on the first topic than you. I am writing to ask whether you think you could afford to attend the dinner and take the part that I have suggested.
>
> Besides the speeches mentioned above there will of course be some brief remarks by a member of the Committee presenting the portrait to the University, a response by President Conant, and then remarks by Dr. Folin himself. Altogether I think it should be a very pleasant occasion. I hope that you may find that you could attend.

Philip Shaffer replied on Oct. 1:

> I am delighted to learn that the portrait of Dr. Folin is a good one. I appreciate highly your letter asking me to be one of those to speak at the dinner on the occasion of presenting the portrait to the University. In spite of the distance, I think this is an invitation and a privilege which I cannot well decline. Of course, I shall be very happy to participate in this tribute to my old Chief.
>
> Lacking experience in affairs of this sort, I shall be very grateful if you will let me know about how much time you expect my remarks to occupy.

The Final Years

The above letter brought the following response by Cannon, on Oct. 4:

> I am delighted that you will be able to come to the "party" to be given as a tribute to Dr. Folin on November 16.
>
> Dr. Christian will speak about Dr. Folin's contributions to practical medicine and you are to have, as I explained, the pleasing opportunity of telling about his contributions to biological chemistry. It seems to me that it would be well for each of you to speak for fifteen or twenty minutes. That would leave about the same length of time for Dr. Folin to talk, and a few minutes more for the presentation of the portrait and its acceptance by President Conant. The speaking would thus take a little over an hour.
>
> Dr. Folin has been laid up with a mild affliction. When I went to his home yesterday and told him about the plans, he was rather overwhelmed by the news, and wanted the affair much more simple. I told him that those who wished to do him honor had some rights and that he would have to be present and take all that was coming to him.
>
> Let me tell you again how pleased I am that you will be able to attend the dinner on November 16.

On Oct. 15, Shaffer wrote:

> Thanks for your letter of October 4. I shall be very glad to be on hand at the dinner for Dr. Folin on November 16. I shall plan to be in Boston on that date unless I hear from you in case the arrangements are changed.

On Oct. 21, 1934, Otto wrote what was apparently his final communication.

> Dear Shaffer,
>
> Your nice little note came yesterday. First of all I must tell you that I have been and still am in more or less trouble. We left suddenly our summer place and Mrs. Folin took me straight to the Brigham hospital about Sept first. I stayed there for ten days and since that time I have been confined to the house, most of the time in bed. I have a nasty combination of cystitis and orchitis. It is not painful and I have no temperature any more but I am supposed to rest and to drink so much water as to make it impractical to go to the school. It is getting to be very tedious and I don't believe now that I shall be in good order even by Nov. 16.
>
> Cannon came in and told me the other day about the arrangements for that date. He said that Christian would speak about my contribution to medicine and yourself about my contributions to biochemistry. I am rather sorry for both of you as well as for myself.
>
> A light tone would be desirable and, of course, would be less trying on me, but I shouldn't over do it before an audience of New Englanders. I should think that a concise sketch of the practical biochemistry as it was when you and I began our work at the McLean is called for since that condition served to give a trend to my work on methods. My portrait includes a colorimeter and a couple of volumetric flasks, and it might therefore fit in pretty well to say something about the introduction of colorimetry into biochemistry. This is about all that I can think of at the moment, but if I think of anything else, perhaps I will write again. I wish it were over!

Teresa is coming from Chicago. Grant and his wife may possibly come also. If they do, I will arrange for you to stay at the Harvard Club; and if they don't, you must stay with us.

It was extremely nice and friendly of you to take on this work for your old friend, there is no one else that would be anywhere nearly as acceptable to me.

With kindest regards to Mrs. Shaffer.

On Oct. 31, 1934, Harry Trimble wrote Hilding Berglund about Otto's last days:

> Last week-end several saw him and came back with the report that he expected to be at the laboratory during the week. Instead on Wednesday [Oct. 24] of the week he again entered the hospital for what was anticipated to be some very minor surgery. I have not had opportunity to learn the details of the operation. I do understand that more was necessary than had been anticipated. The Professor came through the operation satisfactorily, and around 6 pm Mrs. Folin talked with him and then went home for the night. Around 3 am the Professor's blood pressure dropped greatly and a crisis was on. The staff came in and under the emergency treatment he rallied, but by early afternoon a terminal pneumonia appeared, and he grew weaker and weaker until 8 pm when life ceased.
>
> The termination came so suddenly that we were totally unprepared for it. Even though "the Chief" had not been in his office this fall, yet the laboratory seems much emptier. I know that you will join with us in the realization of his loss.

Laura Folin also wrote Hilding Berglund the following message:

> It is my sad duty to tell you that Otto died in the evening of October 25.... Last week Dr. Quinby removed the testicle. Otto stood the operation but in the early morning following it his blood pressure fell, and efforts made to stimulate the heart were not more than temporarily effective.
>
> After services in the Appleton Chapel in the Harvard yard [October 29], we went by automobile to Kearsarge, and laid him in the country graveyard among the hills he loved so well.
>
> Teresa came from Chicago and Grant and his wife came from Detroit. On Saturday morning they will all have gone back to their own work.

On Dec. 3, Harry Trimble wrote to Philip Shaffer (both Trimble and Shaffer, along with SubbaRow, had accompanied the Folin family to the burial in Kearsarge).

> The memorial meeting for Professor Folin was held a week ago last Friday afternoon [November 23, Amphitheater C, Harvard Medical School]. Professor Fiske spoke on the scientific work of Professor Folin. Professor Christian's topic was "Folin and his Influence on Medicine." Professor Cannon acted as presiding officer and talked on the personal characteristics of Professor Folin. The portrait was presented to the Faculty at this time, and Dean Edsell, in accepting the gift, gave a very fine appreciation of the service of Professor Folin to the Medical School.

Among the scores of letters and notes of sympathy that Laura Folin received from relatives, colleagues, friends, students, and alumni around the world, we shall cite only two:

From Stanley Benedict, who attended the chapel service, a telegram:

> My thoughts and deepest sympathy are with you. No words can tell how much I admired him or how deeply I feel his loss.

From P. A. Levene:

> You can understand the depth of my shock for only last March we had a long visit, reminiscing and discussing future plans. Our aims, ambitions, and careers were identical and for a long time, we were two of a very small group. The group has grown large in number, but nobody has taken the place of Dr. Folin in my life. . . . I know how heavy the loss is for you and I wish I knew how to console you.

References

18-1. Folin, O., The micro method for the determination of blood sugar. *N. Engl. J. Med.* **206**, 727–29 (1932).
18-2. Folin, O., The determination of ammonia in blood and other biological fluids. *J. Biol. Chem.* **97**, 141–54 (1932).
18-3. Folin, O., The department of biological chemistry. *Harvard Medical Alumni Bull.* **7**, 47–49 (1933).
18-4. The five epic papers by Fiske and SubbaRow were:
    (a) The colorimetric determination of phosphorus. *J. Biol. Chem.* **66**, 379–400 (1925).
    (b) The nature of the inorganic phosphorus in voluntary muscle. *Science* **65**, 401–3 (1927).
    (c) The isolation and function of phosphocreatine. *Science* **67**, 169–70 (1928).
    (d) Phosphocreatine. *J. Biol. Chem.* **81**, 629–79 (1929).
    (e) Phosphorus compounds of muscle and liver. *Science* **70**, 381–82 (1929).
18-5. Folin, O., Standardized methods for the determination of uric acid in unlaked blood and in urine. *J. Biol. Chem.* **101**, 111–25 (1933).
18-6. Folin, O., The preparation of sodium tungstate free from molybdate, together with a simplified process for the preparation of a correct uric acid reagent (and some comments). *J. Biol. Chem.* **106**, 311–14 (1934).
18-7. Folin, O., The preparation of pure sodium tungstate for use in the preparation of correct uric acid reagent. *Apartado de la Revista de la Sociedad Argentina de Biologia, Supplement*, 261–64 (November 1934).
18-8. Folin, O., Bemerkungen zur Bestimmung von Kreatinin (und Kreatin) im Blut [Remarks on the determination of creatinine (and creatine) in blood]. *Hoppe Seyler's Z. Physiol. Chem.* **228**, 268–72 (1934).

Part II: Methods

Chapter 3. Chicago University (1892-96)

On Urethanes (1)

The purpose of Otto's research thesis is best explained in the introduction of his published dissertation (the italics are the editor's):

Sodium methylate and acetbromamide in methyl alcohol solution do *not* yield a hydroxylamine derivative according to the following equation:

$CH_3CONHBr + NaOCH_3 = CH_3CONHOCH_3 + NaBr$

but a molecular rearrangement, analogous to that observed by Hofmann in the action of alkalies on acid bromamides, occurs in the course of the reaction, and a urethane is obtained in place of the isomeric hydroxylamine ether:

$CH_3CONHBr + NaOCH_3 = CH_3NHCOOCH_3 + NaBr$

A similar rearrangement occurs in the action of sodium methylate on para-nitrobenzbromamide and on succinimide bromide.

The investigations described in this paper were undertaken at the suggestion of Dr. Steiglitz and Dr. Lengfeld for the purpose of determining whether the reaction

$RCOHNHBr + NaOCH_3 = RNHCOOCH_3 + NaBr$

is general, *or if it is possible to prevent the rearrangement and effect direct substitution by varying the positive and negative character of the group R.*

In the course of their investigations on urethanes, Lengfeld and Steiglitz obtained chlorformanilide by the action of phosphorus pentachloride on phenylurethane:

$C_6H_5HNCOOC_2H_5 + PCl_5 = C_6H_5NH \cdot COCl + C_2H_5Cl + POCl_3$

The action of phosphorus pentachloride on substituted urethanes, as well as the action of phosphorus pentachloride and of phosgene on urethane itself was therefore made a part of this investigation in order to test the applicability of the reaction.

It should be stated at the outset that Otto's attempt to prevent the rearrangement was doomed to failure. As a later student of Steiglitz's wrote:

It could not have been understood then that when a halogen atom separates from a nitrogen atom it tends to do so in the positive form (e.g.: Br, leaving the pair of electrons of the covalence with the nitrogen). For this reason the halogen atom could not be replaced by the negative methoxy $CH_3:O:$. Instead of this there is a rearrangement which Steiglitz and Lengfeld recognized in their paper as that of Hofmann, and of Hoogewerf and van

361

Dorp, by means of which an amide group, $CONH_2$, is replaced by an amino group. The simplest illustration of this reaction is the conversion of bromoacetamide to methylamine,

$CH_3:\ddot{N}:\overset{:\ddot{B}r:}{} = HBr + CH_3CO:N:$

The methyl group then shifts to the univalent, highly unsaturated, nitrogen atom and a pair of electrons shifts to give a double covalence between the nitrogen atom and the carbonyl group, forming methyl isocyanate,

$CH_3:\ddot{N}::C::\ddot{O}$

Sodium hydroxide then hydrolyses the isocyanate to methylamine, CH_3NH_2, and sodium carbonate.

The reaction was not formulated in this way by Lengfeld and Stieglitz but the momentary formation of univalent nitrogen atoms was postulated by Stieglitz and his students in many reactions studied by them in later years. They never expressed these by means of electronic formulas, however. (2)

The first part of Otto's research problem was to prepare urethanes from acid benzbromamides (terminology and symbols are part of the thesis) that had been substituted in the ortho, para, and meta positions. The hope was that some benzhydroxamate would be formed as a result of inhibiting the anticipated rearrangement when sodium methylate was reacted with the acid bromamide.

Unsubstituted benzbromamide was the first compound that Otto synthesized; and the urethane was obtained quantitatively according to the reaction:

$C_6H_5CONHBr + NaOCH_3 = \quad C_6H_5NHCOOCH_3 + NaBr$
$\qquad\qquad\qquad\qquad\qquad$ methylphenylcarbamate

Otto then prepared m-nitrobenzbromamide and obtained a 90% yield of the urethane:

$m\text{-}NO_2C_6H_4CONHBr + NaOCH_3 = \quad m\text{-}NO_2C_6H_4NHCOOCH_3 + NaBr$
$\qquad\qquad\qquad\qquad\qquad\qquad$ methyl-m-nitrophenylcarbamate

For the sake of brevity, we shall not include information on the tests of purity, composition, or physical properties, but these included combustion analysis, single and mixed melting points, and solubilities.

The carbamate above was Otto's first new compound, and he confirmed its structure by preparing it synthetically a different way and comparing properties, which proved identical. For this synthesis he used m-nitraniline and methylchloroformate.

The next compound that Otto prepared was o-nitrobenzbromamide, and from it he obtained the urethane, methyl-o-nitrophenylcarbamate, in good yield. When Otto lowered the temperature of his reaction conditions, he must have been momentarily excited to find that a different compound slowly precipitated out, o-nitrophenyl-o-nitrobenzoylurea, $o\text{-}NO_2C_6H_4NHCONHCOC_6H_4NO_2\text{-}(o)$. This compound was identical with one that his colleague Samuel Swartz had produced in another project, and its presence was explained by a series of side reactions not particularly helpful to Otto's problem. It was not the sought-for hydroxamate.

Otto then synthesized another meta-substituted acid bromamide, m-brombenzbromamide, $m\text{-}BrC_6H_4CONHBr$, and obtained the expected urethane, methyl-m-bromphenylcarbamate, $m\text{-}BrC_6H_4NHCOOCH_3$.

Otto did not try the p-nitrobenzbromamide, because this had already been shown by Lengfeld and Stieglitz to produce the urethane.

At this stage, Otto concluded that

> ... as the work just described shows that the rearrangement of the acid bromamide, RCONHBr, occurs almost quantitatively, even when R is strongly negative (R = C_6H_5; o-$NO_2C_6H_4$; $m\text{-}NO_2C_6H_4$; $m\text{-}BrC_6H_4$), the effect of positive radicals was studied."

The first choice was to prepare a dialkylamidobenzamide, $R_2NC_6H_4CONH_2$, to convert into the acid bromamide. To accomplish this synthesis, a few experiments were first made to determine the best route to take. One was to react phosphorus pentachloride with the hydrochloride of p-dimethylamidobenzoic acid:

$$\overset{H}{\underset{Cl}{>}}N(C_6H_5)_2C_6H_4COOH + PCl_5 = \overset{H}{\underset{Cl}{>}}N(C_2H_5)_2C_6H_4COCl + HCl + POCl_3$$

The acid chloride obtained gives, with ammonia, the amide in very pure condition, at 50% yield. This white, crystalline compound was Otto's second new one. He then tried to prepare the amide in greater yield from the ethyl ester of dimethylamidobenzoic acid. The ester proved to be a white solid that Otto purified and characterized. Previous work by others had mistakenly declared the compound to be an oil. Otto's product was, in essence, another new compound. Otto could not, unfortunately, convert the ester into the corresponding amide.

A third approach at preparing the amide was then tried by converting dimethylaniline and two of its derivatives into their corresponding nitriles, but all three were unproductive. Otto did obtain a new compound in small yield, p-dimethylamidobenzonitrile, $p\text{-}(CH_3)_2NC_6H_4CN$.

Otto then prepared the dialkylamidobenzamide from the dialkylamidobenzoic acid by successive action of hydrochloride acid, phosphorus pentachloride, and ammonia, as indicated previously.

The next task was to prepare the corresponding acid bromamide from the amide. He attempted to do this with potassium hypobromite unsuccessfully. He was also unsuccessful with hypochlorous acid. The halogen amide proved too unstable. So Otto was forced to abandon hope of using dialkylamidobenzamide with its positive "radicals" to test the rearrangement. He also tried to make the acid bromamide via m-amidobenzamide and potassium hypobromite, but was luckless, and concluded that the amido and dialkylamido groups on the benzene ring made the ring too reactive. He therefore resorted to preparing the analogous acid bromamide in the aliphatic series instead. In this endeavor he was successful.

Otto synthesized the compound, carboxy-β-amidopropionbromamide in good yield from its corresponding amide. This was then reacted with sodium methylate and the usual rearrangement occurred to produce the urethane:

$CH_3O_2CNHCH_2CH_2CONHBr + NaOCH_3 = CH_3O_2CNHCH_2CH_2NHCO_2CH_3 + NaBr$
   carboxy-β-amidopropion-   dimethyl ester of ethyl-
   bromamide   enecarbamic acid

For this portion of the study Otto was forced to conclude that the reaction

$RCONHBr + NaOCH_3 = RNHCOOCH_3 + NaBr$

was quite general, insofar as his work was concerned.

The second phase of his research dealt with the reaction of phosphorus pentachloride on urethanes according to the equation:

$RNHCOOCH_3 + PCl_5 = RNHCOCl + CH_3Cl + POCl_3$

First he improved the yield of chlorformanilide by 50% over that previously obtained by his mentors Lengfeld and Stieglitz, by modifying the conditions for the reaction:

$C_6H_5NHCOOC_2H_5 + PCl_5 = C_6H_5NHCOCl + C_2H_5Cl + POCl_5$
   phenylurethane   chlorformanilide

Then he reacted PCl_5 with m-nitrophenylurethane and obtained the new compound, chlorform-m-nitranilide in 81% yield:

$$m\text{-}NO_2C_6H_4NHCOOC_2H_5 + PCl_5 = m\text{-}NO_2C_6H_4NHCOCl + C_2H_5Cl + POCl_5$$

The nitranilide (also called m-nitrophenylurea chloride) was further identified by adding it to ammonia to obtain the expected m-nitrophenylurea, $m\text{-}NO_2C_6H_4NHCONH_2$, whose properties were known and that he found comparable. The nitranilide was also heated and extracted to give the m-nitrophenyl isocyanate, $m\text{-}NO_2C_6H_4NCO$, a new compound that Otto characterized by its addition products with methyl and ethyl alcohols, and with ammonia. The methyl alcohol derivative was evidently a new product, m-nitrophenylurethane, $m\text{-}NO_2C_6H_4NHCOOCH_3$.

The isocyanate was heated with m-nitrobenzamide in molecular quantities to produce m-nitrophenyl-m-nitrobenzoyl urea, another new compound, and an isomer of the corresponding ortho derivative that Otto also prepared by the action of sodium methylate on o-nitrobenzbromamide.

$$m\text{-}NO_2C_6H_4NCO + H_2NOCC_6H_4NO_2\text{-}m = NO_2C_6H_4NHCONHCOC_6H_4NO_2\text{-}m$$

Otto then switched to the action of phosphorus pentachloride on m-bromphenylurethane, $m\text{-}BrC_6H_4NHCOOC_2H_5$, and obtained a good yield of m-bromphenylurea chloride, $m\text{-}BrC_6H_4NHCOCl$, that he identified by treating it with strong ammonia to produce m-bromphenylurea, $m\text{-}BrC_6H_4NHCONH_2$, another new compound for the literature. Otto now felt that he could state that the reaction of PCl_5 on *aromatic* urethanes was a general one producing chlorformanilides from which isocyanates can readily be obtained and that this was a better way to synthesize these compounds than to prepare them from aniline and phosgene.

Next he studied the reaction to see if it were also applicable to the urethanes of *aliphatic* amines and to urethane itself. He prepared the urethane by modifying an earlier approach of Liebig and Wohler's using potassium isocyanate and hydrochloric acid in place of isocyanic acid vapors, and reacting this with ethyl alcohol in hydrochloric acid. From this he obtained a 60% yield of urethane. He then treated urethane with phosphorus pentachloride but did not get the expected chlorformamide. More than likely phosphoric acid derivatives of chlorformamide were formed, because of the vulnerability of the amide group. He did not pursue this any further.

Because the phosphorus oxychloride byproduct was difficult to remove when phosphorus pentachloride was used to produce urethanes, the third and final phase of his research project was to try another synthetic route. Otto replaced the pentachloride with phosgene in the hope that chlorformamide would be produced along with easily removed gaseous byproducts:

$$NH_2COOC_2H_5 + COCl_2 = NH_2COCl + CO_2 + C_2H_5Cl$$

When Otto tested these substances, an oil was produced that reacted violently with aniline to produce a white solid with a melting point of 106 °C. The literature, however, indicated that the expected product, phenylallophanic ether, melted at 120 °C. Otto, therefore, prepared this ether for comparison to prove that the oil contained the expected chlorformamide (after loss of CO_2 from $ClONHCOOC_2H_5$). After trying to prepare the ether unsuccessfully via phenyl isocyanate and urethane, he obtained the phenylallophanic ether by using the hydrochloric acid addition product of phenylisocyanate instead:

$$C_6H_5NHCOCl + H_2NCO_2C_2H_5 = C_6H_5NHCONHCO_2C_2H_5 + HCl$$
$$\text{phenylallophanic ether}$$

This ether melted at 106 °C and the mixed melting point with Otto's original product was the same. The ether was not formed when aniline and urethane were mixed together in benzene, and phosgene then added. Hence Otto could conclude that chlorformylurethane was present in the oil, but lost carbon dioxide to form the chlorformamide:

$$NH_2COOC_2H_5 + COCl = C_2H_5O_2CNHCOCl + HCl$$
$$\text{chlorformylurethane}$$

By adding ammonia to the oil the presence of chlorformylurethane was further proved by the production of the allophanic ether:

$$C_2H_5O_2CNHCOCl + NH_3 = C_2H_5O_2CNHCONH_2 + HCl$$

Two other substances were present in the oil, and Otto investigated these. One that Otto extracted and purified was a suspected carbonyldiurethane,

$$CO\begin{smallmatrix}\diagup NHCOOC_2H_5 \\ \diagdown NHCOOC_2H_5\end{smallmatrix}$$

which could be formed by action of the chlorformylurethane on a second molecule of urethane:

$$C_2H_5O_2CNHCOCl + H_2NCO_2C_2H_5 = CO\begin{smallmatrix}\diagup NHCOOC_2H_5 \\ \diagdown NHCOOC_2H_5\end{smallmatrix} + HCl$$

The carbonyldiurethane was a new compound that Otto characterized. The third substance present in the oil was allophanic ether, which has already been mentioned.

The following reactions summarize the action of phosgene on urethane to produce three substances:

$$NH_2COOC_2H_5 + COCl = NH_2COCl + ClC_2H_5 + CO_2 \quad (1)$$
$$NH_2COCl + NH_2COOC_2H_5 = NH_2CONHCOOC_2H_5 + HCl$$

$$COCl_2 + NH_2COOC_2H_5 = ClCONHCOOC_2H_5 + HCl \quad (2)$$

Reaction 1 proceeds to a limited extent, Reaction 2 to a far greater extent.

The chloride produced in Reaction 2 is also quite reactive and acts on a second molecule of urethane to give:

$$C_2H_5OOCNHCOCl + NH_2COOC_2H_5 = CO(NHCO_2H_5)_2 + HCl \quad (3)$$

Otto then improved the conditions for synthesizing the carbonyldiurethane. Next he prepared a pure silver salt of this diurethane and characterized its composition and solubility. With the completion of his study of the silver salt, Otto ended his research, and began the writeup of his results.

The thesis failed to provide new insights into organic chemistry or open up fresh lines of research and was thus neither outstanding nor brilliant. Nevertheless, it offered new and useful information on the preparation and properties of urethanes as it had proposed to do and was published by the University of Chicago.

Startling new discoveries were being reported on atomic structure in this era, but consideration of electronic configuration of organic compounds was in its early stages. Nef, one of the pioneers in this era, had been intrigued when he was with Baeyer in Germany by Baeyer's study of unsaturated valences of carbon. At that time, Nef himself studied tautomerism, especially the keto-enol type, of the nitroparaffins. He later worked on compounds containing bivalent carbon such as the isocyanides and fulminates, and developed theories concerning addition reactions and dissociation.

If Steiglitz had ideas on why Otto's effort could not prevent the Hofmann rearrangement, he chose not to present them, nor did Otto present any of his own. Some conjecture on this aspect of the work might have enhanced the thesis, perhaps significantly.

1. Folin, O. On Urethanes. Dissertation submitted to the Faculties of the Graduate School of Arts, Literature and Science in candidacy for the degree of Doctor of Philosophy, Department of Chemistry, University of Chicago, October 1896. Easton, Pa.: Chemical Publishing Co., 1897.
2. Noyes, W.A., Julius Stieglitz (1867–1937). *Biog. Mem. Natl. Acad. Sci.* **21**, 275–314 (1941).

Chapter 5. Berlin (1897)

Folin's Comparison of Salkowski's Analytical Method for Uric Acid with the Hopkins Method (1)

The Hopkins method (2) was based on two facts: 1) uric acid in water solution or urine is completely precipitated by saturation with ammonium chloride, and 2) uric acid in warm sulfuric acid solution can be accurately titrated with standard permanganate. In the method 30-g of ammonium chloride was added to 100 mL of urine or uric acid solution. Uric acid completely precipitated in the form of the monoammonium salt in less than 2 h. After this time, the precipitate was filtered off, washed once with saturated ammonium chloride solution, and rinsed into an evaporating dish. The uric acid was liberated by addition of hydrochloric acid, filtered after standing 2 h, and the mother liquor was measured. The uric acid was then washed free of chloride and dissolved in 100 mL of water with a little sodium carbonate, and after the addition of 20 mL of conc. sulfuric acid, it was immediately titrated with 0.05 N $KMnO_4$ solution. Each milliliter of this solution corresponded to 3.75 mg of uric acid. To the end result was added 1 mg of uric acid/15 mL of the measured mother liquor.

The 3.75 mg value for the permanganate solution was obtained empirically. The correction factor of 1 mg/15 mL of mother liquor was the soluble uric acid fraction remaining when the uric acid was liberated with HCl.

Otto, as did others, found that results on pure solutions of uric acid with the Hopkins method agreed well with Salkowski's. But later attempts by Hopkins to simplify the method for clinical use were not particularly successful. Otto studied first the conditions for the permanganate titration using solutions of purified uric acid that he had prepared from snake excrement (Schlangenexcrementen) and checked by Kjeldahl nitrogen determination. In general, Otto found little to criticize in the conditions for the titration. Temperature for the permanganate reaction could vary between 50–70 °C, with ideal use between 55–65 °C. This temperature was achieved by the addition of the sulfuric acid to water. Using 20 mL/100 mL of solution raised the temperature to 70–75°C, and perhaps 15 mL of acid was better, but no less. The empirical conversion factor of 3.75 mg of uric acid per milliliter of permanganate consumed was confirmed.

Otto then turned to the precipitation of uric acid. He found that by studying pure solutions of uric acid, the correction factor that Hopkins used for the soluble uric acid in the mother liquor was too low and should have been 1 mg of uric acid/11 mL of mother liquor rather than 1 mg/15 mL, and to this had to be added the amount also that was dissolved in the wash water used in the filtration step. For 50 mL of wash water, Otto found a correction of 3 mg was needed. When these factors were taken into consideration, the Hopkins method gave results closer to Salkowski's. But Hopkins had tried to simplify his method by avoiding the initial precipitation with ammonium chloride; however, the troubling influence of chloride appeared to make this impossible.

Otto thought that perhaps the original ammonium urate precipitate might be titrated directly and the second precipitation eliminated. To achieve this, he tried two other ammonium salts, the carbonate and the acetate, for saturating the uric acid solution. Sodium, potassium, and magnesium salts of urate could be precipitated but others' work showed that they were too soluble for use quantitatively. They would seemingly work, of course, if a little ammonia were added, but that in effect produced the ammonium urate. On the other hand, using ammonium sulfate at saturating quantities coprecipitated too many impurities and ammonium oxalate reacted with permanganate.

With ammonium carbonate used at a saturating concentration of 25 g/100 mL of uric acid solution, Otto found that he could not only get quantitative recovery but that the amount lost through uric acid solubility in the filtrate was 1 mg/100 mL, far less than the Hopkins method with ammonium chloride. The ammonium urate formed was dissolved in 100 mL of hot water, cooled, treated with a small amount of dilute sulfuric acid to remove excess carbonate; then 15 mL of concentrated sulfuric acid was added and the direct titration was performed.

When Otto used ammonium acetate at a saturating concentration of 50–60 g, he found that it contained some free acetic acid that he neutralized by adding a little ammonia. He found that the acetic acid liberated when the sulfuric acid was added for the titration did not interfere with the titration and therefore did not require removal. The results were as quantitative as those with ammonium carbonate.

Hopkins and Ritter had assumed that xanthine bases would interfere with their determination of uric acid, the supposition being that like uric acid they would be oxidized in the permanganate titration. Otto determined not only whether the bases would be titrated by permanganate, but whether they could be precipitated to any extent as ammonium salts in the method. Partly at his own expense, Otto was able to obtain some xanthine, hypoxanthine, and guanine. He found that xanthine and hypoxanthine did not affect the titration but that guanine reacted slowly and without a sharp end point. When the three substances were added to pure uric acid solution, xanthine and hypoxanthine had no effect on the recovery of the acid, whereas guanine in rather high concentration caused a slight increase in values. Otto concluded that guanine's effect was negligible, but anyhow, that its presence in urine was surely minimal, when it was present at all.

Otto, however, wanted to leave no doubts about guanine interference so he tested its precipitation from acetate, carbonate, chloride, and sulfate. Not a trace of guanine was precipitated from acetate even when he reduced the ammonium acetate from saturation to 25 g/100 mL of solution. For the acetate he added just sufficient ammonium hydroxide to ensure alkalinity. He then tested ammonium acetate and carbonate on a few normal and pathological urines and compared the uric acid values found with those of the Salkowski silver method and found them quite comparable.

Two problems for further study emerged. One of the urines tested was icteric and the visual end point was difficult to detect because of the color. A second problem was that the ammonium urate precipitate was too finely divided, making filtration difficult. Although this could be overcome by using a double layer of filter paper, Otto attributed the problem to the use of saturating or high concentrations of ammonium salts and wondered if they were really necessary.

Otto then studied the effect of reducing the concentration of the ammonium salts on the quantitative precipitation of uric acid and found that, at much less than saturating concentration, all of them precipitated uric acid as the ammonium salt equally well. The correction was in all cases 1 mg/100 mL of filtrate.

Otto then showed that he could use any of four ammonium salts (acetate, carbonate, nitrate, sulfate) at 10% concentration with good results. Otto set up a double precipitation method to ensure that guanine, if present, would not be coprecipitated with the urate. He also wanted to try this double precipitation method on the urine of the icteric patient he had previously used, but the patient was no longer available, and another icteric urine was at that moment unobtainable. As far as he was then aware there was no other known substance in urine that would both precipitate with ammonium salts and be oxidized by permanganate.

When neither albumin nor globulin was precipitated from urine by the addition of 10% ammonium sulfate, Otto presented his version of the Hopkins method with the following new or modified features: 10 g of ammonium sulfate was added to 100 mL of urine and allowed to stand 2 h. The precipitated urate was washed with 10% ammonium sulfate until chloride-free, then dissolved and titrated, with the addition of 1 mg to the calculated end result.

Otto had unquestionably made major improvements to the Hopkins procedure. He had markedly simplified it by removing the need for a second precipitation and reducing the analytical time involved; he had found a more reliable correction factor; he had shown that the precipitation did not require saturating concentrations of ammonium salts and that a variety of ammonium salts could be used interchangeably as a uric acid precipitating agent; he had recognized and studied the xanthine bases as potential interferences; and he had shown at least provisionally the clinical applicability of the method to normal and pathological urines. It was, indeed, the Folinische modification!

1. Folin, O., Eine Vereinfachung der Hopkinsche Method zur Bestimmung der Harnsaure im Harn (A simplification of the Hopkins method for determining uric acid in urine). *Hoppe Seyler's Z. Physiol. Chem.* **24**, 224–45 (1897).
2. *J. Pathol. Bacteriol.* (June 1893).

Chapter 10. McLean Hospital (1900–3)

A New Method for Determining Urea in Urine (1)

Existing methods for determining urea in urine required that urine be heated for up to 8 h in strong acid or base to convert a portion of the urea to its ammonium salt. Then the NH_3 produced was distilled into standard acid, and its value was determined titrimetrically with standard base. Before the urea procedure was performed, protein (and presumably other nitrogenous substances) was removed by the use of ether-alcohol (Mörner-Sjöquist) or phosphotungstic acid (Pflüger-Gumlich) methods, and the filtrate used. Otto did not evaluate the Krüger-Wulff reagent that used copper sulfate and sodium bisulfite.

Otto found that urea could be split quantitatively in 30 min with dry magnesium chloride (or in 20 min with calcium chloride). He also found that the reaction could be carried out on urine directly without preparing a filtrate.

This simple method of urea determination was carried out in the following manner: 3 mL of urine, 20 g of $MgCl_2$, and 2 mL of conc. HCl were heated in a 200-mL Erlenmeyer flask with the use of a short reflux condenser (200 mm × 10 mm) until the drops flowing back into the flask made a hissing sound in the mixture. The heating was continued for 25–30 min, diluted cautiously with water, transferred to a liter flask and the NH_3 distilled after the addition of about 7 mL of 20% NaOH solution. Usually about 350 mL had to be distilled (taking about 60 min) before all the NH_3 was separated. The distillate was boiled, cooled, and titrated. Each milliliter of 0.1 N NH_3 of the distillate corresponded to 3 mg or 0.1% urea. The correction for the NH_3 content of the $MgCl_2$, as well as for preformed NH_3 of the urine, had to be separately determined.

The analysis took 90–100 min to carry out. Otto compared its values with those obtained from five urinary filtrates and found good agreement, though there were difficulties with the phosphotungstic acid filtrate in other cases. This much simpler method was, of course, still too long and bulky, and there were many questions about it to be answered. Otto did not perform recovery experiments. (He did not add a known amount of pure urea to urine to see if he could get back what he added.) He could not reach onto the reagent shelf or pick up a chemical catalogue and order a group of substances that might potentially interfere with his test. And, of course, the procedure could only be compared with others that were perhaps crude, rather than with a "standard" one. The Folin urea method, however, was a major step up the staircase of urea analysis.

1. Folin, O., Eine neue Methode zur Bestimmung der Harnstoffe im Harne (A new method for determining urea in urine). *Hoppe-Seyler's Z. Physiol. Chem.* **32**, 504–14 (1901).

A Simple Procedure for Determining Ammonia in Urine (1)

Otto's second paper was a short one on urinary ammonia determination. The urea determination required the determination of preformed NH_3, of course, and there was a need to know the preformed NH_3 anyhow as part of the studies to be carried out on patients. This was done simply by omitting the urea-splitting agents, $MgCl_2$ and HCl, and carrying out the subsequent steps. The urine in the distillation flask was diluted to 400–500 mL with water, alkalinized, and then distilled and collected for 45 min. The content of the distillation flask was then diluted with an equal volume of water, and a second 45-min distillate was collected. The undesired decomposition of urea was assumed to be constant and quite small during the two distillations. The NH_3 value (blank) of the second distillation (derived presumably from urea alone) was then substracted from the NH_3 of the first distillation. A single mixture of urea and ammonium chloride subjected to this procedure using MgO as the alkali gave 8.8 mL of 0.1 NH_3 against an expected value of 8.1 mL. Otto belatedly found that Berthelot, 14 years earlier, had described a similar method, but had not applied it to urine. The only other method that Otto was aware of was Schlösing's, which required several days to perform and which was filled with uncertainty.

1. Folin, O., Ein einfaches Verfahren zur Bestimmung des Ammoniaks im Harne (A simple procedure for determining ammonia in urine). *Hoppe-Seyler's Z. Physiol. Chem.* **32**, 515–17 (1901).

On the Quantitative Determination of Uric Acid in Urine (*1*)

Folin and Shaffer first showed how to improve a step in the Salkowski procedure in which uric acid was precipitated and separated as the silver magnesium compound with H_2S. The removal of silver as sulfide was often incomplete so that repetition of the process was required. To avoid this obstacle, Otto showed that if some copper sulfate and HCl were added to the silver magnesium compound of uric acid before the addition of H_2S, then a clear filtrate was obtained free of silver sulfide. While they offered this improvement they also pointed out that the loss of uric acid in the mother liquor from which it was precipitated varied with the physical state of the crystallized uric acid as did the amount lost in the wash water. Because of these significant variations, they were, in effect, condemning the Salkowski method as fatally flawed. Application of a constant correction for the lost uric acid, even if used, would not solve the problem. Salkowski's method, in other words, gave low values, even if a correction for the lost uric acid were applied. Salkowski had not applied the correction, but he had recently acknowledged that it should be used.

Otto's modification of the Hopkins procedure had been attacked, while the Hopkins and Salkowski procedures that used no corrections were immune. Briefly, the Folin procedure was carried out as follows: To 100 mL of urine, 10 g $(NH_4)_2SO_4$ and a few drops of ammonia were added. After standing 2 h, the solution was filtered and then after conc. H_2SO_4 was added, the separated ammonium urate was directly titrated with $KMnO_4$ solution.

Wörner had stated that 2 h was insufficient for the solution to stand to precipitate all of the uric acid, and that this could be solved by use of a much stronger $(NH_4)_2SO_4$ concentration over a longer standing period. He presented data showing great variability of uric acid values in one urine: 2.94, 5.46, 71.82, and 81.90 in 150 mL. Otto was quite willing to concede that a longer period of standing would increase the yield of uric acid, so that an additional 2–4 mg of uric acid was obtained when the standing was lengthened to 12–24 h. But Otto's concentration of $(NH_4)_2SO_4$ was the correct one, as he had found from experience with the analysis of uric acid in several hundred urine specimens. Wörner's recommendation for using more $(NH_4)_2SO_4$ not only made it difficult to obtain a clear filtrate, but it coprecipitated other substances with uric acid. The huge variability that Wörner showed in his multiple analysis of a single urine specimen was the result of his own modification of the method. It had nothing to do with the Hopkins or Folin procedures.

Jolles stated that ammonium acetate was preferable to sulfate or chloride as the uric acid precipitating medium. He further declared that with ammonium acetate, no further precipitation of uric acid occurred after 2–3 h. He also recommended that the precipitating solution be stirred often with a glass rod. In a later report Jolles recommended that a few drops of NH_3 also be added (as did Otto) to the uric acid precipitating medium along with ammonium acetate, whereas larger amounts of NH_3 should not be used to avoid precipitating phosphates.

Jolles was wrong, Otto pointed out, about ammonium acetate's superiority to sulfate, particularly if the precipitating medium were allowed to stand to the next day. Besides, ammonium phosphate was even better than acetate in producing a clear filtrate in a shorter time. Jolles (as Otto had been) was mistaken to stop the uric acid precipitation at 2–3 h. He should have studied the yields obtained on longer standing. More uric acid would have been found. Stirring finely divided the precipitated ammonium urate so that it became suspended, and both passed through the filter and clogged up its pores. Stirring, not excess NH_3 in the precipitating medium, caused the problem in separating and washing the urate. Finally, Jolles had confirmed that the Folin procedure using the standing period of 2–3 h gave uric acid values on more than 20 urines that were on the average about 4 mg/100 mL of urine higher than the Hopkins or the Ludwig-Salkowski method. This, of course, validated what Otto had observed in his first paper, but it is doubtful that Otto considered this a blessing. In effect, Jolles had not added any new information, nor was his "modification" an improvement in any way.

For some urine specimens, Otto noted, particularly "pathological" ones, it was simply impossible to determine uric acid values close to those obtained by the Salkowski or Hopkins method. Results were sometimes far too high, possibly because mucoid substance in urine present in colloidal solution would coprecipitate with uric acid and be titrated with permanganate. To remove this mucoid material before precipitating the uric acid, Otto and Phil proposed precipitating it with uranium acetate in acid medium. They studied some of the properties of this substance and then offered the following procedure for its removal in the Folin method for uric acid determination:

Five hundred grams of $(NH_4)_2SO_4$, 5 g of uranium acetate, and 60 mL of 10% (v/v) acetic acid were dissolved in 650 mL of water to make a solution of about 1 L. Seventy-five milliliters of this solution was mixed with 300 mL of urine in a 500-mL flask. The mixture was allowed to stand for 5 min to allow the precipitate to form, then filtered through a double thickness of fluted filter paper. One hundred twenty-five milliliters of the filtrate was measured into each of two beakers. To each of these filtrates was now added 5 mL of conc. NH_3, and after stirring the solution was allowed to stand until the next day. The ammonium urate settled to the bottom. The liquid overlayer was then carefully poured into a filter and finally the precipitate in the remaining 10% ammonium sulfate solution was filtered and washed once. Traces of chloride did not interfere with the titration, and the filtration and washing could be accomplished in 20–30 min. During the rinsing of the urate into the beaker they recommended that the filter paper be first taken out of the funnel and opened by punching a hole in the bottom with a glass rod. To the urate suspended in approximately 100 mL of water was added 15 mL of conc. H_2SO_4, and the solution was titrated immediately. Near the end of the titration the $KMnO_4$ solution was allowed to flow in 2-drop portions until the first rose coloring was just visible throughout the solution. Each milliliter of 0.05 N $KMnO_4$ solution equaled 3.75 mg of uric acid.

A note added at this time refuted the claim of those who preferred to isolate and weigh uric acid rather than titrate it, because they considered $KMnO_4$ to be too unstable. Folin and Shaffer correctly pointed out that solutions prepared from pure permanganate were stable for a month and that oxalic acid solutions were also stable for long periods. Finally, a solubility correction of 3 mg uric acid per each 100 mL of urine was added to the uric acid found.

Otto and Phil then compared uncorrected uric acid values obtained by the Folin method on 19 urine specimens, of which 16 were obtained from patients with various disorders, with values obtained by the Salkowski method modified to include the $CuSO_4$ step and found truly excellent correlation.

As a supplement to the article, they presented some further crippling blows to the recent methods of Mörner and of Jolles. Briefly Mörner had determined uric acid with the Kjeldahl procedure but had failed to remove the mucoid-protein substances, making his results too high. The Jolles procedure had removed urea via a hypobromite decomposition prior to oxidation of the uric acid to ammonium sulfate by using heated $KMnO_4$ and H_2SO_4. However, because the hypobromite step was hardly quantitative, urea was a major contaminant.

1. Folin, O., and Shaffer, P.A., Ueber die quantitative Bestimmung de Harnsäure im Harn (On the quantitative determination of uric acid in urine). *Hoppe-Seyler's Z. Physiol. Chem.* **32**, 552–72 (1901).

On Phosphate Metabolism (1)

Urine was collected from a patient wrongly diagnosed as manic depressive. That urine collected daily in three periods during his "good" days was analyzed for the phosphoric acid values present in each period and compared with corresponding periods on the "nervous" days. Phosphate expressed as P_2O_5 was greater on the "nervous" days. This was better shown by the periodicity in the increase in P_2O_5 output calculated on the basis of 100 g output of either nitrogen (N_2) or sulfate (SO_3) that Otto calculated from the "ratio," 100 N_2/P_2O_5 or 100 SO_3/P_2O_5. This periodicity corresponding to the mental state of the patient could not be explained on the basis of diet, because if the patient had eaten less on the "nervous" days, the P_2O_5 output should have gone down but did not. A healthy control ate exactly the same diet as the patient, and the results were almost opposite to those of the patient. In short, the healthy control's P_2O_5 rose on the "good" days of the patient but definitely not on the "nervous" days. What is more, the increase on the patient's nervous days coincided with the onset and duration of the attack, from 10 to 11 a.m. to evening. "The results of the experiments show conclusively that the periodicity in $N_2:P_2O_5$ and $SO_3:P_2O_5$ ratios is due to some kind of periodicity in the metabolism of the patient."

Otto and Phil tried to determine the cause of the periodicity, but could not prove that it was due to increased secretion of HCl in the stomach, to increased absorption of phosphate, to bone disintegration, or to any evidence of altered nerve metabolism.

The analytical method used for phosphate was not discussed in detail in this paper, but it was based on determining phosphorus by titration with a uranium salt solution standardized against monopotassium phosphate, with potassium ferrocyanide as indicator. Nitrogen was determined with the Kjeldahl procedure, and sulfate gravimetrically as $BaSO_4$.

The authors closed the paper with a concluding hypothesis:

> Since the nervous tissues contain considerable quantities of organic phosphorus compounds and since the abnormal phosphate metabolism here described is associated only with peculiar mental symptoms which must in turn be dependent on abnormal processes in the nervous tissues, these are the ones which are subject to an abnormal periodicity in their ability to assimilate "circulating" phosphate. This view is of course advocated only as a working hypothesis which it is hoped may lead to the finding of more facts in connection with the practically unknown field of phosphate metabolism.

1. Folin, O., and Shaffer, P.A. On phosphate metabolism. *Am. J. Physiol.* **7**, 135–51 (1902).

On The Quantitative Determination of Total Sulfates in Urine (*1*)

This method determined sulfate gravimetrically as $BaSO_4$. The urine (50 mL) was clarified by boiling it for 15–20 min with a pinch (0.2 g) of potassium chlorate and 4 mL of conc. HCl. $BaCl_2$ (25 mL of a 60 g/L solution) was added, and the mixture heated but not boiled for 45 min. The mixture was then filtered through ashless filter paper and the $BaSO_4$ washed with hot water (500–700 mL) alternately with NH_4Cl solution (100 mL) for at least 30 min.

After pressing it with another piece of ashless filter paper, the moist filter paper was transferred to a weighed porcelain crucible, placed on a porcelain plate or pipe stem triangle, and 2–3 mL of strong alcohol was poured on the filter and ignited. The burning alcohol dried the paper without sputtering and with little loss of time. Otto cautioned that the paper should continue to burn after the alcohol had disappeared. Otherwise, more alcohol had to be added and the burning repeated.

The SO_3 in the $BaSO_4$ was calculated by multiplying the weight of the $BaSO_4$ found by the factor 0.34293. As this factor was essentially the sum of 0.333 + 0.01, rather than multiply by the factor, it was easier to *divide* the weight of $BaSO_4$ by 3 and by 100, and add the two together.

This method, Otto pointed out, was at least as accurate as that of Baumann-Salkowski, which was subject to two errors that his improvements had eliminated. The B-S method was contaminated with organic matter, especially uric acid that could not be removed by acids, and *moist* $BaSO_4$ could not be washed with alcohol because the first addition of the alcohol would carry some of the precipitate with it through the filter paper. Clarifying the urine was helpful, Otto pointed out, because it promoted rapid settling of the $BaSO_4$ precipitate.

1. Folin, O., On the quantitative determination of total sulphates in urine. *Am. J. Physiol.* **7**, 152–54 (1902).

On the Quantitative Determination of Urea in Urine (1)

While the decomposition of pure solutions of urea to NH_3 could take place quantitatively by 30-min refluxing with $MgCl_2$ and HCl, the procedure with 3 mL of urine was lengthened to 45–60 min. Otto had earlier preferred the shorter time of 30 min out of a fear that other nitrogen compounds could be decomposed by the longer heating period. This fear had proved groundless. By using 45 min or longer Otto probably felt that a greater range of urea concentrations could be quantitatively determined, but again, there were no recovery experiments performed. The 45-min heating caused more frothing in the boiling flask, so he recommended that a small piece of paraffin, about twice the size of a coffee bean, be added.

Otto was not satisfied that the reflux condenser held back enough HCl. For multiple analysis it was inconvenient to add drops of concentrated HCl repeatedly via the condenser to replenish the supply. To remedy this, Otto designed and put into use a safety tube of special construction and size that made the potential loss of NH_3 almost impossible. An illustration of this triple-bulbed device was included. It could be made by any glass blower or purchased from Eimer & Amend in New York.

The possibility that NH_3 could be lost as magnesium ammonium phosphate when the decomposed urea was alkalinized and distilled in the next step was tested in several ways—replacing $MgCl_2$ with $CaCl_2$ prior to removal of phosphates before urea decomposition and tests on pure urea solutions—and there were no indications of loss. The long distillation time was necessary because of the free water present in the heated mixture.

Otto believed that the mechanism for converting urea into NH_3 and CO_2 was due to conversion of urea via cyanic acid, NCOH, into cyanuric acid,

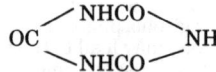

long known to be the product (not biuret) when urea was heated rapidly at high temperature. Cyanuric acid hydrolytically split into NH_3 and CO_2 during the distillation process. This also explained why the distillation could not be achieved in less than 1 h. Interesting to note was the fact that $MgCl_2$ containing water of crystallization could not bring about the hydrolytic cleavage.

In the addendum to the paper, Otto countered criticism of Arnold and Mentzel of his paper on urea analysis. A-M had declared that the least possible volume that could be used in the Folin urea method was 10 mL and had "reproached" him for using 3 mL. By doing so, Folin had introduced a pipetting error of 1.5%. Folin countered that he used not a "well-pipette" but the normal mass pipette of 5-mL volume, graduated into 0.05-mL divisions, which was accurate to 0.0125 mL. This accuracy was readily checked by measuring normal acid with it, and titrating the acid with 0.1 N base. Parallel determinations always gave results within 0.1–0.2 mL of base. Folin did admit that he had used 5 mL of urine lately, but 10 mL was too much because the volume to be titrated was too great for the sensitivity of the indicator used.

A-M also had stated that the Folin method was inaccurate, giving NH_3 yields that were too low. Otto vehemently denied this, but suggested that the cause for low yield was probably the loss of HCl and with it some NH_3 during the refluxing step. He hoped that this paper would now solve this problem. He then stressed the importance of carrying out the heating step properly. Otto used an electric heater that could be set at three separate temperatures. With lamp (illuminating) gas, heated iron disks or sand baths could be used equally well. Heating Erlenmeyer flasks on a wire gauze or directly by a flame was not recommended.

Otto then cited an experiment taken from the June 1902 protocol of his assistant, Phil. A. Shaffer, in which the NH_3 produced by a pure urea solution was determined by both the standard Kjeldahl and the Folin procedures. The results were identical because the conditions of study were what they should have been, in contrast to the work of A-M.

A-M had concluded that Otto's method could not be applied to urine when very accurate results were desired because of interfering substances such as uric acid, hippuric acid, and creatine, which they showed had decomposed, producing some NH_3. But Otto had already tested pure uric acid and it did not decompose to produce any NH_3. He recommended that A-M try purifying their uric acid by twice crystallizing it from boiling water. He also chided them with the fact that they had obtained less NH_3 by heating it for 2 h in the procedure than they did by heating it for only half an hour. When Otto obtained a pure preparation of hippuric acid and used seven times the amount that A-M had used, only a negligible amount of NH_3 was produced in the urea procedure. A-M had probably used impure hippuric acid. The only point in the entire paper that he admitted was undeniable was that he had not studied either creatine or creatinine, and therefore they could potentially interfere. Because A-M had not presented data on the purity of their creatine, their claims of up to 33% of creatine decomposition were unreasonable. On the other hand, assuming that A-M were correct, how much error would this cause? If the daily output of creatine and creatinine were about 1.5 g then Otto calculated that the urea results would be high by about 1% at the most.

1. Folin, O., Ueber die quantitative Bestimmung des Harnstoffs im Harne (On the quantitative determination of urea in urine). *Hoppe-Seyler's Z. Physiol. Chem.* **36**, 333–42 (1902).

A New Method for Determining Ammonia in Urine and Other Animal Fluids (1)

A weak alkali such as Na_2CO_3 or $Ca(OH)_2$ was added to the NH_3-containing liquid, and the liberated NH_3 was expelled into standard acid by a strong airstream at room temperature. Otto, amazed that this simple principle had not been used before, searched the literature and, as expected, found that Boussingault in 1850 had liberated NH_3 from NH_4Cl with $Ca(OH)_2$ and expelled the NH_3 with an airstream at 35–40 °C. Boussingault, who had used a 5-h airstream, had discarded the method as impractical because he had obtained a poor yield of NH_3. Boussingault's problems, Otto noted, were first that he had used far too much NH_4Cl so that he could not collect the NH_3 quantitatively into his absorption flask, and second, the airstream used was too weak to drive out the NH_3 in a sufficiently short time.

Boussingault, however, also described a very accurate procedure for distilling off NH_3 in vacuo, which when properly carried out, as Phil Shaffer had done in Otto's laboratory, provided results more accurate than others. (Shaffer would shortly report his results.)

The following observations were made in support of this new method:

1. The NH_3 in 25–50 mL of the alkalinized liquid had to be driven off with air in the shortest time (about 1–1.5 h) with a moderately strong airstream (600–700 L/h). A useful water-jet blast or a good water-jet pump gave about 2–3 atm water pressure, a sufficiently strong airstream. Otto preferred the water-jet blast, but both systems worked well on 25 mL of urine.

2. To prevent foaming in urine and other protein-containing fluids, 5–10 mL of petroleum or toluene was added. For blood or other strong protein-containing fluids, some methanol was added periodically during the aeration procedure.

3. A series of alkalis were tested for their speed in liberating NH_3 and for their speed in decomposing nitrogen-containing (contaminating) compounds. Others whose work Otto reviewed (Schlösing; von Nencki and Zaleski; and Söldner) had used $Ca(OH)_2$ or $Mg(OH)_2$. Otto tested the carbonates of sodium and potassium and found them better suited to his new method than the hydrates of calcium or magnesium. A 4% (w/v) Na_2CO_3 solution that had been saturated with NaCl decomposed NH_3 salts as fast as a saturated $Ca(OH)_2$ solution without causing the splitting of organic substances that the latter did. This was important in the analysis of blood NH_3 particularly, though weaker solutions of calcium or magnesium hydroxides would work at room temperatures (20–25 °C).

4. The speed of driving off the NH_3 from an alkalinized solution by means of an airstream depended on the pressure of the airstream, the kind and concentration of the alkali used for liberation of NH_3, and the temperature. The same airstream in a solution of 25 mL of 4% Na_2CO_3 quantitatively drove off the same amount of NH_3 at 25–30 °C, in 20 min, as in 1 h at 20 °C. In 4 h at ice temperature (the latter was done on 50 mL of liquid), only half the expected amount of ammonia was obtained.

5. The amount of NH_3 to be absorbed must be within the capacity of the standard acid in the receiving or absorption tube. If so, a continuous chain of receivers could be established for multiple NH_3 determinations. Otto had succeeded in performing four NH_3 determinations simultaneously and without difficulty in about 2 h. To accomplish this more conveniently, Otto designed a new absorption tube that was described in detail in a diagram. The airstream bearing the NH_3 was passed via a perforated bulb through the standard acid contained in a tube, and the "deammoniated" air exited the tube through perforations near its top and then passed through the next NH_3-containing solution.

6. To prevent possible contamination of the airstream by alkali from the alkalinized liquid, it was passed through either a calcium chloride tube or through an adapter containing cotton.

7. Alizarin red was selected as the acid-base indicator. Though less sensitive than others, it was not interfered with by moderate amounts of carbonic acid, ammonia salts, or the organic medium for solution (methanol, petroleum, toluene). Titration was carried out to the red, not the violet color.

Otto found that there was an unknown labile substance that liberated NH_3 during the alkali treatment (as well as by acid reaction) and caused values somewhat too high. The heating caused its generation. Carbonic acid in excess would interfere with the titration and could be removed by warming the receiving tube in warm water for the last 15 min of the airstream.

The procedure to perform the analysis of preformed NH_3 was now as follows:

Twenty-five milliliters of urine was measured into a special cylinder (aerometer) about 45 cm high and 5 cm in diameter though it could be smaller. Then 8–10 g of NaCl, 5–10 mL of petroleum or toluene, and about 1 g of dry Na_2CO_3 were added. A strong airstream of 600–700 mL/h was passed through the urine for 1–1.5 h until all of the NH_3 was driven out at 20–25 °C. The NH_3-containing air was passed through a cotton plug into either the above-described absorption device or through two tubes containing 0.1 N acid and water and the NH_3 content was determined by back-titration with alizarin red indicator and standard alkali. To determine NH_3 in larger volumes (to 100 mL) of urine, the airstream had to be used for a longer period, an extra hour at 20–25 °C.

For historic reasons, we present Otto's first procedure for blood NH_3:

Fifty milliliters of blood was measured into the aerometer cylinder and the cylinder packed well in ice. To this was added 16 g of NaCl, 25 mL of methanol, and 2 g of dried (or 5 g of crystalline) Na_2CO_3. The airstream was used for a 5-h period. The receiver contained 10 mL of 0.05 N acid, and was, as in the case of water, warmed at 30–35 °C during the last 15 min of the airstream to drive off CO_2 before the final titration. After about the first 2 h of the airstream, an additional 25 mL of methanol was added to counteract the foam.

Otto stressed that the blood used must be fresh or results would be too high. But in this study, Otto was unable to make a thorough study of this important point. Otto's studies of blood NH_3 were all carried out on dogs.

The results obtained were much too high, amounting to about 400–500 µg/dL (corrected for a blank run) or 286–357 µmol/L. On the other hand, Otto obtained impossibly high values by methods involving vacuum distillation, obviously signifying destructive production of NH_3. This also indicated the uselessness of vacuum distillation for determining the alkalinity of blood.

Otto's new technique was promising for determining blood NH_3, but was only an initial small step in this important analysis although it was a huge step for urinary NH_3 and greatly simplified the urea procedure.

Philip Shaffer made a thorough study of the various analytical methods for ammonia, and found defects in all of them (Schlösing; Wurster; Nencki and Zaleski; Söldner; and Otto's first method). He then modified the Boussingault vacuum distillation method into a simpler, more defined procedure that could be carried out in 15 min and that agreed remarkably with Otto's latest version (above). Otto's method was more time-consuming but less demanding of equipment and permitted multiple analyses in a "train." But Shaffer's method was a very attractive alternative. The principle involved was exactly the same except where Otto had originated the use of an airstream (or current), Shaffer used a vacuum of about 10 mm Hg and warmed the distilling flask at 50 °C. Otto's receiving vessel was, of course, unique. Both methods used methanol, NaCl, and Na_2CO_3.

Shaffer pursued this work independently and published his results in the *American Journal of Physiology* [8, 330–54 (1903)]. This was his first solo effort, as he entered the graduate school at Harvard.

1. Folin, O., Eine Neue Methode zur Bestimmung des Ammoniaks im Harne und anderen thiereschen Flussigkeiten (A new method for determining ammonia in urine and other animal fluids). *Hoppe-Seyler's Z. Physiol. Chem.* **37**, 161–76 (1902).

On the Quantitative Determination of Urea in Urine (1)

Otto purified commercially obtained creatine by decolorizing it with animal charcoal and crystallizing it from water with the addition of a small amount of alcohol. He verified the purity of this preparation by microscopic examination, by determining its water of crystallization, and by nitrogen analysis. He then put a large amount of creatine through the urea procedure heating it twice as long as required, yet found no trace of NH_3. During the heating step, creatine became converted to creatinine but it also did not interfere. Glycine, which would have formed hippuric acid did not interfere, so Otto assumed that amino acids, in general, did not.

1. Folin, O., Ueber die quantitative Bestimmung des Harnstoffs im Harne (On the quantitative determination of urea in urine). *Hoppe-Seyler's Z. Physiol. Chem.* **37**, 548–50 (1903).

The Acidity of Urine (1)

To determine the total acidity of urine by direct titration powdered potassium oxalate was added to remove the troubling calcium salts without affecting acidity (as Liebig had shown) and to reduce the problem of endpoint detection in the presence of ammonium salts. The method was as follows:

Twenty-five milliliters of urine was pipetted into a 200-mL Erlenmeyer flask. One, or at most 2 drops, of 0.5% phenolphthalein solution and 15–20 g of powdered potassium oxalate were added. The solution was shaken for about 1 min and titrated at once with 0.1 N NaOH until a faint yet distinct pink color was produced. During the titration, the flask was shaken to keep the oxalate strongly dissolved.

Otto found that this procedure worked as well with "pathological" as with "normal" urines.

The determination of mineral acidity of urine was much more complex. The mineral acidity was defined as the excess of the combining equivalence of all the mineral acids of urine above that of the known "bases," sodium, potassium, calcium, magnesium, and ammonia. As Otto stated,

> My own purpose in attempting to determine the excess of mineral acid or alkalinity, was to try to discover whether there is any abnormality in the metabolism of the insane, and whether the urines of the insane show any indication of the occurrence of any unknown organic bases or acids.

Otto thought that he had a method that gave the true excess of mineral acids or bases in urine with as great an accuracy as could be obtained with the current state of knowledge concerning the composition of urine.

Otto's method was borrowed from his knowledge of determining acid in vinegar, and HCl in stomach juice. The determination of mineral acidity required 1) the burning and titrating of the urine as described below, 2) one NH_3 determination, and 3) two sulfate determinations. For historic reasons, we present an abbreviated version of the method:

Pure, dry granular 0.3–0.6 g K_2CO_3 was weighed into a platinum dish, and 25 mL of urine was added. After being evaporated to dryness on a sand bath or in an electric oven, the dry sample was burned at just below red heat for an hour after the visible NH_3 fumes came off. Ten milliliters of peroxide was added; the solution was covered with a watch glass and warmed to decompose the peroxide. The "sputterings" on the watch glass were washed into the dish with a little water, evaporated to dryness for an hour, and then burned as before for 1 h.

The "burning" residue in water was dissolved with the help of an excess of 0.1 N HCl (75–100 mL) and rinsed into an Erlenmeyer flask. Carbonic acid was driven off by boiling the solution, cooled, titrated with 0.1 N NaOH in the presence of a small amount of potassium oxalate and 2 drops of 0.5% phenolphthalein solution.

Because the volume of added alkali or acid was known, the final titration gave the data for calculating the apparent excess of mineral acids or alkalis originally present in the urine. Before the final result was obtained, however, certain other factors had to be taken into account: 1) the alkaline strength of K_2CO_3, 2) the acidity of the peroxide, 3) the SO_3 content of the peroxide, 4) the preformed NH_3 in the urine, 5) the inorganic SO_3 of the urine, and 6) the total SO_3 found in the titrated solution of the urine residue.

Results were calculated as follows: The preformed NH_3, the acidity of peroxide, and the acidity due to organic SO_3 of the urine were subtracted from the apparent excess of acidity found on titrating the burned urine residue. The acidity (in milliliters of 0.1 N) of the organic SO_3 was obtained by subtracting the sum of the SO_3 of the peroxide and the inorganic SO_3 of the urine from the total SO_3 of the urine residue and dividing the amount so obtained by 8.

Otto then provided a sample set of data to illustrate the above calculation and a set of notes on the determination, with some experiments to show its validity.

The organic acidity of urine was obtained by subtracting the mineral acidity from the total acidity to give the total equivalence of organic acid whether free or combined.

Contrary to previous assumptions, Otto found if the organic acids were considered to be in the free form and if only the excess acidity was due to acid phosphates then "acidity due to the phosphates varies within very wide limits, but that the greater part of the acidity of urine in most cases is due to organic acids." Much more attention would therefore have to be given to these organic acids and their origin.

1. Folin, O., The acidity of urine. *Am. J. Physiol.* **9**, 265–78 (1903).

On Rigor Mortis (*1*)

To prove that rigor was not due to coagulation, Otto produced rigor in frog muscles by subjecting them to a temperature of −15 °C to −20 °C. It was not generally known in Otto's time that lowering temperature caused rigor, but it was pointed out by Brücke in 1842, who maintained that rigor so obtained was identical with the usual rigor. Cold rigor, Otto felt, was better adapted to throw light on the process than "normal" or "heat" rigor.

Another fact that Otto used was that frog muscles could be cooled to −7 °C, i.e., frozen solid without losing their irritability and contractility when carefully thawed. Upon lowering the temperature of such frozen muscles to −15 °C, the power to recover contractility was lost, and on thawing, they would be in rigor. Otto wrote,

> It seems clear that the chemical changes involved in the transformation of the frozen but potentially living muscle into the dead and rigid muscle must be relatively very small. Such low temperatures and the absence of a liquid medium acting together would tend to reduce to a standstill all those chemical reactions that are associated with constructive or destructive metabolism, i.e., oxidations, reductions, and ferment [enzyme] or other hydrolytic reactions.

Otto hoped that he could freeze to a standstill those chemical changes, however small, so that they could be measured against the same process in muscles that were not in rigor. At any rate he could determine whether there was a difference in the coagulum they produced, and he could certainly compare their opaqueness or translucency.

Otto obtained saline extracts of frog muscles that had been placed for 2 h at the different temperatures to produce the effects mentioned. In this case, one set of muscles was placed in a freezing mixture cold enough to produce rigor and the control was placed on ice. At the end of 2 h, the control was quick-frozen so that it could be more easily treated along with the "rigor" set by maceration and then extraction into saline solution. This work was all carried out in a "cold" room.

The two saline extracts were treated in various ways. Both produced an equal amount of coagulum on standing or on heating, had identical acidity, and the same total nitrogen (and presumably the same protein) values. Otto had already seen that muscles in cold rigor showed no sign of cloudiness or opaqueness. In fact, they displayed a "perfect translucency" contrary to the coagulation process. The fact that

> the onset of rigidity may be prevented and rigidity already produced may be made to disappear by simply gently bending or pressing a muscle between the fingers would seem far more in harmony with a "contraction theory" than with a coagulation theory of muscle rigor.

Otto's idea that the contraction and the onset of rigor mortis were derived from the same source was correct.

1. Folin, O., On rigor mortis. *Am. J. Physiol.* **9**, 374–79 (1903). [From the Chemical Laboratory of the United States Fish Commission, Woods Hole, and The Chemical Laboratory of the McLean Hospital for the Insane, Waverley, Mass.].

Chapter 11. A Classical Period (1903–7)

Some Metabolism Studies with Special Reference to Mental Disorders (1, 2)

1. *Urea.* Five milliliters of urine was pipetted into a 200-mL Erlenmeyer flask, to which were added 5 mL conc. HCl, 20 g crystalline $MgCl_2$, a piece of paraffin the size of a hazel nut, and finally 2–3 drops of aqueous 1% solution of alizarin red. The safety tube was inserted, and the mixture was boiled on an electric oven until each returning drop from the safety tube produced a very perceptible bump. The heat of the oven was reduced somewhat, and heating continued for a full hour.

The alizarin red was stable and turned bright red in the presence of a trace of alkali. If the boiling contents of the flask became alkaline, a few drops (only) of the acid distillate in the safety tube were shaken back. The contents were transferred to a liter flask with about 700 mL water, and 20 mL of 10% NaOH was added. Distillation was carried out until the flask was dry or until the distillate showed no trace of NH_3 with delicate litmus paper. The distillate was boiled to remove carbonic acid, cooled, and titrated as usual. The free NH_3 of the urine, as well as that of $MgCl_2$, was subtracted from the result obtained. Two to four determinations could be carried on simultaneously.

2. *Uric Acid. Reagent*: 500 g of $(NH_4)_2SO_4$, 5 g of uranium acetate, and 60 mL of acetic acid (10% v/v) in 650 mL of water.

Urine (150 mL) was measured into a tall, narrow beaker or a cylinder, and 37.5 mL of reagent was added. If more urine was available, 200 mL and 50 mL of reagent were used. The solution was allowed to stand for half an hour. After the uranium precipitate settled, the clear supernatant liquid was decanted or siphoned into a beaker. One hundred twenty-five milliliters of this liquid was measured into another beaker, and the mixture was allowed to stand until the next day. The liquid was filtered, and the precipitate was washed with 10% $(NH_4)_2SO_4$ solution until the filtrate was free of chloride. The filter from the funnel was removed, and the precipitate was rinsed from the filter into a beaker. Enough water was added to make about 100 mL, and the precipitate was dissolved in about 15 mL of conc. H_2SO_4. The solution was titrated at once with 0.05 N $KMnO_4$ solution. Each milliliter corresponded to 3.75 mg of uric acid. A correction value of 3 mg for solubility of ammonium urate was added. The end point was the very first pink coloration extending throughout the liquid, obtained by adding 2 drops of $KMnO_4$ while the solution was stirred with a glass rod.

3. *Ammonia.* To obtain a sufficiently rapid air current, a good water vacuum pump, or preferably a force pump was used.

Urine (25 mL) was measured into a tall aerometer-cylinder that was plugged with a double-bored rubber stopper, through which passed one long and one short glass tube, exactly as in an ordinary wash bottle, with the longer tube reaching almost to the bottom of the cylinder. The shorter tube consisted of a $CaCl_2$ tube, and was filled with cotton.

About 1 g of Na_2CO_3 and 5 mL of petroleum were added to the cylinder. The upper end of the $CaCl_2$ tube was connected to another glass tube that led to the bottom of a wide-mouthed 16-oz bottle half-filled with water, 25 mL of 0.1 N acid, and some indicator.

A strong air current was passed through the alkaline urine for 1 h and through the bottle containing the acid. The NH_3 present was determined in the standard acid by titration. The airstream coming from the bottle where the NH_3 was absorbed could be fed into a second similar apparatus, so that two determinations could be done simultaneously.

4. *Creatinine.* Otto had severely criticized the available methods of analyzing for creatinine as unreliable, inaccurate, tedious, time-robbing, and requiring too much urine. He therefore devised a colorimetric method that required 10 mL of urine, that could be performed rapidly, and that was reasonably accurate. Although this method had not yet been published, it had been in constant use for almost a year.

This method not only introduced the use of the colorimeter, but was a masterful stroke in clinical chemistry. For historical reasons, we cite the method verbatim in its entirety.

> The principle of the method is the color reaction which kreatinin gives with alkaline picric acid solution. According to Jaffé, the discoverer of this reaction, it is characteristic for kreatinin and is given by no other normal urinary constituent. The only substances which I have found to give the reaction under the conditions of the experiment are aceto-acetic ether, acetone, and hydrogen-sulphide. Since these can all be removed from the small quantity of urine used in this determination by a few minutes heating with dil. hydrochloric acid and are moreover rarely present, this fact is of but little consequence. A fortunate circumstance discovered in the course of the investigation is the fact that the color produced is identical in suitable concentration with the color of potassium bichromate solutions. This gave a standard for comparison as easy to prepare as it is stable. The half normal solution was selected for the purpose.
>
> In order to make color comparisons accurate enough for quantitative purposes a high-grade colorimeter was necessary and the French instrument of Duboscq was secured from Eimer and Amend "on approval." This instrument proved eminently satisfactory. The construction and workmanship are excellent, the prisms are absolutely flawless, and the height of each solution can easily be adjusted to within an accuracy of a tenth of a millimeter.
>
> Perfectly pure kreatinin was then prepared and the conditions under which its highest colorimetric value is produced was ascertained, and this value was determined in terms of the standard bichromate solution. It was found that 10 mg. kreatinin dissolved in about 10 cc. water gave the highest value five minutes after the addition of 15 cc. saturated picric acid solution and 5 cc. 10 per cent sodic hydrate. The resulting solution diluted at the end of that time to 500 cc. showed about the same color as the 2/n [2/n should have read n/2, or 0.5 N] bichromate solution when viewed in the colorimeter (i.e. by transmitted light) through a depth of 8 mm. To be more accurate 8.1 mm of the kreatinin picrate solution was exactly equal to 8 mm. of the bichromate.
>
> The absorption of light was further found to be quite different for the two solutions, but this fact cannot be discussed here except in so far as to state that this necessitated the making of a definite depth of the bichromate the standard for comparison. 8 millimeters was chosen for the purpose.
>
> On the basis of these facts the kreatinin determination in urine is made as follows:
>
> Half normal potassium bichromate containing 24.55 g. per liter; saturated picric acid solution containing about 12 g. per liter; and 10 per cent sodic hydrate solution, are the reagents needed.
>
> 10 cc. urine are measured into a 500 cc. volumetric flask, 15 cc. picric acid and 5 cc. sodic hydrate are then added, and the mixture is allowed to stand for 5 to 6 minutes. This interval is used to pour a little of the bichromate solution into each of the two cylinders of the colorimeter. The depth of the solution in one of the cylinders is then adjusted accurately to the 8 mm. mark. With the solution in the other cylinder is made a few preliminary colorimetric readings, simply for the sake of insuring greater accuracy in the subsequent readings of the unknown solution. The two bichromate solutions must of course be equal in color, and in taking their readings, no two should differ more than 12 mm. [12 mm. should have read 0.12 mm.] from the true value (8 mm.), leaving out of consideration the very first reading made, which is sometimes less accurate. Four or more readings should be made in each case and an average taken of all but the first. After a while one becomes very sure of the true point, and can take the average of the first two readings.

At the end of five minutes in the 500 cc. flask the contents are diluted up to the 500 cc. mark. The bichromate solution is thoroughly rinsed out of one of the cylinders by means of the unknown solution and several colorimetric readings are then made at once.

The calculation of the results is very simple. If for example it is found that it takes 9.5 mm. of the unknown urine picrate solution to equal the 8 mm. of the bichromate, then the 10 cc. of urine contains $10 \times 8.1/9.5 = 8.4 +$ mg. kreatinin.

The amount of urine taken for the determination is usually 10 cc., but if this should be found to contain more than 15 or less than 5 mg. kreatinin the determination should be repeated with a correspondingly different amount of urine, because outside of these limits the determination is much less accurate. With kreatinin solutions the results are uniformly surprisingly accurate, and I have as yet found no reason for believing that it is not equally reliable, at least for normal urines. The color of the urine does not materially affect the result on account of the very great dilution.

A kreatinin determination can be made by this method in less than 15 minutes.

5. *Specific Gravity.* Current urinometers were all found to be unreliable. Otto had special ones made for him, but did not name the manufacturer or state how he checked their accuracy.

6. *Volume.* Ordinary graduated cylinders were used.

7. *Chlorides.* The Volhard method was used. Otto added a "fair" amount of HNO_3 to ensure a clear filtrate.

8. *Total Sulfates.* Otto used his published method.

9. *Ethereal Sulfates.* Otto modified the currently used method. To 200 mL of urine was added 100 mL of 10% $BaCl_2$ solution; the mixture was allowed to stand overnight at ordinary room temperature. The supernatant liquid was decanted and filtered, if it was not clear. One hundred fifty milliliters of filtrate was transferred to an Erlenmeyer flask and boiled with a few milliliters of strong HCl and enough potassium chlorate to give a colorless solution. The remaining operations were identical with those for the determination of total sulfates.

10. *Neutral Sulfur.* Neutral sulfur was calculated as the difference between the total sulfate and the total sulfur found in the mineral acidity determination.

11. *Phosphates.* Otto cited as the source of this method *Sutton's Volumetric Analysis* (8th ed., p. 316). This method determined phosphates volumetrically by means of a uranium acetate solution, with potassium ferrocyanide as indicator. Chemically pure, crystallized monopotassium phosphate was used to standardize the uranium salt solution. Otto could not determine organic phosphates in urine, which were present only in traces.

12. *Acidity.* The Folin procedure, as published, was used for total, mineral, and organic acid.

13. *Indican.* A Folin modification was used that had not been described previously. The method used 0.01 of a 24-h urinary volume. Indigo was developed by means of Obermayer's reagent, and the indigo blue was extracted with 5 mL of chloroform. The color of the chloroform solution was matched in the colorimeter against Fehling's solution, which was assigned an arbitrary value of 100 units. Otto did not, of course, consider this method quantitative.

14. *Total Nitrogen.* In the Kjeldahl procedure, Otto used for the oxidation step a mixture of copper sulfate, potassium sulfate, and sulfuric acid and a digestion period of about half an hour.

Otto regretted being unable to perform duplicate analysis as a routine, due to the large number of determinations made on each 24-h urine. This lack was countered partly by visual inspection of values that were more readily apparent, because the subjects were on a uniform diet and because *each* subject's urine was analyzed daily for at least five days. Otto pointed out that the nitrogen compounds analyzed represented 92–98% of the total nitrogen in urine.

The full report on each 24-h specimen included the following: volume, specific gravity, total nitrogen, urea, NH_3, creatinine, uric acid, chloride, phosphates, ordinary sulfates, ethereal sulfates, neutral sulfur, total acidity, mineral acidity (or alkalinity), organic acidity, and indican. In addition, the following ratios and percentages were reported: N_2/P_2O_5, N_2/SO_5, N_2/Cl_2, SO_3/P_2O_5, SO_3/ethereal SO_3, SO_3/neutral SO_3. The percentage of total nitrogen was given for urea, NH_3, creatinine, uric acid, and "rest" nitrogen.

Otto paid very strict attention to the diet administered to each subject whose urine was tested. He prepared a liquid diet in his own laboratory. It was used very scrupulously with the help of the nurses. A daily food intake for men consisted of 500 mL of whole milk; 300 mL of cream (18–22%

fat); 450 g of eggs (white and yolk); 20 g of sugar; 200 g of Horlicks Malted Milk; 6 g of salt (75 mL of 10% solution); water to make up to 2 L. In addition, 900 mL of extra water was given. Women received two thirds of the above diet.

Dr. Hoch and his successor, Dr. F. Packard, selected the patients to be tested, and provided the clinical descriptions accompanying the tables of data. A reliable special nurse or sometimes two were put in charge of each case, with sufficient relief nurses to ensure constant attention. At the beginning of the experiment and each morning the weight of the patient was taken. Two kilograms of liquid food prepared in the laboratory, as well as a measuring cylinder for the water, was given to the nurse in charge, and in the course of the day (usually three to five feedings), all the food and 900 mL of water were administered to the patient.

During the first two days on the diet, urine was collected (and discarded) to determine if any difficulties would develop either in the collection or with the patient. The true collection began on the morning of the third day. No fewer than two and usually more than four 24-h urine samples were collected. In some cases the experiments continued for several weeks. Otto noted that Dr. Cowles had many years before devised a special cabinet chair that was particularly useful for quantitative collection of urine. The chair was supplied with a divided pan by means of which urine and feces were collected separately. This device was especially helpful for women patients. Otto regretted very much that the sheer weight of the number of analyses to perform on urine did not allow him time to perform fecal analyses.

Otto performed analyses on seven normal subjects (six nurses and one physician from McLean Hospital) who were on the liquid diet for one week.

There were several observations that Otto made that were highly important, yet did not appear in his written conclusions. Otto was now talking about *discovering the laws that govern the absolute and relative amount of each constituent excreted in the urine.*

The standard diet brought out certain tendencies. Contrary to a recent report in the literature, NH_3 output was not constant. Hence the laws governing the output of NH_3 had to be different from those governing elimination of total nitrogen. It was not true that the volume of urine excreted during a 24-h period was proportional to the liquid intake! The variations Otto noted in his subjects on a constant fluid intake apparently were not related to temperature, moisture, or gain or loss in body weight, and therefore had to be related to water loss via the lungs and skin. Otto had made some undescribed studies on water loss from the lungs in 1902 and had found considerable variation. He thought this loss could have importance in the metabolism of the insane, and he hoped to study it further. The remarkable gain or loss of weight in some of his patients could have been due to elimination of water through the lungs.

The two-part article contained 67 tables, most of which showed the analytical data obtained on the urines of the 46 subjects involved. While there were seven "normal" subjects whose data were summarized, no distinct abnormalities were found in the urines of the mental patients. The periodicity in output of phosphorus in one patient did not lead to a finding concerning his condition. In all, 10 patients with "general paralysis" were tested, but none of the urinary constituents or ratios that seemed abnormal for one or a few were so for all. The greater fluctuations in values obtained on these patients, however, could have association at one stage or another with a metabolic disorder. Claims in the literature of metabolic derangement in mental disease were simply untrue, at least from the standpoint of the constituents in urine that Otto had analyzed. Obviously, in the expectation of uncovering something metabolically different about the patients in his hospital, Otto had to be disappointed. Of course, he was at that time in no position to realize that his results had much greater potential impact for clinical application where salt and water metabolism, acid-base regulation, and nutrition were essential factors to be considered, and that his analytical methods would open up new horizons in the field of biochemistry. He would soon have much more to write about the topic of urine biochemistry.

1. Folin, O., Some metabolism studies. With special reference to mental disorders. *Am. J. Insanity* **60**, 699–732 (1903–4).
2. Folin, O., Some metabolism studies. With special reference to mental disorders. *Am. J. Insanity* **61**, 299–364 (1904–5).

Contribution on the Chemistry of Creatinine and Creatine in Urine (1)

Otto presented not only convincing evidence that creatine existed in urine, but a very precise method to measure it quantitatively. To determine the creatine of any urine, the creatinine (preformed) was measured in the urine in the usual way. To determine the creatine, 10 mL of urine was heated with 5 mL of 1 N HCl for 3 h on a waterbath, quantitatively converting the creatine to creatinine as Otto proved. The creatinine of this "converted" solution was now determined, and from this value was subtracted the value of the preformed creatinine to obtain the milligrams of creatinine derived from creatine. The creatine-derived creatinine was converted to creatine by multiplying by the conversion factor 1.16 (1 mg creatinine is equivalent to 1.16 mg of creatine). Otto found that most urines contained creatine, but occasionally there were urines with no creatine present.

Because pure, commercially available creatinine then was prohibitively costly, Otto presented a fairly simple method for isolating and purifying creatinine from normal human urine (containing 5–20 mg/10 mL) by using the double picrate reaction of Jaffé.

For each liter of urine, 18 g of picric acid was weighed and dissolved in boiling alcohol (100 mL for each 40 g picric acid). This hot solution was poured into the urine with vigorous stirring. The solution was stirred continuously without touching the walls of the vessel until precipitation began. In 30–45 min almost all of the creatine was contained in a heavy sediment contaminated with precipitated uric acid. The liquid was siphoned off as completely as possible, and the sediment was washed on a suction filter with saturated picric acid.

The sediment (double picrate) was then purified by recrystallization four times from hot water. However, this step could be omitted if the dry product was not desired. Optionally, the creatinine could be obtained directly in concentrated solution as follows:

The still moist precipitate was weighed, and about half this weight of $KHCO_3$ was added with about 150 mL of water for each 4 L of urine used, and ground an hour in a large mortar. The carbonate caused no decomposition of creatinine and picric acid as did the strong alkalis and no perceptible transformation of creatinine into creatine. During this treatment the creatinine dissolved, while the equivalent amount of picric acid converted to the difficultly soluble potassium salt. The solution was filtered, and the precipitate was washed with a small quantity of $KHCO_3$ solution.

The carbonate solution containing the creatinine and small amounts of impurity was carefully neutralized with 20% sulfuric acid. To this weakly acidic solution, two volumes of methyl or ethyl alcohol were added, and immediately (without filtering) decolorized with small amounts of animal charcoal, about 50 g/8 L of urine. After a few minutes the solution was filtered to remove the charcoal, and the K_2SO_4 was precipitated by the alcohol. The barely yellow filtrate was allowed to stand for several hours or until the next day, and refiltered.

The creatinine obtained in this way could be used for most practical purposes, but to obtain completely pure creatinine Otto proposed that its double salt with $ZnCl_2$ be prepared and that the $ZnCl_2$ be removed with H_2S and lead hydrate. Details of this procedure will not be given, but once the $ZnCl_2$ was removed, Otto found that there was an admixture of creatine present with the creatinine. He then quantitatively converted the creatine back to creatinine by heating it with dilute sulfuric acid on a waterbath for a 36- to 48-h period, and removed the sulfate with barium chloride. The creatinine was washed with alcohol and recrystallized twice.

1. Folin, O., Beitrag zur Chemie des Kreatinins und Kreatins im Harne (Contribution on the chemistry of creatinine and creatine in urine.) *Hoppe-Seyler's Z. Physiol. Chem.* **41**, 223–42 (1904).

On the Separate Determination of Acetone and Diacetic Acid in Diabetic Urines (1)

Acetone solution or urine (20–25 cc) was measured into an aerometer cylinder and 0.2–0.3 g of oxalic acid or a few drops of 10% phosphoric acid, 8–10 g of sodium chloride (acetone is insoluble in saturated sodium chloride solutions), and a little petroleum were added. The aerometer cylinder was connected with the absorbing bottle (as in the ammonia determination) in which had been placed water and 40% potassium hydroxide solution (about 10 cc of the latter to 150 cc of the former), and an excess of a standardized solution of iodine. The whole apparatus was connected with a Chapman pump and the air current was run through for 20–25 min. (The air current had to be fairly strong but not as strong as for the ammonia determination.) Every trace of acetone would by then be converted into iodoform in the receiving bottle. The contents of the receiving bottle were acidified by addition of conc. HCl (10 cc for each 10 cc of the strong alkali used), and the excess of the iodine was titrated, as in the Messinger method, with standard thiosulfate solution and starch.

Diacetic acid was obtained by subtracting the acetone determined from the total found for the unaerated sample, and by a conversion factor for acetone to diacetic acid.

1. Folin, O., On the separate determination of acetone and diacetic acid in diabetic urines. *J. Biol. Chem.* **3** 177–82 (1907).

Chapter 12. Harvard—Teaching and Profession (1907–12)

Determination of Fat and Fatty Acids in Feces: Highlights (1)

Standard sodium methylate was prepared from metallic Na and absolute alcohol, and standardized against 0.1 N HCl. The ether-HCl mixture was prepared by generating HCl and H_2SO_4 added to powdered NaCl and leading the HCl gas formed into cold, dry ether. After titration, the HCl content was adjusted to 0.1 N with more ether. The indicator in all the titrations was an alcoholic solution of phenolphthalein.

The stool sample was dried, then pulverized thoroughly, and sifted twice through a 40-mesh sieve. For the extraction, 1 g was weighed in a fat-free filter paper, transferred to a fat-free filter-paper thimble, and then inserted into the extraction apparatus. This was attached to a 250-mL Erlenmeyer flask containing 150 mL of the ether-HCl solution, and then boiled for 20 h. The ether was distilled off (along with the dry HCl), 50 mL of low-boiling petroleum was added, and the solution was allowed to stand overnight. This was then filtered, and the filtrate collected into a preweighed, tall 100-mL beaker. After the solvent was boiled off, the residue was then dried at 95 °C for 5 h, cooled, and weighed to give the total weight of neutral fat and fatty acids.

The residue was dissolved in benzene and phenolphthalein was added. The solution was heated almost to boiling and immediately titrated with standardized sodium alcoholate solution until the maximum color of the end point was obtained. Values were reported as stearic acid. Each millimeter of the alcoholate represented 28.4 mg of stearic acid. Stearic acid was the principal fatty acid found in stools.

The authors analyzed six specimens from normal adults and found that the total fat varied from 13 to 30%, and the neutral fat from 5 to 9%. The larger the amount of total fat contained in the feces, the more the fat was present as fatty acids (which include the soaps).

> In the case of an "atrophied" infant the results obtained are even more striking, for here the total "fat" seems to consist almost entirely of fatty acids and soaps. . . . The following figures were obtained: On modified cow's milk, total fat 54.5 per cent, fatty acid 54.6 per cent. On mother's milk, total fat 28.8 per cent, fatty acids 25.8 per cent.

The high fatty acid values were not due to a splitting effect (saponification) by the acid-ether of the neutral fat. Nor could they be explained on the basis of lower molecular weights of the fatty acids present because of the high melting point of the extract—accounted for as an admixture of oleic acid in small concentration.

1. Folin, O., and Wentworth, A.H., A new method for the determination of fat and fatty acids in feces. *J. Biol. Chem.* **7**, 421–26 (1910).

On the Preparation of Cystin (1)

Because genuine hair was hard to get, Folin used oil-free pure wool and boiled it for 3–5 h in an air-reflux condenser with concentrated HCl until the contents no longer gave the biuret reaction for protein. The hot solution of amino acids was then treated with an excess of sodium acetate until the Congo Red reaction was negative for mineral acid. A dark, heavy precipitate containing almost all of the cystine was formed. After standing for several hours, the crude cystine was filtered and washed with cold water. (Note: If the solution stands for a long time, tyrosine usually precipitates from the mother liquor.)

The crude cystine was then boiled in 3–5% HCl, and decolorized with purified boneblack. The filtrate from the boneblack treatment was heated to boiling and the cystine precipitated by the slow addition of a hot, concentrated sodium acetate solution. "Large amounts of colorless cystine consisting of typical hexagonal plates can thus be prepared without difficulty and with very little labor."

1. Folin, O., On the preparation of cystin. *J. Biol. Chem.* **8**, 9–10 (1910).

The Preparation of Creatinine from Urine (1)

Creatinine was precipitated from urine as its picric acid complex. Excess picric acid was removed by treatment with $KHCO_3$. The solution was then acidified with acetic acid. An excess of saturated alcoholic zinc chloride was added to form a double chloride of zinc and creatinine. When this was dissolved in warm 10% H_2SO_4, a new compound creatinine zinc alum, $Kr_2ZnSO_4 \cdot 8H_2O$ (Kr = creatinine) was formed. It was almost quantitatively precipitated from solution by the addition of acetone, or alcohol and ether. The double salt was purified by boiling with boneblack.

Pure creatinine was prepared from the purified creatinine zinc alum by precipitating the sulfate with $BaCl_2$ and using H_2S to remove the zinc. While this method was cumbersome, the yield was almost theoretical. (The previous method had decomposed the double chloride with lead hydroxide and eventually obtained *both* pure creatine and creatinine. The dual product was not a problem because, as the next paper would show, creatine could be converted to creatinine quantitatively.)

1. Folin, O., The preparation of creatinine from urine. *J. Biol. Chem.* **8**, 395–97 (1910).

The Preparation of Creatinine from Creatine (1)

Dry creatine containing water of crystallization, but free from mineral acids and inorganic salts, was autoclaved (at a pressure of 4.5 kg/cm^2 for 3 h resulting in a quantitative conversion to creatinine. (Anhydrous creatine would not convert. Excess water slowed the conversion and caused impurities to develop.) Following a simple purification step the creatinine was obtained in about 90% of theoretical yield, with 99–100% purity.

1. Folin, O., and Denis, W., The preparation of creatinine from creatine. *J. Biol. Chem.* **8**, 399–400 (1910).

Note on the Determination of Ammonia in Urine (1)

Otto's final paper of 1910, submitted in November, was a note acknowledging a point that Steel had repeatedly made, namely that sodium carbonate was inadequate to liberate the ammonia from the triple phosphate ammonium magnesium phosphate in urine, so that values for urinary ammonia would be low. Steel's solution to the problem—the use of a mixture of sodium hydroxide and sodium chloride—was, however, too drastic, Otto pointed out. Besides, he wrote, there was really no problem because the triple phosphate was seldom, if ever, found in urine that was not decomposed or alkaline. When the triple phosphate was present in alkaline urine, it could be dissolved by acidifying the urine and an excess of potassium oxalate added (7–10 g/25 mL of urine) to prevent its re-formation.

1. Folin, O., Note on the determination of ammonia in urine. *J. Biol. Chem.* **8**, 497–98 (1910).

The Determination of Benzoic Acid (*1*)

The paper on benzoic acid was begun because "a recent investigation on cranberries undertaken in this laboratory involved a great many determinations of benzoic acid." Fred Flanders, who was working under a Bullard Fellowship on hippuric acid in urine, had obtained a great deal of experience with a method used by the U.S. Department of Agriculture, Bureau of Chemistry. The method was flawed by a very slow filtration; a long, difficult extraction; a tendency to form emulsions; and a poor titration procedure. By applying the Folin-Wentworth approach for fatty acid determination—extracting with chloroform, removing all acids except benzoic acid from the extract, and titrating with sodium alcoholate—they successfully improved this procedure.

When the authors tested the method on catsup, they found present a higher molecular weight fatty acid that caused artificially high results. The authors added sodium nitrite to overcome an emulsion-forming tendency as well as to reduce contamination from the fatty acid. Filtration was not required. They performed recovery experiments in which known amounts of sodium benzoate were added to catsup and found that the benzoic acid by the analysis was in fair agreement with the amounts added. Although all of the potential contaminants were not removed, the method was much improved over the older one because the analytical time was markedly reduced, the procedure was simpler, and the method more exact.

The quantitative procedure was also modified to provide a quick extract on which to do a qualitative test for the presence of benzoic acid.

1. Folin, O., and Flanders, F.F., The determination of benzoic acid. *J. Am. Chem. Soc.* **33**, 161–66 (1911).

Chapter 13. Willey Denis—The Prodigious Year (1912)

Protein Metabolism from the Standpoint of Blood and Tissue Analysis. I (1)

In their first paper on the fate of amino acids obtained from protein metabolism, Folin and Denis began what would be an extended series of experiments on nitrogenous materials injected into the intestine of cats.

Their general approach was to obtain a "base-line" normal sample of blood from the right carotid artery and the portal vein on a fasted, anesthetized (ether) cat. For urea injection, the blood supply to both kidneys was cut off by ligatures to prevent urea from escaping. The right iliac artery was also ligatured and an outside ligature was applied to the whole leg to prevent urea from entering the leg via the lymph. The small intestine was tied off below the stomach and just above the cecum. One hundred milliliters of the warm solution (urea, 4%) was then injected into the small intestine by means of a large syringe. Samples of blood were then taken at appropriate intervals from the portal vein and left carotid artery. The solution remaining in the intestine was then washed out to be used to determine the amount of the substance absorbed. Finally, pieces of muscle were cut out of the two hind legs for urea determination. In a second experiment, the gracilis muscle was used as a control by removing it under anesthesia from the left hind leg at the beginning of the experiment and from the right hind leg at the end of the experiment. This general surgical approach, with some variations, was used by the research team for testing the fate of urea, glycine, pancreatic digestion mixture (amino acids), and egg albumin in this and in all of their papers that followed. These were to be considered as short-term experiments in that the injected substances were followed for periods of up to about 3 h, but usually shorter.

No details of the colorimetric methods used on blood and tissue for NPN, urea, and ammonia were given in this paper. (For clinical chemistry, these research methods were momentous. As this paper was submitted for publication in January 1912, obviously the experiments with cats began at the latest in the autumn of 1911. This implies that the analytical methods were in progress in early 1911 and very probably in late 1910.)

When urea was injected, the levels of NPN and urea in the portal blood rose within minutes, but they also rose correspondingly in carotid or arterial blood and increased significantly in muscle. This experiment demonstrated the rapid rate of absorption and transport of a nitrogenous substance from the small intestine to presumably all tissues of the body. To determine if this were also true of amino acids, in another experiment glycine (glycocoll) was injected and blood samples (2 mL) were obtained at 0, 6, 18, and 45 min. NPN rose in the portal, carotid, and mesenteric blood and in the muscle tissue. While this experiment showed that the amino acid nitrogen was apparently not immediately split off and converted into urea, Folin and Denis were concerned because no urea formation occurred. When pancreatic digestion mixture containing mixed amino acids was injected, both NPN and urea increased in portal and carotid blood. They did not determine whether the increase in urea was due to urea formation itself or to ammonia not separately determined that might have been preformed in the digestion mixture or produced by deamination. [In an unpublished experiment, the authors found that the ammonia in the portal blood rose markedly (seven times normal) after injecting the pancreatic digestion mixture, indicating that it was preformed.] Finally, when egg albumin was injected a slight increase (6 mg) could be detected in the NPN but none in the urea nitrogen of portal or carotid blood within 90 min (Note: The amount of nitrogen absorbed (65 mg) from the intestine, however, was so small that the experiment may have been invalid). The paper included a table of NPN and urea-nitrogen values for fasting and fed cats and for fresh slaughterhouse beef blood.

If the amino acids were immediately absorbed and dispersed to all tissues of the body, then what happened to them? The authors theorized that the tissues served as a storehouse for such reserve materials, accounting for the several-day lag in reaching a constant nitrogen elimination when extreme changes were made in nitrogen intake. The filling and emptying of a reservoir would account for different results obtained when a substance such as amino acids was fed together with a diet rich or poor in nitrogen.

1. Folin, O., and Denis, W., Protein metabolism from a standpoint of blood and tissue analysis. First paper. *J. Biol. Chem.* **11**, 87–95 (1912).

Protein Metabolism from the Standpoint of Blood and Tissue Analysis. II (*1*)

In the second paper of this series on protein metabolism, Folin and Denis sought an explanation for the very high concentration of ammonia in portal blood. Based on the work of Nencki and Pawlow and their collaborators on dogs with an *Eck fistula* (portacaval shunt), the concept at the time was that food protein was immediately deaminated. Dogs fed meat had "colossal" quantities of ammonia in the portal blood. This work provided an experimental basis for a theory of ammonia toxicity as well as the view that the liver transformed ammonia into urea. Because Kutscher had shown that proteins were digested to amino acids, ammonia was thought to be derived from localized deamination in the intestinal wall.

Folin and Denis sought to test the theory of Nencki and Pawlow by determining the ammonia in the mesenteric veins of both the large and small intestines, in the portal vein, and in the carotid artery of cats treated as in the first paper, but without ligaturing the small intestine. They found, as expected, that the ammonia concentration was higher in both mesenteric veins than in the portal vein and much higher in the vein from the large intestine. The carotid blood was relatively unchanged. When preformed ammonia, however, was injected with the pancreatic digestion mixture (it contained one-fifth ammonia) into the ligatured small intestine, the ammonia appeared within 15 min in increased quantities in both the portal and carotid blood. This procedure showed that ammonia was absorbed rapidly through the intestinal wall, and that the blood's capacity to remove excess ammonia rapidly by way of other tissues was easily exceeded. On the other hand, and most important, when glycine was injected into the ligatured small intestine of a cat who had been well fed, the carotid blood showed no increase in ammonia 30 min later. However, the mesenteric veins of the small and large intestines contained high concentrations of ammonia, with ammonia in portal blood higher than that in the mesenteric vein of the small intestine, and about half the value of the ammonia from the vein of the large intestine. This experiment was repeated using asparagine, which provided a double source of ammonia with the amide as well as the amino group. If the wall of the small intestine had a deamination mechanism in it, then the concentration of ammonia in the mesenteric vein should have been greater than that in the portal blood. However, 24 min after the injection the ammonia value of mesenteric blood and that in portal blood rose only slightly, whereas the ammonia level in the carotid blood showed no significant change.

The "normal" values reported for ammonia nitrogen in the blood of the carotid artery of the cat lay in the range of 30–100 µg/100 mL (21–71 mmol/L), a range within that now accepted for human blood.

1. Folin, O., and Denis, W., Protein metabolism from the standpoint of blood and tissue analysis. Second paper. The origin and significance of the ammonia in the portal blood. *J. Biol. Chem.* **11**, 161–67 (1912).

Protein Metabolism from the Standpoint of Blood and Tissue Analysis. III (*1*)

Folin and Denis's previous experiments had followed changes in concentration over short periods of time, usually less than 1 h. Their next step was to cover periods up to 4 h. First, however, they reported some of the absorption experiments of their first paper. When asparagine was injected into a cat whose small intestine and blood supply to both kidneys were ligatured, the NPN rose in blood everywhere and in muscle, but the values were "minimal," because any asparagine present could not be quantitatively recovered. Urea could not be determined, because asparagine interfered with the method. When tyrosine was injected, it was so poorly absorbed that no rise in urea could be detected anywhere.

To determine the fate of injected creatinine and creatine, they first had to find a way to determine these compounds quantitatively in blood. When blood was added to methyl alcohol to precipitate the proteins, the filtrate contained an inhibitor of the alkaline picrate reaction. The interference, however, could be removed by extraction with ether or chloroform, in which creatinine is insoluble. Although full details of the method were not provided, the authors indicated that rather than use potassium dichromate as standard in the colorimetric analysis, pure creatinine dissolved in 0.1 N acid was indefinitely stable and should be used. The injected creatinine appeared to accumulate in blood in one trial and in muscle in another, and the NPN increased in all sites studied. For creatinine determination, the alcoholic blood or tissue extract was treated the same way as creatinine, but it was first autoclaved to convert the creatine to creatinine, prior to the chloroform extraction to remove the interfering substance. NPN and urea rose everywhere.

Urea formation was then studied over a longer interval of time. Glycine and alanine gave an increase in NPN and urea in all samples tested up to 3 h. This information added little to the fact already demonstrated in the earlier experiment that urea formation had begun markedly within the first 45 min after injection. Results obtained with Witte's peptone were less clear cut, giving an increase everywhere in NPN; the rise in urea in portal blood was not greater than that in the iliac artery, indicating that "the liver had not brought about any demonstrable specialized deaminization."

While the results indicated that absorbed amino acids were accompanied by urea formation, no definite site for this could be selected. Neither deamination nor urea formation was localized. Folin and Denis therefore felt that urea synthesis was a characteristic of all tissues, and probably occurred mostly in muscle (a conclusion that proved wrong later).

1. Folin, O., and Denis, W., Protein metabolism from the standpoint of blood and tissue analysis. Third paper. Further absorption experiments with especial reference to the behavior of creatine and creatinine and to the formation of urea. *J. Biol. Chem.* **12**, 141–62 (1912).

Protein Metabolism from the Standpoint of Blood and Tissue Analysis. IV (*1*)

For these experiments the cats were fasted and given a large dose of castor oil a day before the studies began. Ligatures were placed at the ileocecal valve and at the lower end of the rectum. Blood samples for urea and NPN were drawn at intervals following the injection of urea, glycine, alanine, creatinine, and Witte's peptone. The NPN in the mesenteric vein of the large intestine rose to higher values than in the systemic blood, except with Witte's peptone, which showed little change. Only injected urea gave a significant change in urea concentration. Creatinine was absorbed well and rose to high levels in blood. Portal blood was not sampled.

1. Folin, O., and Denis, W., Protein metabolism from a standpoint of blood and tissue analysis. IV. Absorption from the large intestine. *J. Biol. Chem.* **12**, 253–57 (1912).

Nitrogen Retention in the Blood in Experimental Acute Nephritis of the Cat (1)

Acute nephritis chemically induced by uranium nitrate, by potassium chromate, and by cantharidin was studied as a preliminary step toward examining other types. The cats were injected subcutaneously with each of the poisons, and blood samples were obtained by heart puncture over the next seven to 10 days. Urine samples were also obtained daily to observe the expected harbingers of nephritis, the appearance of albuminuria, and cast formation. NPN and urea nitrogen were measured in the blood sample.

Uranium- and catharidin-induced nephritis, which involved both tubules and glomeruli, caused a marked accumulation of nitrogen in the blood, whereas the chromate-induced nephritis, which involved the tubules almost exclusively, produced only moderate nitrogen retention. Control (normal) cats kept under the same conditions showed only slight variations from day to day. "... the additional involvement of the glomerulus is extremely important in leading to a retention of nitrogenous waste products."

1. Folin, O., Karsner, H.T., and Denis, W., Nitrogen retention in the blood in experimental acute nephritis of the cat. *J. Exp. Med.* **16**, 789–96 (1912).

New Methods For the Determination of Total Non-Protein Nitrogen Urea and Ammonia in Blood (1)

Folin's and Denis's method of drawing blood from animals was interesting, as it varied from the common practice. They used needles (1 mm in diameter and 25 mm long) that had been immersed in a dilute solution of Vaseline dissolved in ether and then allowed to dry on clean paper for a few minutes. (For humans, of course, the needles had to be thoroughly sterilized.) A small pinch of powdered potassium oxalate (anticoagulant) was allowed to run down through the pipette into the tip. The other end of the pipette was connected by another piece of tubing to a mouthpiece consisting of a short tapering glass tube. A pinchcock was placed in the tubing near the pipette. After inserting the needle into the vein or artery, the pinchcock was opened and by mouth suction the pipette was filled with the desired volume without waste or clotting. The analyses were then performed on oxalated "whole" blood.

To isolate non-protein nitrogenous constituents, the proteins were precipitated in acetone-free methanol in the blood-to-alcohol ratio of 2:23 or 5:45. The mixture was allowed to stand for 2 h, and the filtrate was treated with 2 to 3 drops of saturated alcoholic solution of zinc chloride. A clear filtrate was obtained, and 5 mL was used for either the NPN or urea nitrogen determination. While some substances could be recovered fairly quantitatively from this filtrate, other substances could not. To overcome this deficiency, the methanolic precipitate was triturated, washed with methanol, and then refiltered. This technique provided quantitative recovery of urea, glycine, acetamide, and other substances, but not of all non-protein nitrogenous substances including creatine, asparagine, and tyrosine. The shorter method, however, was used for the most part. For muscle analysis trituration of 5 g with methanol was essential. Details will not be presented here.

For the NPN determination, 5 mL of the filtrate was transferred to a Jena test tube (as used in the urine analysis), and a drop of kerosene, and a pebble were added. The methanol was driven off by immersing the tube in a beaker of boiling water for 5–10 min. One milliliter of conc. H_2SO_4 was added and then the remaining filtrate was boiled, cooled, and diluted as in the analysis of urine (presented elsewhere).

Ammonia was removed from this digestion mixture as usual, except that it was passed into a second large test tube containing 1 mL of 0.1 N acid with 2–3 mL of water. This procedure was necessary because of the much smaller concentration of NPN in blood than in urine. The final nesslerized solution could not be diluted to 100 mL, and smaller volumetric flasks could not be used because of splattering; thus large test tubes were used as receivers, and for nesslerization before the solution was transferred to volumetric flasks (usually 25 mL for cat blood determinations). For high NPN concentrations, greater dilutions were made, e.g., to 50 or 100 mL.

"Human blood contains scarcely more than one half as much non-protein nitrogen as cat's blood. In the case of human blood we therefore never draw less than 5 cc. and take 10 cc. of the filtrate for each determination." Usually 7–8 mL of Nessler's reagent (diluted 1:5) was used, unless larger amounts were needed. The nitrogen value was determined colorimetrically using a 1 mg % $(NH_4)_2SO_4$ standard nesslerized in a 100 mL flask and set at 20 mm on the colorimeter.

> ... the smaller the quantity of blood which can be made to give reliable results the greater becomes the usefulness of the method. The work which we have already done on cats could not have been done on such a small animal except by means of these microchemical methods. Finally, small amounts of blood must be used for the urea determination because of disturbing effects of the sugar present.

For blood urea determination, 5 mL of the filtrate of cat's blood or 10 mL of human blood was transferred to a large Jena test tube. A drop of dilute acetic acid and 2–3 drops of kerosene were added, and the tube was closed with a two-hole rubber stopper. Through one hole of the stopper was passed a glass tube drawn out to a capillary several inches long, which reached nearly to the bottom of the tube. Through the other hole a short bent glass tube was passed and connected to a water pump. The tube was placed in warm water and the vacuum was started. Gentle heat and the air current that the vacuum produced through the capillary combined to remove all the alcohol. The stopper was then removed, and the capillary broken off. Then 2 mL of 25% acetic acid, a "temperature indicator," a pebble, and 7 g of dry potassium acetate were added. The urea was decomposed by heating the tube at 153–158 °C for about 8–10 min, exactly as done for urine.

The liberated NH_3 was set free by subsequent air current treatment, collected into another large test tube and nesslerized, and then diluted to 10 mL in a volumetric flask and read in a colorimeter. Usually the tubes for total nitrogen, urea, and the standards were nesslerized at the same time.

The determination of blood NH_3 was beset with difficulties. Folin and Denis recognized that blood decomposed spontaneously particularly in the presence of alkalies capable of setting NH_3 free at all temperatures, even when kept on ice. Because of this decomposition, any delay was harmful, and .when distillation methods are applied, whether in the vacuum or otherwise, the determination becomes little else than a measure of the decomposition." Decomposition was even greater in liver tissue than in blood. ". . . we are of the opinion that there is not a single experiment on record proving that macerated liver tissue splits off by hydrolysis the NH_2 groups from ordinary amino-acids when the latter are added to such tissue."

The speed essential to determining blood ammonia made the method difficult because it was difficult to achieve sensitivity sufficient for microvolumes of blood.

> The Nesslerization process lends itself as does none other to the quantitative estimation of small amounts of ammonia but instead of working with milligrams, as in urine, or with tenths of a milligram, as with blood in the estimation of total nitrogen and urea, it becomes a question of working with hundreths of a milligram.

The Duboscq color could not be read for these color changes. Therefore, two important changes were made in the colorimeter: 1) The light passing through the standard was reduced by installing an iris diaphragm, and the standard was reduced in strength to 0.5 mg % nitrogen; 2) a 100-mm polariscope tube was used to lengthen the light path through the blood ammonia solution replacing the usual cylinder (cup) of the colorimeter. When the solid movable glass prism was removed, the narrow polariscope tube just fit.

Two other precautions were necessary for good blood NH_3 determinations: The use of excessive Nessler's reagent had to be avoided because of its green color; and NH_3-free water was essential for diluting the unknown. For diluting to volume after nesslerization, water treated with saturated bromine water and a few drops of caustic soda was used.

The method for blood ammonia is historic. Ten milliliters of systemic blood or 5 mL of portal or mesenteric blood was transferred to a large Jena test tube. To this were added 2–3 mL of oxalate-carbonate solution (15% potassium oxalate and 10% sodium carbonate) and about 5 mL of toluene. An air current was passed through the mixture for 20–30 min. The liberated NH_3 was collected into another large test tube charged with 5–6 drops of 0.1 N acid and 1 mL of water. The receiver tube was covered with a small stemless funnel.

The NH_3 solution was then nesslerized with 1 mL of diluted (1:5) reagent and transferred to a 10-mL volumetric flask. It was then diluted to the mark, mixed, and poured into the polariscope tube on the colorimeter, and the tube was closed as in ordinary polariscope work. Two standards (0.5 mg and 1.0 mg of nitrogen) were nesslerized simultaneously and made to 100-mL volume. The one closest to the unknown was used.

For colorimetry, the unknown (polariscope tube) remained stationary while the standard solution was adjusted to make the color match. The diaphragm in the standard was adjusted in the standard solution until the right position was obtained. "The colorimeter, as thus used, represents, we believe, a new departure in colorimetry, and we are taking steps to secure the making of such instruments." The diaphragm that Folin and Denis used was an ordinary one from a microscope. It was fastened to the top of the colorimeter platform on which the cylinder stood.

1. Folin, O., and Denis, W., New methods for the determination of total non-protein nitrogen, urea, and ammonia in blood. *J. Biol. Chem.* **11**, 527–36 (1912).

An Apparatus for the Absorption of Fumes (*1*)

Folin and Denis devised an ingenious, simple "fume absorber" to remove the irritating fumes formed from the sulfuric acid used in the new method for total nitrogen determinations in urine (and blood). The device could be easily used at any workbench, without the need for a fume hood, but with the help of a water pump to supply suction.

The apparatus consisted of a volumetric pipette in which the tip was invaginated and the top was bent above the bulb. The gadget was placed on top of the fume-generating digestion tube or Kjeldahl flask, or on top of a stemless funnel placed on a beaker or inverted on an evaporating dish. The invaginated pipette tip was a trap for condensing water that would not flow back into the digestion mixture. For safety and to protect the sink from corrosion, the fume line was passed through a large bottle containing 10% NaOH, though the authors found little need for it.

"We believe that the single absorber at each student's desk might prove a valuable accessory in classroom laboratories where hoods so often are inadequate and ineffective." (Eimer and Amend would manufacture the device and list it in their next catalogue.)

1. Folin, O., and Denis, W., An apparatus for the absorption of fumes. *J. Biol. Chem.* **11**, 503–5 (1912).

On Creatine in the Urine of the Children (*1*)

Both creatine and creatinine analyses were performed on the urine of Folin's children (ages 11, 8⅔, 3⅔). Creatine was present in all urines collected and was apparently increased following the eating of meat.

> We are inclined to believe that the creatine in children's urine does not depend as Rose suggests on a peculiar carbohydrate metabolism but that it is due to an excessively high level of protein consumption (in proportion to mass of muscles in the body).

They reasoned that if this were true, further testing should prove that forced feeding of protein to adults would cause creatine elimination; and reduced protein feeding should cause creatine-free urines in children. The first trials on the Folin children and on Dr. A. H. Wentworth's two children, ages 9½ and 6, on a creatine-free diet, were unsuccessful in totally eliminating creatine output.

> Through the kindness of Professor Wiener we were able to include in our determinations the morning urine obtained from his four healthy and unusually robust children, all of whom are vegetarians and have never eaten food containing creatine—Norbert, 17 years, 3 months; Constance, age 13 years, 10 months; Bertha, age 9 years, 10 months; Fritz, age 6.

Although the morning urines showed much lower concentrations than that of the Folin children, creatine was present, indicating a minimal amount of daily excretion independent of meat-protein intake. The urine of Norbert Weiner, a young adult, showed the most creatine.

1. Folin, O., and Denis., W., On creatine in the urine of children. *J. Biol. Chem.* **11**, 253–56 (1912).

On Phosphotungstic-Phosphomolybdic Compounds as Color Reagents (1)

Folin and Denis systematically combined phosphoric acid with sodium tungstate and with sodium molybdate, as well as with various mixtures of the two,

> ... and by assaying the chromophoric value of each preparation both with uric acid and with phenol derivatives, we have obtained first, a highly sensitive reagent for uric acid which does not react with monohydric phenols, or their derivatives, such as tyrosine, and secondly, a preparation which is probably far more delicate than any color reagent yet known for phenol groups. The first is phosphotungstic acid; the second is a phosphotungstic-phosphomolybdic compound.

No effort was made to isolate these very complex compounds in pure condition.

The uric reagent contained 10% sodium tungstate and 16% phosphoric acid boiled together for about 2 h. To prepare the reagent, 100 g of tungstate and 80 mL of 85% H_3PO_4 were added to 750 mL of water, gently boiled for 2 h using a reflux condenser, cooled, and diluted to 1 L. Two milliliters of this solution gave the maximum color obtainable with 1 mg of tyrosine or uric acid.

The phenol reagent contained 10% sodium tungstate, 2% phosphomolybdic acid, and 10% phosphoric prepared as follows: To 750 mL of water, 100 g of sodium tungstate, 20 g of phosphomolybdic acid, and 50 mL of H_3PO_4 (85%) were added and boiled for 2 h using a reflux condenser, cooled, and diluted to 1 L. Two milliliters of this solution gives the maximum color obtainable with 1 mg of tyrosine or uric acid.

The reagents had to be free of nitrates, because they interfered with the color reaction. In both reactions alkali destroyed the active compound, but the color was produced only in alkaline solution. Thus the alkali used had to be added last, because a reduction of the active compound by uric acid or phenol derivatives occurred in acid solution, and the reduced compound formed gave blue salts when the alkali was added. The blue color was not stable in alkali, making the selection of the alkali critical for quantitative work. Sodium carbonate was selected rather than potassium carbonate, because the latter gave precipitates with the reagents.

The qualitative test was made as follows: 1–2 mL of reagent was mixed in a test tube with about the same volume of the solution to be tested. Then an excess of saturated Na_2CO_3 solution was added (3–10 mL) to produce the color immediately. When only minute traces were present, then solid Na_2CO_3 was used to avoid dilution. When the test was performed with solid Na_2CO_3, it was unmistakably positive in the two respective reagents with solutions containing 1 part of uric acid in 500,000 parts of water, and with 1 part of tyrosine in 1,000,000 parts of water.

The two reagents were tested on a large number of organic compounds. Aliphatic substances, ketones, aldehydes, amines, and indole and its derivatives did not react. The uric acid reagent did not react with ordinary monohydric phenols except those containing an amino group in the benzene ring. The phenol reagent reacted with all oxybenzol compounds and was far more sensitive to tyrosine than Millon's reagent. The reactivity of the two reagents as well as Millon's reagent to monohydric, dihydric, and trihydric phenols and other compounds were presented in three tables. The reagents could not be made reproducibly.

1. Folin, O., and Denis, W., On phosphotungstic-phosphomolybdic compounds as color reagents. *J. Biol. Chem.* **12**, 239–43 (1912).

Tyrosine in Proteins as Determined by a New Colorimetric Method (1)

Folin and Denis wasted no time in applying the phenol reagent to the determination of proteins based on their tyrosine content. There were wide discrepancies in values for the tyrosine content of proteins reported in the literature, e.g., in zein (maize) or vitelline (hen's egg).

T. B. Osborne provided them with a large number of individual proteins with published tyrosine values determined by gravimetric methods, and they obtained a few proteins from their own sources. The method used to hydrolyze the proteins consisted of placing 1 g of the dry protein into a 500-mL Kjeldahl flask, adding 25 mL of 20% HCl, closing the flask with a Hopkins condenser made from a large tube, and boiling for 12 h over a microburner. On cooling, the contents were

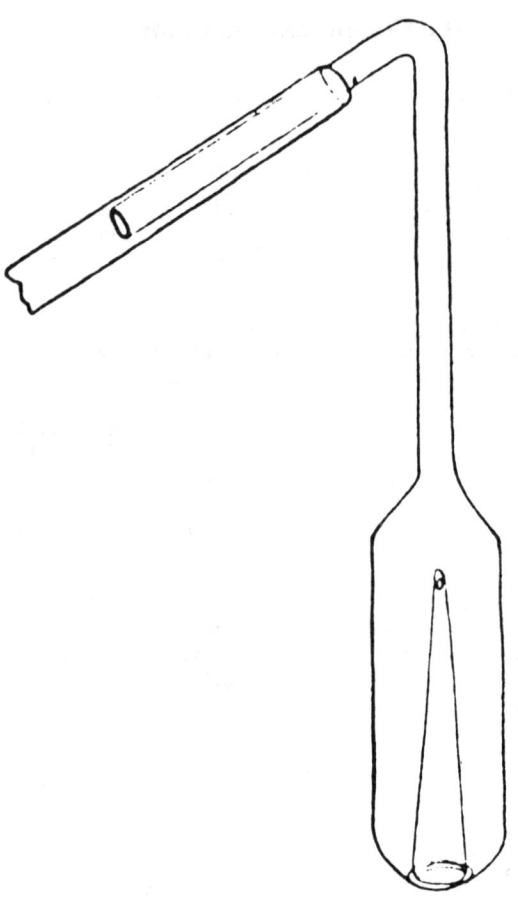

Fume absorber for digestion step in the total nitrogen determination. [*J. Biol. Chem.* **11**, 504 (1912).]

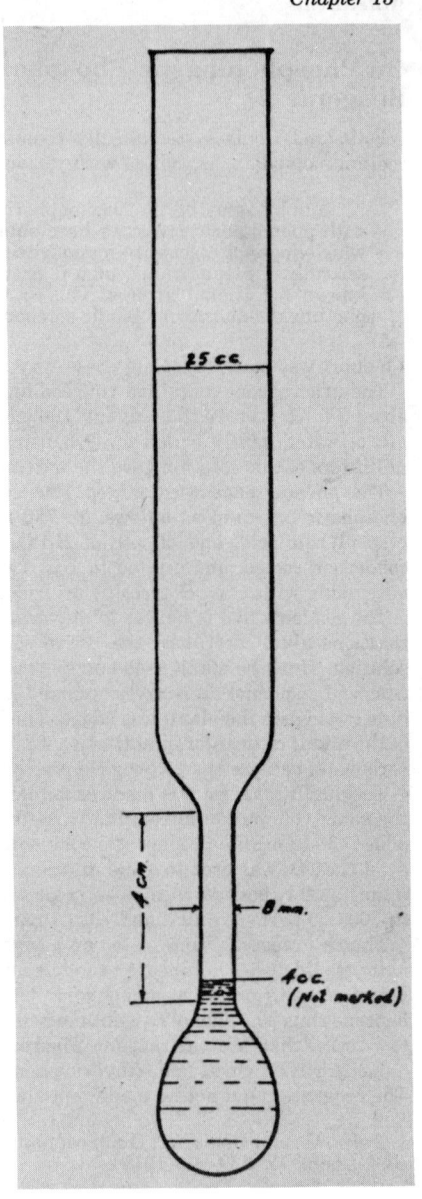

Blood sugar tube. [*J. Biol. Chem.* **41**, 372 (1920).]

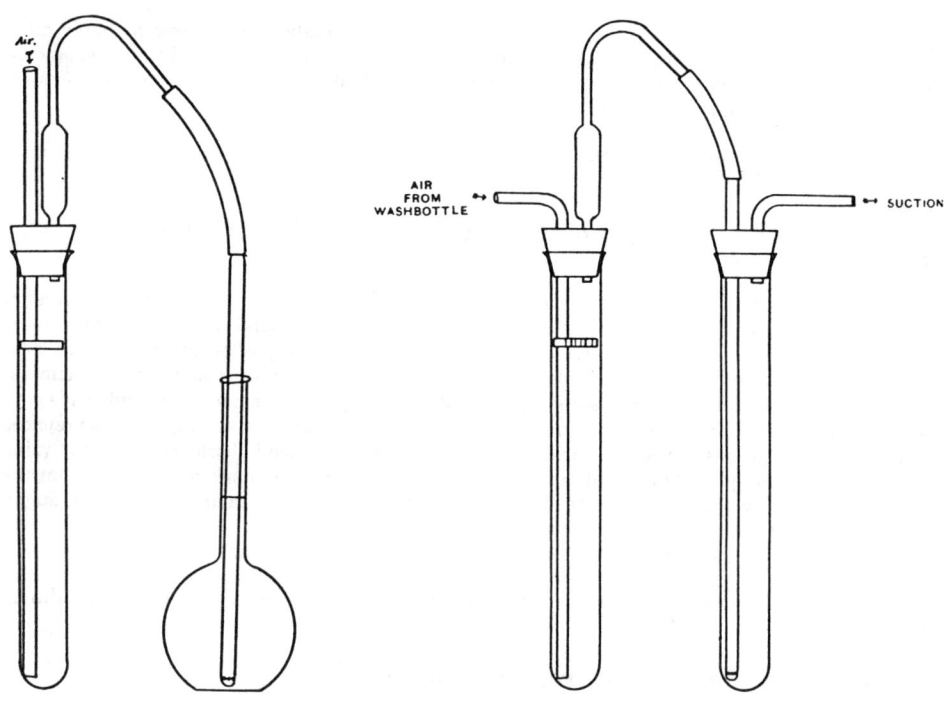

Apparatus used for aeration in determining total nitrogen as NH_3; left, for compressed air; right, for suction. (*Handbuch der Biochemischen Arbeitsmethoden*. E. Abderhalden, Ed. Berlin and Vienna: Urban and Schwartzenberg, 1913; p. 717.)

transferred to a 100-mL volumetric flask, 5 mL of tyrosine (phenol) reagent was added; then after a 5-min wait, 25 mL of a saturated solution of Na_2CO_3 was added. The solution was brought to volume and allowed to stand for 10 min. At the same time a standard of 1 mg of tyrosine was treated in a 100-mL volumetric flask with 5 mL of the phosphotungstic-phosphomolybdic (phenol) reagent and 25 mL of the Na_2CO_3 solution before being diluted to volume. Readings were then made on the Duboscq colorimeter.

Tyrosine values obtained by this method were uniformly higher than those previously reported. They tested various possible reasons for this discrepancy but could not find any defect in the new method.

The above experiment, we think, strongly favors our view that the discrepancies between our colorimetric tyrosine determination and the values for this substance which appear in the literature are due to the fact that it is practically impossible to separate tyrosine quantitatively from the hydrolysis products of any protein.

In short, gravimetric methods could not isolate all of the tyrosine, and therefore gave low results.

That our reagent gives the same color values with tyrosine in peptid combinations as with the same amino-acid in the free condition is proved by the fact that the length of time which the hydrolysis is continued has no influence on the subsequent determinations.

For example, 1 g of sheep's wool boiled with 25% HCl for 1, 5, 12, and 20 h all gave the tyrosine value of 6%.

Not only was this method speedier than the gravimetric method, but it was probably more accurate, and required a much smaller amount of protein (0.1–0.2 g) for analysis. It should be remembered that the method was used on individually purified proteins, and not on mixtures such as serum. That would come later.

1. Folin, O., and Denis, W., Tyrosine in proteins as determined by a new colorimetric method. *J. Biol. Chem.* **12**, 245–51 (1912).

A New Colorimetric Method for the Determination of Vanillin in Flavoring Extracts (*1*)

A direct reaction for vanillin in the extract was not possible because of the probable interference of tannins. To purify the extract, therefore, it was first treated with a lead acetate solution, and the reaction with the phenol reagent carried out on the filtrate, producing a "beautiful deep blue color admirably suited for quantitative colorimetric work." The color was read in the colorimeter against that of a vanillin standard similarly treated with the phenol reagent. Folin and Denis gave a few precautions on the use of the colorimeter. They obtained authentic vanilla extracts prepared from the various grades of beans in the U.S. Food and Drug Inspection Laboratory. After analyzing more than 100 samples they found "in no case has there been any marked deviation from the values obtained by the official method." Extracts of tonka bean, coumarin, acetanilid, sugar, caramel, and glycerine did not interfere.

The new method offered speed and reliability and used a microvolume of sample.

1. Folin, O., and Denis, W., A new colorimetric method for the determination of vanillin in flavoring extracts. *J. Ind. Eng. Chem.* **4**, 1–6 (1912).

On the Determination of Urea in Urine (*1*)

The first topic of the paper, "Observations on my magnesium chloride method and on Benedict's method," answered criticism of Folin's method in the literature and discussed his latest experience with it. He noted that the chief source of error was incomplete decomposition of the urea. "Lack of experience in how to obtain and to maintain this temperature is the chief cause of the failures to get accurate results." He also pointed out the generally unrecognized fact—the time necessary for decomposition of urea depended on the amount of urea to be decomposed—had caused him to change the time of heating from 30 to 90 min.

Two publications had attacked the method as unreliable on the basis of the mass action law: $MgCl_2$ would react with NH_4OH to produce NH_4Cl and hence it would not distill off. This argument was not supported by any data. Otto answered this assertion by argument and by experiment. If the mass action law was a factor, he stated, then the Kjeldahl method could not determine ammonia either because the Na_2SO_4 present would also react with NH_4OH to produce $(NH_4)_2SO_4$, so no NH_3 could be liberated. Experimentally, Otto provided data on standard $(NH_4)_2SO_4$ solutions treated with $MgCl_2$ and distilled. He obtained the expected values in 23 determinations.

The most valid criticism raised against Otto's method was that it required too much skill and experience to carry it out, and it was too lengthy.

> It is interesting to note that this criticism comes chiefly from American laboratories where metabolism experiments for the past few years have been conducted on a wholesale, factory-like basis.
> In his last paper on the estimation of urea S. R. Benedict describes a new method which he believes to be very accurate, giving figures slightly lower than those obtained by means of the magnesium chloride method, though the agreement between the two methods is often as close as two duplicate determinations by the same method. . . . In working on pure products, creatinine, uric acid, and allantoin, Benedict finds that whatever difference there is between the two methods is rather in favor of his new one.

Benedict recommended that NaOH rather than Na_2CO_3 be added for distilling of the NH_3. Folin compared Benedict's values on 16 urines with his values obtained by the $MgCl_2$ method and a new one that he was proposing. The results were substantially the same.

Otto had modified the $MgCl_2$ method slightly by using a 500-mL Kjeldahl flask and a microburner. A large test tube filled with cold water, suspended about half way into the neck of the flask by means of a cork or copper wire, was used as a condenser. After about 350 mL of hot water and the alkali were added, the NH_3 was distilled off directly from the Kjeldahl flask in about an hour.

In the second part of the paper Folin described a new method for decomposing urea with phosphoric acid. He wanted to create a microscale method to accompany the total nitrogen method. "The problem was to decompose one or a few milligrams of urea and either titrate the ammonia with very dilute acid and alkali or to determine it colorimetrically by means of Nessler's solution." The $MgCl_2$ method was not suitable for working on a small scale. He now proposed a phosphoric acid method using 1 mL of urine. The procedure was as follows: One milliliter of urine was measured with an Ostwald pipette into a Jena test tube, to which was added 3 good-sized drops of pure H_3PO_4, one drop of indicator (alizarin red), and a few grains of talcum powder. The mixture was boiled over a free flame until about half the water had escaped (2–3 min). The tube was placed in a bath (paraffin, oil, or H_2SO_4) previously heated to 175–180 °C for 15 min. "The urea is completely decomposed in that time." The contents of the tube were dissolved with 1–2 mL of water and a little heat and 0.5–1.0 mL of KOH solution was added. (*Note*: KOH was better than NaOH because potassium phosphate is more soluble than sodium phosphate.) The NH_3 was removed with a strong air current in 10 min and collected in 25 mL of 0.1 N HCl, and titrated with 0.01 N NaOH using alizarin red as indicator. With the paraffin bath, the time of analysis was 0.5 h. This method compared well with the $MgCl_2$ method and Benedict's method on the 16 urines mentioned earlier, but Folin did not like the idea of using a heating bath for the decomposition.

Otto reported in the third section of the paper that urea could be decomposed by using boiling potassium acetate solution at 158–160 °C. The idea was to eliminate the preliminary boiling off of water as required in the $MgCl_2$ method or in Benedict's procedure.

Because potassium acetate is hygroscopic he dried about a pound at a time in a porcelain dish that was standing on a warm plate at approximately 115 °C for about 24 h.

The new procedure made use of a "temperature indicator" needed during the procedure to ensure maintenance of the 153–160 °C range. The indicator consisted of powdered chloride-iodide of mercury (HgICl) enclosed in a sealed glass bulb not more than 1 mm in diameter. This salt, which is bright red at ordinary temperatures, turns lemon yellow at 118 °C and melts to a clear dark red liquid at 155 °C. It solidifies again at about 148 °C, and resumes its red color gradually in about 24 h. The melting point at 153 °C is fortunately easily obtained and maintained with potassium acetate. Because the acetate begins to cake and solidify at 160–161 °C the combination of HgICl and potassium acetate makes a temperature that is too high or too low readily apparent. In other words, when the temperature of the potassium acetate drops too low, the red indicator solidifies and loses color, while if the temperature is too high, the potassium acetate solidifies.

The temperature indicator salt was prepared by heating molecular portions of $HgCl_2$ and HgI_2 at 150–160 °C for 6–8 h, and then using it. It was kept dry until it was sealed. The temperature indicators could be obtained from Eimer and Amend.

Because the urea was decomposed in a practically saturated potassium acetate solution, the liberated NH_3 was retained in acetic acid. In the presence of so much acetate the acetic acid was an extremely weak acid, barely capable of holding the NH_3. The decomposition of urea occurred in almost neutral medium. In fact, to alizarin red the medium was neutral or alkaline, but certainly not acid.

In the new method, urine was diluted to contain 0.75–1.5 mg of urea nitrogen (usually 1 + 9, 1 + 19, rarely 1 + 4). With an Ostwald pipette, 1 mL of diluted urine was transferred to a large Jena test tube (20 mm × 200 mm) previously charged with 7 g of dry potassium acetate (free from lumps), and 1 mL of 50% acetic acid, a small sand pebble or better, a little powdered zinc (not zinc dust), to prevent bumping during boiling, and a temperature indicator.

The tube was closed with a rubber stopper carrying an empty narrow "$CaCl_2$-tube" (25 cm × 1.5 cm without bulb) as a condenser. The tube and condenser were suspended by a clamp so that they could be easily raised or lowered with reference to the small flame of the microburner. As soon as the acetate was dissolved and the mixture began to boil (about 2 min) the indicator began to melt showing that the desired temperature (153–160 °C) was reached. Gentle, even boiling was continued for 10 min.

After the apparatus was removed from the flame, the contents of the tube were diluted with 5 mL of water (rinsed in with a pipette via the $CaCl_2$ tube). An excess of alkali (2 mL of saturated NaOH or K_2CO_3 solution) was added and the liberated NH_3 driven off by means of a strong air current into a 100-mL measuring flask containing about 35 mL of water and about 2 mL of 1.0 N acid. Ten minutes was sufficient for this. The NH_3 was determined colorimetrically against an $(NH_4)_2SO_4$ standard containing 1 mg of nitrogen as in the procedure for total nitrogen.

In execution the determination of urea as described above is about as simple and free from complications requiring unusual skill or experience as it is possible to make a quantitative method.

Folin completed the section by noting precautions and the rationale for using the reagents.

The fourth division of Folin's paper—shared with Willey Denis—dealt with determining urea in the presence of sugar. Mörner, in 1903, had pointed out how sugar caused low values for urea, but his method for preliminary removal of sugar was tedious. Consequently, determining urea in diabetic urines was still a comparatively long and laborious process. Folin found that sugar would lower urea values by 20–50% in his acetate procedure.

A more systematic investigation of the subject has, however, shown that it is possible by means of this method to meet the unusual conditions which must be fulfilled if urea is to be quantitatively converted into ammonia in the presence of sugar.

The interference by sugar was at the time thought to be due to formation of stable ureids with urea. The solution was perhaps to dilute the urine. Experiments showed that 0.1–0.3 mg of urea nitrogen could be determined in the presence of as much as 2 mg of dextrose.

It is therefore possible by simply diluting diabetic urine until 1 cc (as cited) contains about 0.1 mg of urea nitrogen to determine the urea without any preliminary removal of the sugar when the dextrose-nitrogen ratio (D:N) is as high as 20:1.

In the method, the urine was diluted 20–100-fold. One milliliter of diluted urine was decomposed in the described manner with potassium acetate–acetic acid. The NH_3 was driven off into a second tube containing about 2 mL of water and 0.5 mL of 0.1 N HCl. About 2 mL of water and then 3 mL of dilute (1:5) Nessler's solution were added. The colored solution was rinsed into a 10 mL measuring flask, washed, and diluted to the mark. The solution was transferred to a dry cylinder of the Duboscq colorimeter and measured against the 1-mg nitrogen standard.

Folin and Denis tested the validity of the method by showing the higher values of urea in the presence of sugars at 1:20 dilution when compared with 1:2 dilution. More convincing was the good recovery of urea in urines of known concentration to which dextrose was added. The recovery was quantitative at 1:100 dilution, but much too low at 1:10.

1. Folin, O., On the determination of urea in urine. *J. Biol. Chem.* **11**, 507–22 (1912).

A New Method for the (Colorimetric) Determination of Uric Acid in Urine (*1*)

The authors found that for uric acid the best reagent was a phosphotungstic acid prepared by boiling 100 g of sodium tungstate with 80 mL of 85% phosphoric acid and 750 mL of water for a couple of hours and then diluting it to 1 L. Another investigator (E. Riegler) published a method in 1912 for determining uric acid in urine with phosphomolybdic acid. Folin and Macallum pointed to their own precedence in this work, and the fact that Riegler had not studied the reagent used for its chromophoric value. Obviously it was not optimal. But Otto noted that the phosphomolybdic acid was used for precipitating (protein) purposes. This technique would be of much importance later on, particularly for work on blood.

Because urine contained di- and polyphenol acids in addition to more abundant monophenol derivatives, it was unfortunately impossible to carry out the color reaction directly. However, "we have now worked out a new method for the determination of uric acid which is materially simpler than the method outlined in our first paper."

To remove the impurities in urine that react with the color reagent, a measured volume of urine was evaporated to dryness, and then extracted with ether-methanol. The color reaction was then carried out on the residue, along with a standard, and the values determined from the readings on the Duboscq colorimeter. The procedure was given for preparing stable uric acid standard in lithium carbonate solution.

1. Folin, O., and Macallum, A.B., Jr., A new method for the (colorimetric) determination of uric acid in urine. *J. Biol. Chem.* **13**, 363–69 (1912).

On the Determination of Ammonia in Urine (1)

To carry out the procedure properly for determining ammonia, a good pump was needed that would provide a pressure of 40–45 lb/in^2. The procedure was as follows:

With an Ostwald pipette 1–5 mL of urine containing 0.75–1.5 mg of NH_3 nitrogen was transferred into a test tube. For diabetics, dilution of the urine was sometimes necessary. A few drops of a solution containing 10% K_2CO_3 and 15% potassium oxalate and a few drops of kerosene or heavy crude machine oil (to prevent foaming) were added. A strong air current was passed through the mixture for 10 min (or as long as it took to remove the NH_3) and the NH_3 was collected into a 100-mL measuring flask containing about 20 mL of water and 2 mL of 0.1 N acid. For total nitrogen, it was nesslerized as described and compared with 1 mg of nitrogen obtained from a standard $(NH_4)_2SO_4$ solution similarly nesslerized.

1. Folin, O., and Macallum, A.B., On the determination of ammonia in urine. *J. Biol. Chem.* **11**, 523–25 (1912).

A New Method for the Determination of Hippuric Acid in Urine (1)

The urine sample (100 mL) was alkalinized with NaOH and evaporated to dryness on a waterbath. The greater part of the hippuric acid was hydrolyzed during this process (overnight). The residue was dissolved in water in a Kjeldahl flask and treated with strong HNO_3 with a little copper nitrate catalyst. It was boiled for 4.5 h with a microburner, and this completed the hydrolysis. A Hopkins condenser was used on top of the flask. The benzoic acid was extracted from the solution with chloroform using separatory funnels. The chloroform layer was then titrated with standard sodium alcoholate using phenolphthalein as indicator.

Hippuric acid values were obtained on eight urines, but recovery experiments were not performed.

1. Folin, O., and Flanders, F.F., A new method for the determination of hippuric acid in urine. *J. Biol. Chem.* **11**, 257–63 (1912).

A New Method for the Determination of Total Nitrogen in Urine (1)

Into a Jena glass tube (20–25 mm × 200 mm) were placed 1 mL of appropriately diluted urine, 1 mL of conc. H_2SO_4, 1 g of K_2SO_4, 1 drop of 5% $CuSO_4$ solution, and a small pebble to prevent bumping. The mixture was boiled over a microburner for about 6 min (about 2 min after the solution was colorless) and cooled for about 3 min until viscous, but not solidified. Six milliliters of water and then 3 mL of a saturated solution of NaOH were added. The NH_3 was aspirated by means of a rapid air current into a 100-mL volumetric flask containing about 20 mL water and 2 mL of 0.1 N HCl. The flask was disconnected and contents were diluted to about 60 L. A 1-mg nitrogen standard of $(NH_4)_2SO_4$ was diluted about the same volume in a second flask. Both flasks were nesslerized with 5 mL of Nessler's reagent diluted 1:5 with water immediately before use. The solution was diluted to volume, mixed, and read in the colorimeter, and nitrogen values were calculated.

The authors then presented several pointers and precautions for the procedure. A table was also included giving values obtained on 20 urine samples showing good correlation with values obtained by the Kjeldahl procedure, and by a titration procedure on a microsample.

1. Folin, O., and Farmer, C.J., A new method for the determination of total nitrogen in urine. *J. Biol. Chem.* **11**, 493–501 (1912).

Chapter 14. Developing Clinical Biochemistry (1913–18)

A New (Colorimetric) Method for the Determination of Uric Acid in Blood (1)

For the uric acid determination, blood proteins were coagulated in boiling 0.01 N acetic acid in a 1:5 ratio. Because of the small quantity of uric acid in blood, the procedure required 15–25 mL, and Otto later suggested 20–30 mL. The blood was placed into a preweighed wide-mouth bottle containing the anticoagulant, powdered potassium oxalate; weighed; and then transferred to a 1-L flask containing five times the weight of the blood of 0.01 N acetic acid heated to boiling.

The coagulated material was filtered while hot. Then the precipitate was transferred back to the flask with a spoon or spatula, and 200 mL of boiling water was poured into it and mixed. After standing for 5 min, the mixture was filtered. The combined filtrates were clear. To handle the occurrence of coagulation despite the oxalate, they gave a procedure in which heating (with acetic acid) was not used.

The filtrate was then further acidified with acetic acid, evaporated to about 3 mL, and transferred to a 10-mL centrifuge tube, with washing using dilute $LiCO_3$. Then were added a few drops of silver lactate solution, 2 drops of magnesia mixture, and 10–15 drops of NH_4OH, sufficient to dissolve the AgCl formed. The uric acid was precipitated as the silver salt. After centrifugation and of decanting the supernate, the residue was treated with saturated H_2S solution and a drop of conc. HCl; then the tube was placed into boiling water to remove excess H_2S. A drop of lead acetate was added to ensure that the H_2S was removed (otherwise, it was returned to the bath), and the tube was centrifuged. The supernate containing the free uric acid was transferred to a small beaker, and the uric acid reagent added, followed by saturated Na_2CO_3 solution. Depending on the intensity of the blue color produced, the mixture was transferred to a volumetric flask (25, 50, or 100 mL) and diluted to volume. A standard of 1 mg of uric acid was treated in a separate 100-mL flask with 2 mL of uric acid reagent and 20 mL of the Na_2CO_3, made to volume. The solutions were then read on the colorimeter, and the milligrams of uric acid per 100 g of blood was calculated.

Folin and Denis gave results of a recovery experiment they performed by adding varying quantities of uric acid standard to sheep blood (which contained no uric acid). Recovery was well above 90% of the added values of 1–10 mg at six different concentrations. Results on human blood would be published elsewhere.

1. Folin, O., and Denis, W., A new (colorimetric) method for the determination of uric acid in blood. *J. Biol. Chem.* **13**, 469–75 (1913).

On the Colorimetric Determination of Uric Acid in Urine (1)

The new standard of uric acid in formaldehyde was prepared by first dissolving 1 g of uric acid in a 1-L flask in an excess of Li_2CO_3 solution (200 mL of a 4% w/v solution). Then 40 mL of a 40% (w/v) formaldehyde solution was added. The solution was shaken and allowed to stand a few minutes.

The clear solution was acidified with 20 mL of 1 N acetic acid and diluted to the mark with H_2O. The next day the solution was standardized against a fresh Li_2CO_3 solution of uric acid. The color of 5 mL of this solution corresponded very nearly to that of 1 mg of uric acid.

In the new uric acid method, urine (1–2 mL) was measured into a centrifuge tube (10 mL). Water was added to bring the volume to 5 mL, then in succession, 6 drops of 3% (w/v) silver lactate solution, 2 drops of magnesia mixture, 10–20 drops of NH_4OH to dissolve the AgCl were added, and the solution was centrifuged 1–2 min. The supernate was poured into a beaker. The residue was treated in the tube with 5–6 drops of sat. H_2S solution and 1 drop of conc. HCl; the tube was placed in a beaker of boiling H_2O until all H_2S was expelled. This solution was tested with lead acetate solution. Two milliliters of uric acid reagent and 10 mL of sat. Na_2CO_3 solution were added, and the contents were transferred to a 50-mL flask and diluted to volume. The color was read against a standard made from 5 mL of the standard similarly treated. For urines containing albumin, after removal of the H_2S, 2–10 drops of 10% (w/v) sodium acetate solution were added to the hot solution and centrifuged. The supernate was transferred to the volumetric flask, and the color reaction carried out.

When this method was tested on 19 urines before and after the addition of high concentrations of dextrose, of serum (as a source of protein), or in one case, of egg albumin, the results were identical.

1. Folin, O., and Denis, W., On the colorimetric determination of uric acid in urine. *J. Biol. Chem.* 14, 95–99 (1913).

A New Colorimetric Method for the Determination of Epinephrine (1)

Extracts were made from the suprarenal glands of several animals by the following procedure. The weighed gland was ground in a mortar with fine sand and 0.1 N HCl. The contents were rinsed into an Erlenmeyer flask with more 0.1 N HCl (about 15 mL/2 g of gland) and about as much H_2O and heated to boiling to dissolve the epinephrine (EP). To the boiling mixture a little 10% (w/v) sodium acetate (5 mL for each 15 mL of 0.1 N HCl present) was added and then heated again to boiling. The "albuminous" material came down promptly. The mixture, except the sand, was transferred to a volumetric flask (100 mL for each 2 g of gland), and diluted to the mark with H_2O, then filtered or centrifuged. The final solution was clear, or nearly so. The authors preferred centrifugation to filtration.

To perform the colorimetry, 5 mL of the clear extract was pipetted into a 100-mL flask, and 1 mL of fresh uric acid standard solution (1 mg) was pipetted into another 100-mL flask. To both flasks, 2 mL of uric acid reagent and 20 mL of sat. Na_2CO_3 solution were added. The solution was allowed to stand 2–3 min, then diluted to the mark, and mixed thoroughly. The color comparison was made on the Duboscq colorimeter with the standard set at 20 mm. The EP was first calculated as though it were uric acid, then the value was divided by 3 to obtain the value of EP.

The authors presented values they obtained on fresh suprarenal glands from several animal sources (sheep, lamb, cow, rabbit, dog, monkey, calf). They found that the samples did not preserve well.

> It being Saturday the solutions were transferred to the cold storage room. Two days later they were again assayed colorimetrically and their epinephrine content was then also determined by the blood pressure method.

At 2–3 °C, the samples did not keep well.

1. Folin, O., Cannon, W.B., and Denis, W., A new colorimetric method for the determination of epinephrine. *J. Biol. Chem.* 13, 477–83 (1913).

On the Preparation of Creatine, Creatinine, and Standard Creatinine Solutions (1)

Creatinine in fresh urine was precipitated with picric acid (in alcohol). After an overnight stand the supernate was siphoned off and the remainder of the mixture filtered on a Buchner filter, and the precipitate of creatinine picrate washed with cold water. The precipitate was considered to be composed largely of the double picrate of creatinine and potassium [$C_4H_7N_3O \cdot C_6H_2(NO_2)_3OH \cdot C_6H_2(NO_2)_2OH$] containing 18.55% creatinine. This compound was extraordinarily insoluble.

The dry precipitate was mixed with solid K_2CO_3, then dissolved in water. After standing 1–2 h, it was filtered, and the sediment washed with water. The carbonate in the filtrate was neutralized and then acidified with acetic acid. The addition of a concentrated alcoholic solution of $ZnCl_2$ produced immediately an abundant precipitate of creatinine zinc chloride. In the old procedure Otto had converted this compound to creatinine zinc alum (sulfate) before separating off the creatinine. In the new procedure, he decomposed it with lead hydroxide (prepared from lead nitrate, not acetate) with boiling. After cooling and filtering, the filtrate was clear and free of lead. If not, H_2S was passed through it, and the solution refiltered. Further treatment depended on whether creatine or creatinine was desired.

For creatine, the creatinine in the solution was decomposed at low temperature (80–90 °C) by evaporating the solution to dryness. The residue was dissolved in boiling water and twice the volume of 95% alcohol added. The resulting solution was allowed to stand overnight in a cool place, and creatine precipitated out. It was filtered, and then washed with dilute alcohol. This process produced creatine unmixed with creatinine. About half the creatinine was converted in the first run. The filtrate was then reprocessed for further conversion of creatinine, each time reducing the creatinine by half.

Creatinine could be prepared from creatine or from a mixture of creatine and creatinine (above) by heating in an autoclave at 135–140 °C for 3 h (Folin-Denis procedure of 1910). Rather than finish with a cold alcohol treatment, Otto now preferred pouring a little boiling water over the creatinine obtained, and then precipitating with alcohol. Creatinine prepared this way had a purity of 98–100%.

For preparing standard creatinine solution, Otto recommended that pure creatinine itself be avoided as a standard simply because further purification of creatinine was too laborious. Instead, he pointed out that pure salts of creatinine were quite easily made. He preferred the creatinine zinc chloride over others. The crude material had about 5–8% impurities, but could be purified simply by three crystallizations from acetic acid and alcohol containing some $ZnCl_2$, followed by filtering and washing with alcohol. The product was pure, and 1.6106 g/L equaled 1 mg of creatinine/mL. It was dissolved in 0.1 N HCl in which it was quite stable.

1. Folin O., On the preparation of creatine, creatinine and standard creatinine solutions. *J. Biol. Chem.* **17**, 463–68 (1914).

On the Determination of Creatinine and Creatine in Urine (*1*)

One milligram of creatinine plus 20 cc. of saturated picric acid solution plus 1.5 cc. of 10 per cent sodium hydrate solution (added from a burette) diluted after ten minutes' standing to 100 cc. gives a highly colored, stable solution, one therefore eminently suitable for use as a standard in connection with all ordinary creatinine determinations. With the more common human urines containing from 0.5 to 1.5 mgm. of creatinine in 1 or 2 cc. the process of making the color comparison is so simple that it is hardly possible to make a mistake—provided only that the 10 per cent alkali is measured out with a burette so that the same amount within 0.1 or 0.2 cc. is added to both the standard and the unknown.

One cubic centimeter of the standard creatinine solution is measured into a 100-cc. volumetric flask and 1 cc. of the urine into another; 20 cc. of saturated picric acid solution (measured with a cylinder) are added to each and then the alkali, 1.5 cc. of 10 per cent solution. At the end of ten minutes the flasks are filled up to the mark with tap water and the color of the unknown is determined. It makes little difference whether the standard is set at 10, 15 or 20 mm., 10, 15 or 20 divided by the reading of the unknown gives in milligrams the amount of creatinine present in the volume of urine taken. If the urine reads less than two-thirds or more than one and one-half that of the standard the determination should be repeated with more or with less urine.

Otto provided some added pointers for the procedure. He recommended use of the Ostwald-Folin pipettes for measurement of the urine. When the final color was too strong, less error would occur if the procedure were rerun on diluted urine rather than by diluting the color, or less favorably, one could use a stronger standard. The latter point was one of the advantages of using a creatinine standard rather than bichromate: It could be used at a variety of concentrations. A source of error could arise if bubbles of air were trapped or accumulated under the prism of the colorimeter. This was prevented by emptying the cups and refilling them.

1. Folin, O., On the determination of creatinine and creatine in urine. *J. Biol. Chem.* **17,** 469–73 (1914).

On Determination of Creatinine and Creatine in Blood, Milk, and Tissue (1)

Ten milliliters of blood or milk was measured into a 50-mL flask or into a 50-mL glass-stoppered mixing cylinder that was filled to the mark with saturated picric acid solution, and shaken. One gram of dry picric acid was added and shaken for 5 min. The solution was transferred to centrifuge tubes and the sediment was shaken down. The supernate was poured through a filter. A standard solution was prepared by placing 1 mg of creatinine standard in a 500-mL flask and diluting to volume with saturated picric acid solution. The creatinine content in either the standard or the "unknown" was not to exceed the other by a factor of 1.5.

The volume of alkali added to both standard and unknown had to be identical. Five milliliters of a 10% (w/v) NaOH solution was used for each 100 mL of the picric acid solution present measured from a burette. If a slight turbidity developed, the colored solution was filtered and let stand for 10 min before the color comparison was made in the colorimeter.

The dry picric acid was added to the diluted blood or milk to keep the mixture as close to saturation with picric acid as possible so that when the alkali was added, the darkening that occurred in the absence of creatinine was the same for both standard and unknown. (This step was one that Otto could not eliminate until several years later when protein removal technique was more effective and could be performed prior to the addition of a combined alkaline picrate solution.) In other words, picric acid was serving as both the protein precipitant and as the coloring agent for the alkaline picrate reaction to follow with creatinine.

Otto cautioned that in collecting the blood sample for creatinine analysis excess potassium oxalate (anticoagulant) had to be avoided because picric acid would liberate oxalate as oxalic acid, causing variation in the intensity of the final color obtained. Ten drops of a 20% solution (w/v) was enough anticoagulant for 30 mL of blood. The filtrate resulting otherwise had to be titrated to have the same acidity as that of saturated picric acid solution so that the volume of alkali used was the same. Otto did not find sodium oxalate superior to potassium oxalate.

For determining creatine and creatinine in blood, milk, and exudates, the preliminary precipitation was that used for preformed creatinine, followed by conversion of creatine to creatinine. Ten milliliters of the filtrate was measured into a 25- or 50-mL Erlenmeyer flask or test tube marked at 25 mL, the top was covered with tinfoil, and the flask was autoclaved at about 120 °C for 20 min, and then cooled to room temperature. The solution was rinsed into a 25-mL flask with saturated picric acid solution and diluted to the mark with the same solution.

Depending on the amount of filtrate remaining for determining the preformed creatinine, 10, 15, or 20 mL of the filtrate was used, and 1.25 mL of 10% NaOH was added to develop the color.

Because of variation in the creatine content, three standard solutions of 0.5, 1, and 2 mg per 100 mL of saturated picric acid solution were prepared for comparison. To 20 mL of each of these standards, 1 mL of 10% NaOH was added. For color comparison, the one that most closely matched the color of the unknown was used. The standards were prepared by diluting, respectively, 1, 2, and 4 mL of standard creatinine zinc chloride solution to the mark in a 200-mL flask with saturated picric acid solution.

Finally, Otto gave a procedure for the analysis of creatine in muscle and other tissue, the highlights of which are as follows. Five grams of tissue was cut with scissors or a meat grinder and transferred to a 200-mL Erlenmeyer flask, to which was added 100 mL of 0.5 N H_2SO_4. The flask was covered with tinfoil and autoclaved at 130–135 °C for 30–40 min, and cooled to room temperature. The contents were transferred to a 200-mL flask and diluted to the mark with water, mixed, and filtered. How much 10% NaOH was required to neutralize 10 mL of the filtrate by titration was determined with phenolphthalein used as the indicator. Ten milliliters of the filtrate was pipetted into a 100-mL flask, to which was added 20 mL of saturated picric acid and enough 10% NaOH was added to give 1.5 mL over and above that required to neutralize the H_2SO_4 present. The color comparison was made against a standard of 1 mg of creatinine per milliliter (1.389 g of creatinine zinc chloride per liter) for striated muscle, or half as strong for tissues other than striated muscle. This procedure was suitable for use with small animals and tissue samples of 2 g or less.

1. Folin, O., On the determination of creatinine and creatine in blood, milk and tissues. *J. Biol. Chem.* **17**, 475–81 (1914).

Turbidity Methods for the Determination of Acetone, Acetoacetic Acid, and β-Oxybutyric Acid in Urine (1)

For determining acetone, Otto and Willey combined the Folin air-current method for removing acetone from urine with the Marriott method for determining acetone bodies in blood by use of a turbidity reaction with the Scott-Wilson reagent, an alkaline solution of mercuric cyanide and silver nitrate (below). An extraordinarily insoluble precipitate was formed. Urine containing about 0.5 mg of acetone was acidified, then the acetone aspirated at 35–40 °C into a tube containing 10 mL of a 2% aqueous solution of $NaHSO_3$. The air-current arrangement was about the same as that used for NH_3 determination in urine. The aspiration took about 10 min for 2 mg of acetone from 5 mL of liquid. The bisulfite solution was then transferred to a 100-mL volumetric flask, diluted to 50–60 mL with water, then treated with 15 mL of Scott-Wilson reagent, and diluted to 100 mL. The standard contained 0.5 mg of acetone, 10 mL of the bisulfite solution, and 15 mL of Scott-Wilson reagent that had been diluted to 100 mL. Both flasks were allowed to stand for 12–15 min, and the relative turbidity was measured in the colorimeter.

The Scott-Wilson reagent was prepared by dissolving 10 g of mercuric cyanide in 600 mL of water, and 180 g of NaOH was added in 600 mL of water. The solution was cooled, if warm, then a solution of 2.9 g of $AgNO_3$ dissolved in 400 mL of water was added slowly with constant stirring. The resultant solution was set aside for 3–4 days to allow a small sediment to settle out. The supernate was siphoned off. No filtration was necessary.

Marriott had noted that freshly distilled acetone gave more turbidity with the reagent than undistilled acetone. Folin and Denis found that freshly distilled acetone was stable for several weeks in 0.25 N H_2SO_4. They therefore prepared a standard containing freshly distilled Kahlbaum acetone in the acid so that 10 mL contained 0.5 mg as titrated with the iodine-thiosulfate reaction, and adjusted it to the correct concentration of the acid solution.

They noted that "acetone urines contain from two or three to nine or ten times as much acetoacetic acid as acetone." The older the urine (the longer it stands) the greater the proportion of acetone because of spontaneous breakdown of the acetoacetic acid. Because of the small amount of acetone, it was more convenient to determine the sum of acetone and acetoacetic acid and to measure the preformed acetone in a separate sample. Urines giving a strong ferric chloride reaction usually contained more than 0.5 mg of acetoacetic acid per milliliter and had to be diluted. The amount of urine to use should yield about 0.5 mg of acetone (0.3–0.7 mg). This volume of urine was transferred to a large test tube containing 1 mL of 10% H_2SO_4 and connected in the usual way to a second test tube containing 10 mL of 2% sodium bisulfite solution. The test tube containing the urine was immersed in a beaker of boiling water while an extremely slow air current was passed through it. After 10 min the current was increased and the heating continued for 5 min. The contents of the second tube were then transferred to a 100-mL flask, and acetone was determined as in the method for preformed acetone. For calculation, 1 mg acetone equaled 1.8 mg of acetoacetic acid. The total preformed acetone was subtracted from the total acetone calculated in the 24-h urine, and the remainder multiplied by 1.8 gave the acetoacetic acid concentration.

The β-oxybutyric acid was determined by a modification of the well-known Shaffer oxidation method followed by using the Marriott turbidity method (above) on the acetone formed. Quantitative yields were obtained by oxidizing 1–2 mg of β-oxybutyric acid (β-hydroxybutyric acid;

BHBA) in boiling chromic acid solution. The method was carried out in either of two ways depending on whether or not the urine contained sugar.

For urines containing sugar, the sample was diluted (10–50-fold) to get about 2 mg of BHBA and put into a 500-mL Kjeldahl flask, to which was added 200 mL of water and 5 mL of 10% H_2SO_4. The solution was boiled for 10 min to remove acetone and acetoacetic acid. Twenty-five milliliters of a solution containing 2% potassium bichromate and 35% H_2SO_4 was added, and the solution was distilled slowly for 40–60 min into a flask containing 75–100 mL of water. About 2 g of sodium peroxide was added to the distillate and the distillation repeated to collect about 80 mL into a 100-mL cylinder containing 10 mL of water. This distillation required 10–15 min. The distillate was diluted to 100 mL. Twenty-five to 50 mL was transferred into another 100-mL flask; 25 mL of water and then 15 mL of Scott-Wilson reagent were added and diluted to the mark. The resultant solution was read against the 0.5-mg acetone standard. Note that in this procedure no bisulfite was used. Each milligram of acetone was equivalent to 1.78 mg of β-oxybutyric acid.

[In this method] all the oxidation products obtained from sugar and which react with the Scott-Wilson acetone reagent are destroyed or removed by the second distillation with the alkaline peroxide mixture, and the rather tedious precipitation with basic lead acetate has therefore been omitted.

If the urine contained *no* sugar, the second distillation was omitted. Eighty milliliters of distillate was collected in 40–60 min into a 100-mL measuring cylinder and diluted to volume. One to 2 g of sodium peroxide was added and mixed, producing a solution ready for precipitation of the acetone with the Scott-Wilson reagent.

The authors presented a table of recovery of BHBA from urine to which varying quantities of dextrose and BHBA were added. Recovery ranged from 98–101%.

1. Folin, O., and Denis, W., Turbidity methods for the determination of acetone, acetoacetic acid and β-oxybutyric acid in urine. *J. Biol. Chem.* **18**, 263–71 (1914).

The Quantitative Determination of Albumin in Urine (*1*)

The sulfosalicylic acid method was borrowed from a nephelometric procedure applied by Kober to measure proteins in milk and "digestion" mixtures. Five milliliters of a 25% (w/v) sulfosalicylic acid solution was measured into each of two 100-mL flasks containing about 75 mL of water. To one flask was added 5 mL of standard protein solution (containing 10 mg "albumin") and to the other was added the urine 1 mL at a time (with an Ostwald-Folin pipette) until the turbidity obtained appeared to match the standard. The flask was filled to the mark with water and cautiously inverted a few times to mix. The two solutions were compared in the colorimeter. Both cups of the colorimeter were first adjusted with the standard. With the standard set at 20 mm, the unknown was not to read less than 10 or more than 30 mm.

Calculation: mg/mL of albumin in urine = 200/U × mL urine used

The standard protein solution was prepared by obtaining fresh, normal human or slaughterhouse blood and separating the serum, which had to be free of hemoglobin. Commercial preparations or those provided for bacteriological culture media were not to be used. Twenty-five to 35 mL of the serum was diluted with 15% (w/v) NaCl solution to about 1500 mL, mixed, and then filtered. The nitrogen content was determined by the Kjeldahl method and the protein content calculated using the conversion factor for nitrogen of 6.25 (protein = N × 6.25). The serum was diluted with 15% NaCl to contain 2 mg/mL of protein (albumin). The protein was saturated with 20 mL of chloroform as additional preservative to the salt. When stored in the refrigerator, the standard was stable for several months.

Their gravimetric method differed only in minor details from those described by other people. Ten milliliters of urine was pipetted into a preweighed centrifuge tube. One milliliter of 5% (v/v) acetic acid was added. The tube was placed in a beaker of boiling water for 15 min, removed, and centrifuged for a few minutes. The supernate was poured off and the precipitate stirred up in 10 mL of boiling 0.5% (v/v) acetic acid. The solution was centrifuged, and the supernate poured off. The precipitate was washed with 50% (v/v) alcohol, and the supernate poured off. The precipitate was dried in an air bath at 100–110 °C, cooled in a desiccator, and weighed.

When the turbidimetric and gravimetric methods were compared using 18 urines containing from 0.8–7.1 g of protein, the results correlated well.

1. Folin, O., and Denis, W., The quantitative determination of albumin in urine. *J. Biol. Chem.* **18**, 273–76 (1914).

Impure Picric Acid As a Source of Error in Creatine and Creatinine Determinations (*1*)

To 20 cc. of a saturated (1.2 per cent) solution of picric acid add 1 cc. of 10 per cent sodium hydroxide and let it stand for 15 minutes. The color of the alkaline picrate solution thus obtained must not be more than about twice as deep as the color of the saturated acid solution. If the quality of the picric acid is good, the color of the picrate solution will be no deeper at the end of 24 hours than at the end of 15 minutes provided that organic impurities, dust, etc., be excluded. If the picric acid is unusually pure, the color of the picrate solution will not be more than one and a half times as deep as that of a saturated picric acid solution; i.e., by setting the picric acid solution at 20 mm. in the Duboscq colorimeter, the picrate will give a reading of 13 to 14 mm.

To the fresh blood add four volumes of saturated picric acid solution and about 1 gm. of powdered picric acid for each 10 cc. of blood used. Shake for 10 minutes, and filter. To 10 (or 20) cc. of the filtrate add 1 (or 2) cc. of a solution containing 7 per cent potassium hydroxide and 25 per cent potassium chloride. Let stand for 10 minutes, centrifuge, and compare in the colorimeter with the standard creatinine in picric acid to which has been added a corresponding amount of alkaline potassium solution. The potassium in the alkali precipitates fully 75 per cent of the picric acid present in a saturated solution, and thereby makes the color due to the minute amounts of creatinine present in normal blood distinctly more predominant than when the filtrates remain saturated solutions of picric acid.

In the above modification, as well as in the original method, the blood is diluted fivefold by the addition of four volumes of picric acid solution. The chief reason for this great dilution of the blood filtrates is that it enables one to make a creatinine determination with as little as 2 cc. of blood. When there is no reason for economy of blood, the creatinine concentration of the blood filtrates can advantageously be increased by using only two volumes of picric acid solution (+ solid picric acid) for the precipitation of the proteins. The precipitation in this case requires more time, about an hour. The standard creatinine solution in this case should, of course, be less dilute than when four volumes of picric acid are used for the precipitation.

1. Folin, O., and Doisy, E.A., Impure picric acid as a source of error in creatine and creatinine determinations. *J. Biol. Chem.* **28**, 349–56 (1917).

Copper-Phosphate Mixtures As Sugar Reagents. A Qualitative Test and a Quantitative Titration Method for Sugar in Urine (*1*)

One reagent for the qualitative test for glucose in urine was prepared by dissolving 100 g of Na_3PO_4 (USP), 30 g of Na_2HPO_4, and 50 g of anhydrous Na_2CO_3 in about 1 L of H_2O; 13 g of $CuSO_4$ previously dissolved in about 200 mL of H_2O was added. This solution kept indefinitely and was preferably stored at room temperature.

The test was performed like Benedict's. In a test tube were placed 5 mL of reagent and 5–8 drops of urine (not over 0.5 mL). The tube was boiled for 1 min or heated in a boiling water bath for 3–5 min. Minute traces of sugar were indicated by cuprous oxide precipitates appearing in various grades of turbidity as in Benedict's test. "The test is quite as sensitive and reliable as Benedict's and a trifle more prompt." (Otto and McEllroy offered no proof of this, however. This, of course, was uncharacteristic of Otto, and one is inclined to believe that he was perhaps "digging" his dogged critic and modifier, Stanley Benedict, just a little.)

For the quantitative titration method for sugar in urine, one solution and a dry mixture were used. The solution consisted of 60 g of $CuSO_4 \cdot 5H_2O$ per liter. Five milliliters of this corresponded to 25 mg of dextrose or levulose, 45 mg of anhydrous maltose, or 40.4 mg of anhydrous lactose. The mixture contained 100 g of $Na_2HPO_4 \cdot 12H_2O$, 60 g of dry $Na_2CO_3 \cdot H_2O$, and 30 g of sodium or potassium sulfocyanate ground together in a mortar. Each titration required less than 1 g of the mixture. Both the solution and the mixture kept indefinitely.

The quantitative titration was carried out in a test tube to which 5 mL of the solution and 4–5 g of the dry mixture were added. The urine was added first in a large volume, then in smaller volumes, with a boiling period (open flame) of 1 min between additions for a total boiling time of not less than 4 min or more than 7 min, until the blue copper sulfate color had disappeared. The product formed was white cuprous sulfocyanide (thiocyanide). The quantity of urine consumed could be obtained by drop-counting from a large burette or by measuring from a special 5-mL burette graduated in intervals of 0.02 mL (recommended).

The method was sensitive to sugar concentrations as low as 0.5%, and it was not interfered with by albumin. Otto and McEllroy presented many fine points of technique. They found that the method correlated quite well when 10 urines containing sugar from 0.98–6.66% were tested and values obtained compared with those of Benedict's method and with a polariscope.

1. Folin, O., and McEllroy, W.S., Copper-phosphate mixtures as sugar reagents. A qualitative test and a quantitative titration method for sugar in urine. J. Biol. Chem. 33, 513–19 (1918).

Chapter 15. Hsien Wu—A Major Leap Forward (1919–21)

A System of Blood Analysis (1)

The paper was divided into distinct topics. After the introduction came the question of preparing the protein-free filtrate.

As a working principle or guide in this we have first of all required that the procedure employed must permit the quantitative recovery of at least 10 mg. of uric acid and creatinine when added to 100 cc. of sheep, beef, or chicken blood, and that the total nonprotein nitrogen must certainly be no higher than the figures obtained from a corresponding trichloroacetic filtrate representing a 10 per cent trichloroacetic acid concentration (in the diluted unfiltered blood mixture)—or a corresponding 1.5 per cent m-phosphoric acid filtrate.

Folin and Wu first tried Kahlbaum's phosphotungstic acid and their own prepared sodium phosphotungstate.

In connection with our work on sodium phosphotungstate we have discovered a new protein precipitant which probably has never before been used in blood analysis. We refer to it as [a] new protein precipitant because so far as we have been able to learn, it has never before been used in that capacity. This substance is tungstic acid. Tungstic acid, like sodium phosphotungstate or phosphotungstic acid, must be used in a definite way, but the necessary conditions are not difficult to find. Less than 1 gm. is used for the precipitation from 10 cc. of blood, yet the precipitation is more complete than that produced by 10 gm. of trichloroacetic acid, and the filtrate obtained gives no trouble in connection with any of the determinations so far investigated.

There was no loss of creatinine or uric acid by absorption on the precipitate. As much as 20 mg of uric acid could be added to blood without any loss.

The precipitation of the blood proteins by means of our new reagent is made in the following manner. Transfer a measured amount of blood into a flask having a capacity of fifteen to twenty times that of the volume taken. Dilute the blood with 7 volumes of water and mix. With an appropriate pipette add 1 volume of 10 per cent solution of sodium tungstate ($Na_2WO_4 2H_2O$) and mix. With another suitable pipette add to the contents in the flask (with shaking) 1 volume of ⅔ normal sulfuric acid. Close the mouth of the flask with a rubber stopper and give a few vigorous shakes. If the conditions are right hardly a single air bubble will form as a result of the shaking. Much oxalate or citrate interferes with the coagulation and later with the uric acid determination. 20 mg. of potassium oxalate is ample for 10 cc. of blood. Citrate, except in the minimum amount, is to be avoided. When a blood [sample] is properly coagulated, the color of the coagulum gradually changes from pink to dark brown. If this change does not occur, the coagulation is incomplete, due, in every case we have encountered, to too much oxalate or citrate. In such an emergency the sample may be saved by adding 2 normal sulfuric acid drop by drop, shaking vigorously after each addition and allowing the mixture to stand for a few minutes before adding more, until the coagulation is complete. Pour the mixture on a filter large enough to hold the entire contents of the flask and cover with a watch-glass. If the filtration is begun by pouring the first few cc. of the mixture down the double portion of the filter paper and withholding the remainder till the whole filter has been wet, the filtrates are almost invariably as clear as water from the first drop. If a filtrate is not perfectly clear, the first 2 or 3 cc. may have to be returned to the funnel.

The authors made several observations on the blood protein precipitation: 1) the precipitation was complete within a few seconds, 2) when shaken hard, the sound was like that of shaken mercury, and there was not more than a trace of foam produced, 3) a fine precipitate formed, yet it would not pass through good filter paper or clog its pores, 4) filtration was slow, but the amount of filtrate obtained was equal to that obtained with trichloroacetic acid, 5) when heated for 3–5 min in a water bath the precipitate settled spontaneously. If heated for 2–3 min, the precipitate could be centrifuged out rather than filtered. Folin and Wu did not, however, recommend a heating procedure.

It was important to use pure tungstic acid. In the precipitation all of the tungstic acid was set free, leaving about 10% excess to neutralize the carbonate usually present in commercial tungstates. To test the carbonate in the tungstate, 1 drop of phenolphthalein was added to 10 mL of a 10% (w/v) solution of sodium tungstate and titrated with 0.1 N HCl. The amount of acid required was not to exceed 0.4 mL.

The protein-free filtrate (PFF) was nearly neutral; 10 mL required only about 0.2 mL of 0.1 N NaOH to neutralize it with phenolphthalein as indicator.

The PFF could be preserved longer than 2–3 days if 1–2 drops of toluene or xylene were added. Both were equally effective.

The precipitation worked equally well with the blood of several animals tested: human, cow, sheep, chicken, dog, and rabbit. Tungstic acid filtrates gave lower values of NPN than those obtained with trichloroacetic acid.

The PFF adapted easily to the Folin-Denis NPN method for which improvements were now made.

The acid digestion mixture is made as follows: Mix 300 cc. of phosphoric acid syrup (about 85 per cent H_3PO_4) with 100 cc. of concentrated sulfuric acid. Transfer to a tall cylinder, cover well to exclude the absorption of ammonia, and set aside for sedimentation of calcium sulfate. This sedimentation is very slow, but in the course of a week or so the top part is clear and 50 to 100 cc. can be removed by means of a pipette. (It is not absolutely necessary that the calcium should be thus removed, but it is probably a little safer to have it done.) To 100 cc. of the clear acid add 10 cc. of 6 per cent copper sulfate solution and 100 cc. of water. 2 cc. of this solution are substantially equivalent to 1 cc. of the acid mixture previously described by Folin and Denis. We prefer this diluted acid, first, because the objectionable viscosity of the undiluted reagent is practically eliminated, and second, because we now use for a nitrogen determination only 5 cc. (instead of 10 cc.) of blood filtrate, and 1 cc. of acid (corresponding to 0.5 cc. of the undiluted acid reagent).

The digestion was now carried out in Pyrex ignition test tubes (rather than Jena test tubes) of 75-mL capacity (200 × 25 mm). These tubes were graduated at 35 and 50 mL, which allowed for direct nesslerization. It was noted that the best way to prevent bumping was to start with *dry* tubes and dry quartz pebbles or pieces of granite to keep the pores filled with air.

A purer, more economical Nessler's stock reagent was prepared as follows:

Because of the difficulties encountered in obtaining high grade mercuric iodide, we have devised a new process for making the mercuric potassium iodide solution. This process is as follows: Transfer 150 gm. of potassium iodide and 110 gm. of iodide to a 500 cc. Florence flask; add 100 cc. of water and an excess of metallic mercury, 140 to 150 gm. Shake the flask continuously and vigorously for 7 to 15 minutes or until the dissolved iodine has nearly disappeared. The solution becomes quite hot. When the red iodine solution has begun to become visibly pale, though still red, cool in running water and continue the shaking until the reddish color of the iodine has been replaced by the greenish color of the double iodide. This whole operation usually does not take more than 15 minutes. Now separate the solution from the surplus mercury by decantation and washing with liberal quantities of distilled water. Dilute the solution and washings to a volume of 2 liters. If the cooling is begun in time, the resulting reagent is clear enough for immediate dilution with 10 per cent alkali and water, and the finished solution can at once be used for Nesslerizations.

The Nessler's reagent was prepared from the stock solution as follows:

From the stock solution of mercuric potassium iodide, made according to either of the processes described above, we prepare the final Nessler solution as follows: From completely saturated caustic soda solution containing about 55 gm. of NaOH per 100 cc. decant the clear supernatant liquid and dilute to a concentration of 10 per cent. (It is worth while to determine by titration that a 10 per cent solution has been obtained within an error of not over 5 per cent.) Introduce into a large bottle 3,500 cc. of 10 per cent sodium hydroxide solution, add 750 cc. of the double iodide solution and 750 cc. of distilled water, giving 5 liters of Nessler's solution.

The Nessler solution so obtained contains enough alkali in 15 cc. to neutralize 1 cc. of the diluted phosphoric-sulfuric acid mixture, and to give a suitable degree of alkalinity for the development of the color given by ammonia at a volume of 50 cc.

(In other Nesslerizations, as in urine analysis when there is no acid to be neutralized, 10 cc. of the Nessler reagent per 100 cc. of Nesslerized ammonia solution is the correct amount.)

For historic reasons the methods of this paper will be reproduced. Except for the uric acid procedure, which of these methods is not familiar to those of us performing blood analysis in clinical laboratories in the decade before and the decade after the outbreak of World War II?

Concise Description of Non-Protein Nitrogen Determination. Introduce 5 cc. of the protein-free blood filtrate into a dry 75 cc. test-tube graduated at 35 cc. and at 50 cc. Add 1 cc. of the sulfuric-phosphoric acid mixture. . . . Add a dry quartz pebble and boil vigorously over a microburner until the characteristic dense acid fumes begin to fill the test-tube. This is usually accomplished in from 3 to 7 minutes. When the fumes are unmistakable, cut down the size of the flame so that the contents of the tube are just visibly boiling, and close the mouth of the test-tube with a watch-glass or a very small Erlenmeyer flask. Continue the heating very gently for 2 minutes from the time the fumes began to be unmistakable, even if the solution has become clear and colorless at the end of 20 to 40 seconds. If the oxidations are not visibly finished at the end of 2 minutes the heating must be continued until the solution is nearly colorless. Such cases are very rare; the oxidation is almost invariably finished within the 1st minute. Allow the contents to cool for 70 to 90 seconds, and then add 15 to 25 cc. of water. Cool further, approximately to room temperature, and add water to the 35 cc. mark. Add, preferably with a pipette, 15 cc. of the Nessler solution described above. Insert a clean rubber stopper and mix. If the solution is turbid, centrifuge a portion before making the color comparison with the standard. The standard most commonly required is 0.3 mg. of N (in the form of ammonium sulfate) in a 100 cc. flask. Add to it 2 cc. of the sulfuric-phosphoric acid mixture, about 50 cc. of water, and 30 cc. of Nessler solution. Fill to the mark and mix. The unknown and the standard should be Nesslerized at approximately the same time. If the standard is set at 20 mm. for the color comparison, 20 divided by the reading and multiplied by 30 gives the non-protein nitrogen in mg. per 100 cc. of blood.

Otto had abandoned the direct nesslerization technique for determination of urea in blood because of interference by amino acids and peptones with the small amount of NH_3 liberated from urea. It was not feasible to use more filtrate or to use permutit. The method now chosen was to liberate NH_3 with jack bean urease, aerate to isolate the NH_3, and nesslerize to provide the color for measurement. Alternatively, autoclaving to split the urea and distillation prior to nesslerization could be used, though this method was not preferred. Goodbye to the Folin urea-splitting techniques: $MgCl_2$ + HCl (1901); potassium acetate + heat (1912); H_3PO_4 + heat (1912); acid + autoclaving (1919).

A procedure was presented for preparing an economical urease suspension in dilute NH_3-free alcohol from jack bean flour. The suspension was stable at room temperature for at least 1 week or for 3–5 weeks if stored on ice in the refrigerator. Urease had to be used in a buffer solution. Folin and Wu studied various phosphate mixtures and found the best to be a combination of sodium pyrophosphate and glacial phosphoric acid. This was superior to those investigated by Van Slyke, possibly because it was less injurious to urease than orthophosphates.

The method for urea determination was as follows:

Determination of Urea by Urease Decomposition and Distillation. Transfer 5 cc. of the tungstic acid blood filtrate to a clean and dry Pyrex ignition tube (capacity about 75 cc.). The graduated Pyrex tubes recommended for the non-protein nitrogen determination should never be used for urea determinations, because they have contained Nessler solutions and Nessler solutions leave behind films of mercury compounds which destroy the urease. If those tubes must be used, they should first be washed with nitric acid to remove the mercury films. Add to the blood filtrate two drops of the pyrophosphate solution described above or two drops of a molecular *o*-phosphate solution (⅓ molecular monosodium phosphate plus ⅔ molecular disodium phosphate). Then add 0.5 to 1 cc. of the urease solution described . . . and immerse the test-tube in a beaker of warm water and leave it there for 5 minutes. The temperature of the water is not very important, but should not exceed 55 °C. The warm water can perhaps scarcely be said to be essential, for the hydrolysis is very rapid at room temperature, but we nevertheless much prefer to use it. If no hot water is used, continue the digestion for 10 to 15 minutes, or as much longer as is convenient. The ammonia formed can be conveniently and quickly distilled into 2 cc. of 0.05 normal hydrochloric acid contained in a second test-tube. The second test-tube should not be so heavy as the ordinary test-tubes and should be graduated at 25 cc. A simple and compact arrangement for this distillation is indicated by Fig. 2. The test-tube which serves as a receiver is held in place by means of a rubber stopper in the side of which has been cut a fairly deep notch to permit the escape of air (and some steam). The rubber stopper serving as a holder for the receiver fits quite loosely to the delivery tube by means of which the two test-tubes are connected. The delivery tube must, of course, be so adjusted as to reach below the surface of the hydrochloric acid solution in the receiver before the distillation is begun.

Add to the hydrolyzed blood filtrate a dry pebble, 2 cc. of saturated borax solution, and a drop or two of paraffin oil; insert firmly the rubber stopper carrying both delivery tube and receiver, and boil moderately fast over a microburner for 4 minutes. The size of the flame should never be cut down during the distillation, nor should the boiling be so brisk that the emission of steam from the receiving tube begins before the end of 3 minutes. At the end of 4 minutes slip off the receiver from the rubber stopper, and put it in the position shown in Fig 2. Continue the distillation for 1 more minute and rinse off the lower outside part of the delivery tube with a little water. Cool the distillate with running water, dilute to about 20 cc., and add 2.5 cc. of the Nessler solution described on page 90. Fill to the 25 cc. mark and compare in the colorimeter with a standard containing 0.3 mg. of N in a 100 cc. flask and Nesslerized with 10 cc. of the Nessler solution. The standard and unknown should always be Nesslerized as nearly simultaneously as practicable.

Calculation. Multiply 20 (the height of the standard in mm.) by 15 and divide the colorimetric reading to get the urea nitrogen per 100 cc. of blood. The reasons for this calculation are, of course, to be found in the fact that the standard containing 0.3 mg. of N is diluted to 100 cc., while the unknown, which corresponds to 0.5 cc. of blood, is diluted to only 25 cc.

It is even more important in this distillation than in the non-protein nitrogen digestion that the Pyrex test-tube should not be in a condition that leads to bumping. Dry the tube, or rinse it with alcohol, after each determination.

The determination of creatinine and creatine was as follows:

Determination of Preformed Creatinine. Transfer 25 (or 50) cc. of a saturated solution of purified picric acid to a small, clean flask, add 5 (or 10) cc. of 10 per cent sodium hydroxide, and mix. Transfer 10 cc. of blood filtrate to a small flask or to a test-tube, transfer 5 cc. of the standard creatinine solution described below to another flask, and dilute the standard to 20 cc. Then add 5 cc. of the freshly prepared alkaline picrate solution to the blood filtrate, and 10 cc. to the diluted creatinine solution. Let stand for 8 to 10 minutes and make the color comparison in the usual manner, never omitting first to ascertain that the two fields of the colorimeter are equal when both cups contain the standard creatinine picrate solution. The color comparison should be completed within 15 minutes from the time the alkaline picrate was added; it is therefore never advisable to work with more than three to five blood filtrates at a time.

When the amount of blood filtrate available for the creatinine determination is too small to permit repetition, it is of course advantageous or necessary to start with more than one standard. If a high creatinine should be encountered unexpectedly without several standards ready, the determination can be saved by diluting the unknown with an appropriate amount of the alkaline picrate solution—using for such dilution a picrate solution first diluted with two volumes of water—so as to preserve equality between the standard and the unknown in relation to the concentration of picric acid and sodium hydroxide.

One standard creatinine solution, suitable both for creatinine and for creatine determinations in blood, can be made as follows: Transfer to a liter flask 6 cc. of standard creatinine solution used for urine analysis (which contains 6 mg. of creatinine); add 10 cc. of normal hydrochloric acid, dilute to the mark with water, and mix. Transfer to a bottle and add four or five drops of toluene or xylene. 5 cc. of this solution contain 0.03 mg. of creatinine, and this amount plus 15 cc. of water represents the standard needed for the vast majority of human bloods, for it covers the range of 1 to 2 mg. per 100 cc. In the case of unusual bloods representing retention of creatinine, take 10 cc. of the standard plus 10 cc. of water, which covers the range of 2 to 4 mg. of creatinine per 100 cc. of blood; or 15 cc. of the standard plus 5 cc. of water by which 4 to 6 mg. can be estimated. By taking the full 20 cc. volume from the standard solution at least 8 mg. can be estimated; but when working with such blood it is well to consider whether it may not be more advantageous to substitute 5 cc. of blood filtrate plus 5 cc. of water for the usual 10 cc. of blood filtrate.

Calculation. The reading of the standard in mm. (usually 20) multiplied by 1.5, 3, 4.5, or 6 (according to how much of the standard solution was taken), and divided by the reading of the unknown, in mm., gives the amount of creatinine, in mg. per 100 cc. of blood. In connection with this calculation it is to be noted that the standard is made up to twice the volume of the unknown, so that each 5 cc. of the standard creatinine solution, while containing 0.03 mg., corresponds to 0.015 mg. in the blood filtrate.

Determination of Creatine plus Creatinine. Transfer 5 cc. of blood filtrate to a test-tube graduated at 25 cc. These test-tubes are also used for urea and for sugar determinations. Add 1 cc. of normal hydrochloric acid. Cover the mouth of the test-tube with tin-foil and heat in the autoclave to 130 °C for 20 minutes or, as for the urea hydrolysis, to 155 °C for 10 minutes. Cool. Add 5 cc. of the alkaline picrate solution and let stand for 8 to 10 minutes, then dilute to 25 cc. The standard solution required is 10 cc. of creatinine solution in a 50 cc. volumetric flask. Add 2 cc. of normal acid and 10 cc. of the alkaline picrate solution and after 10 minutes standing dilute to 50 cc. The preparation of the standard must of course have been made first so that it is ready for use when the unknown is ready for the color comparison. The height of the standard, usually 20 mm., divided by the reading of the unknown and multiplied by 6 gives the "total creatinine" in mg. per 100 cc. blood.

In the case of uremic bloods containing large amounts of creatinine 1, 2, or 3 cc. of blood filtrate, plus water enough to make approximately 5 cc., are substitutes for 5 cc. of the undiluted filtrate.

The normal value for "total creatinine" given by this method is about 6 mg. per 100 cc. [530 μmol/L] of blood.

Although Benedict had materially simplified the procedure for uric acid in blood by use of cyanide to dissolve the silver urate, Otto had long doubted its reliability, and did not make use of it in the following modification. But no method for uric acid was useful unless a stable uric acid standard was made available. This feat was achieved as follows:

Before describing the determination of uric acid in tungstic acid blood filtrates we wish to describe the preparation of a new standard solution of uric acid—a solution the keeping quality of which we now, after 18 months of constant use, consider much superior to any other as yet devised. The solvent is 10 per cent sodium sulfite, and the keeping quality of the solution depends on the fact that the sulfite keeps the solution free from dissolved oxygen. The solution is prepared as follows:

Make 1 to 3 liters of a 20 per cent solution of sodium sulfite, let stand overnight, and filter. Dissolve 1 gm. of uric acid in 125 to 150 cc. of 0.4 per cent lithium carbonate solution and dilute to a volume of 500 cc. Transfer 50 cc., corresponding to 100 mg. of uric acid, to each of a series of volumetric liter flasks. Add 200 to 300 cc. of water, then 500 cc. of filtered 20 per cent sodium sulfite solution, and finally make up to volume, and mix well. Fill a series of 200 cc. bottles, and stopper very tightly with rubber stoppers. The solution in a bottle which is opened daily will keep for at least 3 to 4 months. Our records kept for one larger bottle so used show that no measurable loss of uric acid had occurred at the end of 6 months. In unopened bottles we expect the uric acid to keep for many years.

The surplus 20 per cent sulfite solution should be diluted to [a] concentration of 10 per cent and should then be transferred to another series of small, tightly stoppered bottles. This sulfite is added to the unknown in order to offset the sulfite content of the standard.

The solutions required for the determination of uric acid, and the procedure, were as follows:

1. The standard uric acid sulfite solution already described (3 cc. used for each series of determinations).
2. A 10 per cent sodium sulfite solution, also described (2 cc. used for each determination).
3. A 5 per cent sodium cyanide solution, to be added from a burette (2.5 to 5 cc. used for each series of determinations).
4. A 10 per cent solution of sodium chloride in 0.1 normal hydrochloric acid (10 to 20 cc. used for each series of determinations).
5. The uric acid reagent prepared according to Folin and Denis. A still stronger reagent is obtained by heating the sodium tungstate (100 gm.) and the phosphoric acid (80 cc.) plus water (700 cc.) for 24 hours, instead of 2 hours; but the advantage gained, about 20 per cent, is not needed. Dilute the solution to 1 liter.
6. A solution of 5 per cent silver lactate in 5 per cent lactic acid (4 to 5 cc. needed for each determination).

In our new method for the determination of uric acid the latter is precipitated directly from the filtrate, without any previous concentration. 20 cc. of filtrate corresponding to 2 cc. of blood are used. In describing the process we shall have to introduce a slight variation from the way we actually do it. This variation is due to the fact that we use a larger centrifuge than most laboratories possess and by means of which we are able to use 30 cc. test-tubes for the precipitation. Using the small 15 cc. centrifuge tubes, it is necessary either to precipitate 10 cc. of filtrate in each of two tubes or to make the precipitation in two 10 cc. installments.

To 10 cc. of blood filtrate in each of two centrifuge tubes add 2 cc. of a 5 per cent solution of silver lactate in 5 per cent lactic acid, and stir with a very fine glass rod. Centrifuge; add a drop of silver lactate to the supernatant solution, which should be almost perfectly clear and should not become turbid when the last drop of silver solution is added. Remove the supernatant liquid by decantation as completely as possible. Add to each tube 1 cc. of a solution of 10 per cent sodium chloride in 0.1 normal hydrochloric acid and stir thoroughly with the glass rod. Then add 5 or 6 cc. of water, stir again, and centrifuge once more. By this chloride treatment the uric acid is set free from the precipitate. Transfer the two supernatant liquids by decantation to a 25 cc. volumetric flask. Add 1 cc. of a 10 per cent solution of sodium sulfite, 0.5 cc. of a 5 per cent solution of sodium cyanide, and 3 cc. of a 20 per cent solution of sodium carbonate. Prepare simultaneously two standard uric acid solutions as follows:

Transfer to one 50 cc. volumetric flask 1 cc., and to another 50 cc. flask 2 cc. of the standard uric acid sulfite solution described above. To the first flask add also 1 cc. of 10 per cent sodium sulfite solution. Then add to each flask 4 cc. of the acidified sodium chloride solution, 1 cc. of the sodium cyanide solution, and 6 cc. of the sodium carbonate solution. Dilute with water to about 45 cc. When the two standard solutions and the unknown have been prepared as described, they are ready for the addition of the uric acid reagent of Folin and Denis. Add 0.5 cc. of this reagent to the unknown and 1 cc. to each of the standards, and mix. Let stand for 10 minutes, fill to the mark with water, mix, and make the color comparison.

Calculation. In connection with the calculation it is to be noted (a) that the blood filtrate taken corresponds to 2 cc. of blood, (b) that the standard is diluted to twice the volume of the unknown, and (c) that the standard used contains 0.1 or 0.2 mg. of uric acid. The blood filtrate from blood containing 2.5 mg. of uric acid will be just equal in color to the weaker standard. 20 times 2.5 divided by the reading of the unknown gives, therefore, the uric acid content of the blood when the weaker standard is set at 20 mm.

Originally Folin and Wu were going to adapt Benedict's picrate method for determining blood sugar to their tungstic acid filtrate.

But a few exploratory experiments showed that an intense and stable color reaction can be obtained by the application of the phenol reagent of Folin and Denis to cuprous oxide. The color obtained from a given quantity of sugar is far more intense than that obtained by the alkaline picrate reaction; so that a small fraction of a mg. of dextrose (1 or 2 cc. of blood filtrate) is all that is required for a determination of the blood sugar.

The method employed a weakly alkaline copper tartrate solution, extremely sensitive to traces of sugar, yet unaffected by creatinine or uric acid in concentrations corresponding to 50 mg each per 100 mL blood. "We are therefore inclined to regard our method as more accurate than any method as yet proposed for the determination of sugar in blood."

The picrate methods (Benedict's latest, or the Myers modification) gave results substantially higher than those of the new method. This was attributed to error introduced because dextrose in pure solution reacted at a different rate from dextrose in filtrate.

Such quantitative variations are not encountered in our process when equal amounts of dextrose in the form of pure sugar and of blood filtrate are heated, except that the reduced copper is, of course, more extensively precipitated and visible in the pure sugar solution. It need scarcely be stated that added sugar is quantitatively recovered by our method.

The solutions used and the procedure for the new Folin-Wu method are presented below:

Solutions Needed for Determination of Sugar in Blood

1. *Standard Sugar Solution.* Dissolve 1 gm. of pure anhydrous dextrose in water and dilute to a volume of 100 cc. Mix, add a few drops of xylene or toluene, and bottle. If pure dextrose is not available, a standard solution of invert sugar made from cane sugar is equally useful. Transfer exactly 1 g. of cane sugar to a 100 cc. volumetric flask; add 20 cc. of normal hydrochloric acid and let the mixture stand over night at room temperature (or rotate the flask and contents continuously for 10 minutes in a water bath kept at 70 °C). Add 1.68 gm. of sodium bicarbonate and about 0.2 gm. of sodium acetate, to neutralize the hydrochloric acid. Shake a few minutes to remove most of the carbonic acid and fill to the 100 cc. mark with water. Then add 5 cc. more of water (1 gm. of cane sugar yields 1.05 gm. of invert sugar) and mix. Transfer to a bottle; add a few drops of xylene or toluene, shake well, and stopper tightly. The stock solution made in either way keeps indefinitely. Dilute 5 cc. to 500 cc., giving a solution 10 cc. of which contain 1 mg. of dextrose or invert sugar. Add some xylene. Use 2 cc. for each determination.

2. *Alkaline Copper Solution.* Dissolve 40 gm. of anhydrous sodium carbonate in about 400 cc. of water and transfer to a liter flask. Add 7.5 gm. of tartaric acid and when the latter has dissolved add 4.5 gm. of crystallized copper sulfate; mix, and make up to a volume of 1 liter. If the carbonate used is impure, a sediment may be formed in the course of a week or so. If this happens, decant the clear solution into another bottle.

3. *Phosphotungstic-phosphomolybdic Acid.* Transfer to a large flask 25 gm. of molybdenum trioxide (MoO_3) or 34 gm. of ammonium molybdate ($NH_4)_2(MoO_4)$; add 140 cc. of 10 per cent sodium hydroxide and about 150 cc. of water. Boil for 20 minutes to drive off the ammonia (molybdic acid sometimes contains large amounts of ammonia as impurity). Add to the solution 100 gm. of sodium tungstate, 50 cc. of 85 per cent phosphoric acid, and 100 cc. of concentrated hydrochloric acid. Dilute to a volume of 700 to 800 cc.; close the mouth of the flask with a funnel and watch-glass. Boil gently for not less than 4 hours, adding hot water from time to time to replace that lost during the boiling. Cool and dilute to 1 liter. This solution is identical with the phenol reagent of Folin and Denis. For use in connection with the determination of blood sugar dilute 1 volume (100 cc.) of the reagent with one-half volume (50 cc.) of water and one-half volume (50 cc.) of concentrated hydrochloric acid.

Saturated Sodium Carbonate Solution. The determination of blood sugar is carried out as follows: Heat a beaker of water to vigorous boiling. Transfer 2 cc. of the tungstic acid blood filtrate to a test-tube (20 mm. × 200 mm.) graduated at 25 cc. The graduated test-tubes used as receivers when distilling off the ammonia in urea determinations (p. 95) are suitable for this work. Transfer 2 cc. of the dilute standard sugar solution to another similar test-tube. Add to each tube 2 cc. of the alkaline copper tartrate solution. Heat in the boiling water for 6 minutes. Remove the test-tube and add at once (without cooling), preferably from a graduated pipette, 1 cc. of the strongly acidified and diluted phenol reagent. This should be done as nearly simultaneously as possible; it is not advisable to use one standard for a set of more than four determinations. The purpose of the added hydrochloric acid in the reagent is to dissolve the cuprous oxide. Mix, cool, and add 5 cc. of saturated sodium carbonate solution. An intense blue color is gradually developed which will remain unaltered for several days. Dilute the contents of both test-tubes to the 25 cc. mark, and after at least 5 minutes make the color comparison in the usual manner.

The depth of the standard (in mm.) multiplied by 100 and divided by the reading of the unknown gives the sugar content, in mg., per 100 cc. of blood.

The copper solution is adjusted to give proportionate reductions with 0.12 to 0.4 mg. of dextrose. This covers the range of hypoglycemic and hyperglycemic bloods. But in extreme cases it is better to use 3 or 1 cc. of the filtrate, instead of 2 cc., adding water to the standard or to the unknown so as to equalize the concentration of the alkaline copper.

1. Folin, O., and Wu, H., A system of blood analysis. *J. Biol. Chem.* **38**, 81–110 (1919).

Chapter 16. Metabolic Studies; MLIC (1922–26)

Colorimetric Methods for the Separate Determination of Tyrosine, Tryptophane, and Cystine in Proteins (1)

The phenol reaction was tested on 17 pure amino acids and only tyrosine and tryptophane gave the blue color. Otto admitted that he and Denis had erred in stating that indole would not react. It was found that tyrosine gave 58% of the color of tryptophane and that boiling 25% HCl did not affect tryptophane.

To determine tyrosine, they first separated the tryptophane via the Hopkins-Cole reaction (precipitation) with mercuric sulfate. This precipitate was centrifuged off. An aliquot of the supernate was treated with sodium cyanide in sodium carbonate solution to bind the mercury, and the tyrosine was then reacted with the phenol reagent to produce the color for measurement. For tryptophane, a similar procedure was used on the tryptophane–mercuric sulfate precipitate. The standards were similarly treated and precipitated in advance. Cystine was determined with the Folin-Denis uric acid reagent, the only known amino acid to react with it. Cystine was converted to cysteic acid with a large excess of sulfite, in the presence of sodium carbonate. Color was developed with the uric acid reagent.

For determining the three amino acids in a protein, the protein had to be hydrolyzed in acid medium to preserve cystine, whereas tryptophane could only be recovered well from alkaline hydrolysis. Tyrosine survived quantitatively under both conditions. For alkaline hydrolysis, therefore, the purified dry protein was boiled over a microburner in a Kjeldahl flask (fitted with a Hopkins condenser) for 40–48 h using a solution of $Ba(OH)_2$. The barium was precipitated with H_2SO_4, the solution heated to remove H_2S, and filtered. The filtrate was then ready for the treatment with the Hopkins-Cole reagent to precipitate the tryptophane. For cystine determination the protein was hydrolyzed for 12 h with 20% H_2SO_4 with the same reflux apparatus, and an aliquot was then treated with Na_2CO_3 and 20% sodium sulfite solution, prior to addition of the uric acid reagent.

A comparison of values obtained on a number of pure proteins showed excellent agreement with the literature values for tyrosine and tryptophane. Proteins particularly rich in cystine (horn, wool, human hair), but not those with lesser amounts (e.g., gliadin, glutenin, and fibrin), gave comparable values.

1. Folin, O., and Looney, J.M., Colorimetric methods for the separate determination of tyrosine, tryptophane, and cystine in proteins. *J. Biol Chem.* **51**, 421–34 (1922).

The Uric Acid Problem. An Experimental Study on Animals and Man, Including Gouty Subjects (1)

After a thorough review of the literature, the authors decided that they would try to determine whether uric acid (UA) was in part destroyed by the human organism. This had to be carried out by use of intravenous (IV) injections of UA because UA taken orally or via a stomach tube was not found in the urine. They would also try to determine whether UA was taken up by tissues.

The first experiments were carried out on dogs. Following an IV injection of UA in dogs there was a huge rise in UA (as would be predicted) in plasma and in the kidneys and urine. The distribution of UA in other tissues, particularly muscle and liver, was unremarkable. Even more important was the very rapid fall (within minutes) in the concentration of plasma UA, far greater than could be attributed to the kidneys alone. Obviously in the dog, there was "destruction" of UA and blood analysis could be used to monitor the fate of the injected material.

Where does the destruction of UA occur? By determining UA in the venous and arterial blood supply of muscle, no difference in UA concentration was found (as would be expected if the muscles destroyed UA) 3 min after an IV dose was injected. Removing either the kidneys or the pancreas did not stop the destruction of UA.

There was no support in their work for the idea that the liver was the chief site of destruction of uric acid. But Otto was particularly careful about the liver this time! As previously mentioned, it was in a publication by Mann and co-workers from the Mayo Clinic that additional evidence was found that urea was synthesized in the liver. Mann et al. also showed that the values of blood uric acid promptly rose (and urea fell) soon after the liver was extirpated [Bollman, J.L., Mann, F.C., and Magath, T.B., Studies on the physiology of the liver. VIII. Effect of total removal of the liver on the formation of urea. *Am. J. Physiol.* **69**, 371–92 (1924)]. The work also showed that UA rose when both the liver and the kidneys were removed. The Harvard trio found that following injection of UA, plasma UA values dropped when the kidneys alone were removed.

They brought in a surgeon, E. C. Cutler, to prepare an Eck fistula in a dog, a procedure that created a shunt between the portal vein and the vena cava. It was estimated that such a dog should retain 10% of its capacity to destroy circulating UA as compared to a healthy dog, yet the rate of destruction of the injected UA remained the same. Could the destruction of UA during the time of the fistula have transferred to another organ? Injecting UA directly into the mesenteric vein of a dog did not produce UA values lower than the dog with the Eck fistula. Finally, when UA was injected into an isolated loop of the jejunum of a dog (289 mg UA absorbed in 51 min), there was a small but significant rise in plasma UA, which would not have happened if the liver were destroying it.

One wonders if Otto (and his co-workers) determined the urea nitrogen in the blood of the dog with the Eck fistula. The NPN was measured and rose significantly in a matter of less than 30 min. Values were very high after 26 h. No elaboration was made on the statement, "Such nitrogen accumulations are always obtained in dogs allowed to survive the uric acid injections." Because the UA was falling dramatically during this period the rise in NPN was certainly attributable in large part to synthesis of urea. This would indicate the failure of the Eck fistula to solve the problem of the liver's involvement in uric acid "destruction." Extirpation was the only way to settle the issue, and this was not done. Indisputable, however, was the loss in uric acid not attributable to elimination from the kidneys. Our investigators concluded, therefore, that the loss occurred directly in circulating blood. Blood evidently picked up a UA-oxidizing agent (perhaps from liver) that was used up almost as rapidly as it entered the blood. They decided this agent must be so unstable that it disappeared when blood was drawn and tested for this ability to transform UA. The agent could not, therefore, be an enzyme.

The rate of destruction of UA varied between species, but it occurred in all of those tested: dog, including the Dalmatian; cat; goat; rabbit; and duck. Destruction was in all cases related to the level of UA in blood, and was much slower in herbivorous than in carnivorous animals.

Other factors concerned in UA removal were examined. There was an endogenous and an exogenous uric acid metabolism, which were experimentally approachable through purine-free diets. Excretion of UA was controlled by the absorbing and concentrating power of the kidneys, particularly in birds, apart from its rate of elimination.

The fate of uric acid injected into healthy and gouty subjects was tested. The results showed that in six normal men tested on low-purine and low- and high-protein diets, the excretion of UA was 30–90% of the injected dose. The destruction, therefore, was from 10 to 70%, and averaged about 50%. The excretion period lasted from 1 to 4 days. "The unique and characteristic high levels of UA in normal human blood are due to a lack of responsiveness on the part of the human kidney." This was attributed to the absorbing power of the kidney. High-protein diet promoted this absorbing power, and therefore lowered the concentration of plasma UA. Because destruction of UA related to its concentration in plasma, more endogenous UA was excreted on a high-protein than on a low-protein diet.

Injection of UA into nine gouty subjects did not reveal any differences from the normal except a transient retention of NPN products other than UA. The authors believed that the comparative lack of responsiveness to plasma UA concentration by the kidneys was exaggerated in gout, accounting for the higher levels found.

1. Folin, O., Berglund H., and Derick, C., The uric acid problem. An experimental study on animals and man, including gouty subjects. *J. Biol. Chem.* **60**, 361–471 (1924).

Afterword

For the 1935–36 academic year, Cyrus Fiske was promoted to full professor of biological chemistry, but contrary to his own expectations, an outsider, A. Baird Hastings, Ph.D., was appointed Hamilton Kuhn Professor of Biological Chemistry, in Otto's place. Frederick W. Klemper was appointed a research fellow; otherwise the staff remained the same.

One can only conjecture concerning the decision by the administration to seek a new chairman outside of the department, but only one fact seems plausible: that the decision not to choose Fiske was probably made before the search began. His promotion indicated the respect that he had earned as a premiere investigator.

A. Baird Hastings wrote Philip Shaffer on Nov. 12, 1935:

> I have wanted you to know with what humility and hesitation I have decided to assume the position left vacant by Dr. Folin's passing. . . . It was a very difficult decision for me to make. . . . I cannot fill Dr. Folin's shoes. I do not like to be thought of as his successor. I have simply come here to work; and hope that with the passing of years, I will learn to teach and guide those who come through this Department on their way to the practice of medicine, or a career of investigation.
>
> My new colleagues in the School and in the Department are being very kind and helpful; and although I find myself a bit bewildered at times, I feel that I have already learned a lot. Your letter of good wishes helps a lot, and I should like to think that when I need some counsel, I might appeal to you.

Laura Folin lived to age 89, her mind sharp and clear to the end. In the summer or autumn of 1935 she sold the Brookline home and moved into an apartment on Brookline Avenue near the Beth Israel Hospital. She retained the home in Kearsarge and it remains in the family's possession to this date. Daughter Teresa, after pediatric training in Chicago, married a budding young surgeon, Jonathan E. Rhoads, in 1936 in Cambridge. By 1940, Teresa had her third of six children. When the clouds of war approached in the summer of 1941, Laura moved to Philadelphia to be near Teresa and Jonathan where she lived out her life. Grant continued successfully in his machine tool business in Detroit. There were a total of nine grandchildren.

Following her death on Jan. 21, 1961, Laura was buried in Kearsarge beside Otto and Joanna. Teresa Folin Rhoads died on May 12, 1987.

As the 11-year old Joanna had written to Laura in 1911,

> ... the Ledges and Moat were half covered with those fleecy white clouds that give the appearance of snow drifts. The sun's first rays were striking the northern part of Moat, causing such wonderful light and shadow among the clouds, the like of which I have never seen. Through those banks of clouds, portions of the sky could be seen; it was that exquisite blue, that is rarely seen, even in the sky.

Chronological Highlights of Otto Folin's Career

1897
- Modified the Hopkins method for quantitative determination of urinary uric acid.

1900
- Became the first clinical biochemist in the U.S., McLean Hospital, Waverley, Mass.

1900–4
- Developed quantitative analytical methods for urinary constituents: urea, NH_3, creatinine, creatine, uric acid, total nitrogen, phosphorus, chloride, total sulfate, ethereal sulfate, acidity. Did first experiments with protein precipitants, ether-alcohol, phosphotungstate, uranium acetate, picric acid.
- Introduced aeration technique for NH_3 determination and release of NH_3 from mildly alkaline carbonate solution (1902) and applied the technique to urea, acetone, and diacetic acid. First attempted blood NH_3 determination.
- Introduced alkaline picrate (Jaffé) reaction for creatinine analysis.
- Introduced visual colorimeter (Duboscq) into clinical biochemistry (1903) and concomitantly, modern colorimetric analysis.

1902–5
- Launched major study on the quantitative chemical composition of 24-h urines from subjects on uniform diets and presented 67 tables of data.
- Developed laws governing the output of urinary constituents, e.g., constancy of creatinine output, relation of total nitrogen to its nitrogenous constituents.
- Developed theory of endogenous and exogenous protein metabolism from studies with subjects on protein-restricted diets.
- Established minimum daily protein requirement and its relation to nitrogen and sulfate excretion.

1905
- Appointed as collaborating editor (board of directors) of the newly founded *Journal of Biological Chemistry* (edited by Abel and Herter); served continuously as chairman and member of its editorial committee.

1906
- Found that creatine metabolism differed from creatinine. Postulated special role of creatine in muscle, and restated it in 1909, paving the way for Fiske and SubbaRow's later discovery of phosphocreatine.
- Helped found the American Society of Biological Chemistry, became member of first Council, and served for many terms thereafter.

1907
- Quantified acetone and diacetic acid in diabetic urine; pointed out that diacetic acid was far more prevalent than acetone.
- Appointed Associate Professor of Biochemistry, Harvard Medical School; the sole non-MD on the faculty.

1908
- Issued first call for hospitals to hire biochemists (Harvey Society lecture).
- Elected third president, ASBC.

1909
- Promoted to Hamilton Kuhn Professor of Biological Chemistry, Harvard Medical School.

1910
- Analyzed fat and fatty acids in stools.
- Prepared purified creatinine from urine and from creatine.

1911
- Improved determination of benzoic acid.

1912
- Introduced modern quantitative chemical analysis of blood and tissues.
- Introduced phosphotungstate for uric acid analysis.
- Introduced phosphomolybdate for detection of phenols.
- Demonstrated that the large intestine was the source of NH_3 in portal blood, attributed to the putrefactive action of bacteria.
- Showed that amino acids were absorbed as such from the small and large intestines, and not first synthesized into protein prior to absorption.
- Showed that chemically induced acute nephritis in cats led to nitrogen retention (NPN, urea) in blood.
- Introduced specific new techniques and reagents for blood and urine uric acid analysis: e.g., Nessler's reagent for NH_3, purified phosphotungstic acid, a newly designed "fume absorber" for NH_3 determination; Ostwald-Folin pipettes.
- Confirmed urinary creatine excretion by children.
- Measured tyrosine content of purified proteins; vanillin in vanilla extract.
- Showed that nitrogenous digestion products were absorbed from the stomach.

1913
- Published normal values for uric acid, NPN, and urea nitrogen in human blood. Compared NPN and urea nitrogen values with PSP excretion, and interpreted their use as indicators of kidney function.
- Introduced colorimetric method for epinephrine in extracts.
- Measured decrease in blood uric acid in gouty patients treated with a uricosuric drug.

1914
- Introduced sulfosalicylic acid for urine protein determination, with stable serum protein as standard; used picric acid as protein precipitant.
- Introduced turbidimetry using the Duboscq colorimeter.
- Prepared a stable, pure creatinine standard.
- Studied creatine in muscle and its possible presence in organic combination with protoplasm.
- Improved methods for determining acetone, acetoacetic acid, and β-oxybutyric acid in urine.
- Published monograph on use and abuse of preservatives in foods.

1916
- Published first of five editions of a laboratory manual.
- First used metaphosphoric acid as protein precipitant.

1917
- Purified picric acid for creatinine determination.
- Introduced permutit (sodium aluminum silicate) as an NH_3 adsorbent for quantifying urinary NH_3.

1919
- Introduced Folin-Wu tungstic acid for protein removal from blood, and a "system" of analysis on the filtrate.
- Studied the protein, fat, and lactose content of animal milk.

1920
- Introduced the Folin-Wu method for blood sugar.
- Chaired a committee that prepared a report on the teaching of biochemistry in medical schools.
- First used Lloyd's reagent to remove creatinine and other substances from urine; used benzoic acid as preservative for glucose standards.
- Studied the fate of ingested carbohydrates in blood, urine, red blood cells.
- Introduced β-naphthoquinonesulfonic acid for determining amino acids.
- Improved colorimetric methods for tyrosine, tryptophane, and cystine in proteins.
- Updated the analysis of uric acid in blood.

1923–34
- Directed laboratory and later served as consultant (with S. R. Benedict) in research at the Metropolitan Life Insurance Co. of N.Y. on improved measurement of urinary sugar and protein, glucose tolerance tests, tests of kidney function, and standardization of methods.

1927
- Improved methods for tyrosine and tryptophane in protein.

1928
- Introduced first of several micromethods for skin-puncture-derived blood, the Prussian blue method for blood sugar; introduced gum ghatti to maintain color stability; new techniques in clinical microchemistry.

1929
- Improved the micromethod for sugar; introduced the use of a light filter in colorimetry.
- Improved cystine analysis in protein.

1930
- Removed interferences caused by red blood cells by introducing the use of protein-free filtrates made from unlaked blood; compared unlaked blood values with plasma.
- Provided micromethods for NPN, urea, and uric acid.
- Studied the effect of eating on NPN constituents, 2.5 h after breakfast.

1932
- Improved the blood NH_3 determination as a long-time challenge to analysis, despite its lack of clinical orientation at that time.

1932–34
- Significantly improved uric acid and creatinine determinations.

With the permission of the Nobel Committee for Physiology or Medicine at the Karolinska Institute, Stockholm, Anders Kallner (Department of Clinical Chemistry, Karolinska Hospital) made the following information available.

Otto Folin was nominated six times for the Nobel Prize in Physiology or Medicine between 1920 and 1932. Listed below are the years in which he was nominated and those who nominated him:

1920	Philip Shaffer, the first of the Folin students to receive his Ph.D.
1924	C. J. V. Pettibone, one of Folin's students at Harvard
1924	H. Steudel, Berlin
1927	V. J. Harding, Montreal
1931	A. D. Hirschfelder, Minneapolis, and F. W. Schultz, Chicago
1932	Victor Myers, who nominated Folin to share the prize with Stanley Benedict

Incredibly, the Nobel Committee for Physiology or Medicine judged Folin's work to consist primarily of modifications of old techniques, hence insufficiently original, and of low specificity. His contributions and knowledge of purine and protein metabolism were deemed too premature for adequate evaluation.

General References

1. Bearman, D. and Edsall, J.T., Eds. *Archival Sources for the History of Biochemistry and Molecular Biology*. Philadelphia: The American Philosophical Society, 1980.
2. Beecher, H.K., and Altschule, M.D. *Medicine at Harvard, The First Hundred Years*. Hanover, N.H.: University Press of New England, 1977.
3. Forbes, H.M. *West Virginia History—A Bibliography and Guide to Research*. Morgantown, W.Va.: West Virginia University Press, 1981.
4. Garraty, J.A. and Gay, P., Eds. *The Columbia History of the World*. New York: Harper and Row, 1972.
5. Storr, R.J. *Harper's University, the Beginnings*. Chicago: University of Chicago Press, 1966.
6. Aub, J.C., and Hapgood, R.K. *Pioneer in Modern Medicine, David Linn Edsall of Harvard*. Boston: Harvard Medical Alumni Association, 1970.

Folin Obituary Notices and Biographical Sketches

1934-35

1. Otto Folin. *J. Biol. Chem.* **107**, 606–7 (1934).
2. Dr. Otto Folin. *Minn. Alumni Weekly* **34**, 171 (Nov. 10, 1934).
3. Svedberg, A.A., Otto Folin (Obituary). *Nord. Med. Tidsk.* **8**, 1668–69 (Dec. 8, 1934).
4. Shaffer, P.A., Otto Folin (Obituary). *Science* **81**, 35–37 (1935).
5. Christian, H.A., Tribute to Professor Folin. *Science* **81**, 37–38 (1935).
6. Christian, H.A., Otto Knut Olof Folin, Ph.D., M.D. (hon.). *Harvard Med. Alumni Bull.* 22 (Jan. 1935)
7. Trimble, H., Otto Folin—A life work in biochemistry. *Am. Swed. Monthly* **29**, 9–11 (1935).
8. Otto Folin (Obituary). *Med. J. Aust.* **1**, 69 (1935).

1936-51

9. Berglund, H., Otto Folin. Papers from the IV Medical Service of St. Erik's Hospital, Stockholm, Sweden, Alb. Bonniers Boktryckeri, Stockholm, 1937.

10. Haagensen, C.D., and Lloyd, Wyndham, E.B., In: *A Hundred Years of Medicine*. pp. 65–67. New York: Sheridan House, 1943.
11. Folin, O. In: *Who Was Who in America*. Vol. 1, 1897–1942. p. 409. Chicago: Marquis, 1943.
12. Christian, H.A., Folin, Otto Knut Olof. In: H.E. Starr, Ed. *Dictionary of American Biography*. Vol. 21 (Supplement 1), pp. 306–8. New York: Charles Scribner's Sons, 1944.

1952–88

13. Shaffer, P.A., Otto Folin (1867–1934). *Biogr. Mem. Natl. Acad. Sci., U.S.A.* **27**, 47–82 (1952).
14. Marble, A., Otto Folin—Benefactor of diabetics through biochemistry. *Diabetes* **2**, 503–5 (1953).
15. Shaffer, P., Otto Folin (1867–1934). *J. Nutr.* **52**, 1–11 (1954).
16. Van Liere, E.J., Otto Knut Olof Folin (1867–1934). *W. Va. Med. J.* **59**, 41–42 (1963).
17. Forbes, W.H., Recollections of Otto Folin. *Harvard Med. Alumni Bull.* **46**, 8–10 (1971).
18. Leicester, H.M., Folin, Otto. *Dictionary Sci. Biogr.* **5**, 53 (1972).
19. Hastings, A.B., The clinical-chemical interface of medical science: Its development in this century. *Ann. Clin. Lab. Sci.* **4**, 213–21 (1974).
20. Meites, S., Otto Folin's decade in Minnesota, 1882–1892: A brief review. *Clin. Chem.* **28**, 2173–77 (1982).
21. Meites, S., The first call for clinical chemists in the United States. *Clin. Chem.* **29**, 1852–53 (1983).
22. Meites, S., Otto Folin's medical legacy. *Clin. Chem.* **31**, 1402–4 (1985).
23. Büttner, J., and Habrich, C., *Roots of Clinical Chemistry*. pp. 99–102. Darmstadt, West Germany: Git Verlag, GMBH, 1988.